London Mathematical Society Lecture Note Series: 371

Conformal Fractals: Ergodic Theory Methods

FELIKS PRZYTYCKI
Polish Academy of Sciences

MARIUSZ URBAŃSKI
University of North Texas

CAMBRIDGE
UNIVERSITY PRESS

CAMBRIDGE UNIVERSITY PRESS
Cambridge, New York, Melbourne, Madrid, Cape Town, Singapore,
São Paulo, Delhi, Dubai, Tokyo, Mexico City

Cambridge University Press
The Edinburgh Building, Cambridge CB2 8RU, UK

Published in the United States of America by
Cambridge University Press, New York

www.cambridge.org
Information on this title: www.cambridge.org/9780521438001

First published 2010

A catalogue record for this publication is available from the British Library

Library of Congress Cataloguing in Publication Data

ISBN 978-0-521-43800-1 Paperback

LONDON MATHEMATICAL SOCIETY LECTURE NOTE SERIES

Managing Editor: Professor M. Reid, Mathematics Institute, University of Warwick, Coventry CV4 7AL, United Kingdom

The titles below are available from booksellers, or from Cambridge University Press at www.cambridge.org/mathematics

221 Harmonic approximation, S. J. GARDINER
222 Advances in linear logic, J.-Y. GIRARD, Y. LAFONT & L. REGNIER (eds)
223 Analytic semigroups and semilinear initial boundary value problems, K. TAIRA
224 Computability, enumerability, unsolvability, S. B. COOPER, T. A. SLAMAN & S. S. WAINER (eds)
225 A mathematical introduction to string theory, S. ALBEVERIO *et al*
226 Novikov conjectures, index theorems and rigidity I, S. C. FERRY, A. RANICKI & J. ROSENBERG (eds)
227 Novikov conjectures, index theorems and rigidity II, S. C. FERRY, A. RANICKI & J. ROSENBERG (eds)
228 Ergodic theory of Z^d-actions, M. POLLICOTT & K. SCHMIDT (eds)
229 Ergodicity for infinite dimensional systems, G. DA PRATO & J. ZABCZYK
230 Prolegomena to a middlebrow arithmetic of curves of genus 2, J. W. S. CASSELS & E. V. FLYNN
231 Semigroup theory and its applications, K. H. HOFMANN & M. W. MISLOVE (eds)
232 The descriptive set theory of Polish group actions, H. BECKER & A. S. KECHRIS
233 Finite fields and applications, S. COHEN & H. NIEDERREITER (eds)
234 Introduction to subfactors, V. JONES & V. S. SUNDER
235 Number theory: Séminaire de théorie des nombres de Paris 1993–94, S. DAVID (ed)
236 The James forest, H. FETTER & B. GAMBOA DE BUEN
237 Sieve methods, exponential sums, and their applications in number theory, G. R. H. GREAVES *et al* (eds)
238 Representation theory and algebraic geometry, A. MARTSINKOVSKY & G. TODOROV (eds)
240 Stable groups, F. O. WAGNER
241 Surveys in combinatorics, 1997, R. A. BAILEY (ed)
242 Geometric Galois actions I, L. SCHNEPS & P. LOCHAK (eds)
243 Geometric Galois actions II, L. SCHNEPS & P. LOCHAK (eds)
244 Model theory of groups and automorphism groups, D. M. EVANS (ed)
245 Geometry, combinatorial designs and related structures, J. W. P. HIRSCHFELD *et al* (eds)
246 p-Automorphisms of finite p-groups, E. I. KHUKHRO
247 Analytic number theory, Y. MOTOHASHI (ed)
248 Tame topology and O-minimal structures, L. VAN DEN DRIES
249 The atlas of finite groups - Ten years on, R. T. CURTIS & R. A. WILSON (eds)
250 Characters and blocks of finite groups, G. NAVARRO
251 Gröbner bases and applications, B. BUCHBERGER & F. WINKLER (eds)
252 Geometry and cohomology in group theory, P. H. KROPHOLLER, G. A. NIBLO & R. STÖHR (eds)
253 The q-Schur algebra, S. DONKIN
254 Galois representations in arithmetic algebraic geometry, A. J. SCHOLL & R. L. TAYLOR (eds)
255 Symmetries and integrability of difference equations, P. A. CLARKSON & F. W. NIJHOFF (eds)
256 Aspects of Galois theory, H. VÖLKLEIN, J. G. THOMPSON, D. HARBATER & P. MÜLLER (eds)
257 An introduction to noncommutative differential geometry and its physical applications (2nd edition), J. MADORE
258 Sets and proofs, S. B. COOPER & J. K. TRUSS (eds)
259 Models and computability, S. B. COOPER & J. TRUSS (eds)
260 Groups St Andrews 1997 in Bath I, C. M. CAMPBELL *et al* (eds)
261 Groups St Andrews 1997 in Bath II, C. M. CAMPBELL *et al* (eds)
262 Analysis and logic, C. W. HENSON, J. IOVINO, A. S. KECHRIS & E. ODELL
263 Singularity theory, W. BRUCE & D. MOND (eds)
264 New trends in algebraic geometry, K. HULEK, F. CATANESE, C. PETERS & M. REID (eds)
265 Elliptic curves in cryptography, I. BLAKE, G. SEROUSSI & N. SMART
267 Surveys in combinatorics, 1999, J. D. LAMB & D. A. PREECE (eds)
268 Spectral asymptotics in the semi-classical limit, M. DIMASSI & J. SJÖSTRAND
269 Ergodic theory and topological dynamics of group actions on homogeneous spaces, M. B. BEKKA & M. MAYER
271 Singular perturbations of differential operators, S. ALBEVERIO & P. KURASOV
272 Character theory for the odd order theorem, T. PETERFALVI. Translated by R. SANDLING
273 Spectral theory and geometry, E. B. DAVIES & Y. SAFAROV (eds)
274 The Mandelbrot set, theme and variations, T. LEI (ed)
275 Descriptive set theory and dynamical systems, M. FOREMAN, A. S. KECHRIS, A. LOUVEAU & B. WEISS (eds)
276 Singularities of plane curves, E. CASAS-ALVERO
277 Computational and geometric aspects of modern algebra, M. ATKINSON *et al* (eds)
278 Global attractors in abstract parabolic problems, J. W. CHOLEWA & T. DLOTKO
279 Topics in symbolic dynamics and applications, F. BLANCHARD, A. MAASS & A. NOGUEIRA (eds)
280 Characters and automorphism groups of compact Riemann surfaces, T. BREUER
281 Explicit birational geometry of 3-folds, A. CORTI & M. REID (eds)
282 Auslander–Buchweitz approximations of equivariant modules, M. HASHIMOTO
283 Nonlinear elasticity, Y. B. FU & R. W. OGDEN (eds)
284 Foundations of computational mathematics, R. DEVORE, A. ISERLES & E. SÜLI (eds)
285 Rational points on curves over finite fields, H. NIEDERREITER & C. XING
286 Clifford algebras and spinors (2nd Edition), P. LOUNESTO
287 Topics on Riemann surfaces and Fuchsian groups, E. BUJALANCE, A. F. COSTA & E. MARTÍNEZ (eds)
288 Surveys in combinatorics, 2001, J. W. P. HIRSCHFELD (ed)
289 Aspects of Sobolev-type inequalities, L. SALOFF-COSTE
290 Quantum groups and Lie theory, A. PRESSLEY (ed)
291 Tits buildings and the model theory of groups, K. TENT (ed)
292 A quantum groups primer, S. MAJID
293 Second order partial differential equations in Hilbert spaces, G. DA PRATO & J. ZABCZYK
294 Introduction to operator space theory, G. PISIER
295 Geometry and integrability, L. MASON & Y. NUTKU (eds)
296 Lectures on invariant theory, I. DOLGACHEV
297 The homotopy category of simply connected 4-manifolds, H.-J. BAUES
298 Higher operads, higher categories, T. LEINSTER (ed)

299 Kleinian groups and hyperbolic 3-manifolds, Y. KOMORI, V. MARKOVIC & C. SERIES (eds)
300 Introduction to Möbius differential geometry, U. HERTRICH-JEROMIN
301 Stable modules and the D(2)-problem, F. E. A. JOHNSON
302 Discrete and continuous nonlinear Schrödinger systems, M. J. ABLOWITZ, B. PRINARI & A. D. TRUBATCH
303 Number theory and algebraic geometry, M. REID & A. SKOROBOGATOV (eds)
304 Groups St Andrews 2001 in Oxford I, C. M. CAMPBELL, E. F. ROBERTSON & G. C. SMITH (eds)
305 Groups St Andrews 2001 in Oxford II, C. M. CAMPBELL, E. F. ROBERTSON & G. C. SMITH (eds)
306 Geometric mechanics and symmetry, J. MONTALDI & T. RATIU (eds)
307 Surveys in combinatorics 2003, C. D. WENSLEY (ed.)
308 Topology, geometry and quantum field theory, U. L. TILLMANN (ed)
309 Corings and comodules, T. BRZEZINSKI & R. WISBAUER
310 Topics in dynamics and ergodic theory, S. BEZUGLYI & S. KOLYADA (eds)
311 Groups: topological, combinatorial and arithmetic aspects, T. W. MÜLLER (ed)
312 Foundations of computational mathematics, Minneapolis 2002, F. CUCKER *et al* (eds)
313 Transcendental aspects of algebraic cycles, S. MÜLLER-STACH & C. PETERS (eds)
314 Spectral generalizations of line graphs, D. CVETKOVIC, P. ROWLINSON & S. SIMIC
315 Structured ring spectra, A. BAKER & B. RICHTER (eds)
316 Linear logic in computer science, T. EHRHARD, P. RUET, J.-Y. GIRARD & P. SCOTT (eds)
317 Advances in elliptic curve cryptography, I. F. BLAKE, G. SEROUSSI & N. P. SMART (eds)
318 Perturbation of the boundary in boundary-value problems of partial differential equations, D. HENRY
319 Double affine Hecke algebras, I. CHEREDNIK
320 L-functions and Galois representations, D. BURNS, K. BUZZARD & J. NEKOVÁR (eds)
321 Surveys in modern mathematics, V. PRASOLOV & Y. ILYASHENKO (eds)
322 Recent perspectives in random matrix theory and number theory, F. MEZZADRI & N. C. SNAITH (eds)
323 Poisson geometry, deformation quantisation and group representations, S. GUTT *et al* (eds)
324 Singularities and computer algebra, C. LOSSEN & G. PFISTER (eds)
325 Lectures on the Ricci flow, P. TOPPING
326 Modular representations of finite groups of Lie type, J. E. HUMPHREYS
327 Surveys in combinatorics 2005, B. S. WEBB (ed)
328 Fundamentals of hyperbolic manifolds, R. CANARY, D. EPSTEIN & A. MARDEN (eds)
329 Spaces of Kleinian groups, Y. MINSKY, M. SAKUMA & C. SERIES (eds)
330 Noncommutative localization in algebra and topology, A. RANICKI (ed)
331 Foundations of computational mathematics, Santander 2005, L. M PARDO, A. PINKUS, E. SÜLI & M. J. TODD (eds)
332 Handbook of tilting theory, L. ANGELERI HÜGEL, D. HAPPEL & H. KRAUSE (eds)
333 Synthetic differential geometry (2nd Edition), A. KOCK
334 The Navier–Stokes equations, N. RILEY & P. DRAZIN
335 Lectures on the combinatorics of free probability, A. NICA & R. SPEICHER
336 Integral closure of ideals, rings, and modules, I. SWANSON & C. HUNEKE
337 Methods in Banach space theory, J. M. F. CASTILLO & W. B. JOHNSON (eds)
338 Surveys in geometry and number theory, N. YOUNG (ed)
339 Groups St Andrews 2005 I, C. M. CAMPBELL, M. R. QUICK, E. F. ROBERTSON & G. C. SMITH (eds)
340 Groups St Andrews 2005 II, C. M. CAMPBELL, M. R. QUICK, E. F. ROBERTSON & G. C. SMITH (eds)
341 Ranks of elliptic curves and random matrix theory, J. B. CONREY, D. W. FARMER, F. MEZZADRI & N. C. SNAITH (eds)
342 Elliptic cohomology, H. R. MILLER & D. C. RAVENEL (eds)
343 Algebraic cycles and motives I, J. NAGEL & C. PETERS (eds)
344 Algebraic cycles and motives II, J. NAGEL & C. PETERS (eds)
345 Algebraic and analytic geometry, A. NEEMAN
346 Surveys in combinatorics 2007, A. HILTON & J. TALBOT (eds)
347 Surveys in contemporary mathematics, N. YOUNG & Y. CHOI (eds)
348 Transcendental dynamics and complex analysis, P. J. RIPPON & G. M. STALLARD (eds)
349 Model theory with applications to algebra and analysis I, Z. CHATZIDAKIS, D. MACPHERSON, A. PILLAY & A. WILKIE (eds)
350 Model theory with applications to algebra and analysis II, Z. CHATZIDAKIS, D. MACPHERSON, A. PILLAY & A. WILKIE (eds)
351 Finite von Neumann algebras and masas, A. M. SINCLAIR & R. R. SMITH
352 Number theory and polynomials, J. MCKEE & C. SMYTH (eds)
353 Trends in stochastic analysis, J. BLATH, P. MÖRTERS & M. SCHEUTZOW (eds)
354 Groups and analysis, K. TENT (ed)
355 Non-equilibrium statistical mechanics and turbulence, J. CARDY, G. FALKOVICH & K. GAWEDZKI
356 Elliptic curves and big Galois representations, D. DELBOURGO
357 Algebraic theory of differential equations, M. A. H. MACCALLUM & A. V. MIKHAILOV (eds)
358 Geometric and cohomological methods in group theory, M. R. BRIDSON, P. H. KROPHOLLER & I. J. LEARY (eds)
359 Moduli spaces and vector bundles, L. BRAMBILA-PAZ, S. B. BRADLOW, O. GARCÍA-PRADA & S. RAMANAN (eds)
360 Zariski geometries, B. ZILBER
361 Words: Notes on verbal width in groups, D. SEGAL
362 Differential tensor algebras and their module categories, R. BAUTISTA, L. SALMERÓN & R. ZUAZUA
363 Foundations of computational mathematics, Hong Kong 2008, F. CUCKER, A. PINKUS & M. J. TODD (eds)
364 Partial differential equations and fluid mechanics, J. C. ROBINSON & J. L. RODRIGO (eds)
365 Surveys in combinatorics 2009, S. HUCZYNSKA, J. D. MITCHELL & C. M. RONEY-DOUGAL (eds)
366 Highly oscillatory problems, B. ENGQUIST, A. FOKAS, E. HAIRER & A. ISERLES (eds)
367 Random matrices: High dimensional phenomena, G. BLOWER
368 Geometry of Riemann surfaces, F. P. GARDINER, G. GONZÁLEZ-DIEZ & C. KOUROUNIOTIS (eds)
369 Epidemics and rumours in complex networks, M. DRAIEF & L. MASSOULIÉ
370 Theory of p-adic distributions, S. ALBEVERIO, A. YU. KHRENNIKOV & V. M. SHELKOVICH
371 Conformal fractals, F. PRZYTYCKI & M. URBAŃSKI
372 Moonshine: The first quarter century and beyond, J. LEPOWSKY, J. MCKAY & M. P. TUITE (eds)
373 Smoothness, regularity, and complete intersection J. MAJADAS & A. RODICIO

F. P. dedicates this book to the memory of his parents,
Jakub and Róża, and to his wife Jolanta Słomińska.

M. U. dedicates this book to his parents,
Anna and Henryk, and his wife Irena.

Contents

Introduction

This book is an introduction to the theory of iteration of expanding and non-uniformly expanding holomorphic maps and topics in geometric measure theory of the underlying invariant fractal sets. Probability measures on these sets yield information on Hausdorff and other fractal dimensions and properties. The book starts with a comprehensive chapter on abstract ergodic theory, followed by chapters on uniform distance-expanding maps and *thermodynamical formalism*. This material is applicable in many branches of dynamical systems and related fields, far beyond the applications in this book.

Popular examples of the fractal sets to be investigated are Julia sets for rational functions on the Riemann sphere. The theory, which was initiated by Gaston Julia [1918] and Pierre Fatou [1919–1920], has become very popular since the publication of Benoit Mandelbrot's book [Mandelbrot 1982] with beautiful computer generated illustrations. Top mathematicians have since made spectacular progress in the field over the last 30 years.

Consider, for example, the map $f(z) = z^2$ for complex numbers z. Then the unit circle $S^1 = \{|z| = 1\}$ is f-invariant, $f(S^1) = S^1 = f^{-1}(S^1)$. For $c \approx 0, c \neq 0$ and $f_c(z) = z^2 + c$, there still exists an f_c-invariant set $J(f_c)$ called the *Julia set* of f_c, close to S^1, homeomorphic to S^1 via a homeomorphism h satisfying the equality $f \circ h = h \circ f_c$. However, $J(f_c)$ has a fractal shape. For large c the curve $J(f_c)$ pinches at infinitely many points; it may pinch everywhere to become a dendrite, or even crumble to become a Cantor set.

These sets satisfy two main properties, standard attributes of 'conformal fractal sets':

1. Their fractal dimensions are strictly larger than the topological dimension.

2. They are conformally 'self-similar': that is, arbitrarily small pieces have shapes similar to large pieces *via* conformal mappings, here *via* iteration of f.

To measure fractal sets invariant under holomorphic mappings, one applies probability measures corresponding to equilibria in the thermodynamical formalism. This is a beautiful example of the interlacing of ideas from mathematics and physics.

The following *prototype lemma* [Bowen, 1975, Lemma 1.1], resulting from Jensen's inequality applied to the function logarithm, stems from the thermodynamical formalism.

Lemma. *(Finite Variational Principle) For given real numbers* $\phi_1, \dots, \phi_n$ *the quantity*

$$F(p_1, \dots p_n) = \sum_{i=1}^{n} -p_i \log p_i + \sum_{i=1}^{n} p_i \phi_i$$

has maximum value $P(\phi_1, ...\phi_n) = \log \sum_{i=1}^{n} e^{\phi_i}$ *as* $(p_1, \dots, p_n)$ *ranges over the simplex* $\{(p_1, \dots, p_n) : p_i \geq 0, \sum_{i=1}^{n} p_i = 1\}$ *and the maximum is attained only at*

$$\hat{p}_j = e^{\phi_j} \Big(\sum_{i=1}^{n} e^{\phi_i}\Big)^{-1}.$$

We can read $\phi_i, p_i, i = 1, \dots, n$ as a function (*potential*), resp. probability distribution, on the finite space $\{1, \dots, n\}$. The proof follows from the strict concavity of the logarithm function.

Let us further follow Bowen [1975]. The quantity

$$S = \sum_{i=1}^{n} -p_i \log p_i$$

is called the *entropy* of the distribution $(p_1, \dots, p_n)$. The maximizing distribution $(\hat{p}_1, .., \hat{p}_n)$ is called the *Gibbs* or *equilibrium state.* In statistical mechanics $\phi_i = -\beta E_i$, where $\beta = 1/kT$, T is the temperature of an external 'heat source' and k is a physical (Boltzmann) constant. The quantity $E = \sum_{i=1}^{n} p_i E_i$ is the average energy. The Gibbs distribution thus maximizes the expression

$$S - \beta E = S - \frac{1}{kT} E$$

or, equivalently, minimizes the so-called *free energy* $E - kTS$. Nature prefers states with low energy and high entropy. It minimizes free energy.

The idea of the Gibbs distribution as a limit of distributions on finite spaces of configurations of states (spins, for example) of interacting particles over increasing to infinite, bounded parts of the lattice $\mathbb{Z}^d$ was first introduced in statistical mechanics by Bogolyubov and Hacet [1949] where it plays a fundamental role. It was applied in dynamical systems to study Anosov flows and hyperbolic diffeomorphisms at the end of the 1960s by Ja. Sinai, D. Ruelle and R. Bowen. For more historical remarks see [Ruelle 1978a] or [Sinai 1982]. This theory met the notion of entropy S, borrowed from information theory and introduced by Kolmogorov as an invariant of a measure-theoretic dynamical system.

Later, the usefulness of these notions to the geometric dimensions became apparent. It was already present in [Billingsley 1965], but papers by Bowen [1979] and McCluskey & Manning [1983] were also crucial.

In order to illustrate the idea, consider the following example. Let $T_i : I \to I$, $i = 1, \dots, n > 1$, where $I = [0, 1]$ is the unit interval, $T_i(x) = \lambda_i x + a_i$, where λ_i, a_i are real numbers chosen in such a way that all the sets $T_i(I)$ are pairwise disjoint and contained in I. Define the limit set Λ as follows:

$$\Lambda = \bigcap_{k=0}^{\infty} \bigcup_{(i_0, \dots, i_k)} T_{i_0} \circ \dots \circ T_{i_k}(I) = \bigcup_{(i_0, i_1 \dots)} \lim_{k \to \infty} T_{i_0} \circ \dots \circ T_{i_k}(x),$$

the latter union taken over all infinite sequences $(i_0, i_1, \dots)$, the former over sequences of length $k+1$. By our assumptions $|\lambda_j| < 1$: hence the limit exists, and does not depend on x.

It occurs that its Hausdorff dimension is equal to the only number α for which

$$|\lambda_1|^\alpha + \cdots + |\lambda_n|^\alpha = 1.$$

Λ is a Cantor set. It is self-similar with small pieces similar to large pieces with the use of linear (more precisely, affine) maps $(T_{i_0} \circ \cdots \circ T_{i_k})^{-1}$. We call such a Cantor set *linear*. We can distribute a measure μ by setting $\mu(T_{i_0} \circ \cdots \circ T_{i_k}(I)) = (\lambda_{i_0} \dots \lambda_{i_k})^\alpha$. Then for each interval $J \subset I$ centred at a point of Λ, its diameter raised to the power α is comparable to its measure μ (this is immediate for the intervals $T_{i_0} \circ \cdots \circ T_{i_k}(I)$). (A measure with this property for all small balls centred at a compact set, in a Euclidean space of any dimension, is called a *geometric measure*.) Hence $\sum(\operatorname{diam} J)^\alpha$ is bounded away from 0 and ∞ for all economical (of multiplicity not exceeding 2) covers of Λ by intervals J.

Note that for each k the measure μ restricted to the space of unions of $T_{i_0} \circ \cdots \circ T_{i_k}(I)$, each such interval viewed as one point, is the Gibbs distribution, where we set $\phi((i_0, \dots, i_k)) = \phi_\alpha((i_0, \dots, i_k)) = \sum_{l=0,\dots,k} \alpha \log \lambda_{i_l}$. The number α is the unique zero of the *pressure function* $\mathrm{P}(\alpha) = \frac{1}{k+1} \log \sum_{(i_0,\dots,i_k)} e^{\phi_\alpha((i_0,\dots,i_k))}$. In this special affine example this is independent of k. In the general non-linear case to define pressure one considers the limit as k goes to ∞.

The family T_i and compositions is an example, very popular in recent years, of *Iterated Function Systems* [Barnsley 1988]. Note that on a neighbourhood of each $T_i(I)$ we can consider $\hat{T} := T_i^{-1}$. Then Λ is an invariant repeller for the distance-expanding map $\hat{T}$.

The relations between dynamics, dimension and geometric measure theory start in our book with the theorem that the Hausdorff dimension of an expanding repeller is the unique zero of the adequate pressure function for sets built with the help of $C^{1+\varepsilon}$ usually non-linear maps in $\mathbb{R}$ or conformal maps in the complex plane $\mathbb{C}$ (or in $\mathbb{R}^d, d > 2$; in this case conformal maps must be Möbius, i.e. a composition of inversions and symmetries, by Liouville's theorem).

This theory was developed for non-uniformly hyperbolic maps or flows in the setting of smooth ergodic theory: see [Katok & Hasselblatt 1995], [Mañé 1987]. Let us also mention [Ledrappier & Young 1985]. See [Pesin 1997] for recent developments. The advanced chapters of our book are devoted to this theory, but we restrict ourselves to complex dimension 1. So the maps are non-uniformly expanding, and the main technical difficulties are caused by critical points, where we have strong contraction, since the derivative by definition is equal to 0 at critical points.

A direction not developed in this book is conformal iterated function systems with infinitely many generators T_i. They occur naturally as return maps in many important constructions, for example for rational maps with parabolic periodic points, or in the *induced expansion* construction for polynomials [Graczyk & Świątek 1998]. See also the recent [Przytycki & Rivera-Letelier 2007]. Beautiful examples are provided by infinitely generated Kleinian groups. For a measure-theoretic background see [Young 1999].

The systematic treatment of iterated function systems with infinitely many generators can be found in [Mauldin & Urbanski 1996] and [Mauldin & Urbański 2003], for example. Recently this has been rigorously explored in the iteration of entire and meromorphic functions.

Below is a short description of the content of the book.

Chapter 1 contains some introductory definitions and basic examples. It is a continuation of this Introduction.

Chapter 2 is an introduction to abstract ergodic theory: here T is a probability measure-preserving transformation. The reader will find proofs of the fundamental theorems: the Birkhoff Ergodic Theorem and the Shannon–McMillan–Breiman Theorem. We introduce entropy and measurable partitions, and discuss canonical systems of conditional measures in Lebesgue spaces, the notion of *natural extension* (inverse limit in the appropriate category). We follow here Rokhlin's Theory [Rokhlin 1949], [Rokhlin 1967]: see also [Kornfeld, Fomin & Sinai 1982]. Next, to prepare for applications for finite-to-one rational maps, we sketch Rokhlin's theory on countable-to-one endomorphisms, and introduce the notion of the Jacobian: see also [Parry 1969]. Finally we discuss mixing properties (K-property, exactness, Bernoulli) and probability laws: the Central Limit Theorem (abbr. CLT), the Law of Iterated Logarithm (LIL), the Almost Sure Invariance Principle (ASIP) for the sequence of functions (random variables on our probability space) $\phi \circ T^n, n = 0, 1, \dots$.

Chapter 3 is devoted to ergodic theory and thermodynamical formalism for general continuous maps on compact metric spaces. The main point here is the so called Variational Principle for pressure: compare with the Finite Variational Principle lemma, above. We also apply functional analysis in order to explain the Legendre transform duality between entropy and pressure. We follow here [Israel 1979] and [Ruelle 1978a]. This material is applicable in *large deviations* and *multifractal analysis*, and is directly related to the uniqueness question of Gibbs states.

In Chapters 2 and 3 we often follow the beautiful book by Peter Walters [Walters 1982].

In Chapter 4 *distance-expanding maps* are introduced. Analogously to Axiom A diffeomorphisms [Smale 1967], [Bowen 1975] or endomorphisms [Przytycki 1976] and [Przytycki 1977], we outline a topological theory: spectral decomposition, specification, Markov partition, and start a 'bounded distortion' play with Hölder continuous functions.

In Chapter 5 thermodynamical formalism and mixing properties of Gibbs measures for open distance-expanding maps T and Hölder continuous potentials ϕ are studied. To a large extent we follow [Bowen 1975] and [Ruelle 1978a]. We prove the existence of Gibbs probability measures (states): m with Jacobian being $\exp(-\phi)$ up to a constant factor, and T-invariant $\mu = \mu_\phi$ equivalent to m. The idea is to use the *transfer operator* $\mathcal{L}_\phi(u)(x) = \sum_{y \in T^{-1}(x)} u(y) \exp \phi(y)$ on the Banach space of Hölder continuous functions u. We prove the exponential convergence $\xi^{-n}\mathcal{L}_\phi^n(u) \to (\int u\,dm)u_\phi$, where ξ is the eigenvalue with the largest absolute value and u_ϕ the corresponding eigenfunction. One obtains $u_\phi = dm/d\mu$. We deduce CLT, LIL and ASIP, and the Bernoulli property for the natural extension.

We provide three different proofs of the uniqueness of the invariant Gibbs measure. The first, and simplest, follows [Keller 1998], the second relies on the Finite Variational Principle, and the third on the differentiability of the pressure function in adequate function directions.

Finally we prove Ruelle's formula:

$$d^2P(\phi + tu + sv)/dt\,ds|_{t=s=0}$$
$$= \lim_{n\to\infty} \frac{1}{n} \int \left(\sum_{i=0}^{n-1} (u \circ T^i - \int u\,d\mu_\phi) \right) \cdot \left(\sum_{i=0}^{n-1} (v \circ T^i - \int v\,d\mu_\phi) \right) d\mu_\phi.$$

This expression for $u = v$ is equal to σ^2 in CLT for the sequence $u \circ T^n$ and measure μ_ϕ.

(In the book we use the letter T to denote a measure-preserving transformation. Maps preserving an additional structure, continuous, smooth or holomorphic for example, are usually denoted by f or g.)

In Chapter 6 (Section 6.1) a metric space with the action of a distance-expanding map f is embedded in a smooth manifold, and it is assumed that the map extends smoothly (or only continuously) to a neighbourhood. Similarly with hyperbolic sets [Katok & Hasselblatt 1995] we discuss basic properties. The intrinsic property of f being an open map on X occurs equivalent to X being repeller for the extension.

We call a repeller X with smoothly extended dynamics a *Smooth Expanding Repeller* (SER).

If an extension is conformal, we say (X, f) is a *conformal expanding repeller* (CER). In Section 6.2 we discuss some distortion theorems and holomorphic motion to be used later in Section 6.4, and in Chapter 9 to prove the analytic dependence of 'pressure' and the Hausdorff dimension of CER on a parameter.

In Section 6.3 we prove that for CER the density $u_\phi = dm/d\mu$ for measures of harmonic potential is real-analytic (and extends so on a neighbourhood of X). This will be used in Chapter 9 for the potential being $-\log|f'|$, in which case μ is equivalent to a Hausdorff measure in the maximal dimension (geometric measure).

In Chapter 7 we provide in detail D. Sullivan's theory classifying $C^{r+\varepsilon}$ line Cantor sets via a *scaling function*, sketched in [Sullivan 1988], and discuss the realization problem [Przytycki & Tangerman 1996]. We also discuss applications for Cantor-like closures of postcritical sets for infinitely renormalizable *Feigenbaum* quadratic-like maps of interval. The infinitesimal geometry of these sets occurs independent of the map, which is one of the famous Coullet–Tresser–Feigenbaum universalities.

In Chapter 8 we provide definitions of various 'fractal dimensions': Hausdorff, box and packing. We also consider Hausdorff measures with gauge functions different from t^α. We prove the 'Volume Lemma' linking, roughly speaking, (global) dimension with local dimensions.

In Chapter 9 we develop the theory of conformal expanding repellers, and relate pressure to the Hausdorff dimension.

Section 9.2 provides a brief exposition of multifractal analysis of the Gibbs measure μ of a Hölder potential on CER X. We rely mainly on [Pesin 1997]. In particular, we discuss the function $F_\mu(\alpha) := \mathrm{HD}(X_\mu(\alpha))$, where $X_\mu(\alpha) := \{x \in X : d(x) = \alpha\}$ and $d(x) := \lim_{r\to 0} \log \mu(B(x,r))/\log r$. The decomposition $X = \bigcup_\alpha (X_\mu(\alpha)) \cup \hat{X}$, where the limit $d(x)$, called the local dimension, does not exist for $x \in \hat{X}$, is called the local dimension spectrum decomposition.

Next we follow the easy (uniform) part of [Przytycki, Urbański & Zdunik 1989] and [Przytycki, Urbański & Zdunik 1991]. We prove that for CER (X, f) and Hölder continuous $\phi : X \to R$, for $\kappa = \mathrm{HD}(\mu_\phi)$, the Hausdorff dimension of the Gibbs measure μ_ϕ (infimum of Hausdorff dimensions of sets of full measure), either $\mathrm{HD}(X) = \kappa$ the measure μ_ϕ is equivalent to Λ_κ, the Hausdorff measure in dimension κ, and is a *geometric measure*, or μ_ϕ is singular with respect to Λ_κ and the right gauge function for the Hausdorff measure to be compared to μ_ϕ is $\Phi(\kappa) = t^\kappa \exp(c\sqrt{\log 1/t \log\log\log 1/t})$. In the proof we use LIL. This theorem is used to prove a dichotomy for the harmonic measure on a Jordan curve ∂, bounding a domain Ω, which is a repeller for a conformal expanding map. Either ∂ is real-analytic, or the harmonic measure is comparable to the Hausdorff measure with gauge function $\Phi(1)$. This yields information about the lower and upper growth rates of $|R'(r\zeta)|$, for $r \nearrow 1$, for almost every ζ with $|\zeta| = 1$ and univalent function R from the unit disc $|z| < 1$ to Ω. This is a dynamical counterpart of Makarov's theory of boundary behaviour for general simply connected domains [Makarov 1985].

We prove, in particular, that for $f_c(z) = z^2 + c, \quad c \neq 0, \quad c \approx 0$ it holds that $1 < \mathrm{HD}(J(f_c)) < 2$.

We show how to express another interesting function in the language of pressure: $\int_{|\zeta|=1} |R'(r\zeta)|^t \, |d\zeta|$ for $r \nearrow 1$.

Finally, we apply our theory to the boundary of the von Koch 'snowflake' and more general Carleson fractals.

Chapter 10 is devoted to Sullivan's rigidity theorem, saying that if two non-linear expanding repellers $(X, f), (Y, g)$ are Lipschitz conjugate (or more generally if there exists a measurable conjugacy that transforms a geometric measure on X to a geometric measure on Y), then the conjugacy extends to a conformal one. This means that measures classify non-linear conformal repellers. This fact, announced in [Sullivan 1986] with only a sketch of the proof, is proved here rigorously for the first time.

(This chapter is one of the oldest chapters in this book; we already made it available in 1991 and many papers have since followed.)

In Chapter 11 we start to deal with non-uniform expanding phenomena. At the heart of this chapter is the proof of the formula $\mathrm{HD}(\mu) = \mathrm{h}_\mu(f)/\chi_\mu(f)$ for an arbitrary f-invariant ergodic measure μ of positive Laypunov exponent $\chi_\mu := \int \log |f'| \, d\mu$.

(The phrase 'non-uniform expanding' is used just to say that we consider (typical points of) an ergodic measure with positive Lyapunov exponent. In higher dimensions one uses the name 'non-uniform hyperbolic' for measures with all Lyapunov exponents non-zero.)

It is so roughly because a small disc around z, whose n-th image is large, has diameter of order $|(f^n)'(z)|^{-1} \approx \exp(-n\chi_\mu)$ and measure $\exp(-n\,\mathrm{h}_\mu(f))$ (the Shannon–McMillan–Breiman theorem is involved here).

Chapter 12 is devoted to conformal measures: that is, probability measures with Jacobian Const $\exp(-\phi)$ or more specifically $|f'|^\alpha$ in a non-uniformly expanding situation, in particular for any rational mapping f on its Julia set J. It is proved that there exists a minimal exponent $\delta(f)$ for which such a measure exists, and that $\delta(f)$ is equal to each of the following quantities:

Dynamical dimension $\mathrm{DD}(J) := \sup\{\mathrm{HD}(\mu)\}$, where μ ranges over all ergodic f-invariant measures on J of positive Lyapunov exponent.

Hyperbolic dimension $\mathrm{HyD}(J) := \sup\{\mathrm{HD}(Y)\}$, where Y ranges over all Conformal Expanding Repellers in J, or CERs that are Cantor sets.

It is an open problem whether for every rational mapping $\mathrm{HyD}(J) = \mathrm{HD}(J) =$ the box dimension of J, but for many non-uniformly expanding mappings these equalities hold. It is often easier to study the continuity of $\delta(f)$ with respect to a parameter, than study the Hausdorff dimension directly. So one obtains information about the continuity of dimensions due to the above equalities.

Section 12.5 presents a recent approach via pressure for the potential function $-t\log|f'|$, yielding a simple proof of the equalities of the above dimensions, see [Przytycki, Rivera-Letelier & Smirnov 2004].

A large part of this book was written in the years 1990–1992, and was lectured to graduate students by each of us in Warsaw, Yale and Denton. We neglected to finish writing, but recently the methods in Chapter 12, relating hyperbolic dimension to minimal exponent of conformal measure, were unexpectedly used to study the dependence on ε of the dimension of the Julia set for $z^2 + 1/4 + \varepsilon$, for $\varepsilon \to 0$ and other parabolic bifurcations, by A. Douady, P. Sentenac and M. Zinsmeister [1997] and by C. McMullen [1996]. So we decided to make final efforts. Meanwhile good books have appeared on some topics of our book: let us mention [Falconer 1997], [Zinsmeister 1996], [Boyarsky & Góra 1997], [Pesin 1997], [Keller 1998], [Baladi 2000] but a lot of important material in our book is new or has been made more easily accessible.

Acknowledgements. We are indebted to Krzysztof Barański for help with figures and Pawel Góra for Figure 2.1. The first author acknowledges the support of consecutive Polish KBN and MNiSW grants; the recent one is N201022233.

1

Basic examples and definitions

Let us start with definitions of dimensions. We shall come back to them in a more systematic way in Chapter 8.

Definition 1.1. Let (X, ρ) be a metric space. We denote by the *upper (lower) box dimension* of X the quantity

$$\overline{\mathrm{BD}}(X) \text{ (or } \underline{\mathrm{BD}}(X)) := \limsup(\liminf)_{r\to 0} \frac{\log N(r)}{-\log r},$$

where $N(r)$ is the minimal number of balls of radius r that cover X.

Sometimes the names *capacity* or *Minkowski dimension* or *box-counting dimension* are used. The name 'box dimension' comes from the situation where X is a subset of a Euclidean space $\mathbb{R}^d$. Then one can consider only $r = 2^{-n}$, and $N(2^{-n})$ can be replaced by the number of dyadic boxes $[\frac{k_1}{2^{-n}}, \frac{k_1+1}{2^{-n}}] \times \cdots \times [\frac{k_d}{2^{-n}}, \frac{k_d+1}{2^{-n}}]$, $k_j \in \mathbb{Z}$ intersecting X.

If $\overline{\mathrm{BD}}(X) = \underline{\mathrm{BD}}(X)$ we call the quantity the *box dimension* and denote it by $\overline{\mathrm{BD}}(X)$.

Definition 1.2. Let (X, ρ) be a metric space. For every $\kappa > 0$ we define $\Lambda_\kappa(X) = \lim_{\delta\to 0} \inf\{\sum_{i=1}^\infty (\operatorname{diam} U_i)^\kappa\}$, where the infimum is taken over all countable covers $(U_i, i = 1, 2, \dots)$ of X by sets of diameter not exceeding δ. $\Lambda_\kappa(Y)$ defined as above on all subsets $Y \subset X$ is called the κ-th outer *Hausdorff measure.*

It is easy to see that there exists $\kappa_0 : 0 \le \kappa_0 \le \infty$ such that for all $\kappa : 0 \le \kappa < \kappa_0$ $\Lambda_\kappa(X) = \infty$ and for all $\kappa : \kappa_0 < \kappa$ $\Lambda_\kappa(X) = 0$. The number κ_0 is called the *Hausdorff dimension* of X.

Note that if in this definition we replace the assumption: sets of diameter not exceeding δ by equal δ, and $\lim_{\delta\to 0}$ by lim inf or lim sup, we obtain the box dimension.

A standard example to compare the two notions is the set $\{1/n, n = 1, 2, \dots\}$ in $\mathbb{R}$. Its box dimension is equal to $1/2$, and the Hausdorff dimension is 0. If one considers $\{2^{-n}\}$ instead one obtains both dimensions as 0. Also, linear Cantor sets, as introduced in the Introduction, have their Hausdorff and box dimensions equal. The reason for this is self-similarity.

Example 1.3. Shift spaces. For every natural number d consider the space Σ^d of all infinite sequences $(i_0, i_1, \dots)$ with $i_n \in \{1, 2, \dots, d\}$. Consider the metric

$$\rho((i_0, i_1, \dots), (i'_0, i'_1, \dots)) = \sum_{n=0}^{\infty} \lambda^n |i_n - i'_n|$$

for an arbitrary $0 < \lambda < 1$. Sometimes it is more convenient to use the metric

$$\rho((i_0, i_1, \dots), (i'_0, i'_1, \dots)) = \lambda^{-\min\{n: i_n \neq i'_n\}},$$

equivalent to the previous one. Consider $\sigma : \Sigma^d \to \Sigma^d$ defined by $\sigma((i_0, i_1, \dots) = (i_1, \dots)$. The metric space (Σ^d, ρ) is called the *one-sided shift space* and the map σ the *left shift.* Often, if we do not specify metric but are interested only in the Cartesian product topology in $\Sigma^d = \{1, \dots, d\}^{\mathbb{Z}^+}$, we use the name *topological shift space.*

One can consider the space $\tilde{\Sigma}^d$ of all two sides infinite sequences $(\dots, i_{-1}, i_0, i_1, \dots)$. This is called the *two-sided shift space.*

Each point $(i_0, i_1, \dots) \in \Sigma^d$ determines its forward trajectory under σ, but is equipped with a Cantor set of backward trajectories. Together with the topology determined by the metric $\sum_{n=-\infty}^{\infty} \lambda^{|n|} |i_n - i'_n|$ the set $\tilde{\Sigma}^d$ can be identified with the *inverse limit* (in the topological category) of the system $\dots \to \Sigma^d \to \Sigma^d$ where all the maps $\to$ are σ.

Note that the limit Cantor set Λ in the Introduction, with all $\lambda_i = \lambda$, is Lipschitz homeomorphic to Σ^d, with the homeomorphism h mapping $(i_0, i_1, \dots)$ to $\bigcap_k T_{i_0} \circ \dots \circ T_{i_k}(I)$. Note that for each $x \in \Lambda$, $h^{-1}(x)$ is the sequence of integers $(i_0, i_1, \dots)$ such that for each k, $\hat{T}^k(x) \in T_{i_k}(I)$. This is called a *coding sequence.* If we allow the end points of $T_i(I)$ to overlap, and in particular $\lambda = 1/d$ and $a_i = (i-1)/d$, then $\Lambda = I$ and $h^{-1}(x) = \sum_{k=0}^{\infty} (i_k - 1) d^{-k-1}$.

One generalizes the one (or two) -sided shift space, sometimes called the *full shift space*, by considering the set Σ_A for an arbitrary $d \times d$ matrix $A = (a_{ij}$ with $a_{ij} = 0$ or 1 defined by

$$\Sigma_A = \{(i_0, i_1, \dots) \in \Sigma^d : a_{i_t i_{t+1}} = 1 \quad \text{for every } t = 0, 1, \dots\}.$$

By the definition $\sigma(\Sigma_A) \subset \Sigma_A$. Σ_A with the mapping σ is called a *topological Markov chain.* Here the word *topological* is substantial; otherwise it is customary to think of a finite number of states stochastic process – see Example 1.9.

Example 1.4. Adding machine. A complementary dynamics on Σ^d above is given by the map $T((i_0, i_1, \dots)) = (1, 1, \dots, 1, i_k + 1, i_k + 1, \dots)$, where k is the least integer for which $i_k < d$. Finally $(d, d, d, \dots) + 1 = (1, 1, 1, \dots)$. (This is of course compatible with standard adding, except that here the sequences are infinite to the right and the digits run from 1 to d, rather than from 0 to $d-1$.) Notice that unlike the previous example, with an abundance of periodic trajectories, here each T-trajectory is dense in Σ^d (such a dynamical system is called *minimal*).

Example 1.5. Iteration of rational maps. Let $f : \overline{\mathbb{C}} \to \overline{\mathbb{C}}$ be a holomorphic mapping of the Riemann sphere $\overline{\mathbb{C}}$. Then it must be rational, i.e. the ratio of two polynomials. We assume that the topological degree of f is at least 2. The *Julia set* $J(f)$ is defined as follows:

$J(f) = \{z \in \overline{\mathbb{C}} : \forall U \ni z, U$ open, the family of iterates $f^n = f \circ \cdots \circ f|_U$, n times, for $n = 1, 2, \ldots$ is not *normal* in the sense of Montel $\}$.

A family of holomorphic functions $f_t : U \to \overline{\mathbb{C}}$ is called *normal* (in the sense of Montel) if it is pre-compact: that is, from every sequence of functions belonging to the family one can choose a sub-sequence uniformly convergent (in the spherical metric on the Riemann sphere $\overline{\mathbb{C}}$) on all compact subsets of U.

$z \in J(f)$ implies for example, that for every $U \ni z$ the family $f^n(U)$ covers all $\overline{\mathbb{C}}$ but at most two points. Otherwise by Montel's theorem $\{f^n\}$ would be normal on U.

Another characterization of $J(f)$ is that $J(f)$ is the closure of repelling periodic points, namely those points $z \in \overline{\mathbb{C}}$ for which there exists an integer n such that $f^n(z) = z$ and $|(f^n)'(z)| > 1$.

There are only a finite number of attracting periodic points, $|(f^n)'(z)| < 1$: they lie outside $J(f)$, which is an uncountable 'chaotic, expansive (repelling)' Julia set. The lack of symmetry between attracting and repelling phenomena is caused by the non-invertibility of f.

It is easy to prove that $J(f)$ is compact, completely invariant: $f(J(f)) = J(f) = f^{-1}(J(f))$, either nowhere dense or equal to the whole sphere (to prove this use Montel's theorem).

For polynomials, the set of points whose images under iterates $f^n, n = 1, 2, \ldots$, tend to ∞, *basin of attraction to* ∞, is connected and completely invariant. Its boundary is the Julia set.

Check that all these general definitions and statements are compatible with the discussion of $f(z) = f_c(z) = z^2 + c$ in the Introduction. As an introduction to this theory we recommend, for example, the books [Beardon 1991], [Carleson & Gamelin 1993], [Milnor 1999] and [Steinmetz 1993].

Figures 1.1–1.3 are computer pictures exhibiting some Julia sets: rabbit, basilica[1] and Sierpiński's carpet of their mating (see [Bielefeld 1990]).

A Julia set can have Hausdorff dimension arbitrarily close to 0 (but not 0) and arbitrarily close to 2 or even exactly 2 (but not the whole sphere). More precisely: a Julia set is always closed and either the whole sphere or nowhere is dense. Recently examples have been found of quadratic polynomials f_c with a Julia set of positive Lebesgue measure (with c in the cardioid; Example 6.1.10): see [Buff & Cheritat 2008]. See also http://picard.ups-tlse.fr/adrien2008/Slides/Cheritat.pdf

[1]The name was proposed by Benoit Mandelbrot [Mandelbrot 1982], impressed by the Basilica San Marco in Venice plus its reflection in flooded Piazza.

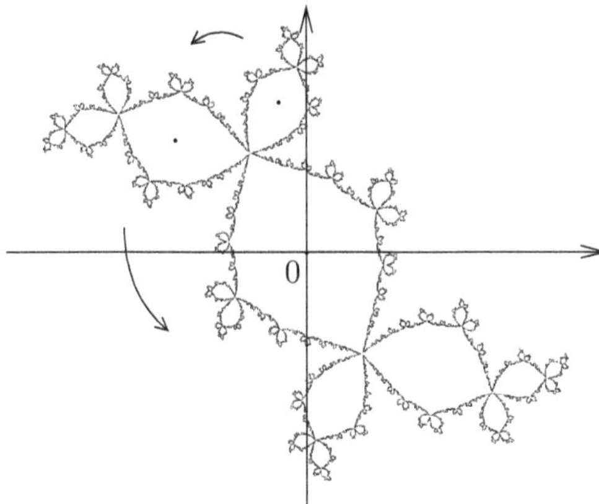

Figure 1.1 Douady's rabbit. Here $f(z) = z^2 + c$, where $c \approx -0.123 + 0.749i$ is a root of $c^3 + 2c^2 + c + 1 = 0$: see [Carleson & Gamelin 1993]. The three distinguished points constitute a period 3 orbit. The arrows hint at the action of f.

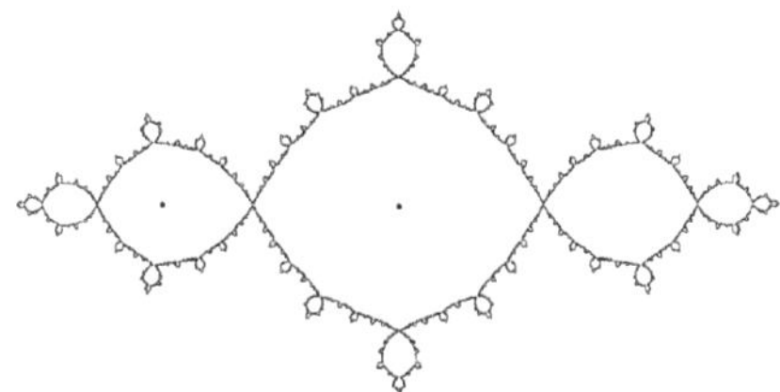

Figure 1.2 Basilica. For decreasing c this shape appears at $c = -3/4$ with thicker components. $f(z) = z^2 - 1$. The critical point 0 is attracting of period 2.

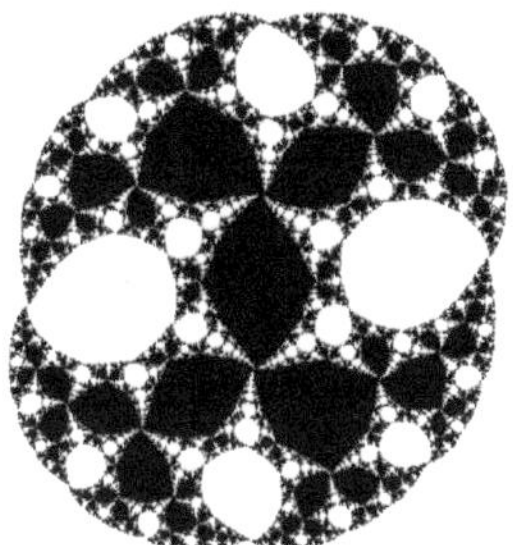

Figure 1.3 The (outer) basilica mated with the rabbit. Here $f(z) = \frac{z^2+c}{z^2-1}$, where $c = \frac{1+\sqrt{-3}}{2}$. Black is attracted to a period 3 orbit, white to period 2. The Julia set is the boundary between black and white.

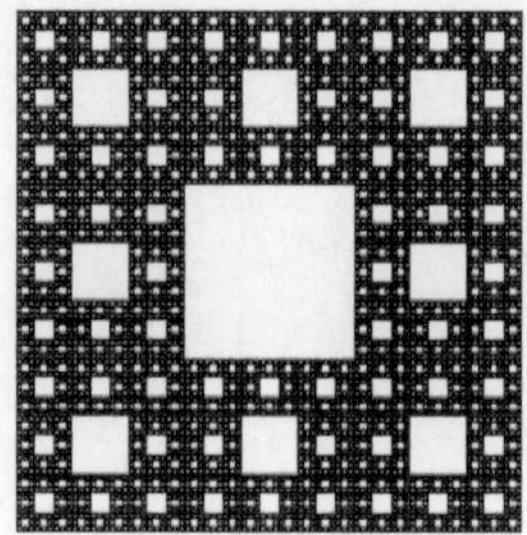
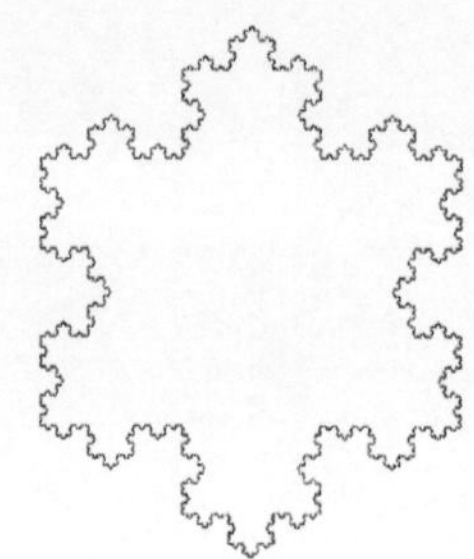

Figure 1.4 Sierpiński gasket, Sierpiński carpet and the boundary of a von Koch snowflake.

Example 1.6. Complex linear fractals. The linear Cantor set construction in $\mathbb{R}$ described in the Introduction can be generalized to conformal linear Cantor and other fractal sets in $\overline{\mathbb{C}}$:

Let $U \subset \mathbb{C}$ be a bounded connected domain and $T_i(z) = \lambda_i z + a_i$, where λ_i, a_i are complex numbers, $i = 1, \ldots, n > 1$. Assume that closures $\mathrm{cl}\, T_i(U)$ are pairwise disjoint and contained in U. The limit Cantor set Λ is defined in the same way as in the Introduction.

In Chapter 10, Example 10.2.8, we shall note that it cannot be the Julia set for a holomorphic extension of $\hat{T} = T_i^{-1}$ on $T_i(U)$ for each i, to the whole sphere $\overline{\mathbb{C}}$.

If we allow that the boundaries of $T_i(U)$ intersect or intersect ∂U we obtain other interesting examples (Figure 1.4).

Example 1.7. Action of Kleinian groups. Beautiful examples of fractal sets arise as limit sets of the action of Kleinian groups on $\overline{\mathbb{C}}$.

Let Ho be the group of all *homographies*, namely the rational mappings of the Riemann sphere of degree 1, i.e. of the form $z \mapsto \frac{az+b}{cz+d}$, where $ad - bc \neq 0$, for complex numbers a, b, c, d. Every discrete subgroup of Ho is called a *Kleinian group.* If all the elements of a Kleinian group preserve the unit disc $\mathbb{D} = \{|z| < 1\}$, the group is called *Fuchsian.*

Consider, for example, a regular hyperbolic $4n$-gon in $\mathbb{D}$ (equipped with the hyperbolic metric) centred at 0 (Figure 1.5). Denote the consecutive sides by $a_i^j, i = 1, \ldots, n, j = 1, \ldots, 4$ in the lexicographical order $a_1^1, \ldots a_1^4, a_2^1, \ldots$. Each side is contained in the corresponding circle C_i^j intersecting $\partial\mathbb{D}$ at right angles. Denote the disc bounded by C_i^j by D_i^j.

It is not hard to see that the closures of D_i^j and D_i^{j+2} are disjoint for each i and $j = 1, 2$.

Let $g_i^j, j = 1, 2$ be the unique homography preserving $\mathbb{D}$ mapping a_i^j to a_i^{j+2} and D_i^j to the complement of $\mathrm{cl}\, D_i^{j+2}$. It is easy to see that the family $\{g_i^j\}$ generates a Fuchsian group G. For an arbitrary Kleinian group G, the *Poincaré limit set* $\Lambda(G) = \bigcup \lim_{k\to\infty} g_k(z)$, the union taken over all

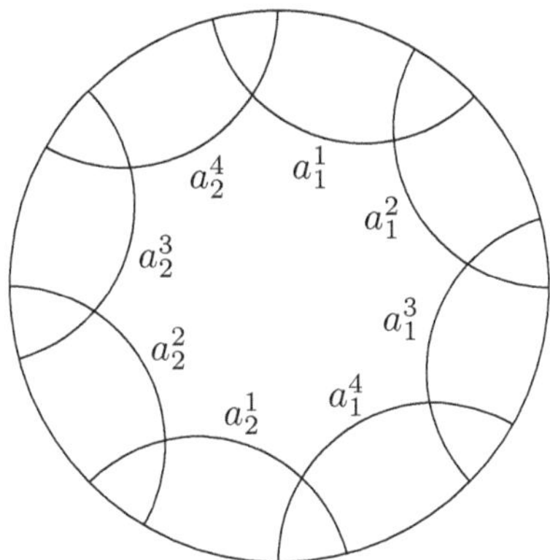

Figure 1.5 Regular hyperbolic octagon

sequences of pairwise different $g_k \in G$ such that $g_k(z)$ converges, where z is an arbitrary point in $\overline{\mathbb{C}}$. It is not hard to prove that $\Lambda(G)$ does not depend on z.

For the above example $\Lambda(G) = \partial\mathbb{D}$. If we change g_i^j slightly (the circles C_i^j change slightly), then either $\Lambda(G)$ is a circle S (all new C_i^j intersect S at the right angle), or it is a fractal Jordan curve. The phenomenon is similar to the case of the maps $z \mapsto z^2 + c$ described in the Introduction and in more detail in Section 9.5. For details see [Bowen 1979], [Bowen & Series 1979] and [Sullivan 1982].

If all the closures of the discs $D_i^j, i = 1, \ldots, n, j = 1, \ldots, 4$ become pairwise disjoint, $\Lambda(G)$ becomes a Cantor set (the group is called then a *Schottky group* or a *Kleinian group of Schottky type*).

Example 1.8. Higher dimensions. Though the book is devoted to one-dimensional real and complex iteration and arising fractals, Chapters 2–4 apply to general situations. A basic example is Smale's horseshoe. Take a square $K = [0,1] \times [0,1]$ in the plane $\mathbb{R}^2$ and map it affinely to a strip by squeezing in the horizontal direction and stretching in the vertical, for example $f(x,y) = (\frac{1}{5}x + \frac{1}{4}, 3y - \frac{1}{8})$, and bend the strip by a new affine map g, which maps the rectangle $[\frac{1}{5}, \frac{2}{5}] \times [\frac{7}{4}, \frac{23}{8}]$ to $[\frac{3}{5}, \frac{4}{5}] \times [-\frac{1}{8}, 1]$. The resulting composition $T = g \circ f$ maps K to a 'horseshoe': see [Smale 1967, p. 773]

The map can be easily extended to a C^∞-diffeomorphism of $\overline{\mathbb{C}}$ by mapping a 'stadium' extending K to a bent 'stadium', and mapping its complement to the respective complement (Figure 1.6). The set Λ^K of points not leaving K under action of $T^n, n = \ldots, -1, 0, 1, \ldots$ is the cartesian product of two Cantor sets. This set is T-invariant, 'uniformly hyperbolic'. In the horizontal direction we have contraction; in the vertical direction uniform expansion. The situation is different from the previous examples of Σ^d or linear Cantor sets, where we had uniform expansion in all directions.

Smale's horseshoe is a universal phenomenon. It is always topologically present for an iterate of a diffeomorphism f having a *transversal homoclinic*

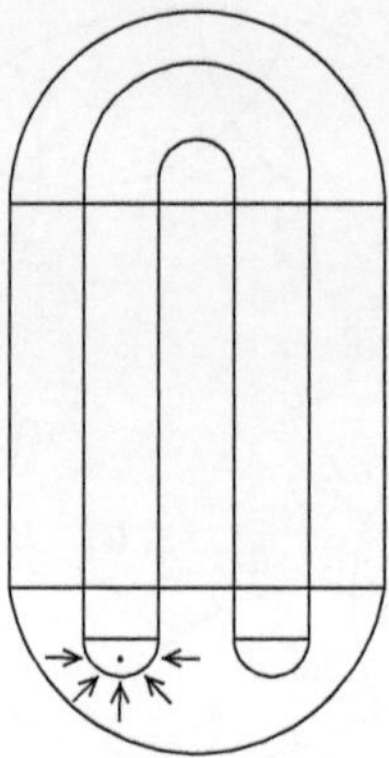

Figure 1.6 Horseshoe, stadium extension

point q for a saddle p (Figure 1.7). The latter says that the stable and unstable manifolds $W^s(p) := \{y : f^n(y) \to p\}, W^u(p) := \{y : f^{-n}(y) \to p\}$ as $n \to \infty$ intersect transversally at q. For more details on hyperbolic sets see [Katok & Hasselblatt 1995]. Compare heteroclinic intersections in Chapter 4, Exercise 4.8.

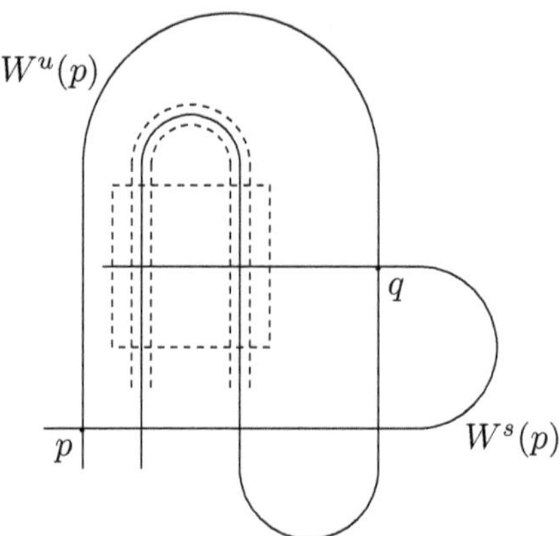

Figure 1.7 Homoclinic point

Note that $T|_{\Lambda^K}$ is *topologically conjugate* to the left shift σ on the two-sided shift space $\tilde{\Sigma}^2$: that is, there exists a homeomorphism $h : \Lambda^K \to \tilde{\Sigma}^2$ such that $h \circ T = \sigma \circ h$. Compare h in Example 1.3. T on Λ^K is the inverse limit of the mapping $\hat{T}$ on the Cantor set described in the Introduction, similar to the inverse limit $\tilde{\Sigma}^2$ of σ on Σ^2. The philosophy is that hyperbolic systems appear as inverse limits of expanding systems.

A partition of a hyperbolic set Λ into local stable (unstable) sets, $W^s(x) = \{y \in \Lambda : (\forall n \geq 0)\rho(f^n(x), f^n(y)) \leq \varepsilon(x)\}$ for a small positive measurable function ε, is an illustration of an abstract ergodic theory measurable partition ξ such that $f(\xi)$ is finer than ξ, $f^n(\xi), n \to \infty$ converges to the partition into points, and the conditional entropy $\mathrm{H}_\mu(f(\xi)|\xi)$ is maximal possible, equal to the entropy $\mathrm{h}_\mu(f)$: all this holds for an ergodic invariant measure μ.

The inverse limit of the system $\cdots \to S^1 \to S^1$ where all the maps are $z \mapsto z^2$, is called a *solenoid*. It has a group structure $(\dots, z_{-1}, z_0)\cdot(\dots, z'_{-1}, z'_0) = (\dots, z_{-1}\cdot z'_{-1}, z_0 \cdot z'_0)$, which is a trajectory if both factors are, since the map $z \mapsto z^2$ is a homomorphism of the group S^1. Topologically the solenoid can be represented as the attractor A of the mapping of the solid torus $\mathbb{D} \times S^1$ into itself $f(z,w) = (\frac{1}{3}z + \frac{1}{2}w, w^2)$. Its Hausdorff dimension is equal in this special example to $1 + \mathrm{HD}(A \cap \{w = w_0\}) = 1 + \frac{\log 2}{\log 3}$ for an arbitrary w_0, as Cantor sets $A \cap \{w = w_0\}$ have Hausdorff dimensions $\frac{\log 2}{\log 3}$. These are linear Cantor sets, discussed in the Introduction.

Especially interesting is the question of the Hausdorff dimension of A if $z \mapsto \frac{1}{3}z$ is replaced by $z \mapsto \phi(z)$ not conformal; but this higher-dimensional problem goes beyond the scope of our book.

If the map $z \mapsto z^2$ in the definition of a solenoid is replaced by an arbitrary rational mapping, then if f is expanding on the Julia set, the solenoid is locally the cartesian product of an open set in $J(f)$ and the Cantor set of all possible choises of backward trajectories. If, however, there are critical points in $J(f)$ (or converging under the action of f^n to parabolic points in $J(f)$), the solenoid (inverse limit) is more complicated: see [Lyubich & Minsky 1997] and more recent papers for an attempt to describe it, together with a neighbourhood composed of trajectories outside $J(f)$. We shall not discuss this in this book.

Example 1.9. Bernoulli shifts and Markov chains. For every positive numbers $p_1, \dots, p_d$ such that $\sum_{i=1}^d p_i = 1$, one introduces on the Borel subsets of Σ^d (or $\tilde{\Sigma}^d$) a probability measure μ by extending to the σ-algebra of all Borel sets the function $\mu(C_{i_0,i_1,\dots,i_t}) = p_0 p_1 \dots p_t$, where $C_{i_0,i_1,\dots,i_t} = \{(i'_0, i'_1, \dots) : i'_s = i_s \text{ for every } s = 0, 1, \dots, t\}$. Each such C is called a *finite cylinder.*

The space Σ^d with left shift σ and measure μ is called a one-sided *Bernoulli shift.*

On a topological Markov chain $\Sigma_A \subset \Sigma^d$ with $A = (a_{ij})$ and an arbitrary $d \times d$ matrix $M = p_{ij}$ such that $\sum_{j=1}^d p_{ij} = 1$ for every $i = 1, \dots, d$, $p_{ij} \geq 0$ and $p_{ij} = 0$ if $a_{ij} = 0$, one can introduce a probability measure μ on all Borel subsets of Σ_A by extending $\mu(C_{i_0,i_1,\dots,i_t}) = p_{i_0} p_{i_0 i_1} \cdots p_{i_{t-1} i_t}$. Here $(p_1, \dots, p_d)$ is an eigenvector of M^*, namely $\sum_i p_i p_{ij} = p_j$, such that $p_i \geq 0$ for every $i = 1, \dots, d$ and $\sum_{i=1}^d = 1$.

The space Σ_A with left shift σ and measure μ is called a one-sided *Markov chain.*

Note that μ is σ-invariant. Indeed,

$$\mu\Big(\bigcup_i (C_{i,i_0,\dots,i_t})\Big) = \sum_i p_i p_{i i_0} p_{i_0 i_1} \cdots p_{i_{t-1} i_t} = p_{i_0} p_{i_0 i_1} \cdots p_{i_{t-1} i_t} = \mu(C_{i_0,\dots,i_t}).$$

As in the topological case, if we consider $\tilde{\Sigma}^d$ rather than Σ^d, we obtain two-sided Bernoulli shifts and two-sided Markov chains.

Example 1.10. Tchebyshev polynomial. Let us consider the mapping $T : [-1,1] \to [-1,1]$ of the real interval $[-1,1]$ defined by $T(x) = 2x^2 - 1$. In the co-ordinates $z \mapsto 2z$ it is just a restriction to an invariant interval of the mapping $z \mapsto z^2 - 2$, already discussed in the Introduction. The interval $[-1,1]$ is the Julia set of T.

Notice that this map is the factor of the mapping $z \mapsto z^2$ on the unit circle $\{|z| = 1\}$ in $\mathbb{C}$ by the orthogonal projection P to the real axis. Since the length measure l is preserved by $z \mapsto z^2$, its projection is preserved by T. Its density with respect to the Lebesgue measure on $[-1,1]$ is proportional to $(dP/dl)^{-1}$, and after normalization is equal to $\frac{1}{\pi}\frac{1}{\sqrt{1-x^2}}$. This measure satisfies many properties of Gibbs invariant measures, discussed in Chapter 5, though T is not expanding; it has a critical point at 0. This T is the simplest example of a non-uniformly expanding map, to which the advanced parts of the book are devoted. See also Figures 2.1 and 2.2 in Section 2.2.

2

Measure-preserving endomorphisms

2.1 Measure spaces and the Martingale Theorem

We assume that the reader is familiar with the basic elements of measure and integration theory. For a complete treatment see, for example, [Halmos 1950] or [Billingsley 1979]. We start with some basics to introduce the notation and terminology.

A family $\mathcal{F}$ of subsets of a set X is said to be a *σ-algebra* if the following conditions are satisfied:

$$X \in \mathcal{F}, \tag{2.1.1}$$

$$A \in \mathcal{F} \Rightarrow A^c \in \mathcal{F} \tag{2.1.2}$$

and

$$\{A_i\}_{i=1}^{\infty} \subset \mathcal{F} \Rightarrow \bigcup_{i=1}^{\infty} A_i \in \mathcal{F}. \tag{2.1.3}$$

It follows from this definition that $\emptyset \in \mathcal{F}$: that the σ-algebra $\mathcal{F}$ is closed under countable intersections and under subtractions of sets. If (2.1.3) is assumed only for finite subfamilies of $\mathcal{F}$ then $\mathcal{F}$ is called an *algebra*. The elements of the σ-algebra $\mathcal{F}$ will frequently be called *measurable* sets.

Notation 2.1.1. *For any family $\mathcal{F}_0$ of subsets of X, we denote by $\sigma(\mathcal{F}_0)$ the least σ-algebra that contains $\mathcal{F}_0$, and we call it the σ-algebra generated by $\mathcal{F}_0$.*

A function on a σ-algebra $\mathcal{F}$, $\mu : \mathcal{F} \to [0, \infty]$, is said to be *σ-additive* if for any countable subfamily $\{A_i\}_{i=1}^{\infty}$ of $\mathcal{F}$ consisting of mutually disjoint sets we have

$$\mu\Big(\bigcup_{i=1}^{\infty} A_i\Big) = \sum_{i=1}^{\infty} \mu(A_i). \tag{2.1.4}$$

We say then that μ is a *measure*. If we consider in (2.1.4) only finite families of sets, we say μ is *additive*. The two notions of *additivity* and of *σ-additivity* make sense for a σ-algebra as well as for an algebra, provided that in the case of an algebra one considers only families $\{A_i\} \subset \mathcal{F}$ such that $\bigcup A_i \in \mathcal{F}$. The simplest consequences of the definition of measure are the following:

$$\mu(\emptyset) = 0; \tag{2.1.5}$$

$$\text{if } A, B \in \mathcal{F} \text{ and } A \subset B \text{ then } \mu(A) \le \mu(B); \tag{2.1.6}$$

if $A_1 \subset A_2 \subset \dots$ and $\{A_i\}_{i=1}^{\infty} \subset \mathcal{F}$ then

$$\mu\Big(\bigcup_{i=1}^{\infty} A_i\Big) = \sup_i \mu(A_i) = \lim_{i\to\infty} \mu(A_i). \tag{2.1.7}$$

We say that the triple $(X, \mathcal{F}, \mu)$ with a σ-algebra $\mathcal{F}$ and μ a measure on $\mathcal{F}$ is a *measure space*. In this book we shall always assume, unless otherwise stated, that μ is a *finite measure*: that is, $\mu : \mathcal{F} \to [0, \infty)$. By (2.1.6) this equivalently means that $\mu(X) < \infty$. If $\mu(X) = 1$, the triple $(X, \mathcal{F}, \mu)$ is called a *probability space* and μ a *probability measure*.

We say that $\phi : X \to \mathbb{R}$ is a measurable function, if $\phi^{-1}(J) \in \mathcal{F}$ for every interval $J \subset \mathbb{R}$, equivalently for every Borel set $J \subset \mathbb{R}$ (compare Section 2.2). We say that ϕ is μ-integrable if $\int |\phi|\, d\mu < \infty$. We write $\phi \in L^1(\mu)$. More generally, for every $1 \le p < \infty$ we write $(\int |\phi|^p\, d\mu)^{1/p} = \|\phi\|_p$, and we say that ϕ belongs to $L^p(\mu) = L^p(X, \mathcal{F}, \mu)$. If $\inf_{\mu(E)=0} \sup_{X\setminus E} |\phi| < \infty$, we say that $\phi \in L^\infty$ and denote the latter expression by $\|\phi\|_\infty$. The numbers $\|\phi\|_p, 1 \le p \le \infty$ are called L^p-norms of ϕ. We usually identify in this chapter functions that differ only on a set of μ-measure 0. After these identifications the linear spaces $L^p(X, \mathcal{F}, \mu)$ become Banach spaces with the norms $\|\phi\|_\infty$.

We say that a property $q(x)$, $x \in X$, is satisfied for μ almost every $x \in X$ (abbr: a.e.), or μ-a.e., if $\mu(\{x : q(x) \text{ is not satisfied}\}) = 0$. We can consider q as a subset of X with $\mu(X \setminus q) = 0$.

We shall often use in this book the following two facts.

Theorem 2.1.2 (Monotone Convergence Theorem). *Suppose $\phi_1 \le \phi_2 \le \dots$ is an increasing sequence of integrable, real-valued functions on a probability space $(X, \mathcal{F}, \mu)$. Then $\phi = \lim_{n\to\infty} \phi_n$ exists a.e. and $\lim_{n\to\infty} \int \phi_n\, d\mu = \int \phi\, d\mu$. (We allow $+\infty$'s here.)*

Theorem 2.1.3 (Dominated Convergence Theorem). *If $(\phi_n)_{n=1}^{\infty}$ is a sequence of measurable real-valued functions on a probability space $(X, \mathcal{F}, \mu)$ and $|\phi_n| \le g$ for an integrable function g, and $\phi_n \to \phi$ a.e., then ϕ is integrable and $\lim_{n\to\infty} \int \phi_n\, d\mu = \int \phi\, d\mu$.*

Recall now that if $\mathcal{F}'$ is a sub-σ-algebra of $\mathcal{F}$ and $\phi : X \to \mathbb{R}$ is a μ-integrable function, then there exists a unique (mod 0) function, usually denoted by $E(\phi|\mathcal{F}')$, such that $E(\phi|\mathcal{F}')$ is $\mathcal{F}'$-measurable and

$$\int_A E(\phi|\mathcal{F}')\, d\mu = \int_A \phi\, d\mu \tag{2.1.8}$$

for all $A \in \mathcal{F}'$. $E(\phi|\mathcal{F}')$ is called the *conditional expectation value* of the function ϕ with respect to the σ-algebra $\mathcal{F}'$. Sometimes we shall use for $E(\phi|\mathcal{F}')$ the simplified notation $\phi_{\mathcal{F}'}$.

For $\mathcal{F}$, generated by a finite partition $\mathcal{A}$ (cf. Section 2.3), one can think of $E(\phi|\sigma(\mathcal{A}))$ as constant on each $A \in \mathcal{A}$ equal to the average $\int_A \phi\, d\mu/\mu(A)$.

The existence of $E(\phi|\mathcal{F}')$ follows from the well-known Radon–Nikodym Theorem, which says that if $\nu \ll \mu$, with both measures defined on the same σ-algebra $\mathcal{F}'$ (where $\nu \ll \mu$ means that ν is *absolutely continuous* with respect to μ, i.e. $\mu(A) = 0 \Rightarrow \nu(A) = 0$ for all $A \in \mathcal{F}'$), then there exists a unique (mod 0) $\mathcal{F}'$-measurable, μ-integrable function $\Phi = d\nu/d\mu : X \to \mathbb{R}^+$, called the Radon–Nikodym derivative, such that for every $A \in \mathcal{F}'$

$$\int_A \Phi\, d\mu = \nu(A).$$

To deduce (2.1.8) we set $\nu(A) = \int_A \phi\, d\mu$ for $A \in \mathcal{F}'$. The trick is that we restrict μ from $\mathcal{F}$ to $\mathcal{F}'$: that is, we apply the Radon–Nikodym Theorem for $\nu \ll \mu|_{\mathcal{F}'}$.

If $\phi \in L^p(X, \mathcal{F}, \mu)$ then $E(\phi|\mathcal{F}') \in L^p(X, \mathcal{F}', \mu)$ for all σ-algebras $\mathcal{F}'$ with L^p norms uniformly bounded. More precisely, the operators $\phi \mapsto E(\phi|\mathcal{F}')$ are linear projections from $L^p(X, \mathcal{F}, \mu)$ to $L^p(X, \mathcal{F}', \mu)$, with L^p-norms equal to 1 (see Exercise 2.7).

For a sequence $(\mathcal{F}_n))_{n=1}^\infty$ of σ-algebras contained in $\mathcal{F}$, denote by $\bigvee_{n=1}^\infty \mathcal{F}_n$ the smallest σ-algebra containing $\bigcup_{n=1}^\infty \mathcal{F}_n$ The latter union is usually not a σ-algebra, but only an algebra (if the sequence is ascending). According to Notation 2.1.1, $\bigvee_{n=1}^\infty \mathcal{F}_n = \sigma(\bigcup_{n=1}^\infty \mathcal{F}_n)$. Compare Section 2.6, where complete σ-algebras of this form are considered in Lebesgue spaces.

We end this section with the following version of the **Martingale Convergence Theorem**.

Theorem 2.1.4. *If $(\mathcal{F}_n : n \geq 1)$ is either an ascending or a descending sequence of σ-algebras contained in $\mathcal{F}$, then for every $\phi \in L^p(\mu)$, $1 \leq p < \infty$, we have*

$$\lim_{n\to\infty} E(\phi|\mathcal{F}_n) = E(\phi|\mathcal{F}'), \quad \text{a.e. and in } L^p,$$

where $\mathcal{F}'$ is equal either to $\bigvee_{n=1}^\infty \mathcal{F}_n$ or to $\bigcap_{n=1}^\infty \mathcal{F}_n$ respectively.

Recall that a sequence of μ-measurable functions $\psi_n : X \to \mathbb{R}$, $n = 1, 2, \dots$ is said to *converge in measure* μ to ψ if for every $\varepsilon > 0$, $\lim_{n\to\infty} \mu(\{x \in X : |\psi_n(x) - \psi(x)| \geq \varepsilon\}) = 0$.

In this book we denote by $1\!\!1_A$ the indicator function of A, namely the function equal to 1 on A and to 0 outside A.

Remark 2.1.5. For the existence of $\mathcal{F}'$ and the convergence in L^p in Theorem 2.1.4, no monotonicity is needed. It is sufficient to assume that for every $A \in \mathcal{F}$ the limit $\lim E(1\!\!1_A|\mathcal{F}_n)$ in measure μ exists.

We shall not provide here any proof of Theorem 2.1.4 in the full generality (but see Exercise 2.5). However, let us at least provide a proof of Theorem 2.1.4

(and of Remark 2.1.5 in the case $\lim E(1\!\!1_A|\mathcal{F}_n) = 1\!\!1_A$) for the L^2-convergence for functions $\phi \in L^2(\mu)$. This is sufficient, for example, to prove the important Theorem 2.8.6 (proof 2) later on in this chapter.

For any ascending sequence $(\mathcal{F}_n)$ we have the equality

$$L^2(X, \mathcal{F}', \mu) = \overline{\bigcup_n L^2(X, \mathcal{F}_n, \mu)}. \tag{2.1.9}$$

Indeed, for every $B, C \in \mathcal{F}$ write $B \div C = (B \setminus C) \cup (C \setminus B)$, the so-called *symmetric difference* of sets B and C. Note that for every $B \in \mathcal{F}'$ there exists a sequence $B_n \in \mathcal{F}_n$, $n \geq 1$, such that $\mu(B \div B_n) \to 0$.

This follows, for example, from Carathéodory's argument: see the comments after the statement of Theorem 2.7.2. We have $\mu(B)$ equal to the outer measure of B constructed from μ restricted to the algebra $\bigcup_{n=1}^{\infty} \mathcal{F}_n$. In the Remark 2.1.5 case, where we assumed $\lim E(1\!\!1_A|\mathcal{F}_n) = 1\!\!1_A$, this is immediate.

Hence $L^2(X, \mathcal{F}_n, \mu) \ni 1\!\!1_{B_n} \to 1\!\!1_B$ in $L^2(X, \mathcal{F}, \mu)$. Finally, to get (2.7.2), use the fact that every function $f \in L^2(X, \mathcal{F}', \mu)$ can be approximated in the space $L^2(X, \mathcal{F}', \mu)$ by the step functions, i.e. finite linear combinations of indicator functions.

Therefore, since $E(\phi|\mathcal{F}_n)$ and $E(\phi|\mathcal{F}')$ are orthogonal projections of ϕ to $L^2(X, \mathcal{F}_n, \mu)$ and $L^2(X, \mathcal{F}', \mu)$ respectively (exercise), we obtain $E(\phi|\mathcal{F}_n) \to E(\phi|\mathcal{F}')$ in L^2.

For a decreasing sequence $\mathcal{F}_n$ use the equality $L^2(X, \mathcal{F}', \mu) = \bigcap_n L^2(X, \mathcal{F}_n, \mu)$.

2.2 Measure-preserving endomorphisms; ergodicity

Let $(X, \mathcal{F}, \mu)$ and $(X', \mathcal{F}', \mu')$ be measure spaces. A transformation $T : X \to X'$ is said to be *measurable* if $T^{-1}(A) \in \mathcal{F}$ for every $A \in \mathcal{F}'$. If, moreover, $\mu(T^{-1}(A)) = \mu'(A)$ for every $A \in \mathcal{F}'$, then T is called *measure preserving*. We write $\mu' = \mu \circ T^{-1}$ or $\mu' = T_*(\mu)$.

We call $(X', \mathcal{F}', \mu')$ *a factor* (or quotient) of $(X, \mathcal{F}, \mu)$, and $(X, \mathcal{F}, \mu)$ *an extension* of $(X, \mathcal{F}, \mu)$.

If a measure-preserving map $T : X \to X'$ is invertible, and the inverse T^{-1} is measurable, then clearly T^{-1} is also measure preserving. Therefore T is an *isomorphism* in the category of measure spaces.

If $(X, \mathcal{F}, \mu) = (X', \mathcal{F}', \mu')$ we call T a *measure-preserving endomorphism*; we shall also say that the measure μ is T–invariant, or that T preserves μ. In the case of $(X, \mathcal{F}, \mu) = (X', \mathcal{F}', \mu')$ an isomorphism T is called an *automorphism*.

If T and T' are endomorphisms of $(X, \mathcal{F}, \mu)$ and $(X', \mathcal{F}', \mu')$ respectively, and $S : X \to X'$ is a measure-preserving transformation from $(X, \mathcal{F}, \mu)$ to $(X', \mathcal{F}', \mu')$ such that $F' \circ S = S \circ F$, then we call $T' : X \to X'$ a factor of $T : X \to X$ and $T : X \to X$ an extension of $T : X' \to X'$.

For every μ-measurable ϕ we define $U_T(\phi) = \phi \circ T$.

U_T is sometimes called the Koopman operator. We have the following easy proposition:

Proposition 2.2.1. *For $\phi \in L^1(X', \mathcal{F}', \mu')$ we have $\int \phi \circ T \, d\mu = \int \phi \, d\mu \circ T^{-1}$. Moreover, for each p the adequate restriction of the Koopman operator $U_T : L^p(X', \mathcal{F}', \mu') \to L^p(X, \mathcal{F}, \mu)$ is an isometry to the image, surjective if and only if T is an isomorphism.*

The isometry operator U_T has been widely explored to understand measure-preserving endomorphisms T. Especially convenient has been $U_T : L^2(\mu) \to L^2(\mu)$, the isometry of the Hilbert space $L^2(\mu)$. Notice that it is an isomorphism (that is *unitary*) if and only if T is an automorphism. For more properties see Exercise 2.23.

We shall now prove the following very useful fact, in which the finiteness of measure is a crucial assumption.

Theorem 2.2.2 (Poincaré Recurrence Theorem). *If $T : X \to X$ is a (finite!) measure-preserving endomorphism, then for every mesurable set A*

$$\mu\big(\{x \in A : T^n(x) \in A \ \text{ for infinitely many } n\text{'s}\}\big) = \mu(A).$$

Proof. Let

$$N = N(T, A) = \{x \in A : T^n(x) \notin A \ \forall n \geq 1\}.$$

We shall first show that $\mu(N) = 0$. Indeed, N is measurable since $N = A \cap \bigcap_{n \geq 1} T^{-n}(X \setminus A)$. If $x \in N$, then $T^n(x) \notin A$ for all $n \geq 1$ and, in particular, $T^n(x) \notin N$, which implies that $x \notin T^{-n}(N)$, and consequently $N \cap T^{-n}(N) = \emptyset$ for all $n \geq 1$. Thus all the sets N, $T^{-1}(N)$, $T^{-2}(N), \ldots$ are mutually disjoint, since if $n_1 \leq n_2$ then

$$T^{-n_1}(N) \cap T^{-n_2}(N) = T^{-n_1}(N \cap T^{-(n_2 - n_1)}(N)) = \emptyset.$$

Hence

$$\infty > \mu\left(\bigcup_{n=0}^{\infty} T^{-n}(N)\right) = \sum_{n=0}^{\infty} \mu(T^{-n}(N)) = \sum_{n=0}^{\infty} \mu(N).$$

Therefore $\mu(N) = 0$. Now set $k \geq 1$ and put

$$N_k = \{x \in A : T^n(x) \notin A \ \forall n \geq k\}.$$

Then $N_k \subset N(T^k, A)$ and therefore from what has been proved above it follows that $\mu(N_k) \leq \mu(N(T^k, A)) = 0$. Thus

$$\mu\big(\{x \in A : T^n(x) \in A \ \text{ for only finitely many } n\text{'s}\}\big) = 0.$$

The proof is complete. ♣

Definition 2.2.3. A measurable transformation $T : X \to X$ of a measure space $(X, \mathcal{F}, \mu)$ is said to be *ergodic* if for any measurable set A

$$\mu(T^{-1}(A) \div A) = 0 \quad \Rightarrow \quad \mu(A) = 0 \text{ or } \mu(X \setminus A) = 0.$$

Recall the notation $B \div C = (B \setminus C) \cup (C \setminus B)$.

Note that we did not assume in the definition of ergodicity that μ is T-invariant (nor that μ is finite). Suppose that for every E of measure 0 the set $T^{-1}(E)$ is also of measure 0. (In Chapter 5 we call this property of μ with respect to T *backward quasi-invariance.* In the literature the name *non-singular* is also used.) Then in the definition of ergodicity one can replace $\mu(T^{-1}(A) \div A) = 0$ by $T^{-1}(A) = A$. Indeed, having A as in the definition, one can define $A' = \bigcap_{n=0}^{\infty} \bigcup_{m=n}^{\infty} T^{-m}(A)$. Then $\mu(A') = \mu(A)$ and $T^{-1}(A') = A'$. If we assume that the latter implies $\mu(A') = 0$ or $\mu(X \setminus A') = 0$, then $\mu(A) = 0$ or $\mu(X \setminus A) = 0$.

Remark 2.2.4. If T is an isomorphism then T is ergodic if and only if T^{-1} is ergodic.

Let $\phi : X \to \mathbb{R}$ be a measurable function. For any $n \geq 1$ we define

$$S_n\phi = \phi + \phi \circ T + \ldots + \phi \circ T^{n-1}. \tag{2.2.1}$$

Let $\mathcal{I} = \{A \in \mathcal{F} : \mu(T^{-1}(A) \div A) = 0\}$. We call $\mathcal{I}$ the σ-algebra of T-invariant (mod 0) sets. Note that every $\psi : X \to \mathbb{R}$, measurable with respect to $\mathcal{I}$, is T-invariant (mod 0): that is, $\psi \circ T = \psi$ on the complement of a set of measure μ equal to 0.

Indeed, let $A = \{x \in X : \psi(x) \neq \psi \circ T(x)\}$, and suppose $\mu(A) > 0$. Then there exists $a \in \mathbb{R}$ such that either $A_a^+ = \{x \in A : \psi(x) < a, \psi \circ T(x) > a\}$ or $A_a^- = \{x \in A : \psi(x) > a, \psi \circ T(x) < a\}$ has positive μ-measure. In the case of A^+ we have $\psi \circ T > a$ on $T^{-1}(A_a^+)$. We conclude that $\psi > a$ and $\psi < a$ on $A_a^+ \cap T^{-1}(A_a^+)$ simultaneously, which contradicts $A_a^+ \cap T^{-1}(A_a^+) = \mu(A_a^+) > 0$. The case of A^- can be dealt with similarly.

Theorem 2.2.5 (Birkhoff's Ergodic Theorem). *If $T : X \to X$ is a measure-preserving endomorphism of a probability space $(X, \mathcal{F}, \mu)$ and $\phi : X \to \mathbb{R}$ is an integrable function, then*

$$\lim_{n\to\infty} \frac{1}{n} S_n\phi(x) = E(\phi|\mathcal{I}) \quad \textit{for } \mu\textit{-a.e. } x \in X\,.$$

If, in addition, T is ergodic, then

$$\lim_{n\to\infty} \frac{1}{n} S_n\phi(x) = \int \phi \, d\mu, \quad \textit{for } \mu\textit{-a.e. } x \tag{2.2.2}$$

We say that the *time average* exists for μ-almost every $x \in X$. If T is ergodic, we say that the *time average equals the space average.*

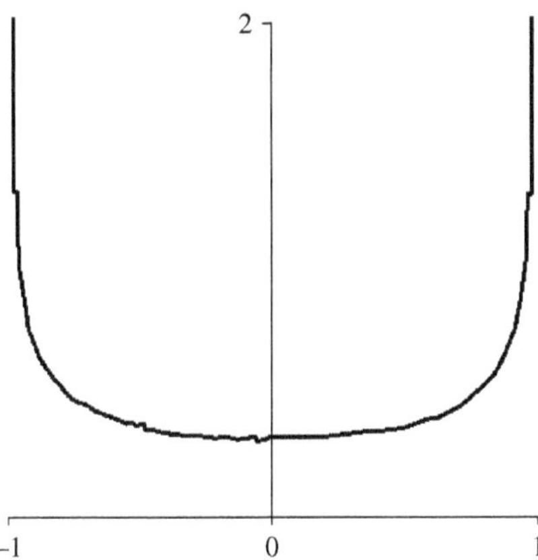

Figure 2.1 The plotted density of an invariant measure for $T(x) = 2x^2 - 1$.

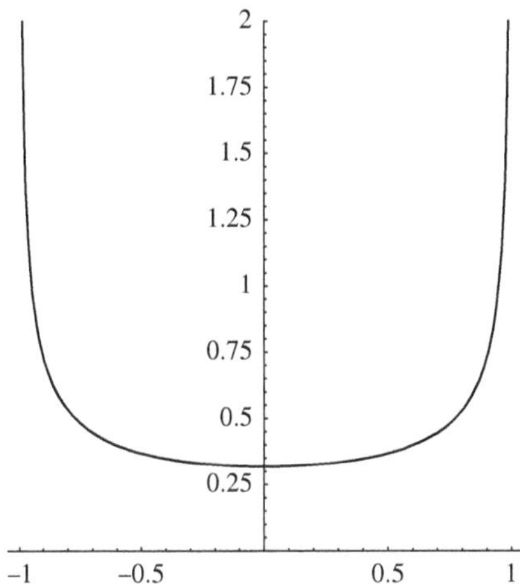

Figure 2.2 The density of an invariant measure for $T(x) = 2x^2 - 1$.

If $\phi = 1\!\!1_A$, the indicator function of a measurable set A, then we deduce that for μ-a.e. $x \in X$ the frequency of hitting A by the forward trajectory of x is equal to the measure (probability) of A: that is,

$$\lim_{n\to\infty} \#\{0 \le j < n : T^j(x) \in A\}/n = \mu(A). \tag{2.2.3}$$

This means, for example, that if we choose a point in X being a bounded invariant part of Euclidean space at random, its sufficiently long forward trajectory fills X, with the density being approximately the density of μ with respect to the Lebesgue measure, provided μ is equivalent to the Lebesgue measure.

In Figure 2.1, for a randomly chosen $x \in [-1, 1]$, the trajectory $T^j(x), j = 0, 1, \dots, n$, for $T(x) = 2x^2 - 1$ is plotted. See Example 1.8. The interval $[-1, 1]$ is

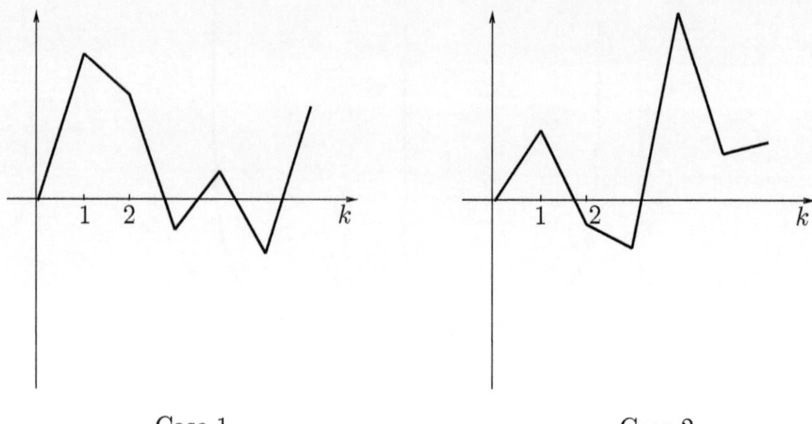

Figure 2.3 Graph of $k \mapsto \sum_{i=0}^{k-1} f \circ T^i(x), k = 1, 2, \ldots$. Case 1: $F_{n+1}(x) = f(x)$ (i.e. $F_n(T(x)) \leq 0$). Case 2: $F_{n+1}(x) = f(x) + F_n(T(x))$ (i.e. $F_n(T(x)) \geq 0$).

divided into $k = 100$ equal pieces. The computer calculated the number of hits of each piece for $n = 500\,000$. The resulting graph indeed resembles the graph of $\frac{1}{\pi}\sqrt{1-x^2}$ (Figure 2.2), which is the density of the invariant probability measure equivalent to the length measure. Compare Figures I.13 and I.14 in [Collet & Eckmann 1980].

As a corollary of Birkhoff's Ergodic Theorem, one can obtain von Neumann's Ergodic Theorem. This says that if $\phi \in L^p(\mu)$ for $1 \leq p < \infty$, then the convergence to $E(\phi|\mathcal{I})$ holds in L^p. This is not difficult: see for example [Walters 1982].

Proof of Birkhoff's Ergodic Theorem. Let $f \in L^1(\mu)$ and $F_n = \max\{\sum_{i=0}^{k-1} f \circ T^i : 1 \leq k \leq n\}$, for $n = 1, 2, \ldots$. Then for every $x \in X$, $F_{n+1}(x) - F_n(T(x)) = f(x) - \min(0, F_n(T(x))) \geq f(x)$ and is monotone decreasing, since F_n is monotone increasing. The two cases under min are illustrated in Figure 2.3.

Define

$$A = \left\{x : \sup_n \sum_{i=0}^{n} f(T^i(x)) = \infty\right\}.$$

Note that $A \in \mathcal{I}$. If $x \in A$, then $F_{n+1}(x) - F_n(T(x))$ monotonously decreases to $f(x)$ as $n \to \infty$. The Dominated Convergence Theorem implies, then, that

$$0 \leq \int_A (F_{n+1} - F_n)\, d\mu = \int_A (F_{n+1} - F_n \circ T)\, d\mu \to \int_A f d\mu. \tag{2.2.4}$$

(We thus get $\int_A f\, d\mu \geq 0$, which is a variant of the so-called Maximal Ergodic Theorem: see Exercise 2.3.)

Notice that $\frac{1}{n}\sum_{k=0}^{n-1} f\circ T^k \le F_n/n$: so, outside A, we have

$$\limsup_{n\to\infty}\frac{1}{n}\sum_{k=0}^{n-1} f\circ T^k \le 0. \tag{2.2.5}$$

Therefore, if the conditional expectation value $f_{\mathcal{I}}$ of f is negative a.e., that is if $\int_C f\,d\mu = \int_C f_{\mathcal{I}}\,d\mu < 0$ for all $C\in\mathcal{I}$ with $\mu(C)>0$, then, as $A\in\mathcal{I}$, (2.2.4) implies that $\mu(A)=0$. Hence (2.2.5) holds a.e. Now if we let $f=\phi-\phi_{\mathcal{I}}-\varepsilon$, then $f_{\mathcal{I}}=-\varepsilon<0$. Note that $\phi_{\mathcal{I}}\circ T=\phi_{\mathcal{I}}$ implies that

$$\frac{1}{n}\sum_{k=0}^{n-1} f\circ T^k = \Big(\frac{1}{n}\sum_{k=0}^{n-1}\phi\circ T^k\Big) - \phi_{\mathcal{I}} - \varepsilon.$$

So (2.2.5) yields

$$\limsup_{n\to\infty}\frac{1}{n}\sum_{k=0}^{n-1}\phi\circ T^k \le \phi_{\mathcal{I}}+\varepsilon \quad \text{a.e.}$$

Replacing ϕ by $-\phi$ gives

$$\liminf_{n\to\infty}\frac{1}{n}\sum_{k=0}^{n-1}\phi\circ T^k \ge \phi_{\mathcal{I}}-\varepsilon \quad \text{a.e.}$$

Thus $\lim_{n\to\infty}\frac{1}{n}\sum_{k=0}^{n-1}\phi\circ T^k = \phi_{\mathcal{I}}$ a.e. ♣

Recall that at the end opposite to the absolute continuity (see Section 2.1) there is the notion of singularity. Two probability measures μ_1 and μ_2 on a σ-algebra $\mathcal{F}$ are called mutually *singular*, $\mu_1\perp\mu_2$, if there exist disjoint sets $X_1, X_2\in\mathcal{F}$ with $\mu_i(X_i)=1$ for $i=1,2$.

Theorem 2.2.6. *If $T: X\to X$ is a map measurable with respect to a σ-algebra $\mathcal{F}$, and if μ_1 and μ_2 are two different T-invariant probability ergodic measures on $\mathcal{F}$, then μ_1 and μ_2 are mutually singular.*

Proof. Since μ_1 and μ_2 are different, there exists a measurable set A such that

$$\mu_1(A)\ne\mu_2(A). \tag{2.2.6}$$

By Theorem 2.2.5 (Birkhoff's Ergodic Theorem) applied to μ_1 and μ_2 there exist sets $X_1, X_2\in\mathcal{F}$ satisfying $\mu_i(X_i)=1$ for $i=1,2$ such that for every $x\in X_i$

$$\lim_{n\to\infty}\frac{1}{n}S_n 1\!\!1_A(x)=\mu_i(A).$$

Thus in view of (2.2.6) the sets X_1 and X_2 are disjoint. The proof is complete. ♣

Proposition 2.2.7. *If $T: X\to X$ is a measure-preserving endomorphism of a probability space $(X,\mathcal{F},\nu)$, then ν is ergodic if and only if there is no T-invariant probability measure on $\mathcal{F}$ absolutely continuous with respect to ν and different from ν.*

Proof. Suppose that ν is ergodic and μ is a T-invariant probability measure on $\mathcal{F}$ with $\mu \ll \nu$. Then μ is also ergodic. Otherwise there would exist $A \in \mathcal{F}$ such that $T^{-1}(A) = A$ and $\mu(A), \mu(X \setminus A) > 0$ so $\nu(A), \nu(X \setminus A) > 0$: thus ν would not be ergodic. Hence, by Theorem 2.2.6, $\mu = \nu$.

Suppose in turn that ν is not ergodic, and let $A \in \mathcal{F}$ be a T-invariant set such that $0 < \nu(A) < 1$. Then the conditional measure on A is also T-invariant, but simultaneously it is distinct from ν and absolutely continuous with respect to ν. The proof is complete. ♣

Observe now that the space $M(\mathcal{F})$ of probability measures on $\mathcal{F}$ is a convex set: i.e. the convex combination $\alpha\mu + (1-\alpha)\nu$, $0 \le \alpha \le 1$, of two such measures is again in $M(\mathcal{F})$. The subspace $M(\mathcal{F}, T)$ of $M(\mathcal{F})$ consisting of T-invariant measures is also convex.

Recall that a point in a convex set is said to be *extreme* if and only if it cannot be represented as a convex combination of two distinct points with corresponding coefficient $0 < \alpha < 1$. We shall prove the following theorem.

Theorem 2.2.8. *The ergodic measures in $M(\mathcal{F}, T)$ are exactly the extreme points of $M(\mathcal{F}, T)$.*

Proof. Suppose that $\mu, \mu_1, \mu_2 \in M(\mathcal{F}, T)$, $\mu_1 \neq \mu_2$ and $\mu = \alpha\mu_1 + (1-\alpha)\mu_2$ with $0 < \alpha < 1$. Then $\mu_1 \neq \mu$ and $\mu_1 \ll \mu$. Thus, in view of Proposition 2.2.7, the measure μ is not ergodic.

Suppose in turn that μ is not ergodic, and let $A \in \mathcal{F}$ be a T-invariant set such that $0 < \mu(A) < 1$. Recall that, given $B \in \mathcal{F}$ with $\mu(B) > 0$, the *conditional measure* $A \mapsto \mu(A|B)$ is defined by $\mu(A \cap B)/\mu(B)$. Thus the conditional measures $\mu(\cdot|A)$ and $\mu(\cdot|A^c)$ are distinct, T-invariant and $\mu = \mu(A)\mu(\cdot|A) + (1 - \mu(A)\mu(\cdot|A^c)$. Consequently μ is not an extreme point in $M(\mathcal{F}, T)$. The proof is complete. ♣

In Section 2.8 we shall formulate a theorem on decomposition into ergodic components that will clarify the situation better. This will correspond to the Choquet Theorem in functional analysis: see Section 3.1.

2.3 Entropy of partition

Let $(X, \mathcal{F}, \mu)$ be a probability space. A *partition* of $(X, \mathcal{F}, \mu)$ is a subfamily (*a priori* may be uncountable) of $\mathcal{F}$ consisting of mutually disjoint elements whose union is X.

If $\mathcal{A}$ is a partition and $x \in X$, then the only element of $\mathcal{A}$ containing x is denoted by $\mathcal{A}(x)$ or, if $x \in A \in \mathcal{A}$, by $A(x)$.

If $\mathcal{A}$ and $\mathcal{B}$ are two partitions of X, we define their *join* or *joining*:

$$\mathcal{A} \vee \mathcal{B} = \{A \cap B : A \in \mathcal{A},\ B \in \mathcal{B}\}.$$

We write $\mathcal{A} \le \mathcal{B}$ if and only if $\mathcal{B}(x) \subset \mathcal{A}(x)$ for every $x \in X$, which in other words means that each element of the partition $\mathcal{B}$ is contained in an element of the partition $\mathcal{A}$ or equivalently $\mathcal{A} \vee \mathcal{B} = \mathcal{B}$. We sometimes say in this case that $\mathcal{B}$ is *finer* than $\mathcal{A}$, or that $\mathcal{B}$ is a *refinement* of $\mathcal{A}$.

Now we introduce the notion of entropy of a countable (finite or infinite) partition, and we collect its basic elementary properties. Define the function $k : [0,1] \to [0,\infty]$, putting

$$k(t) = \begin{cases} -t \log t & \text{for } t \in (0,1] \\ 0 & \text{for } t = 0 \end{cases} \tag{2.3.1}$$

Check that the function k is continuous. Let $\mathcal{A} = \{A_i : 1 \le i \le n\}$ be a countable partition of X, where $n \ge 1$ is a finite integer or ∞. In the sequel we shall usually write ∞.

The *entropy* of $\mathcal{A}$ is the number

$$\mathrm{H}(\mathcal{A}) = \sum_{i=1}^{\infty} -\mu(A_i) \log \mu(A_i) = \sum_{i=1}^{\infty} k(\mu(A_i)). \tag{2.3.2}$$

If $\mathcal{A}$ is infinite, $\mathrm{H}(\mathcal{A})$ may happen to be infinite as well as finite.

Define

$$I(x) = I(\mathcal{A})(x) := -\log \mu(\mathcal{A}(x)). \tag{2.3.3}$$

This is called an *information function*. Intuitively $I(x)$ is information on an object x given by the experiment $\mathcal{A}$ in the logarithmic scale. Therefore the entropy in (2.3.2) is the integral (the average) of the information function.

Note that $\mathrm{H}(\mathcal{A}) = 0$ for $\mathcal{A} = \{X\}$, and that if $\mathcal{A}$ is finite – say, consisting of n elements – then $0 \le \mathrm{H}(\mathcal{A}) \le \log n$ and $\mathrm{H}(\mathcal{A}) = \log n$ if and only if $\mu(A_1) = \mu(A_2) = \ldots = \mu(A_n) = 1/n$. This follows from the fact that the logarithmic function is strictly concave.

In this section we deal only with one fixed measure μ. If, however, we need to consider more measures simultaneously (see for example Chapter 3), we shall use instead the notation $\mathrm{H}_\mu(A)$ for $\mathrm{H}(A)$. We shall use also the notation $I_\mu(x)$ for $I(x)$.

Let $\mathcal{A} = \{A_i : i \ge 1\}$ and $\mathcal{B} = \{B_j : j \ge 1\}$ be two countable partitions of X. The *conditional entropy* $\mathrm{H}(\mathcal{A}|\mathcal{B})$ of $\mathcal{A}$ given $\mathcal{B}$ is defined as

$$\begin{aligned} \mathrm{H}(\mathcal{A}|\mathcal{B}) &= \sum_{j=1}^{\infty} \mu(B_j) \sum_{i=1}^{\infty} -\frac{\mu(A_i \cap B_j)}{\mu(B_j)} \log \frac{\mu(A_i \cap B_j)}{\mu(B_j)} \\ &= \sum_{i,j} -\mu(A_i \cap B_j) \log \frac{\mu(A_i \cap B_j)}{\mu(B_j)}. \end{aligned} \tag{2.3.4}$$

The first equality, defining $\mathrm{H}(\mathcal{A}|\mathcal{B})$, can be viewed as follows. One considers each element B_j as a probability space with conditional measure $\mu(A|B_j) = \mu(A)/\mu(B_j)$ for $A \subset B_j$ and calculates the entropy of the partition of the set B_j into $A_i \cap B_j$. Then one averages the result over the space of B_j's. (This will be generalized in Definition 2.8.3.)

For each x denote $-\log\mu(A(x)|B(x)) = -\log\frac{\mu(A(x)\cap B(x))}{\mu(B(x))})$ by $I(x)$ or $I(\mathcal{A}|\mathcal{B})(x)$. The second equality in (2.3.4) can be rewritten as

$$\mathrm{H}(\mathcal{A}|\mathcal{B}) = \int_X I(\mathcal{A}|\mathcal{B})\,d\mu. \tag{2.3.5}$$

Note, by the way, that if $\tilde{\mathcal{B}}$ is the σ-algebra consisting of all unions of elements of $\mathcal{B}$ (i.e. generated by $\mathcal{B}$), then $I(x) = -\log\mu((\mathcal{A}(x)\cap\mathcal{B}(x))|\mathcal{B}(x)) = -\log E(1\!\!1_{\mathcal{A}(x)}|\tilde{\mathcal{B}})(x)$; compare (2.1.8).

Note finally that for any countable partition $\mathcal{A}$ we have

$$\mathrm{H}(\mathcal{A}|\{X\}) = \mathrm{H}(\mathcal{A}). \tag{2.3.6}$$

Some further basic properties of the entropy of partitions are collected in the following.

Theorem 2.3.1. *Let $(X,\mathcal{F},\mu)$ be a probability space. If $\mathcal{A}$, $\mathcal{B}$ and $\mathcal{C}$ are countable partitions of X, then:*

$$\mathrm{H}(\mathcal{A}\vee\mathcal{B}|\mathcal{C}) = \mathrm{H}(\mathcal{A}|\mathcal{C}) + \mathrm{H}(\mathcal{B}|\mathcal{A}\vee\mathcal{C}) \tag{a}$$
$$\mathrm{H}(\mathcal{A}\vee\mathcal{B}) = \mathrm{H}(\mathcal{A}) + \mathrm{H}(\mathcal{B}|\mathcal{A}) \tag{b}$$
$$\mathcal{A}\le\mathcal{B} \Rightarrow \mathrm{H}(\mathcal{A}|\mathcal{C}) \le \mathrm{H}(\mathcal{B}|\mathcal{C}) \tag{c}$$
$$\mathcal{B}\le\mathcal{C} \Rightarrow \mathrm{H}(\mathcal{A}|\mathcal{B}) \ge \mathrm{H}(\mathcal{A}|\mathcal{C}) \tag{d}$$
$$\mathrm{H}(\mathcal{A}\vee\mathcal{B}|\mathcal{C}) \le \mathrm{H}(\mathcal{A}|\mathcal{C}) + \mathrm{H}(\mathcal{B}|\mathcal{C}) \tag{e}$$
$$\mathrm{H}(\mathcal{A}|\mathcal{C}) \le \mathrm{H}(\mathcal{A}|\mathcal{B}) + \mathrm{H}(\mathcal{B}|\mathcal{C}). \tag{f}$$

Proof. Let $\mathcal{A} = \{A_n : n\ge 1\}$, $\mathcal{B} = \{B_m : m\ge 1\}$, and $\mathcal{C} = \{C_l : l\ge 1\}$. Without loss of generality we can assume that all these sets are of positive measure.
(a) By (2.3.4) we have

$$\mathrm{H}(\mathcal{A}\vee\mathcal{B}|\mathcal{C}) = -\sum_{i,j,k}\mu(A_i\cap B_j\cap C_k)\log\frac{\mu(A_i\cap B_j\cap C_k)}{\mu(C_k)}.$$

But

$$\frac{\mu(A_i\cap B_j\cap C_k)}{\mu(C_k)} = \frac{\mu(A_i\cap B_j\cap C_k)}{\mu(A_i\cap C_k)}\frac{\mu(A_i\cap C_k)}{\mu(C_k)}$$

unless $\mu(A_i\cap C_k) = 0$. But then the left-hand side vanishes, and we need not consider it. Therefore

$$\begin{aligned}
\mathrm{H}(\mathcal{A}\vee\mathcal{B}|\mathcal{C}) &= -\sum_{i,j,k}\mu(A_i\cap B_j\cap C_k)\log\frac{\mu(A_i\cap C_k)}{\mu(C_k)}\\
&\quad -\sum_{i,j,k}\mu(A_i\cap B_j\cap C_k)\log\frac{\mu(A_i\cap B_j\cap C_k)}{\mu(A_i\cap C_k)}\\
&= -\sum_{i,k}\mu(A_i\cap C_k)\log\frac{\mu(A_i\cap C_k)}{\mu(C_k)} + \mathrm{H}(\mathcal{B}|\mathcal{A}\vee\mathcal{C})\\
&= \mathrm{H}(\mathcal{A}|\mathcal{C}) + \mathrm{H}(\mathcal{B}|\mathcal{A}\vee\mathcal{C}).
\end{aligned}$$

(b) Put $\mathcal{C} = \{X\}$ and apply (2.3.6) in (a).
(c) By (a)

$$\mathrm{H}(\mathcal{B}|\mathcal{C}) = \mathrm{H}(\mathcal{A} \vee \mathcal{B}|\mathcal{C}) = \mathrm{H}(\mathcal{A}|\mathcal{C}) + \mathrm{H}(\mathcal{B}|\mathcal{A} \vee \mathcal{C}) \geq \mathrm{H}(\mathcal{A}|\mathcal{C}).$$

(d) Since the function k defined by (2.3.1) is strictly concave, we have for every pair i, j that

$$k\Big(\sum_l \frac{\mu(C_l \cap B_j)}{\mu(B_j)} \frac{\mu(A_i \cap C_l)}{\mu(C_l)}\Big) \geq \sum_l \frac{\mu(C_l \cap B_j)}{\mu(B_j)} k\left(\frac{\mu(A_i \cap C_l)}{\mu(C_l)}\right). \tag{2.3.7}$$

But since $\mathcal{B} \leq \mathcal{C}$, we can write above $C_l \cap B_j = C_l$: hence the left-hand side is equal to $k\left(\frac{\mu(A_i \cap B_j)}{\mu(B_j)}\right)$, and we conclude with

$$k\left(\frac{\mu(A_i \cap B_j)}{\mu(B_j)}\right) \geq \sum_l \frac{\mu(C_l \cap B_j)}{\mu(B_j)} k\left(\frac{\mu(A_i \cap C_l)}{\mu(C_l)}\right).$$

(Note that until now we have not used the specific form of the function k.)

Finally, multiplying both sides of (2.3.7) by $\mu(B_j)$, using the definition of k and summing over i and j, we get

$$\begin{aligned} -\sum_{i,j} \mu(A_i \cap B_j) \log \frac{\mu(A_i \cap B_j)}{\mu(B_j)} &\geq -\sum_{i,j,l} \mu(C_l \cap B_j) \frac{\mu(A_i \cap C_l)}{\mu(C_l)} \log \frac{\mu(A_i \cap C_l)}{\mu(C_l)} \\ &= -\sum_{i,l} \mu(C_l) \frac{\mu(A_i \cap C_l)}{\mu(C_l)} \log \frac{\mu(A_i \cap C_l)}{\mu(C_l)}, \end{aligned}$$

or equivalently $\mathrm{H}(\mathcal{A}|\mathcal{B}) \geq \mathrm{H}(\mathcal{A}|\mathcal{C})$.

Formula (e) follows immediately from (a) and (d), and formula (f) can be proved by a straightforward calculation (its consequences are discussed in Exercise 2.17). ♣

2.4 Entropy of an endomorphism

Let $(X, \mathcal{F}, \mu)$ be a probability space, and let $T : X \to X$ be a measure-preserving endomorphism of X. If $\mathcal{A} = \{A_i\}_{i \in I}$ is a partition of X, then by $T^{-1}\mathcal{A}$ we denote the partition $\{T^{-1}(A_i)\}_{i \in I}$. Note that for any countable $\mathcal{A}$

$$\mathrm{H}(T^{-1}\mathcal{A}) = \mathrm{H}(\mathcal{A}). \tag{2.4.1}$$

For all $n \geq m \geq 0$ denote the partition $\bigvee_{i=0}^n T^{-i}\mathcal{A} = \mathcal{A} \vee T^{-1}(\mathcal{A}) \vee \cdots \vee T^{-n}(\mathcal{A}) = \bigvee_{i=m}^n T^{-i}(\mathcal{A})$ by A_m^n. For $m = 0$ we shall sometimes use the notation $\mathcal{A}^n$.

Lemma 2.4.1. *For any countable partition $\mathcal{A}$,*

$$\mathrm{H}(\mathcal{A}^n) = \mathrm{H}(\mathcal{A}) + \sum_{j=1}^n \mathrm{H}(\mathcal{A}|\mathcal{A}_1^j). \tag{2.4.2}$$

Proof. We prove this formula by induction. If $n = 0$, it is a tautology. Suppose it is true for $n - 1 \geq 0$. Then with the use of Theorem 2.3.1(b) and (2.4.1) we obtain

$$\begin{aligned} \mathrm{H}(\mathcal{A}^n) &= \mathrm{H}(\mathcal{A}_1^n \vee \mathcal{A}) = \mathrm{H}(\mathcal{A}_1^n) + \mathrm{H}(\mathcal{A}|\mathcal{A}_1^n) \\ &= \mathrm{H}(\mathcal{A}^{n-1}) + \mathrm{H}(\mathcal{A}|\mathcal{A}_1^n) = \mathrm{H}(\mathcal{A}) + \sum_{j=1}^{n} \mathrm{H}(\mathcal{A}|\mathcal{A}_1^j). \end{aligned}$$

Hence (2.4.2) holds for all n. ♣

Lemma 2.4.2. *The sequences* $\frac{1}{n+1}\mathrm{H}(\mathcal{A}^n)$ *and* $\mathrm{H}(\mathcal{A}|\mathcal{A}_1^n)$ *are monotone decreasing to a limit* $\mathrm{h}(T, \mathcal{A})$.

Proof. The sequence $\mathrm{H}(\mathcal{A}|\mathcal{A}_1^n)$, $n = 0, 1, \ldots$ is monotone decreasing, by Theorem 2.3.1(d). Therefore the sequence of averages is also monotone decreasing to the same limit; furthermore, it coincides with the limit of the sequence $\frac{1}{n+1}\mathrm{H}(\mathcal{A}^n)$ by (2.4.2). ♣

The limit $\frac{1}{n+1}\mathrm{H}(\mathcal{A}^n)$ whose existence has been shown in Lemma 2.4.2 is known as the *(measure–theoretic) entropy of T with respect to the partition $\mathcal{A}$*, and is denoted by $\mathrm{h}(T, \mathcal{A})$, or by $\mathrm{h}_\mu(T, \mathcal{A})$ if one wants to indicate the measure under consideration. Intuitively this means the limit rate of the growth of average (integral) information (in a logarithmic scale), under consecutive experiments, for the number of those experiments tending to infinity.

Remark. Write $a_k := \mathrm{H}(\mathcal{A}^{k-1})$. In order to prove the existence of the limit $\frac{1}{n+1}\mathrm{H}(\mathcal{A}^n)$, instead of relying on (2.4.2) and the monotonicity, we could use the estimate

$$a_{n+m} = \mathrm{H}(\mathcal{A}^{n+m-1}) \leq \mathrm{H}(\mathcal{A}^{n-1}) + \mathrm{H}(\mathcal{A}_n^{n+m-1}) = a_n + \mathrm{H}(\mathcal{A}^{m-1}) = a_n + a_m.$$

following from Theorem 2.3.1(e) and from (2.4.1), and apply the following:

Lemma 2.4.3. *If* $\{a_n\}_{n=1}^\infty$ *is a sequence of real numbers such that* $a_{n+m} \leq a_n + a_m$ *for all* $n, m \geq 1$ *(any such a sequence is called subadditive), then* $\lim_{n\to\infty} a_n$ *exists and equals* $\inf_n a_n/n$. *The limit could be* $-\infty$, *but if the* a_n*'s are bounded below, then the limit will be non-negative.*

Proof. Fix $m \geq 1$. Each $n \geq 1$ can be expressed as $n = km + i$ with $0 \leq i < m$. Then

$$\frac{a_n}{n} = \frac{a_{i+km}}{i+km} \leq \frac{a_i}{km} + \frac{a_{km}}{km} \leq \frac{a_i}{km} + \frac{ka_m}{km} = \frac{a_i}{km} + \frac{a_m}{m}.$$

If $n \to \infty$ then also $k \to \infty$ and therefore $\limsup_{n\to\infty} \frac{a_n}{n} \leq \frac{a_m}{m}$. Thus $\limsup_{n\to\infty} \frac{a_n}{n} \leq \inf \frac{a_m}{m}$. Now the inequality $\inf \frac{a_m}{m} \leq \liminf_{n\to\infty} \frac{a_n}{n}$ completes the proof. ♣

Notice that there exists a subadditive sequence $(a_n)_{n=1}^\infty$ such that the corresponding sequence a_n/n is not eventually decreasing. Indeed, it suffices to observe that each sequence consisting of 1's and 2's is subadditive, and to consider such a sequence having infinitely many 1's and 2's. If for an $n > 1$ we have $a_n = 1$ and $a_{n+1} = 2$, we have $\frac{a_n}{n} < \frac{a_{n+1}}{n+1}$.

One can consider $a_n + Cn$ for any constant $C > 1$, making the example strictly increasing.

Exercise. Prove that Lemma 2.4.3 remains true under the weaker assumption that there exists $c \in R$ such that $a_{n+m} \le a_n + a_m + c$ for all n and m.

The basic elementary properties of the entropy $\mathrm{h}(T, \mathcal{A})$ are collected in the next theorem.

Theorem 2.4.4. *If $\mathcal{A}$ and $\mathcal{B}$ are countable partitions of finite entropy then*

$$\mathrm{h}(T, \mathcal{A}) \le \mathrm{H}(\mathcal{A}) \tag{a}$$

$$\mathrm{h}(T, \mathcal{A} \vee \mathcal{B}) \le \mathrm{h}(T, \mathcal{A}) + \mathrm{h}(T, \mathcal{B}) \tag{b}$$

$$\mathcal{A} \le \mathcal{B} \Rightarrow \mathrm{h}(T, \mathcal{A}) \le \mathrm{h}(T, \mathcal{B}) \tag{c}$$

$$\mathrm{h}(T, \mathcal{A}) \le \mathrm{h}(T, \mathcal{B}) + \mathrm{H}(\mathcal{A}|\mathcal{B}) \tag{d}$$

$$\mathrm{h}(T, T^{-1}(\mathcal{A})) = \mathrm{h}(T, \mathcal{A}) \tag{e}$$

$$\text{If } k \ge 1 \text{ then } \mathrm{h}(T, \mathcal{A}) = \mathrm{h}\big(T, \mathcal{A}^k\big) \tag{f}$$

$$\text{If } T \text{ is invertible and } k \ge 1, \text{ then } \mathrm{h}(T, \mathcal{A}) = \mathrm{h}\big(T, \bigvee_{i=-k}^{k} T^i(\mathcal{A})\big) \tag{g}$$

The standard proof (see for example [Walters 1982]) based on Theorem 2.3.1 and formula (2.3.2) is left for the reader as an exercise. Let us prove only item (d).

$$\begin{aligned}
\mathrm{h}(T, \mathcal{A}) &= \lim_{n\to\infty} \frac{1}{n} \mathrm{H}(\mathcal{A}^{n-1}) = \lim_{n\to\infty} \frac{1}{n} \Big(\mathrm{H}(\mathcal{A}^{n-1}|\mathcal{B}^{n-1}) + \mathrm{H}(\mathcal{B}^{n-1})\Big) \\
&\le \lim_{n\to\infty} \frac{1}{n} \sum_{j=0}^{n-1} H(T^{-j}(\mathcal{A})|\mathcal{B}^{n-1}) + \lim_{n\to\infty} \frac{1}{n} \mathrm{H}(\mathcal{B}^{n-1}) \\
&\le \lim_{n\to\infty} \frac{1}{n} \sum_{j=0}^{n-1} H(T^{-j}(\mathcal{A})|T^{-j}(\mathcal{B})) + \mathrm{h}(T, \mathcal{B}) \le \mathrm{H}(\mathcal{A}|\mathcal{B}) + \mathrm{h}(T, \mathcal{B}).
\end{aligned}$$

Here is one more useful fact, stronger than Theorem 2.4.4(c):

Theorem 2.4.5. *If $T : X \to X$ is a measure-preserving endomorphism of a probability space $(X, \mathcal{F}, \mu)$, $\mathcal{A}$ and $\mathcal{B}_m, m = 1, 2, \dots$ are countable partitions with finite entropy, and $\mathrm{H}(\mathcal{A}|\mathcal{B}_m) \to 0$ as $m \to \infty$, then*

$$\mathrm{h}(T, \mathcal{A}) \le \liminf_{m\to\infty} \mathrm{h}(T, \mathcal{B}_m).$$

In particular, for $\mathcal{B}_m := \mathcal{B}^m = \bigvee_{j=0}^{m} T^{-j}(\mathcal{B})$, one obtains $\mathrm{h}(T, \mathcal{A}) \le \mathrm{h}(T, \mathcal{B})$.

Proof. By Theorem 2.4.4(d), we get for every positive integer m that

$$\mathrm{h}(T,\mathcal{A}) \le \mathrm{H}(\mathcal{A}|\mathcal{B}_m) + \mathrm{h}(T,\mathcal{B}_m).$$

Letting $m \to \infty$ this yields the first part of the assertion. If $\mathcal{B}_m = \mathcal{B}^m$, then $\mathrm{h}(T,\mathcal{B}^m) = \mathrm{h}(T,\mathcal{B})$, by Theorem 2.4.4(f), and the second part of the theorem follows as well. ♣

The *(measure-theoretic) entropy of an endomorphism* $T : X \to X$ is defined as

$$\mathrm{h}_\mu(T) = \mathrm{h}(T) = \sup_{\mathcal{A}}\{\mathrm{h}(T,\mathcal{A})\}, \tag{2.4.8}$$

where the supremum is taken over all finite (or countable of finite entropy) partitions of X. See Exercise 2.21.

It is clear from the definition that the entropy of T is an isomorphism invariant.

Later on (see Theorem 2.8.7, Remark 2.8.9, Corollary 2.8.10 and Exercise 2.18) we shall discuss the cases where $\mathrm{H}(\mathcal{A}|\mathcal{B}_n) \to 0$ for every $\mathcal{A}$ (finite or of finite entropy). This will allow us to write $\mathrm{h}_\mu(T) = \lim_{m\to\infty} \mathrm{h}(T,\mathcal{B}_m)$ or $\mathrm{h}(T) = \mathrm{h}(T,\mathcal{B})$.

The following theorem is very useful.

Theorem 2.4.6. *If $T : X \to X$ is a measure-preserving endomorphism of a probability space $(X,\mathcal{F},\mu)$, then*

$$\mathrm{h}(T^k) = k\,\mathrm{h}(T) \textit{ for all } k \ge 1, \tag{a}$$

$$\textit{If } T \textit{ is invertible then } \mathrm{h}(T^{-1}) = \mathrm{h}(T). \tag{b}$$

Proof. (a) Fix $k \ge 1$. Since

$$\lim_{n\to\infty} \frac{1}{n}\mathrm{H}\Big(\bigvee_{j=0}^{n-1} T^{-kj}\Big(\bigvee_{i=0}^{k-1} T^{-i}\mathcal{A}\Big)\Big) = \lim_{n\to\infty} \frac{k}{nk}\mathrm{H}\Big(\bigvee_{i=0}^{nk-1} T^{-i}\mathcal{A}\Big) = k\,\mathrm{h}(T,\mathcal{A})$$

we have $\mathrm{h}\big(T^k, \bigvee_{i=0}^{k-1} T^{-i}\mathcal{A}\big) = k\,\mathrm{h}(T,\mathcal{A})$. Therefore

$$k\,\mathrm{h}(T) = k \sup_{\mathcal{A}\text{ finite}} \mathrm{h}(T,\mathcal{A}) = \sup_{\mathcal{A}} \mathrm{h}\Big(T^k, \bigvee_{i=0}^{k-1} T^{-i}\mathcal{A}\Big) \le \sup_{\mathcal{B}} \mathrm{h}(T^k,\mathcal{B}) = \mathrm{h}(T^k). \tag{2.4.3}$$

On the other hand, by Theorem 2.4.4(c), we get $\mathrm{h}(T^k,\mathcal{A}) \le \mathrm{h}\big(T^k, \bigvee_{i=0}^{k-1} T^{-i}\mathcal{A}\big) = k\,\mathrm{h}(T,\mathcal{A})$, and therefore $\mathrm{h}(T^k) \le k\,\mathrm{h}(T)$. The result follows from this and (2.4.3).

(b) In view of (2.4.1), for all finite partitions $\mathcal{A}$ we have

$$\mathrm{H}\Big(\bigvee_{i=0}^{n-1} T^{i}\mathcal{A}\Big) = \mathrm{H}\Big(T^{-(n-1)} \bigvee_{i=0}^{n-1} T^{i}\mathcal{A}\Big) = \mathrm{H}\Big(\bigvee_{i=0}^{n-1} T^{-i}\mathcal{A}\Big).$$

This completes the proof. ♣

Let us end this section with the following theorem, to be used for example in Section 3.6.

Theorem 2.4.7. *If μ and ν are two probability measures on $(X, \mathcal{F})$, both preserved by an endomorphism $T : X \to X$, then for every $a : 0 < a < 1$ and the measure $\rho = a\mu + (1-a)\nu$ we have*

$$\mathrm{h}_\rho(T) = a\,\mathrm{h}_\mu(T) + (1-a)\nu(T).$$

In other words, the mapping $\mu \mapsto \mathrm{h}_\mu$ is affine.

The proof can be found in [Denker, Grillenberger & Sigmund, 1976, Proposition 10.13] or [Walters 1982, Theorem 8.1]. We leave it to the reader as an exercise.

Hint: Prove first that for every $A \in \mathcal{F}$ we have

$$0 \le k(\rho(A)) - ak(\mu(A)) - (1-a)k(\nu(A)) \le -(a\log a)\mu(A) - ((1-a)\log(1-a))\nu(A),$$

using the concavity of the function $k(t) = -t\log t$: see (2.3.1). Summing this up over $A \in \mathcal{A}$ for a finite partition $\mathcal{A}$, obtain

$$0 \le \mathrm{H}_\rho(\mathcal{A}) - a\,\mathrm{H}_\mu(\mathcal{A}) - (1-a)\,\mathrm{H}_\nu(\mathcal{A}) \le \log 2.$$

Apply this to partitions $\mathcal{A}^n$ and use Theorem 2.4.6(a).

Remark. This theorem can be easily deduced from the ergodic decomposition theorem (Theorem 2.8.11) for Lebesgue spaces: see Exercise 2.16. In the setting of Chapter 3, for Borel measures on a compact metric space X, one can refer also to Choquet's Theorem 3.1.11.

2.5 Shannon–McMillan–Breiman Theorem

Let $(X, \mathcal{F}, \mu)$ be a probability space, let $T : X \to X$ be a measure-preserving endomorphism of X, and let $\mathcal{A}$ be a countable finite entropy partition of X.

Lemma 2.5.1 (Maximal inequality). *For each $n = 1, 2, \dots$ let $f_n = I(\mathcal{A}|\mathcal{A}_1^n)$ and $f^* = \sup_{n\ge1} f_n$. Then for each $\lambda \in \mathbb{R}$ and each $A \in \mathcal{A}$*

$$\mu(\{x \in A : f^*(x) > \lambda\}) \le e^{-\lambda}. \tag{2.5.1}$$

Proof. For each $A \in \mathcal{A}$ and $n = 1, 2, \dots$ let $f_n^A = -\log E(1\!\!1_A|\mathcal{A}_1^n)$. Of course $f_n = \sum_{A\in\mathcal{A}} 1\!\!1_A f_n^A$. Denote

$$B_n^A = \{x \in X : f_1^A(x), \dots, f_{n-1}^A(x) \le \lambda, f_n^A(x) > \lambda\}.$$

Since $B_n^A \in \mathcal{F}(\mathcal{A}_1^n)$, the σ-algebra generated by $\mathcal{A}_1^n$,

$$\mu(B_n^A \cap A) = \int_{B_n^A} 1\!\!1_A\,d\mu = \int_{B_n^A} E(1\!\!1_A|\mathcal{A}_1^n)\,d\mu = \int_{B_n^A} e^{-f_n^A}\,d\mu \le e^{-\lambda}\mu(B_n^A).$$

Therefore

$$\mu(\{x \in A : f^*(x) > \lambda\}) = \sum_{n=1}^{\infty} \mu(B_n^A \cap A) \le e^{-\lambda} \sum_{n=1}^{\infty} \mu(B_n^A) \le e^{-\lambda}.$$

♣

Corollary 2.5.2. *The function f^* is integrable and $\int f^*\, d\mu \le \mathrm{H}(\mathcal{A}) + 1$.*

Proof. Of course $\mu(\{x \in A : f^* > \lambda\}) \le \mu(A)$, so $\mu(\{x \in A : f^*(x) > \lambda\}) \le \min\{\mu(A), e^{-\lambda}\}$. So, by Lemma 2.5.1,

$$\begin{aligned}
\int_X f^*\, d\mu &= \sum_{A\in\mathcal{A}} \int_A f^*\, d\mu = \sum_{A\in\mathcal{A}} \int_0^{\infty} \mu\{x \in A : f^*(x) > \lambda\}\, d\lambda \\
&\le \sum_{A\in\mathcal{A}} \int_0^{\infty} \min\{\mu(A), e^{-\lambda}\}\, d\lambda \\
&= \sum_{A\in\mathcal{A}} \Big(\int_0^{-\log\mu(A)} \mu(A)\, d\lambda + \int_{-\log\mu(A)}^{\infty} e^{-\lambda}\, d\lambda \Big) \\
&= \sum_{A\in\mathcal{A}} \Big(-\mu(A)(\log\mu(A)) + \mu(A) \Big) = \mathrm{H}(\mathcal{A}) + 1.
\end{aligned}$$

♣

Note that if $\mathcal{A}$ is finite, then the integrability of f^* follows from the integrability of $f^*|_A$ for each A, following immediately from Lemma 2.5.1. The difficulty with infinite $\mathcal{A}$ is that there is no $\mu(A)$ factor on the right-hand side of (2.5.1).

Corollary 2.5.3. *The sequence $(f_n)_{n=1}^{\infty}$ converges a.e. and in L^1.*

Proof. $E(\mathbb{1}_A|\mathcal{A}_1^n)$ is a martingale to which we can apply Theorem 2.1.4. This gives convergence a.e.: hence convergence a.e. of each f_n^A, and hence of f_n. Now convergence in L^1 follows from Corollary 2.5.2 and the Dominated Convergence Theorem. ♣

Theorem 2.5.4 (Shannon–McMillan–Breiman). *Suppose that $\mathcal{A}$ is a countable partition of finite entropy. Then there exist limits*

$$f = \lim_{n\to\infty} I(\mathcal{A}|\mathcal{A}_1^n) \quad \text{and} \quad f_{\mathcal{I}}(x) = \lim_{n\to\infty} \frac{1}{n} \sum_{i=0}^{n-1} f(T^i(x)) \text{ for a.e. } x$$

and

$$\lim_{n\to\infty} \frac{1}{n+1} I(\mathcal{A}^n) = f_{\mathcal{I}} \quad \text{a.e. and in } L^1. \tag{2.5.2}$$

Furthermore,

$$\mathrm{h}(T,\mathcal{A}) = \lim_{n\to\infty} \frac{1}{n+1} \mathrm{H}(\mathcal{A}^n) = \int f_{\mathcal{I}}\, d\mu = \int f\, d\mu. \tag{2.5.3}$$

The limit f will gain a new interpretation in (2.8.6), in the context of Lebesgue spaces, where the notion of information function I will be generalized.

Proof. First note that the sequence $f_n = I(\mathcal{A}|\mathcal{A}_1^n), n = 1, 2, \ldots$ converges to an integrable function f by Corollary 2.5.3. (Caution: though the integrals of f_n decrease to the entropy, Lemma 2.4.2, it is usually not true that f_n decrease.) Hence the a.e. convergence of time averages to $f_{\mathcal{I}}$ holds by Birkhoff's Ergodic Theorem. It will suffice to prove (2.5.2), since then (2.5.3), the second equality, holds by integration and the last equality by Birkhoff's Ergodic Theorem, the convergence in L^1.

(In fact (2.5.3) already follows from Corollary 2.5.3. Indeed, $\lim_{n\to\infty} \frac{1}{n+1} \mathrm{H}(\mathcal{A}^n) = \lim_{n\to\infty} \mathrm{H}(\mathcal{A}|\mathcal{A}_1^n) = \lim_{n\to\infty} \int I(\mathcal{A}|\mathcal{A}_1^n)\, d\mu = \int \lim_{n\to\infty} I(\mathcal{A}|\mathcal{A}_1^n)\, d\mu = \int f\, d\mu$.)

Let us now establish some identities (compare Lemma 2.4.1). Let $\{\mathcal{A}_n : n \geq 0\}$ be a sequence of countable partitions. Then we have

$$\begin{aligned} I\Big(\bigvee_{i=0}^{n} \mathcal{A}_i\Big) &= I\Big(\mathcal{A}_0 | \bigvee_{i=1}^{n} \mathcal{A}_i\Big) + I\Big(\bigvee_{i=1}^{n} \mathcal{A}_i\Big) \\ &= I\Big(\mathcal{A}_0 | \bigvee_{i=1}^{n} \mathcal{A}_i\Big) + I\Big(\mathcal{A}_1 | \bigvee_{i=2}^{n} \mathcal{A}_i\Big) + \cdots + I(\mathcal{A}_n). \end{aligned}$$

In particular, it follows from the above formula that for $\mathcal{A}_i = T^{-i}\mathcal{A}$ we have

$$\begin{aligned} I(\mathcal{A}^n) &= I(\mathcal{A}|\mathcal{A}_1^n) + I(T^{-1}\mathcal{A}|\mathcal{A}_2^n) + \ldots + I(T^{-n}\mathcal{A}) \\ &= I(\mathcal{A}|\mathcal{A}_1^n) + I(\mathcal{A}|\mathcal{A}_1^{n-1}) \circ T + \ldots I(\mathcal{A}) \circ T^n \\ &= f_n + f_{n-1} \circ T + f_{n-2} \circ T^2 + \ldots + f_0 \circ T^n, \end{aligned}$$

where $f_k = I(\mathcal{A}|\mathcal{A}_1^k)$, $f_0 = I(\mathcal{A})$. Now

$$\left| \frac{1}{n+1} I(\mathcal{A}^n) - f_{\mathcal{I}} \right| \leq \left| \frac{1}{n+1} \sum_{j=0}^{n} (f_{n-i} \circ T^i - f \circ T^i) \right| + \left| \frac{1}{n+1} \sum_{j=0}^{n} f \circ T^i - f_{\mathcal{I}} \right|.$$

Since by Birkhoff's Ergodic Theorem the latter term converges to zero both almost everywhere and in L^1, it suffices to prove that for $n \to \infty$

$$\frac{1}{n+1} \sum_{i=0}^{n} g_{n-i} \circ T^i \to 0 \ \text{ a.e. and in } L^1, \tag{2.5.4}$$

where $g_k = |f - f_k|$.

Now, since T is measure preserving, for every $i \geq 0$

$$\int g_{n-i} \circ T^i d\mu = \int g_{n-i} d\mu.$$

Thus $\frac{1}{n} \sum_{i=0}^{n} \int g_{n-i} \circ T^i\, d\mu = \frac{1}{n} \sum_{i=0}^{n} \int g_{n-i}\, d\mu \to 0$, since $f_k \to f$ in L^1 by Corollary 2.5.3. Thus we have established the L^1 convergence in (2.5.4).

Now, let $G_N = \sup_{n>N} g_n$. Of course G_N is monotone decreasing, and since $g_n \to 0$ a.e. (Corollary 2.5.3), we get $G_N \searrow 0$ a.e.. Moreover, by Corollary 2.5.2, $G_0 \le \sup_n f_n + f \in L_1$.

For arbitrary $N < n$ we have

$$\frac{1}{n+1}\sum_{i=0}^{n} g_{n-i}\circ T^i = \frac{1}{n+1}\sum_{i=0}^{n-N-1} g_{n-i}\circ T^i + \frac{1}{n+1}\sum_{i=n-N}^{n} g_{n-i}\circ T^i$$
$$\le \frac{1}{n+1}\sum_{i=0}^{n-N-1} G_N\circ T^i + \frac{1}{n+1}\sum_{i=n-N}^{n} G_0\circ T^i.$$

Hence for $K_N = G_0 + G_0\circ T + \ldots + G_0\circ T^N$

$$\limsup_{n\to\infty}\frac{1}{n+1}\sum_{i=0}^{n} g_{n-i}\circ T^i \le (G_N)_{\mathcal{I}} + \limsup_{n\to\infty}\frac{1}{n+1}K_N\circ T^{n-N} = (G_N)_{\mathcal{I}} \quad \text{a.e.},$$

where $(G_N)_{\mathcal{I}} = \lim_{n\to\infty}\frac{1}{n+1}\sum_{i=0}^{n} G_N\circ T^i$ by Birkhoff's Ergodic Theorem.

Now $(G_N)_{\mathcal{I}}$ decreases with N because G_N decreases, and

$$\int (G_N)_{\mathcal{I}}\, d\mu = \int G_N d\mu \to 0,$$

because G_N are non-negative uniformly bounded by $G_0 \in L^1$ and tend to 0 a.e.

Hence $(G_N)_{\mathcal{I}} \to 0$ a.e. Therefore

$$\limsup_{n\to\infty}\frac{1}{n+1}\sum_{i=0}^{n} g_{n-i}\circ T^i \to 0 \quad \text{a.e.}$$

establishing the missing a.e. convergence in (2.5.4). ♣

As an immediate consequence of (2.5.2) and (2.5.3) for T ergodic, along with $f_{\mathcal{I}} = \int f_{\mathcal{I}}\, d\mu$, we get the following:

Theorem 2.5.5 (Shannon–McMillan–Breiman, ergodic case). *If $T : X \to X$ is ergodic and $\mathcal{A}$ is a countable partition of finite entropy, then*

$$\lim_{n\to\infty}\frac{1}{n}I(\mathcal{A}^{n-1})(x) = \mathrm{h}_\mu(T,\mathcal{A}) \quad \textit{for a.e. } x \in X.$$

The left-hand side expression in the above equality can be viewed as a *local entropy* at x. The theorem says that at a.e. x the local entropy exists and is equal to the entropy (compare comments after (2.3.2) and Lemma 2.4.2).

2.6 Lebesgue spaces, measurable partitions and canonical systems of conditional measures

Let $(X, \mathcal{F}, \mu)$ be a probability space. We consider only *complete* measures (probabilities), such that every subset of a measurable set of measure 0 is measurable.

If a measure is not complete we can always consider its completion, that is, to add to $\mathcal{F}$ all sets A for which there exists $B \in \mathcal{F}$ with $A \div B$ contained in a set in $\mathcal{F}$ of measure 0.

Notation 2.6.1. *Consider $\mathcal{A}$, an arbitrary partition of X, not necessarily countable nor consisting of measurable sets. We denote by $\tilde{\mathcal{A}}$ the sub σ-algebra of $\mathcal{F}$ consisting of those sets in $\mathcal{F}$ that are unions of whole elements (fibres) of $\mathcal{A}$.*

Note that in the case where $\mathcal{A} \subset \mathcal{F}$, we have $\tilde{\mathcal{A}} \supset \sigma(\mathcal{A})$, the latter defined in Notation 2.1.1, but the inclusion can be strict. For example, if $\mathcal{A} \subset \mathcal{F}$ is the partition of X into points, then $\sigma(\mathcal{A})$ consists of all countable sets and their complements in X, and $\tilde{\mathcal{A}} = \mathcal{F}$. Obviously $\tilde{\mathcal{A}} \supset \{\emptyset, X\}$.

Definition 2.6.2. The partition $\mathcal{A}$ is called *measurable* if it satisfies the following separation property.

> There exists a sequence $\mathbf{B} = (B_n)_{n=1}^{\infty}$ of subsets of $\tilde{\mathcal{A}}$ such that for any two distinct $A_1, A_2 \in \mathcal{A}$ there is an integer $n \geq 1$ such that either
>
> $$A_1 \subset B_n \text{ and } A_2 \subset X \setminus B_n$$
>
> or
>
> $$A_2 \subset B_n \text{ and } A_1 \subset X \setminus B_n.$$

Since each element of the measurable partition $\mathcal{A}$ can be represented as an intersection of countably many elements B_n or their complements, each element of $\mathcal{A}$ is measurable. Let us stress, however, that the measurability of all elements of $\mathcal{A}$ is not sufficient for $\mathcal{A}$ to be a measurable partition (see Exercise 2.7). The sequence $\mathbf{B}$ is called a *basis* for $\mathcal{A}$.

Remark 2.6.3. A popular definition of an uncountable measurable partition $\mathcal{A}$ is that there exists a sequence of finite partitions (recall that this means: finite partitions into measurable sets) $\mathcal{A}_n, n = 0, 1, \ldots$, such that $\mathcal{A} = \bigvee_{n=0}^{\infty} \mathcal{A}_n$. Here (unlike later on) the join $\bigvee$ is in the set-theoretic sense, i.e. as $\{A_{n_1} \cap A_{n_2} \cap \cdots : A_{n_i} \in \mathcal{A}_{n_i}, i = 1, \ldots\}$. Clearly it is equivalent to the separation property in Definition 2.6.2.

Notice that for any measurable map $T : X \to X'$ between probability measure spaces, if $\mathcal{A}$ is a measurable partition of X', then $T^{-1}(\mathcal{A})$ is a measurable partition of X.

Now we pass to a very useful class of probability spaces: *Lebesgue spaces.*

Definition 2.6.4. We call a sequence $\mathbf{B} = (B_n)_{n=1}^{\infty}$ of subsets of $\mathcal{F}$ the *basis* of $(X, \mathcal{F}, \mu)$ if the two following conditions are satisfied:

(i) $\mathbf{B}$ ensures the separation property in Definition 2.6.2 for $\mathcal{A} = \varepsilon$, the partition into points, (i.e. $\mathbf{B}$ is a basis for ε);

(ii) for any $A \in \mathcal{F}$ there exists a set $C \in \sigma(\mathbf{B})$ such that $C \supset A$ and $\mu(C \setminus A) = 0$.

(Recall again, Notation 2.1.1, that $\sigma(\mathbf{B})$ denotes the smallest σ-algebra containing all the sets $B_n \in \mathbf{B}$. Rokhlin used the name Borel σ-algebra.)

A probability space $(X, \mathcal{F}, \mu)$ having a basis is called *separable.*

Now let $\varepsilon = \pm 1$ and $B_n^{(\varepsilon)} = B_n$ if $\varepsilon = 1$ and $B_n^{(\varepsilon)} = X \setminus B_n$ if $\varepsilon = -1$. To any sequence of numbers $\varepsilon_n, n = 1, 2, \dots$ there corresponds the intersection $\bigcap_{n=1}^{\infty} B_n^{(\varepsilon_n)}$. By (i) every such intersection contains no more than one point.

A probability space $(X, \mathcal{F}, \mu)$ is said to be *complete* with respect to a basis $\mathbf{B}$ if all the intersections $\bigcap_{n=1}^{\infty} B_n^{(\varepsilon_n)}$ are non-empty. The space $(X, \mathcal{F}, \mu)$ is said to be *complete* (mod 0) with respect to a basis $\mathbf{B}$ if X can be included as a subset of full measure into a certain measure space $(\overline{X}, \overline{\mathcal{F}}, \overline{\mu})$ that is complete with respect to its own basis $\overline{\mathbf{B}} = (\overline{B}_n)$ satisfying $\overline{B}_n \cap X = B_n$ for all n.

It turns out that a space that is complete (mod 0) with respect to one basis is also complete (mod 0) with respect to every other basis.

Definition 2.6.5. A probability space $(X, \mathcal{F}, \mu)$ complete (mod 0) with respect to one of its bases is called a *Lebesgue space.*

Exercise. If $(X_1, \mathcal{F}_1, \mu_1)$ and $(X_2, \mathcal{F}_2, \mu_2)$ are two probability spaces with complete measures, such that $X_1 \subset X_2$, $\mu_2(X_2 \setminus X_1) = 0$ and $\mathcal{F}_1 = \mathcal{F}_2|_{X_1}, \mu_1 = \mu_2|_{\mathcal{F}_1}$ (where $\mathcal{F}_2|_{X_1} := \{A \cap X_1 : A \in \mathcal{F}_2\}$), then the first space is Lebesgue if and only if the second is.

It is not difficult to check (see Exercise 2.9) that $(X, \mathcal{F}, \mu)$ is a Lebesgue space if and only if $(X, \mathcal{F}, \mu)$ is isomorphic to the unit interval (equipped with classical Lebesgue measure) together with countably many atoms.

Theorem 2.6.6. *Assume that $T : X \to X'$ is a measurable injective map from a Lebesgue space $(X, \mathcal{F}, \mu)$ onto a separable space $(X', \mathcal{F}', \mu')$, and pre-images of the sets of measure 0 (or positive) are of measure 0 (or positive). Then the space $(X', \mathcal{F}', \mu')$ is Lebesgue, and T^{-1} is a measurable map.*

Note that, in particular, a measurable, measure-preserving, injective map between Lebesgue spaces is an isomorphism. If $X = X', \mathcal{F} \supset \mathcal{F}', \mathcal{F} \neq \mathcal{F}'$ and $(X', \mathcal{F}', \mu')$ is separable, then the above implies that $(X, \mathcal{F}, \mu)$ is not Lebesgue.

Now let $(X, \mathcal{F}, \mu)$ be a Lebesgue space and $\mathcal{A}$ be a measurable partition of X. We say that a property holds for almost all atoms of $\mathcal{A}$ if and only if the union of atoms for which it is satisfied is measurable, and of full measure. The following fundamental theorem holds:

Theorem 2.6.7. *For almost all $A \in \mathcal{A}$ there exists a Lebesgue space $(A, \mathcal{F}_A, \mu_A)$ such that the following conditions are satisfied:*

(1) *If $B \in \mathcal{F}$, then $B \cap A \in \mathcal{F}_A$ for almost all $A \in \mathcal{A}$.*

(2) *$\mathcal{F}$-measurable for all $B \in \mathcal{F}$, where $A(x)$ is the element of $\mathcal{A}$ containing x.*

(3)
$$\mu(B) = \int_X \mu_{A(x)}(B \cap A(x))\, d\mu(x). \tag{2.6.1}$$

Remark 2.6.8. One can consider the quotient (factor) space $(X/\mathcal{A}, \mathcal{F}_{\mathcal{A}}, \mu_{\mathcal{A}})$ with $X/\mathcal{A}$ being defined just as $\mathcal{A}$ and with $\mathcal{F}_{\mathcal{A}} = p(\tilde{\mathcal{A}})$ (see Notation 2.6.1 for the tilde), and $\mu_{\mathcal{A}}(B) = \mu(p^{-1}(B))$, for the projection map $p(x) = A(x)$. It can be proved that the factor space is again a Lebesgue space. Then $x \mapsto \mu_A(x)(B \cap A(x))$ is $\mathcal{F}_{\mathcal{A}}$-measurable, and the property (2.6.1) can be rewritten in the form

$$\mu(B) = \int_{X/\mathcal{A}} \mu_A(B \cap A)\, d\mu_{\mathcal{A}}(A). \tag{2.6.2}$$

Remark 2.6.9. If a partition $\mathcal{A}$ is finite or countable, then the measures μ_A are just the conditional measures given by the formulae $\mu_A(B) = \mu(A \cap B)/\mu(A)$.

Remark 2.6.10. (2.6.1) can be rewritten for every μ-integrable function ϕ, or non-negative μ-measurable ϕ, if we allow $+\infty$-ies, as

$$\int \phi\, d\mu = \int_X \left(\int_{A(x)} \phi|_{A(x)}\, d\mu_{A(x)} \right) d\mu(x). \tag{2.6.3}$$

This is a version of Fubini's Theorem.

The family of measures $\{\mu_A : A \in \mathcal{A}\}$ is called the *canonical system of conditional measures* with respect to the partition $\mathcal{A}$. It is unique (mod 0) in the sense that any other system μ'_A coincides with it for almost all atoms of $\mathcal{A}$.

The method of construction of the system μ_A is via conditional expectations values with respect to the σ-algebra $\tilde{\mathcal{A}}$. Having chosen a basis (B_n) of the Lebesgue space $(X, \mathcal{F}, \mu)$, for every finite intersection

$$B = \bigcap_i B_{n_i}^{(\varepsilon_{n_i})} \tag{2.6.4}$$

one considers $\phi_B := E(1\!\!1_B|\tilde{\mathcal{A}})$, which can be treated as a function on the factor space $X/\mathcal{A}$, unique on a.e. $A \in \mathcal{A}$, and such that for all $Z \in \tilde{\mathcal{A}}$

$$\mu(B \cap Z) = \int_{p(Z)} \phi_B(A)\, d\mu_{\mathcal{A}}(A).$$

Clearly $(B_n \cap A)_{n=1}^{\infty}$ is a basis for all A. It is not hard to prove that for a.e. A, for each B from our countable family (2.6.4), $\phi_B(A)$ as a function of B generates a Lebesgue space on A, with $\mu_A(B) := \phi_B(A)$. Uniqueness of ϕ_B yields additivity.

Theorem 2.6.11. *If $T : X \to X'$ is a measurable map of a Lebesgue space $(X, \mathcal{F}, \mu)$ onto a Lebesgue space $(X', \mathcal{F}', \mu')$, then the induced map from $(X/\zeta, \mathcal{F}_\zeta, \mu_\zeta)$ for $\zeta = T^{-1}(\varepsilon)$, to $(X', \mathcal{F}', \mu')$ is an isomorphism.*

Proof. This follows immediately from the fact that the factor space is a Lebesgue space, and from Theorem 2.6.6. ♣

In what follows we consider partitions (mod 0): that is, we identify two partitions if they coincide, restricted to a measurable subset of full measure. For these classes of equivalence we use the same notation $\leq, \geq$ as in Section 2.3. They define a partial order. If $\mathcal{A}_\tau$ is a family of measurable partitions of a measure space (unlike in previous sections, the family may be uncountable), then by its product $\mathcal{A} = \bigvee_\tau \mathcal{A}_\tau$ we mean the measurable partition $\mathcal{A}$ determined by the following two conditions:

(i) $\mathcal{A} \geq \mathcal{A}_\tau$ for every τ;

(ii) if $\mathcal{A}' \geq \mathcal{A}_\tau$ for every τ and $\mathcal{A}'$ is measurable, then $\mathcal{A}' \geq \mathcal{A}$.

Similarly, replacing $\geq$ by $\leq$, we define the intersection $\bigwedge_\tau \mathcal{A}_\tau$.

The product and intersection exist in a Lebesgue space (i.e. the partially ordered structure is complete). They of course generalize the notions dealt with in Section 2.4. Clearly, for a countable family of measurable partitions $\mathcal{A}_\tau$ the above $\bigvee$ and the set-theoretic one coincide (the assumption that the space is Lebesgue and the reasoning (mod 0) is not needed). In Exercise 2.13 we give some examples.

There is a natural one-to-one correspondence between the measurable partitions (mod 0) of a Lebesgue space $(X, \mathcal{F}, \mu)$ and the complete σ-sub-algebras of $\mathcal{F}$, that is, such σ-algebras $\mathcal{F}' \subset \mathcal{F}$ that the measure μ restricted to $\mathcal{F}'$ is complete. This correspondence is defined by assigning to each $\mathcal{A}$ the σ-algebra $\mathcal{F}(\mathcal{A})$ of all sets that coincide (mod 0) with the sets of $\tilde{\mathcal{A}}$ (defined at the beginning of this section). To operations on the measurable partitions (mod 0) there correspond operations on the corresponding σ-algebras. Namely, if $\mathcal{A}_\tau$ is a family of measurable partitions (mod 0), then

$$\mathcal{F}(\bigvee_\tau \mathcal{A}_\tau) = \bigvee_\tau \mathcal{F}(\mathcal{A}_\tau), \qquad \mathcal{F}(\bigwedge_\tau \mathcal{A}_\tau) = \bigwedge_\tau \mathcal{F}(\mathcal{A}_\tau).$$

Here $\bigwedge_\tau \mathcal{F}(\mathcal{A}_\tau) = \bigcap_\tau \mathcal{F}(\mathcal{A}_\tau)$ is the set-theoretic intersection of the σ-algebras, and $\bigvee_\tau \mathcal{F}(\mathcal{A}_\tau)$ is the set-theoretic intersection of all the σ-algebras that contain all $\mathcal{F}(\mathcal{A}_\tau)$.

For any measurable partition $\mathcal{A}$ and any μ-integrable function $\phi : X \to \mathbb{R}$ write

$$E(f|\mathcal{A})(x) := \int f|_{\mathcal{A}(x)} \, d\mu_{\mathcal{A}(x)} \text{ a.e.} \tag{2.6.5}$$

Note that by the definition of the canonical system of conditional measures and by the definition of conditional expectation value, for any measurable partition $\mathcal{A}$ we get the identity

$$E(f|\mathcal{A}) = E(f|\mathcal{F}(\mathcal{A})). \tag{2.6.6}$$

A sequence of measurable partitions $\mathcal{A}_n$ is called *(monotone) increasing* or *ascending* if for all $n_1 \leq n_2$ we have $\mathcal{A}_{n_1} \leq \mathcal{A}_{n_2}$. It is called *(monotone) decreasing* or *descending* if for all $n_1 \leq n_2$ we have $\mathcal{A}_{n_1} \geq \mathcal{A}_{n_2}$.

For a monotone increasing (decreasing) sequence of measurable partitions $\mathcal{A}_n$ and $\mathcal{A} = \bigvee_n \mathcal{A}_n$ ($\mathcal{A} = \bigwedge_n \mathcal{A}_n$) we write $\mathcal{A}_n \nearrow \mathcal{A}$ (or $\mathcal{A}_n \searrow \mathcal{A}$). In the language of measurable partitions of a Lebesgue space, owing to (2.6.6), the Martingale Convergence Theorem 2.1.4 can be expressed as follows:

Theorem 2.6.12. *If $\mathcal{A}_n \nearrow \mathcal{A}$ or $\mathcal{A}_n \searrow \mathcal{A}$, then for every integrable function f,* $E(f|\mathcal{A}_n) \to E(f|\mathcal{A})$ μ a.s.

2.7 Rokhlin natural extension

We shall prove here the following very useful theorem:

Theorem 2.7.1. *For every measure-preserving endomorphism T of a Lebesgue space $(X, \mathcal{F}, \mu)$ there exists a Lebesgue space $(\tilde{X}, \tilde{\mathcal{F}}, \tilde{\mu})$ with measure-preserving transformations $\pi_n : \tilde{X} \to X, n \le 0$ satisfying $T \circ \pi_{n-1} = \pi_n$, which is an inverse limit of the system $\cdots \xrightarrow{T} X \xrightarrow{T} X$.*

Moreover there exists an automorphism $\tilde{T}$ of $(\tilde{X}, \tilde{\mathcal{F}}, \tilde{\mu})$ satisfying

$$\pi_n \circ \tilde{T} = T \circ \pi_n \tag{2.7.1}$$

for every $n \le 0$.

Recall that in *category theory* [Lang, 1965, Ch. 1], for a sequence (system) of objects and morphisms $\cdots \xrightarrow{M_{n-1}} O_{n-1} \xrightarrow{M_n} \cdots \xrightarrow{M_0} O_0$ an object O equipped with morphisms $\pi_n : O \to O_n$ is called an *inverse limit* if $M_n \circ \pi_{n-1} = \pi_n$ and for every other O' equipped with morphisms $\pi'_n : O' \to O_n$ satisfying $M_n \circ \pi'_{n-1} = \pi'_n$ there exists a unique morphism $M : O' \to O$ such that $\pi_n \circ M = \pi'_n$ for every $n \le 0$.

In particular, if all O_n are the same ($= O_0$), and additionally $M_1 : O_0 \to O_0$ is chosen, then for $\pi'_n := M_{n+1} \circ \pi_n : O \to O_0$, $n \le 0$ there exists $M : O \to O$ such that $\pi_n \circ M = \pi'_n = M_{n+1} \circ \pi_n$ for every n. It is easy to see that M is an automorphism.

In Theorem 2.7.1 the objects are probability spaces or probability spaces with complete probabilities, and morphisms are measure-preserving transformations or measure-preserving transformations up to sets of measure 0. (We have thus multiple meanings of Theorem 2.7.1.)

Thus the first part of Theorem 2.7.1 produces $\tilde{T}$ satisfying (2.7.1) automatically, via the category theory definition. The automorphism $\tilde{T}$ is called Rokhlin's *natural extension* of T; compare the terminology at the beginning of Section 2.2. This is a 'minimal' extension of T to an automorphism.

One can consider $\cdots \xrightarrow{T_{n-1}} X_n \xrightarrow{T_n} X_n \xrightarrow{T_{n+1}} \ldots$ in place of $\cdots \xrightarrow{T} X \xrightarrow{T} X$ for all $n \in \mathbb{Z}$ in the statement of Theorem 2.7.1. We have chosen a simplified version with all T_n equal to T to simplify the notation, and since only such a version will be used in this book.

In the proof of Theorem 2.7.1 we shall use the following.

Theorem 2.7.2 (Extension of Measure). *Every probability measure ν (σ-additive) on an algebra $\mathcal{G}_0$ of subsets of a set X can be uniquely extended to a measure on the σ-algebra $\mathcal{G}$ generated by $\mathcal{G}_0$.*

This theorem can be proved with the use of the well-known Carathéodory's construction [Carathéodory, 1927, Ch. V]. We define the outer measure:

$$\nu_e(A) = \inf\{\nu(B) : B \in \mathcal{G}_0, A \subset B\}$$

for every $A \subset X$.

We say that A is *Carathéodory measurable* if for every $E \subset X$ the outer measure ν_e satisfies

$$\nu_e(E) = \nu_e(E \cap A) + \nu_e(E \setminus A).$$

The family of these sets turns out to be a σ-algebra containing $\mathcal{G}_0$, and hence containing $\mathcal{G}$.

For a general definition of outer measures and a sketch of the theory see Chapter 8.

Proof of Theorem 2.7.1. Denote $\Pi = X^{\mathbb{Z}_-}$, the set theory Cartesian product of a countable number of X's, or more precisely the space of sequences (x_n) of points in X indexed by non-positive integers. For each $i \leq 0$ denote by $\pi_i : \Pi \to X$ the projection to the i-th coordinate, $\pi_i((x_n)_{n\in\mathbb{Z}_-}) = x_i$.

We start by producing the inverse limit in the set-theoretic category. Set

$$\tilde{X} = \{(x_n)_{n\in\mathbb{Z}_-} : T(x_n) = x_{n+1} \ \forall n < 0\}. \tag{2.7.2}$$

The mappings π_n in the statement of Theorem 2.7.1 will be the restrictions of the π_n's defined above to $\tilde{X}$.

We shall endow Π with a σ-algebra $\mathcal{F}_\Pi$ and probability measure μ_Π, whose restrictions to $\tilde{X}$ will yield the inverse limit $(\tilde{X}, \tilde{\mathcal{F}}, \tilde{\mu})$. The measure μ_Π will occur to be 'supported' on $\tilde{X}$.

For each $n \leq 0$ consider the σ-algebra $\mathcal{G}_n = \pi_n^{-1}(\mathcal{F})$. Let $\mathcal{F}_{\Pi,0}$ be the smallest algebra of subsets of Π containing all σ-algebras $\mathcal{G}_n$. It is easy to see that $\mathcal{F}_{\Pi,0}$ consists of finite unions of pairwise disjoint 'cylinders' $\bigcap_{i=n}^0 \pi_i^{-1}(C_i))$, considered for arbitrary finite sequences of sets $C_i \in \mathcal{F}, i = n, ..., 0$ for an arbitrary $n \in \mathbb{Z}_-$. Define

$$\mu_\Pi\Big(\bigcap_{i=n}^{0} \pi_i^{-1}(C_i)\Big) := \mu\Big(\bigcap_{i=n}^{0} T^{-(i-n)}(C_i)\Big). \tag{2.7.3}$$

We extend the definition to finite unions of disjoint cylinders A_k by $\mu_\Pi(\bigcup_k A_k) := \sum_k \mu_\Pi(A_k)$.

To ensure that μ_Π is well defined it is sufficient to prove the compatibility condition:

$$\mu_\Pi\Big(\bigcap_{i=n}^{0} \pi_i^{-1}(C_i)\Big) + \mu_\Pi\Big(\Big(\bigcap_{i:n\leq i\leq 0, i\neq j} \pi_i^{-1}(C_i)\Big) \cap \pi_j^{-1}(C'_j)\Big)$$
$$= \mu_\Pi\Big(\Big(\bigcap_{i:n\leq i\leq 0, i\neq j} \pi_i^{-1}(C_i)\Big) \cap \pi_j^{-1}(C_j \cup C'_j)\Big), \tag{2.7.4}$$

for all sequences $C_i \in \mathcal{F}, i = n, ..., 0$ and $C'_j \in \mathcal{F}$ disjoint from C_j. Fortunately (2.7.4) follows immediately from (2.7.3) and from the additivity of the measure μ_Π.

The next step is to observe that μ_Π is σ-additive on the algebra $\mathcal{F}_{\Pi,0}$. For this end we use the assumption that $(X, \mathcal{F}, \mu)$ is a Lebesgue space.[1] We just assume that X is a full Lebesgue measure subset of the unit interval $[0, 1]$, with classical Lebesgue measure and atoms, and the σ-algebra of Lebesgue measurable sets $\mathcal{F}$: see Exercise 2.9. Now it is sufficient to apply the textbook fact that for every Lebesgue measurable set $C \subset [0, 1]$ and $\varepsilon > 0$ there exists a compact set $P \subset C$ with $\mu(C \setminus P) < \varepsilon$. (This is also often proved by the Lebesgue measure construction via Carathéodory's outer measure.) Compare the notion of regularity of measure in Section 3.1.

Consider Π endowed with the product topology, compact by Tichonov's Theorem. Then all π_i are continuous, and for every $\varepsilon > 0$ and for every cylinder $A = \bigcap_{i=n}^{0} \pi_i^{-1}(C_i))$ we can find a compact cylinder $K = \bigcap_{i=n}^{0} \pi_i^{-1}(P_i))$, with compact $P_i \subset C_i$, such that $\mu_\Pi(A \setminus K) < \varepsilon$. This follows from the definition (2.7.3) and the T-invariance of μ. The same immediately follows for finite unions of cylinders.

To prove the σ-additivity of μ_Π on $\mathcal{F}_{\Pi,0}$ it is sufficient to prove that for every descending sequence of sets $A_k \in \mathcal{F}_{\Pi,0}, i = 1, 2, \ldots$ if

$$\bigcap_k A_k = \emptyset \text{ then } \mu_\Pi(A_k) \to 0. \tag{2.7.5}$$

Suppose to the contrary that there exists $\varepsilon > 0$ such that $\mu_\Pi(A_k) \geq \varepsilon$ for every k. For each k, consider a compact set $K_k \subset A_k$ such that $\mu(A_k \setminus K_k) \leq \varepsilon 2^{-k-1}$. Then all $L_m := \bigcap_{k=1}^{m} K_k$ are non-empty, since $\mu_\Pi(L_m) \geq \varepsilon/2$. Hence $\bigcap_{k=1}^{\infty} A_k \supset \bigcap_{k=1}^{\infty} L_k \neq \emptyset$ as $(L_k)_{k=1}^{\infty}$ is a descending family of non-empty compact sets. Thus we have proved that μ_Π is σ-additive on $\mathcal{F}_{\Pi,0}$.

The measure μ_Π extends to σ-additive measure on a σ-algebra generated by $\mathcal{F}_{\Pi,0}$ by Theorem 2.7.2. Set this extension to be our $(\Pi, \mathcal{F}_\Pi, \mu_\Pi)$.

Now we shall prove that the set $\Pi \setminus \tilde{X}$ is μ_Π-measurable, and that $\mu_\Pi(\Pi \setminus \tilde{X}) = 0$. To this end we shall take care that the compact sets $K = K_k$ lie in $\tilde{X}$. Denote

$$\tilde{X}^n := \{(x_i)_{i \in \mathbb{Z}_-} : T(x_i) = x_{i+1} \ \forall n \leq i < 0\}. \tag{2.7.6}$$

Let us recall that $A = \bigcap_{i=n}^{0} \pi_i^{-1}(C_i))$. Note that $\pi_n^{-1}(T^{-(i-n)}(C_i)) \cap \tilde{X}^n \subset \pi^{-1}(C_i)$, but they have the same measure μ_Π, by the formula (2.7.3). Let P_n be a compact subset of $C'_n := \bigcap_{i=n}^{0} T^{-(i-n)}(C_i)$ such that $\mu_\Pi(C'_n \setminus P_n) < \varepsilon$ and T^j restricted to P_n is continuous for all $j = 1, ..., n$. This is possible by Luzin's Theorem.

Then all $T^j(P_n)$ are compact sets, and in particular are μ-measurable. Hence each $Q_n := \bigcap_{i=n}^{0} \pi_i^{-1}(T^{i-n}(P_n))$ belongs to $\mathcal{F}_{\Pi,0}$, and in particular it is μ_Π-measurable. It is contained in $\tilde{X}^n$, but need not be contained in $\tilde{X}$. To cope with

[1] This is a substantial assumption, overlooked by some authors: see the Bibliographical notes at the end of this chapter.

this problem, express A as $A_N = \bigcap_{i=N}^{0} \pi_i^{-1}(C_i))$ for N arbitrarily large, setting $C_i = X$ for $i : N \le i < n$. Then find Q_N for A_N and $\varepsilon_n = \varepsilon 2^{-N-1}$ and finally set $Q = \bigcap Q_N$. The set Q is μ_Π-measurable (even compact), contained in $\tilde{X}$ and $\mu_\Pi(A \setminus Q) \le \varepsilon$.

For $A = \Pi$ we conclude with μ_Π-measurable $Q \subset \tilde{X}$ with $\mu_\Pi(\Pi \setminus Q) \le \varepsilon$, for an arbitrary $\varepsilon > 0$. Hence $\mu_\Pi(\Pi \setminus \tilde{X} = 0$.

Note that the measure μ_Π is complete (i.e. all subsets of measurable sets of measure 0 are measurable) by Carathéodory's construction. Now we prove that it is a Lebesgue space. Let $\mathbf{B} = (B_l)_{l=1}^{\infty}$ be a basis of $(X, \mathcal{F}, \mu)$. Then clearly the family $\mathbf{B}_\Pi := \{\pi_n^{-1}(B_l) : l \ge 1, n \le 0\}$ is a basis of the partition ε in Π. The family $\mathbf{B}_\Pi$ generates the σ-algebra $\mathcal{F}_\Pi$ in the sense of Definition 2.6.4(ii)), because $\mathbf{B}$ generates $\mathcal{F}$ in this sense, and by Carathéodory's outer measure construction. The probability space $(\Pi, \mathcal{F}_\Pi, \mu_\Pi)$ is complete with respect to $\mathbf{B}_\Pi$ since $(X, \mathcal{F}, \mu)$ is complete with respect to $\mathbf{B}$, and by the Cartesian product definition.

Finally, let us restrict all the objects to $\tilde{X}$. In particular, $\tilde{F} := \{A \cap \tilde{X} : A \in \mathcal{F}_\Pi\}$, $\tilde{\mu}$ is the restriction of μ_Π to $\tilde{F}$, and $\tilde{\mathbf{B}} := \{B \cap \tilde{X} : B \in \mathbf{B}_\Pi\}$. The resulting probability space $(\tilde{X}, \tilde{\mathcal{F}}, \tilde{\mu})$ is complete (mod 0) with respect to $\tilde{\mathbf{B}}$. Therefore it is a Lebesgue space (see Definition 2.6.5): the extension required by the definition is just $(\Pi, \mathcal{F}_\Pi, \mu_\Pi)$ with the basis $\mathbf{B}_\Pi$.

Suppose $(X', \mathcal{F}', \mu')$ is any Lebesgue measure space, with measure-preserving transformations $\pi'_n : X' \to X, n \le 0$ satisfying $T \circ \pi'_{n-1} = \pi'_n$. Then define $M : X' \to \tilde{X}$ by

$$M(x') = (\dots \pi'_{n-1}(x'), \pi'_n(x'), \dots, \pi'_0(x')). \tag{2.7.7}$$

We get $\pi_n \circ M = \pi'_n$ by definition. We leave the proof of the measurability of M to the reader.

The uniqueness of M follows from the fact that if $M(x') = (\dots y_n, \dots, y_0)$ for $y_j \in X$, then from $\pi_n \circ M = \pi'_n$ μ'-a.e., we get $y_n = \pi'_n(x')$ a.e. ♣

Remark 2.7.3. $\tilde{X}$ can be interpreted as the space of all backward trajectories for T. The map $\tilde{T} : \tilde{X} \to \tilde{X}$ can be defined by the formula

$$\tilde{T}((x_n)_{n \in \mathbb{Z}_-}) = (\dots, x_{-2}, x_{-1}, x_0, T(x_0)). \tag{2.7.8}$$

$\tilde{X}$ could be defined in (2.7.2) as the space of full trajectories $\{(x_n)_{n \in \mathbb{Z}}; T(x_n) = x_{n+1}\}$. Then (2.7.8) is the *shift to the left.*

The formula (2.7.8) holds because $\tilde{T}$, defined by it, satisfies (2.7.1), and because of the uniqueness of maps $\tilde{T}$ satisfying (2.7.1)

Remark 2.7.4. Alternatively, to compact sets in $\tilde{X}^n$ we could find for all $n \le 0$ sets $E_{n,i} \supset \tilde{X}^n$, with $\tilde{\mu}_\Pi(E_{n,i} \setminus \tilde{X}^n) \to 0$ as $i \to \infty$, which are unions of cylinders $\bigcap_{i=-n}^{0} \pi_i^{-1}(C_i)$. This agrees with the following general fact:

If a sequence of sets Σ generates a σ-algebra $\mathcal{G}$ with a measure ν on it (see Definition 2.6.4(ii)), then for every $A \in \mathcal{G}$ there exists $C \supset A$ with $\nu(C \setminus A) = 0$ such that $C \in \Sigma'_{d\sigma\delta}$: that is, C is a countable intersection of countable unions of finite intersections of sets belonging to Σ or their complements. Exercise: Prove this general fact, using Carathéodory's outer measure constructed on measurable sets.

Remark 2.7.5. Another way to prove Theorem 2.7.1 is to construct $\tilde{\mathcal{F}}$ and $\tilde{\mu}$ on $\tilde{X}$ already at the beginning. One also defines $\tilde{\mu}$ by the formula (2.7.3).

More specifically, for the maps π_n restricted now to $\tilde{X}$, consider $\mathcal{G}_n = \pi_n^{-1}(\mathcal{F})$. Note that this is an ascending sequence of σ-algebras with growing $|n|$, because $\pi_n^{-1}(A) = \pi_{n-1}^{-1}(T^{-1}(A))$ for every $A \in \mathcal{F}$. Write $\tilde{\mathcal{F}}_0 = \bigcup_{n\le 0} \mathcal{G}_n$. This is an algebra. For every $A \in \mathcal{F}$ and $n \le 0$ define $\tilde{\mu}(\pi_n^{-1}(A)) := \mu(A)$. This is well defined, because if $C = \pi_n^{-1}(A_1) = \pi_m^{-1}(A_2)$ for $A_1, A_2 \in \mathcal{F}$ and $n < m$, then $A_1 = T^{-(m-n)}(A_2)$. Since T preserves μ, we have $\mu(A_1) = \mu(A_2)$. This measure is σ-additive on the algebra $\tilde{\mathcal{F}}_0$, since we managed to approximate 'from below' each of its elements by a compact set: see the proof of Theorem 2.7.1. Hence we find $\tilde{\mathcal{F}}$ and $\tilde{\mu}$ on $\tilde{X}$ by Carathéodory's theory.

Unfortunately, the measure space $(\tilde{X}, \tilde{\mathcal{F}}, \tilde{\mu})$ is usually not complete with respect to the basis $\tilde{\mathbf{B}}$, constructed in the proof of Theorem 2.7.1. To make it complete (mod 0) we need to extend it, and the only way we know how to acomplish this is to construct the space $(\Pi, \mathcal{F}_\Pi, \mu_\Pi)$.

We end this section with another version of Theorem 2.7.1. First the following definition:

Definition 2.7.6. Suppose that T is an automorphism of a Lebesgue space $(X, \mathcal{F}, \mu)$. Let ζ be a measurable partition. Assume it is forward invariant: that is, $T(\zeta) \ge \zeta$, or equivalently $T^{-1}(\zeta) \le \zeta$. Then ζ is said to be *exhausting* if $\bigvee_{n\ge 0} T^n(\zeta) = \varepsilon$.

Theorem 2.7.7. *For every measure-preserving endomorphism T of a Lebesgue space $(X, \mathcal{F}, \mu)$ there exist a Lebesgue space $(\tilde{X}, \tilde{\mathcal{F}}, \tilde{\mu})$, its automorphism $\tilde{T}$, and a forward invariant for $\tilde{T}$ exhausting measurable partition ζ, such that $(X, \mathcal{F}, \mu) = (\tilde{X}/\zeta, \tilde{\mathcal{F}}_\zeta, \tilde{\mu}/\zeta)$ the factor space (cf. Remark 2.6.8), and T is a factor of $\tilde{T}$, that is, $T \circ p = p \circ \tilde{T}$ for the projection $p : \tilde{X} \to X$.*

Proof. Take $(\tilde{X}, \tilde{\mathcal{F}}, \tilde{\mu})$ and $\tilde{T}$ from Theorem 2.7.1. Set $\zeta := \pi_0^{-1}(\varepsilon)$. By (2.7.1) and $T^{-1}(\varepsilon) \le \varepsilon$ we get $\tilde{T}^{-1}(\zeta) \le \zeta$.

If $\varepsilon' = \bigvee_{n\ge 0} T^n(\zeta)$ is not the partition of $\tilde{X}$ into points, then $\tilde{T}/\varepsilon'$ is an automorphism of $(\tilde{X}/\varepsilon', \tilde{\mathcal{F}}_{\varepsilon'}, \tilde{\mu}_{\varepsilon'})$. Moreover, if we denote by p' the projection from $\tilde{X}$ to $\tilde{X}/\varepsilon'$, then we can write $\pi_n = \pi'_n \circ p'$ for some maps π'_n for every $n \le 0$. By the definition of inverse limit, p' has an inverse, which is impossible.

The last part, that $\bigvee_{n\ge 0} T^n(\zeta)$ is the partition of $\tilde{X}$ into points, has also an immediate proof following directly from the form of $\tilde{X}$ in (2.7.2). Indeed, for $n \ge 0$ the element of $T^n(\zeta)$ containing $\tilde{x} = (\dots, x_{-2}, x_{-1}, x_0)$ is the n-th image of the element of ζ containing $\tilde{T}^{-n}(\tilde{x})$ i.e. containing $(\dots, x_{-n-1}, x_{-n})$. So it is equal to $\{(\dots, x'_{-n-1}, x'_{-n}, \dots, x'_0) \in \tilde{X} : x'_{-n} = x_{-n})\}$. Intersecting over $n \to \infty$, we obtain $\{\tilde{x}\}$. ♣

2.8 Generalized entropy; convergence theorems

This section contains generalizations of entropy notions (introduced in Section 2.3) to the case of all measurable partitions. The triple $(X, \mathcal{F}, \mu)$ is assumed to be a Lebesgue space.

Definition 2.8.1. If $\mathcal{A}$ is a measurable partition of X, then its (generalized) entropy is defined as follows:

$$\mathrm{H}(\mathcal{A}) = \infty \text{ if } \mathcal{A} \text{ is not a countable partition (mod 0);}$$
$$H(\mathcal{A}) = \textstyle\sum_{A \in \mathcal{A}} -\mu(A) \log \mu(A) \text{ if } \mathcal{A} \text{ is a countable partition (mod 0).}$$

Lemma 2.8.2. *If $\mathcal{A}_n$ and $\mathcal{A}$ are measurable partitions of X and $\mathcal{A}_n \nearrow \mathcal{A}$, then $\mathrm{H}(\mathcal{A}_n) \nearrow \mathrm{H}(\mathcal{A})$.*

Proof. Write $\mathrm{H}(\mathcal{A}) = \int I(\mathcal{A})\, d\mu$, where $I(\mathcal{A})(x) = -\log \mu(\mathcal{A}(x))$ is the information function (compare Section 2.4). We set $\log 0 = -\infty$: hence $I(\mathcal{A})(x) = \infty$ if $\mu(\mathcal{A}(x)) = 0$. Write the same for $\mathcal{A}_n$. As $\mu(\mathcal{A}_n(x)) \searrow \mu(\mathcal{A}(x))$ for a.e. x, the convergence in our lemma follows from the Monotone Convergence Theorem. ♣

Definition 2.8.3. If $\mathcal{A}$ and $\mathcal{B}$ are two measurable partitions of X, then the (generalized) *conditional entropy* $\mathrm{H}(\mathcal{A}|\mathcal{B}) = \mathrm{H}_\mu(\mathcal{A}|\mathcal{B})$ of partition $\mathcal{A}$ subject to $\mathcal{B}$ is defined by the following integral:

$$\mathrm{H}_\mu(\mathcal{A}|\mathcal{B}) = \int_{X/\mathcal{B}} \mathrm{H}_{\mu_B}(\mathcal{A}|B)\, d\mu_{\mathcal{B}}(B), \tag{2.8.1}$$

where $\mathcal{A} \cap B$ is the partition $\{A \cap B : A \in \mathcal{A}\}$ of B and $\{\mu_B, B \in \mathcal{B}\}$ forms a canonical system of conditional measures: see Section 2.7. For the integral in (2.8.1) to be well defined we have to know that the function $B \mapsto \mathrm{H}_{\mu_B}(\mathcal{A}|B)$, $B \in \mathcal{B}$ is measurable. In order to see this, choose a sequence of finite partitions $\mathcal{A}_n \nearrow \mathcal{A}$ (see Remark 2.6.3). Each conditional entropy function $\mathrm{H}_{\mu_B}(\mathcal{A}_n|B)$ is measurable as a function of B in the factor space $(X/\mathcal{B}, \mathcal{F}_{\mathcal{B}}, \mu_{\mathcal{B}})$, and hence of course as a function on $(X, \mathcal{F}, \mu)$, since it is a finite sum of measurable functions:

$$B \mapsto -\mu_B(A \cap B) \log \mu_B(A \cap B) \text{ for } A \in \mathcal{A}.$$

Since $\mathcal{A}_n|B \nearrow \mathcal{A}|B$ for a.e. B, we obtain, by using Lemma 2.8.2, that $\mathrm{H}_{\mu_B}(\mathcal{A}_n|B) \to \mathrm{H}_{\mu_B}(\mathcal{A}|B)$. Hence $\mathrm{H}_{\mu_B}(\mathcal{A}|B)$ is measurable, so our definition of $\mathrm{H}_\mu(\mathcal{A}|\mathcal{B})$ makes sense (we allow ∞'s here).

Of course, (2.8.1) can be rewritten in the form

$$\int_X \mathrm{H}_{\mu_{\mathcal{B}(x)}}(\mathcal{A}|\mathcal{B}(x))\, d\mu(x), \tag{2.8.2}$$

with $\mathrm{H}_{\mu_B}(\mathcal{A}|B)$ understood as a constant function on each $B \in \mathcal{B}$ (compare (2.6.1) with (2.6.2)). As in Section 2.3, we can write

$$\mathrm{H}_\mu(\mathcal{A}|\mathcal{B}) = \int_X I(\mathcal{A}|\mathcal{B})\, d\mu, \tag{2.8.3}$$

where $I(\mathcal{A}|\mathcal{B})$ is the *conditional information function*:

$$I(\mathcal{A}|\mathcal{B})(x) := -\log \mu_{\mathcal{B}(x)}(\mathcal{A}(x) \cap \mathcal{B}(x)).$$

Indeed, $I(\mathcal{A}|\mathcal{B})$ is non-negative and μ-measurable as being equal to $\lim_{n\to\infty} I(\mathcal{A}_n|\mathcal{B})$ (a.e.), so (2.8.3) follows from (2.6.3).

Lemma 2.8.4. *If $\{\mathcal{A}_n : n \geq 1\}$ and $\mathcal{A}$ are measurable partitions, $\mathcal{A}_n \searrow \mathcal{A}$ and $\mathrm{H}(\mathcal{A}_1) < \infty$, then $\mathrm{H}(\mathcal{A}_n) \searrow \mathrm{H}(\mathcal{A})$.*

Proof. The proof is similar to the proof of Lemma 2.8.2. ♣

Theorem 2.8.5. *If $\mathcal{A}, \mathcal{B}$ are measurable partitions and $\{\mathcal{A}_n : n \geq 1\}$ is an ascending (descending and $\mathrm{H}(\mathcal{A}_1|\mathcal{B}) < \infty$) sequence of measurable partitions converging to $\mathcal{A}$, then*

$$\lim_{n\to\infty} \mathrm{H}(\mathcal{A}_n|\mathcal{B}) = \mathrm{H}(\mathcal{A}|\mathcal{B}) \tag{2.8.4}$$

and the convergence is respectively monotone.

Proof. Applying Lemmas 2.8.2 and 2.8.4 we get the monotone convergence $\mathrm{H}_{\mu_B}(\mathcal{A}_n|B) \to \mathrm{H}_{\mu_B}(\mathcal{A}|B)$ for almost all $B \in X/\mathcal{B}$. Thus the integrals in the Definition 2.8.3 converge by the Monotone Convergence Theorem. ♣

Theorem 2.8.6. *If $\mathcal{A}, \mathcal{B}$ are measurable partitions and $\{\mathcal{B}_n : n \geq 1\}$ is a descending (ascending and $\mathrm{H}(\mathcal{A}|\mathcal{B}_1) < \infty$) sequence of measurable partitions converging to $\mathcal{B}$, then*

$$\lim_{n\to\infty} \mathrm{H}(\mathcal{A}|\mathcal{B}_n) = \mathrm{H}(\mathcal{A}|\mathcal{B}) \tag{2.8.5}$$

and the convergence is respectively monotone.

Proof 1. Assume first that $\mathcal{A}$ is finite (or countable with finite entropy). Then the a.e. convergence $I(\mathcal{A}|\mathcal{B}_n) \to I(\mathcal{A}|\mathcal{B})$ follows from the Martingale Convergence Theorem (more precisely from Theorem 2.6.12), applied to $f = 1\!\!1_A$, the indicator functions of $A \in \mathcal{A}$.

Now it is sufficient to prove that $\sup_n I(\mathcal{A}|\mathcal{B}_n) \in L^1$ in order to use the Dominated Convergence Theorem (compare Corollary 2.5.3) and (2.8.3)). One can repeat the proofs of Lemma 2.5.1 (for ascending $\mathcal{B}_n$) and Corollary 2.5.2.

The monotonicity of the sequence $\mathrm{H}(\mathcal{A}|\mathcal{B}_n)$ relies on Theorem 2.3.1(d). However, for infinite $\mathcal{B}_n$ one needs to approximate $\mathcal{B}_n$ by finite (or finite entropy) partitions. For details see [Rokhlin 1967, Sec. 5.12].

For $\mathcal{A}$ measurable, represent $\mathcal{A}$ as $\lim_{j\to\infty} \mathcal{A}_j$ for an ascending sequence of finite partitions $\mathcal{A}_j, j = 1, 2, \ldots$; then refer to Theorem 2.8.5. In the case of a descending sequence $\mathcal{B}_n$ the proof is straightforward. In the case of ascending $\mathcal{B}_n$ use

$$\mathrm{H}(\mathcal{A}|\mathcal{B}_n) - \mathrm{H}(\mathcal{A}_j|\mathcal{B}_n) = H(\mathcal{A}|(\mathcal{A}_j \vee \mathcal{B}_n)) \leq H(\mathcal{A}|(\mathcal{A}_j \vee \mathcal{B}_1)) = \mathrm{H}(\mathcal{A}|\mathcal{B}_1) - \mathrm{H}(\mathcal{A}_j|\mathcal{B}_1).$$

This implies that the convergence as $j \to \infty$ is uniform with respect to n: hence in the limit $\mathrm{H}(\mathcal{A}|\mathcal{B}_n) \to H(\mathcal{A}|\mathcal{B})$. ♣

Proof 2. For $\mathcal{A}$ finite (or countable with finite entropy) there is a simpler way to prove (2.8.5). By Theorem 2.1.4, for every $A \in \mathcal{A}$, the sequence $E(1\!\!1_A|\mathcal{F}(\mathcal{B}_n))$ converges to $E(1\!\!1_A|\mathcal{F}(\mathcal{B}))$ in L^2. Hence, for every $A \in \mathcal{A}$, the sequence $\mu_{\mathcal{B}_n(x)}(A\cap \mathcal{B}_n(x))$ converges to $\mu_{\mathcal{B}(x)}(A\cap\mathcal{B}(x))$ in measure μ. By continuity of the function $k(t) = -t\log t$ (compare Section 2.3), this implies the convergence

$$k(\mu_{\mathcal{B}_n(x)}(A\cap\mathcal{B}_n(x))) \to k(\mu_{\mathcal{B}(x)}(A\cap\mathcal{B}(x)))$$

in measure μ (we do not assume $x \in A$ here). Summing over all $A \in \mathcal{A}$ we obtain the convergence $\mathrm{H}_{\mu_{\mathcal{B}_n(x)}}(\mathcal{A}|\mathcal{B}_n(x)) \to \mathrm{H}_{\mu_{\mathcal{B}(x)}}(\mathcal{A}|\mathcal{B}(x))$ in measure μ. These functions are uniformly bounded by $\log \#\mathcal{A}$ (or by $\mathrm{H}(\mathcal{A})$) and non-negative: hence we get the convergence in L^1 and in consequence, owing to (2.8.2), we obtain (2.8.5). (Note that we have not used the a.e. convergence coming from Theorem 2.1.4, but only the convergence in L^2, which has been proved there.) ♣

Observe that we can now rewrite the definition of the entropy $\mathrm{h}_\mu(T,\mathcal{A})$ from Section 2.5 as follows:

$$\mathrm{h}_\mu(T,\mathcal{A}) = \mathrm{H}(\mathcal{A}|\mathcal{A}^-), \quad \text{where } \mathcal{A}^- := \bigvee_{n=1}^{\infty} T^{-n}(\mathcal{A}). \tag{2.8.6}$$

A countable partition $\mathcal{B}$ is called *a countable (one-sided) generator* for an endomorphism of a Lebesgue space if $\mathcal{B}^m \nearrow \varepsilon$. Because of Theorem 2.8.6 we obtain the following facts, useful for computing the entropy for concrete examples.

Theorem 2.8.7. *(a) If $\mathcal{B}_m$ is a sequence of finite partitions of a Lebesgue space, such that $\mathcal{B}_m \nearrow \varepsilon$, then, for any endomorphism $T : X \to X$, $\mathrm{h}(T) = \lim_{m\to\infty} \mathrm{h}(T,\mathcal{B}_m)$.*

(b) If $\mathcal{B}$ is a countable one-sided generator with finite entropy for an endomorphism T of a Lebesgue space, then $\mathrm{h}(T) = \mathrm{h}(T,\mathcal{B})$.

Proof. By Theorem 2.8.6 for every finite partition $\mathcal{A}$ we have $\lim_{m\to\infty}\mathrm{H}(\mathcal{A}|\mathcal{B}_m) = \mathrm{H}(\mathcal{A}|\varepsilon) = 0$. Hence, in view of Theorem 2.4.5, $\mathrm{h}(T) = \lim_{m\to\infty}\mathrm{h}(T,\mathcal{B}_m)$. This proves (a). Theorem 2.4.5, together with the definition of a generator, also proves (b). ♣

Remark 2.8.8. For T being an automorphism, one considers two-sided countable (and in particular finite) generators, i.e. partitions of $\mathcal{B}$ for which $\bigvee_{n=-\infty}^{\infty} T^n(\mathcal{B}) = \varepsilon$. Then, as in the one-sided case, finiteness of $\mathrm{H}(\mathcal{B})$ implies that $\mathrm{h}(T) = \mathrm{h}(T,\mathcal{B})$.

Remark 2.8.9. In both Theorem 2.8.6 and Theorem 2.8.7(a), the assumption of monotonicity of $\mathcal{B}_m$ can be weakened. Assume, for example, that $\mathcal{A}$ is finite and $\mathcal{B}_m \to \varepsilon$, in the sense that for every measurable set Y, $E(1\!\!1_Y|\mathcal{B}_m) \to 1\!\!1_Y$ in measure, as in Remark 2.1.5. Then $\mathrm{H}(\mathcal{A}|\mathcal{B}_m) \to 0$: hence $\mathrm{h}(T) = \lim_{m\to\infty}\mathrm{H}(T,\mathcal{B}_m)$.

Indeed, for $\mathrm{H}(\mathcal{A}|\mathcal{B}_m) \to 0$ just repeat Proof 2 of Theorem 2.8.6. The convergence in measure μ of $\mu_{\mathcal{B}_n(x)}(A \cap \mathcal{B}_n(x)))$ to $\mu_{\varepsilon(x)}(A \cap \varepsilon(x)))$ means that $E(1\!\!1_A|\mathcal{B}_n) \to 1\!\!1_A$, which has just been assumed.

Corollary 2.8.10. *Assume that X is a compact metric space, and that $\mathcal{F}$ is the σ-algebra of Borel sets (generated by open sets). If $\sup_{B\in\mathcal{B}_m}(\mathrm{diam}(B)) \to 0$ as $m \to \infty$, then $\mathrm{h}(T) = \lim_{m\to\infty} \mathrm{H}(T, \mathcal{B}_m)$.*

Proof. It is sufficient to check $E(1\!\!1_A|\mathcal{B}_m) \to 1\!\!1_A$ in measure. First note that for every $\delta > 0$ there exist an open set U and a closed set K such that $K \subset A \subset U$ and $\mu(U \setminus K) \le \delta$. This property is called the *regularity* of our measure μ, and is true for every finite measure on the σ-algebra of Borel sets of a metric space (compactness is not needed here). This can be proved by Carathéodory's argument: compare the proof of Theorem 2.1.4. That is, we construct the outer measure with the help of open sets, as in the sketch of the proof of Theorem 2.7.2 (where we used $\mathcal{G}_0$), and we notice that since each closed set is an intersection of a descending sequence of open sets, we shall have the same outer measure if, in the construction of the outer measure, we use the algebra generated by open sets. Now we can refer to Theorem 2.7.2.

Now, owing to the compactness of X, and hence K, for m large enough the set $A' := \bigcup\{B \in \mathcal{B}_m : B \cap K \ne \emptyset\}$ contains K and is contained in U: hence $\mu(A \div A') \le \delta$. This implies that

$$\begin{aligned}\int_X |E(1\!\!1_A|\mathcal{B}_m) - 1\!\!1_A|\,d\mu &= \int_{X\setminus(A\cup A')} E(1\!\!1_A|\mathcal{B}_m)\,d\mu \\ &\quad + \int_{A\div A'} |E(1\!\!1_A|\mathcal{B}_m) - 1\!\!1_A|\,d\mu + \int_{A\cap A'} 1\!\!1_A - E(1\!\!1_A|\mathcal{B}_m)\,d\mu \\ &\le \frac{\delta}{\mu(X\setminus A')}\mu(X\setminus(A\cup A')) + \delta \\ &\quad + \Big(1 - \frac{\mu(A\cap A')}{\mu(A')}\Big)\mu(A\cap A') \le 3\delta.\end{aligned}$$

Hence $\mu\{x : |E(1\!\!1_A|\mathcal{B}_m) - 1\!\!1_A| \ge \sqrt{3\delta}\} \le \sqrt{3\delta}$. ♣

For a simpler proof, omitting Theorem 2.8.6, see Exercise 2.18.

We end this section with the *ergodic decomposition theorem* and the adequate entropy formula. Compare this with the Choquet Representation Theorem: Theorem 3.1.11 and Theorem 3.1.13.

Let T be a measure-preserving endomorphism of a Lebesgue space. A measurable partition $\mathcal{A}$ is said to be T-invariant if $T(A) \subset A$ for almost every $A \in \mathcal{A}$. The induced map $T_A = T|_A : A \to A$ is a measurable endomorphism of the Lebesgue space $(A, \mathcal{F}_A, \mu_A)$. One calls T_A a *component* of T.

Theorem 2.8.11. *(a) There exists a finest measurable partition $\mathcal{A}$ (mod 0) into T-invariant sets (called the ergodic decomposition). Almost all of its components are ergodic.*

(b) $h(T) = \int_{X/\mathcal{A}} h(T_A)\,d\mu_{\mathcal{A}}(A)$.

Proof. Part (a) will not be proved. Let us mention only that the ergodic decomposition partition corresponds (see Section 2.6) to the completion of $\mathcal{I}$, the σ-subalgebra of $\mathcal{F}$ consisting of T invariant sets in $\mathcal{F}$ (compare Theorem 2.2.5).

To prove part (b) notice that for every T-invariant measurable partition $\mathcal{A}$, for every finite partition ξ and almost every $A \in \mathcal{A}$, writing ξ_A for the partition $\{s \cap A : s \in \xi\}$, we obtain

$$\mathrm{h}(T_A, \xi_A) = H(\xi_A|\xi_A^-) = \int_A I_{\mu_A}(\xi_A|\xi_A^-)\, d\mu_A.$$

Notice next that the latter information function is equal a.e. to $I_\mu(\xi|\xi^- \vee \mathcal{A})$ restricted to A. Hence

$$\int_{X/\mathcal{A}} h(T_A)\, d\mu_{\mathcal{A}}(A) = \int_{X/\mathcal{A}} d\mu_{\mathcal{A}} \int_A I_{\mu_A}(\xi_A|\xi_A^-)\, d\mu_A$$

$$= \int_X I_\mu(\xi|\xi^- \vee \mathcal{A})\, d\mu = \mathrm{H}(\xi|\xi^- \vee \mathcal{A}) = \mathrm{h}(T, \xi).$$

The latter equality follows from an approximation of $\mathcal{A}$ by finite T-invariant partitions $\eta \nearrow \mathcal{A}$ and from

$$\mathrm{H}(\xi|\xi^- \vee \eta) = \mathrm{H}(\xi \vee \eta|\xi^- \vee \eta^-) = \lim_{n\to\infty} \frac{1}{n} \mathrm{H}((\xi \vee \eta)^n)$$

$$= \lim_{n\to\infty} \frac{1}{n} \mathrm{H}(\xi^n \vee \eta) = \lim_{n\to\infty} \frac{1}{n} \mathrm{H}(\xi^n) = H(T, \xi).$$

Now let ξ_n be a sequence of finite partitions such that $\xi_n \nearrow \varepsilon$. Then $\mathrm{h}(T, \xi_n) \nearrow \mathrm{h}(T)$ and $\mathrm{h}(T_A, (\xi_n)_A) \nearrow \mathrm{h}(T_A)$. So $\mathrm{h}(T, \xi_n) = \int_{X/\mathcal{A}} h(T_A, \xi_n)$ $d\mu_{\mathcal{A}}(A)$, and the Lebesgue Monotone Convergence Theorem proves (b) ♣

2.9 Countable-to-one maps, Jacobian and entropy of endomorphisms

We start with a formulation of

Theorem 2.9.1 (Rokhlin's fundamental theorem of cross-sections). *Suppose that $\mathcal{A}$ and $\mathcal{B}$ are two measurable partitions of a Lebesgue space $(X, \mathcal{F}, \mu)$ such that $\mathcal{A} \cap B$ (see Definition 2.8.3) is countable (mod 0 with respect to μ_B) for almost every $B \in \mathcal{B}$. Then there exists a countable partition $\gamma = \{\gamma_1, \gamma_2, \dots\}$ of X (mod 0) such that each $\gamma_j \in \gamma$ intersects almost every B at not more than one point, which is then an atom of μ_B: in particular,*

$$\mathcal{A} \vee \mathcal{B} = \gamma \vee \mathcal{B} \ \ (mod\ 0).$$

Furthermore, if $\mathrm{H}(\mathcal{A}|\mathcal{B}) < \infty$, *then γ can be chosen so that*

$$\mathrm{H}(\gamma) < H(\mathcal{A}|\mathcal{B}) + 3\sqrt{H(\mathcal{A}|\mathcal{B})} < \infty.$$

Definition 2.9.2. Let $(X, \mathcal{F}, \mu)$ be a Lebesgue space. Let $T : X \to X$ be a measurable endomorphism. We say that T is *essentially countable to one* if the measures μ_A of a canonical system of conditional measures for the partition $\mathcal{A} := T^{-1}(\varepsilon)$ are purely atomic (mod 0 with respect to μ_A), for almost all $A \in \mathcal{A}$.

Lemma 2.9.3. *If T is essentially countable to one and preserves μ, then there exists a measurable set $Y \subset X$ of full measure such that $T(Y) \subset Y$, and:*

1. $T^{-1}(x) \cap Y$ is countable for every $x \in Y$, i.e. $T|_Y$ is countable to one. Moreover, for almost every $x \in Y$, $T^{-1}(x) \cap Y$ consists only of atoms of the conditional measure $\mu_{T^{-1}(x)}$;

2. $T(B)$ is measurable if $B \subset Y$ is measurable;

3. $T|_Y$ is forward quasi-invariant: that is, $\mu(B) = 0$ for $B \subset Y$ implies $\mu(T(B)) = 0$.

Proof. Let Y' be the union of atoms mentioned in Definition 2.9.2. We can write, because of Theorem 2.9.1, $Y' = \bigcup_j \gamma_j$, so Y' is measurable. Set $Y = \bigcap_{n=0}^{\infty} T^{-n}(Y')$. Denote the partition $T^{-1}(\varepsilon)$ in Y by ζ. Property 1 follows from the construction. To prove 2 we use the fact that $(Y/\zeta, \mathcal{F}_\zeta, \mu_\zeta)$ is a Lebesgue space and the factor map $T_\zeta : Y_\zeta \to X$ is an automorphism (Th. 2.6.11). So, for measurable $B \subset Y$, the set

$$\{A \in \zeta : \mu_A(B \cap A) \neq 0\} = \{A \in \zeta : A \cap B \neq \emptyset\} \tag{2.9.1}$$

is measurable by Theorem 2.6.7(2), and therefore its image under T_ζ, equal to $T(B)$, is measurable. If $\mu(B) = 0$, then the set in (2.9.1) has measure μ_ζ equal to 0: hence, as T_ζ is an isomorphism, we obtain the result that $T(B)$ is measurable and of measure 0. ♣

The key property in the above proof is the equality (2.9.1). Without assuming that μ_A are purely atomic there could exist B of measure 0 with $C := \{A \in \zeta : \mu_A(B \cap A) \neq 0\}$ not measurable in $\mathcal{F}_\zeta$.

To have such a situation, just consider a non-measurable $C \subset Y/\zeta$. Consider the disjoint union $D := C \cup Y$ and denote the embedded C by C'. Finally, defining measure on D, put $\mu(C') = 0$ and μ on the embedded Y. Define $T(c') = T(c)$ for $C \ni c$ and c' being the image of c under the above-mentioned embedding. Thus C' is measurable, of measure 0, whereas $T(C')$ is not measurable, because C is not measurable and T_ζ is an isomorphism.

Definition 2.9.4. Let $(X, \mathcal{F}, \mu)$ and $(X', \mathcal{F}', \mu')$ be probability measure spaces. Let $T : X \to X'$ be a measurable homomorphism. We say that a real, non-negative, measurable function J is a *weak Jacobian* if there exists E of measure 0 such that for every measurable $A \subset X \setminus E$ on which T is injective, the set $T(A)$ is measurable and $\mu(T(A)) = \int_A J \, d\mu$. We say J is a *strong Jacobian*, or just a *Jacobian*, if the above holds without assuming $A \subset X \setminus E$.

We say that T is *forward quasi-invariant* if $(\mu(A) = 0) \Rightarrow (\mu'(T(A)) = 0)$. Notice that, if T is forward quasi-invariant, then automatically a weak Jacobian is a strong Jacobian.

Proposition 2.9.5. *Let $(X, \mathcal{F}, \mu)$ be a Lebesgue space and $T : X \to X$ be a measurable, essentially countable to one, endomorphism. Then there exists a weak Jacobian $J = J_\mu$. It is unique (mod 0). For T restricted to Y (Lemma 2.9.3) J is a strong Jacobian.*

Proof. Consider the partition $\gamma = \{\gamma_1, \gamma_2, \dots\}$ given by Theorem 2.9.1 with $\mathcal{A} = \varepsilon$ and $\mathcal{B} = T^{-1}(\varepsilon)$. Then for each j the map $T|_{\gamma_j \cap Y}$ is injective. Moreover, by Lemma 2.9.3, $T|_{\gamma_j \cap Y}$ maps measurable sets onto measurable sets and is forward quasi-invariant. Therefore J exists on each $\gamma_j \cap Y$ by the Radon–Nikodym theorem.

By the presentation of each $A \subset Y$ as $\bigcup_{j=1}^\infty A \cap \gamma_j$ the function J satisfies the assertion of the proposition. The uniqueness follows from the uniqueness of the Jacobian in the Radon-Nikodym theorem on each $\gamma_j \cap Y$. ♣

Theorem 2.9.6. *Let $(X, \mathcal{F}, \nu)$ be a Lebesgue space. Let $T : X \to X$ be a ν-preserving endomorphism, essentially countable to one. Then its Jacobian J_ν, strong on Y defined in Lemma 2.9.3, and weak on X, has logarithm equal to $I_\nu(\varepsilon|T^{-1}(\varepsilon))$. (We do not need to assume here that $T(Y) \subset Y$. I stands for the information function: see Sections 2.4 and 2.8.)*

Proof. Consider T already restricted to Y. Let $Z \subset Y$ be an arbitrary measurable set such that T is one–to–one on it. For each $y \in Y$ denote by $A(y)$ the element of $\zeta = T^{-1}(\varepsilon)$ containing y. We obtain

$$\begin{aligned}
\nu(T(Z)) = \nu\Big(T^{-1}(T(Z))\Big) &= \int_{T^{-1}\left(T(Z)\right)} 1 \, d\nu(y) \\
&= \int_{T^{-1}\left(T(Z)\right)} \left(\int_{A(y)} 1\!\!1_Z(x)/\nu_{A(y)}\{x\}) \, d\nu_{A(y)}(x) \right) d\nu(y) \\
&= \int_{T^{-1}\left(T(Z)\right)} (1\!\!1_Z(y)/\nu_{A(y)}\{y\}) \, d\nu(y) = \int_Z (1/\nu_{A(y)}\{y\}) \, d\nu(y).
\end{aligned}$$

Therefore $J_\nu(y) = 1/\nu_{A(y)}\{y\})$, and its logarithm is equal to $I_\nu(\varepsilon|T^{-1}(\varepsilon))(y)$. ♣

Theorem 2.9.6 gives rise to the so-called Rokhlin entropy formula:

Theorem 2.9.7. *Let $(X, \mathcal{F}, \mu)$ be a Lebesgue space. Let $T : X \to X$ be a measure μ-preserving endomorphism, essentially countable to one. Suppose that on each component A of the ergodic decomposition (cf. Theorem 2.8.11) the restriction T_A has a countable one-sided generator of finite entropy. Then*

$$h_\mu(T) = \mathrm{H}_\mu(\varepsilon|T^{-1}(\varepsilon)) = \int I_\mu(\varepsilon|T^{-1}(\varepsilon)) \, d\mu = \int \log J_\mu \, d\mu.$$

Proof. The third equality follows from Theorem 2.9.6; the second equality is the definition of the conditional entropy: see Sec. 2.8. To prove the first equality we can assume, owing to Theorem 2.8.11, that T is ergodic. Then, for ζ, a countable one-sided generator of finite entropy, with the use of Theorems 2.8.5 and 2.8.7(b), we obtain

$$\mathrm{H}(\varepsilon|T^{-1}(\varepsilon)) = \mathrm{H}(\varepsilon|\zeta^-) = \lim_{n\to\infty} \mathrm{H}(\zeta^n|\zeta^-) = \mathrm{H}(\zeta|\zeta^-) = \mathrm{h}(T,\zeta) = \mathrm{h}(T).$$

♣

Remark. The existence of a countable one-sided generator, without demanding finite entropy, is a general, and not very difficult, fact. That is, the following holds:

Theorem 2.9.8. *Let $(X, \mathcal{F}, \mu)$ be a Lebesgue space. Let $T : X \to X$ be a μ-preserving aperiodic endomorphism, essentially countable to one. Then there exists a countable one-sided generator, that is, a countable partition ζ such that $\zeta \vee \zeta^- = \varepsilon$ (mod 0).*

Aperiodic means that there exists no B of positive measure and a positive integer n so that $T^n|_B = \mathrm{id}$. For the proof see [Rokhlin, 1967, Sec. 10.12–13] or [Parry 1969]. To construct ζ one uses the partition γ ascribed to ε and $T^{-1}(\varepsilon)$ according to Theorem 2.9.1 and the so-called *Rokhlin towers.*

The existence of a one-sided generator with finite entropy is in fact equivalent to $H(\varepsilon|\varepsilon^-) = \mathrm{h}(T) < \infty$. The proof of the implication to the right is contained in the proof of Theorem 2.9.7. The reverse implication (the construction of the partition) is not easy: it uses in particular the estimate in Theorem 2.9.1.

The existence of a one-sided generator with finite entropy is a strong property. It may fail even for *exact* endomorphisms: see Section 2.10 and Exercise 2.22. Nor does its existence imply exactness (Exercise 2.22). On the contrary, for automorphisms, two-sided generators, even finite, always exist, provided the map is aperiodic.

2.10 Mixing properties

In this section we examine briefly some mixing properties of a measure-preserving endomorphism that are stronger than ergodicity. A measure-preserving endomorphism is said to be *weakly mixing* if and only if for every two measurable sets A and B

$$\lim_{n\to\infty} \frac{1}{n} \sum_{j=0}^{n-1} |\mu(T^{-j}(B) \cap A) - \mu(A)\mu(B)| = 0.$$

To see that a weakly mixing transformation is ergodic, suppose that $T^{-1}(B) = B$. Then $T^{-k}(B) = B$ for all $k \geq 0$, and consequently for every n,

$$\frac{1}{n} \sum_{j=0}^{n-1} |\mu(T^{-j}(B) \cap B) - \mu(B)\mu(B)| = |\mu(B) - \mu(B)^2| \to 0.$$

Thus $\mu(B) - \mu(B)^2 = 0$, and therefore $\mu(B) = 0$ or 1.

A measure-preserving endomorphism is said to be *mixing* if and only if for every two measurable sets A and B

$$\lim_{n\to\infty} \mu(T^{-n}(B)\cap A) - \mu(A)\mu(B) = 0.$$

Clearly, every mixing transformation is weakly mixing. The property equivalent to the mixing property is the following: for all square integrable functions f and g,

$$\lim_{n\to\infty} \int f(g\circ T^n)\,d\mu = \int f\,d\mu \int g\,d\mu.$$

Indeed, the former property follows from the latter if we substitute the indicator functions $1\!\!1_A, 1\!\!1_B$ in place of f, g respectively. To prove the opposite implication, note that with the help of the Hölder inequality it is sufficient to restrict our considerations to simple functions $f = \sum_i a_i 1\!\!1_{A_i}$ and $g = \sum_j a_j 1\!\!1_{A_j}$, where (A_i) and (B_j) are arbitrary finite partitions. Then

$$\begin{aligned}&\left|\int f(g\circ T^n)\,d\mu - \int f\,d\mu \int g\,d\mu\right| \\ &= \left|\sum_{i,j} a_i b_j(\mu(A_i\cap T^{-n}(B_j)) - \mu(A_i)\mu(B_j))\right| \to 0\end{aligned}$$

because every summand converges to 0 as $n\to\infty$.

In the sequel we shall also deal with stronger mixing properties. An endomorphism is called *K-mixing* if, for every measurable set A and every finite partition $\mathcal{A}$,

$$\lim_{n\to\infty} \sup_{B\in\mathcal{F}(\mathcal{A}_n^\infty)} |\mu(A\cap B) - \mu(A)\mu(B)| = 0.$$

Recall that $\mathcal{F}(\mathcal{A}_n^\infty)$ for $n\ge 0$ means the complete σ-algebra assigned to the partition $\mathcal{A}_n^\infty = \bigvee_{j=n}^\infty T^{-j}(\mathcal{A})$. The following theorem provides us with alternative definitions of the K-mixing property in the case when T is an automorphism.

Theorem 2.10.1. *Let $(X,\mathcal{F},\mu)$ be a Lebesgue space and $T: X\to X$ be its measure-preserving automorphism. Then the following conditions are equivalent:*

(a) T is K-mixing.

(b) For every finite partition $\mathcal{A}$ Tail($\mathcal{A}$) $:= \bigwedge_{n=0}^\infty \bigvee_{k=n}^\infty T^{-k}(\mathcal{A})$ is equal to the trivial partition $\nu = \{X\}$ (mod 0).

(c) For every finite partition $\mathcal{A}\neq\nu$, $\mathrm{h}_\mu(T,\mathcal{A}) > 0$ (T has completely positive entropy).

(d) There exists a forward invariant exhausting measurable partition α (i.e. satisfying $T^{-1}(\alpha)\le\alpha$, $T^n(\alpha)\nearrow\varepsilon$: see Definition 2.6.4) such that $T^{-n}(\alpha)\searrow\nu$.

The property Tail($\mathcal{A}$) $=\nu$ is a version of the 0-1 Law. An automorphism satisfying (d) is usually called a *K-automorphism.* The symbol K comes from the name Kolmogorov. Each partition satisfying the properties of α in (d) is called a K-partition.

Remark. The properties (a)–(c) make sense for endomorphisms, and they are equivalent (proofs are the same as for automorphisms). Moreover, they hold for an endomorphism if and only if they hold for its natural extension.

Proof. (a part of) To show the reader what Theorem 2.10.1 is about, let us prove some implications:

(a)⇒(b) Let $A \in \mathcal{F}(\mathrm{Tail}(\mathcal{A}))$ for a finite partition $\mathcal{A}$. Then, for every $n \geq 0$, $A \in \mathcal{F}(\bigvee_{k=n}^{\infty} T^{-k}(\mathcal{A}))$. Hence, by K-mixing, $\mu(A \cap A) - \mu(A)\mu(A) = 0$, and therefore $\mu(A) = 0$ or 1.

(b)⇒(c) Suppose (b) and assume $\mathrm{h}(T, \mathcal{A}) = 0$ for a finite partition $\mathcal{A}$. Then $\mathrm{H}(\mathcal{A}|\mathcal{A}^-) = 0$: hence $I(\mathcal{A}|\mathcal{A}^-) = 0$ a.s. (see Section 2.8): hence $\mathcal{A} \leq \mathcal{A}^-$. Thus

$$\bigvee_{k=0}^{\infty} T^{-k}(\mathcal{A}) = \bigvee_{k=1}^{\infty} T^{-k}(\mathcal{A}) \text{ and } \bigvee_{k=m}^{\infty} T^{-k}(\mathcal{A}) = \bigvee_{k=n}^{\infty} T^{-k}(\mathcal{A})$$

for every $m, n \geq 0$. So $\nu = \bigwedge_{n=0}^{\infty} \bigvee_{k=n}^{\infty} T^{-k}(\mathcal{A}) = \bigvee_{k=0}^{\infty} T^{-k}(\mathcal{A}) \geq \mathcal{A}$. So $\mathcal{A} = \nu$, the trivial partition. Thus for every non-trivial partition $\mathcal{A}$ we have $\mathrm{h}(T, \mathcal{A}) > 0$.

(b)⇒(d) (in the case where there exists a finite two-sided generator $\mathcal{B}$, meaning that $\bigvee_{n=-\infty}^{\infty} T^n(\mathcal{B}) = \varepsilon$). Note that $\alpha = \bigvee T_{n=0}^{\infty} T^{-n}(\mathcal{B})$ is exhausting. ♣

Let us finish this section with the following useful definition:

Definition 2.10.2. A measure-preserving endomorphism is said to be *exact* if

$$\bigwedge_{n=0}^{\infty} T^{-n}(\varepsilon) = \nu.$$

(Recall that ε is the partition into points, and ν is the trivial partition $\{X\}$.)

Exercise. Prove that exactness is equivalent to the property that $\mu_e(T^n(A)) \to 1$ for every A of positive measure (μ_e is the outer measure generated by μ), or to the property that $\mu(T^n(A)) \to 1$ provided $\mu(A) > 0$ and the sets $T^n(A)$ are measurable.

The property of being *exact* implies the natural extension to be a K-automorphism (in Theorem 2.10.1(d) set for α the lift of ε). The converse is of course false. The automorphisms of spaces that are not one-atom spaces are not exact. Observe, however, that if T is an automorphism and α is a measurable partition satisfying (d), then the factor mapping of T/α on X/α is exact.

Exercise. Prove that T is the natural extension of T/α.

Recall finally (Section 2.9) that, even for exact endomorphisms, $\mathrm{h}(\varepsilon|T^{-1}(\varepsilon))$ can be strictly less than $\mathrm{h}(T)$.

2.11 Probability laws and Bernoulli property

Let $(X, \mathcal{F}, \mu)$ be a probability space and, whenever it is needed, a Lebesgue space, and let $T : X \to X$ be an endomorphism that preserves μ. Let f and g be real-valued square-integrable functions on X. For every positive integer n, the n-th *correlation* of the pair f, g, is the number

$$C_n(f,g) := \int f \cdot (g \circ T^n)\, d\mu - \int f\, d\mu \int g\, d\mu.$$

provided the above integrals exist. Note that owing to the T-invariance of μ we can also write

$$C_n(f,g) = \int (f - Ef)((g - Eg) \circ T^n)\, d\mu,$$

where $Ef = \int f\, d\mu$ and $Eg = \int g\, d\mu$.

Keep $g : X \to \mathbb{R}$ a square-integrable function. The limit

$$\sigma^2 = \sigma^2(g) = \lim_{n\to\infty} \frac{1}{n} \int \Big(\sum_{j=0}^{n-1} g \circ T^j - nEg \Big)^2 d\mu \tag{2.11.1}$$

is called the *asymptotic variance* or *dispersion* of g, provided it exists. Write $g_0 = g - Eg$. Then we can rewrite the above formula as

$$\sigma^2 = \lim_{n\to\infty} \frac{1}{n} \int \Big(\sum_{j=0}^{n-1} g_0 \circ T^j \Big)^2 d\mu. \tag{2.11.2}$$

Another useful expression for the asymptotic variance is the following:

$$\sigma^2(g) = \int g_0^2\, d\mu + 2 \sum_{j=1}^{\infty} \int g_0 \cdot (g_0 \circ T^j)\, d\mu. \tag{2.11.3}$$

The convergence of the series of correlations $C_n(g,g)$ in (2.11.3) easily implies that $\sigma^2(g)$ from this formula is equal to σ^2, defined in (2.11.1): compare the computation in the proof of Theorem 2.11.3 later on.

We say that the *Law of Iterated Logarithm*, LIL, holds for g if $\sigma^2(g)$ exists (i.e. the above series converges) and

$$\limsup_{n\to\infty} \frac{\sum_{j=0}^{n-1} g \circ T^j - nEg}{\sqrt{n \log\log n}} = \sqrt{2\sigma^2} \quad \mu\text{- almost surely.} \tag{2.11.4}$$

μ almost surely (a.s.) means μ almost everywhere (a.e.). This is the language of probability theory.

We say that the *Central Limit Theorem*, CLT, holds, if for all $r \in \mathbb{R}$, in the case $\sigma^2 \neq 0$,

$$\mu\left(\left\{ x \in X : \frac{\sum_{j=0}^{n-1} g \circ T^j - nEg}{\sqrt{n}} < r \right\} \right) \to \frac{1}{\sigma\sqrt{2\pi}} \int_{-\infty}^{r} e^{-t^2/2\sigma^2}\, dt, \tag{2.11.5}$$

and in the case $\sigma^2 = 0$ the convergence holds for $r \neq 0$ with 0 on the right-hand side for $r < 0$ and 1 for $r > 0$.

The LIL and CLT for $\sigma^2 \neq 0$ are often, and this is the case in Theorem 2.11.1 below, a consequence of the *Almost Sure Invariance Principle*, ASIP, which says that the sequence of random variables g, $g \circ T$, $g \circ T^2$, *centred at the expectation value* (that is, if $Eg = 0$), is approximated by the rate $n^{1/2-\varepsilon}$ with some $\varepsilon > 0$, depending on δ in Theorem 2.11.1 below, by a martingale difference sequence and a respective Brownian motion.

Theorem 2.11.1. *Let $(X, \mathcal{F}, \mu)$ be a probability space and T an endomorphism-preserving μ. Let $\mathcal{G} \subset \mathcal{F}$ be a σ-algebra. Write $\mathcal{G}_m^n := \bigvee_{j=m}^n T^{-j}(\mathcal{G})$ (notation from Section 2.6) for all $m \le n \le \infty$, and suppose that the following property, called ϕ-mixing, holds:*

There exists a sequence $\phi(n), n = 0, 1, \dots$ of positive numbers satisfying

$$\sum_{n=1}^{\infty} \phi^{1/2}(n) < \infty, \tag{2.11.6}$$

such that for every $A \in \mathcal{G}_0^m$ and $B \in \mathcal{G}_n^\infty$, $0 \le m \le n$, we have

$$|\mu(A \cap B) - \mu(A)\mu(B)| \le \phi(n - m)\mu(A). \tag{2.11.7}$$

Now consider a $\mathcal{G}_0^\infty$ measurable function $g : X \to \mathbb{R}$ such that

$$\int |g|^{2+\delta} \, d\mu < \infty \quad \textit{for some } \delta > 0,$$

and that for all $n \ge 1$

$$\int |h - E(h|\mathcal{G}_0^n)|^{2+\delta}) \, d\mu \le K n^{-s}, \quad \textit{for } K > 0, \ s > 0 \textit{ large enough.} \tag{2.11.8}$$

(A concrete formula for s, depending on δ, can be given.)

Then g satisfies CLT and LIL.

LIL for $\sigma^2 \neq 0$ is a special case, for $\psi(n) = \sqrt{2 \log \log n}$, of the following property for a square integrable function $g : X \to \mathbb{R}$ for which σ^2 exists, provided $\int g \, d\mu = 0$: for every real positive non-decreasing function ψ:

$$\mu\left(\left\{x \in X : \sum_{j=0}^{n} g(T^j(x)) > \psi(n)\sqrt{\sigma^2 n} \text{ for infinitely many } n\right\}\right) = 0 \text{ or } 1$$

according to whether the integral $\int_1^\infty \frac{\psi(t)}{t} \exp(-\frac{1}{2}\psi^2(t)) \, dt$ converges or diverges.

As we have already remarked, this theorem, for $\sigma^2 \neq 0$, is a consequence of the ASIP and similar conclusions for standard Brownian motion. We do not give the proofs here. For the ASIP and further references see [Philipp & Stout, 1975, Chapters 4 and 7]. Let us discuss only the existence of σ^2. This follows from the following consequence of (2.11.7): for α, β square integrable real-valued functions on X, α measurable with respect to $\mathcal{G}_0^m$ and β measurable with respect to $\mathcal{G}_n^\infty$, we have

$$\left| \int (\alpha\beta\, d\mu - E\alpha\, E\beta)\, d\mu \right| \le 2(\phi(n-m))^{1/2} \|\alpha\|_2 \|\beta\|_2. \tag{2.11.9}$$

The proof of this inequality is not difficult, but is tricky, with the use of the Hölder inequality: see [Ibragimov 1962] or [Billingsley 1968]. It is sufficient to work with simple functions $\alpha = \sum_i a_i 1\!\!1_{A_i}, \beta = \sum_j a_j 1\!\!1_{A_j}$ for finite partitions (A_i) and (B_j), as in dealing with mixing properties in Section 2.10. Note that if instead of (2.11.7) we have the stronger

$$|\mu(A \cap B) - \mu(A)\mu(B)| \le \phi(n-m)\mu(A)\mu(B), \tag{2.11.10}$$

as will be the case in Chapter 5, then we very easily obtain in (2.11.9) the estimate by $\phi(n-m)\|\alpha\|_1\|\beta\|_1$, by the same computation as for mixing in Section 2.10.

We may assume that g is centred at the expectation value. Write $g = k_n + r_n = E(g|\mathcal{G}_0^{[n/2]}) + (g - E(g|\mathcal{G}_0^{[n/2]})$. We have

$$\begin{aligned}\left|\int g(g\circ T^n)\, d\mu\right| &\le \left|\int k_n(k_n\circ T^n)\, d\mu\right| + \left|\int k_n(r_n\circ T^n)\, d\mu\right| \\ &\quad + \left|\int r_n(k_n\circ T^n)\, d\mu\right| + \left|\int r_n(r_n\circ T^n)\, d\mu\right| \\ &\le 2(\phi(n-[n/2]))^{1/2}\|k_n\|_2^2 + 2\|k_n\|_2\|r_n\|_2 + \|r_n\|_2^2 \\ &\le 2(\phi(n-[n/2]))^{1/2}\|k_n\|_2^2 + 2K[n/2]^{-s}\|k_n\|_2 + K[n/2]^{-2s},\end{aligned}$$

the first summand estimated according to (2.11.9). For $s > 1$ we thus obtain convergence of the series of correlations.

Let us go back to the discussion of the ϕ-mixing property. If $\mathcal{G}$ is associated to a finite partition that is a one-sided generator, ϕ-mixing with $\phi(n) \to 0$ as $n \to \infty$ (that is, weaker than (2.11.6)), implies K-mixing (see Section 2.10). Indeed, B is the same in both definitions, whereas A in K-mixing can be approximated by sets belonging to $\mathcal{G}_0^m$. We leave the details to the reader.

Intuitively, both notions mean that any event B in the remote future weakly depends on the present state A: that is, $|\mu(B) - \mu(B|A)|$ is small.

In applications $\mathcal{G}$ will be usually associated to a finite or countable partition.

In Theorems 2.11.1 the case $\sigma^2 = 0$ is easy. It relies on Theorem 2.11.3 below. Let us first introduce the following fundamental definition:

Definition 2.11.2. Two functions $f, g : X \to \mathbb{R}$ (or $\mathbb{C}$) are said to be *co-homologous* in a space $\mathcal{K}$ of real (or complex) -valued functions on X (or f is said to be co-homologous to g), if there exists $h \in \mathcal{K}$ such that

$$f - g = h \circ T - h. \tag{2.11.11}$$

If f, g are defined mod 0, then (2.11.11) is understood a.s. This formula is called a *cohomology equation.*

Theorem 2.11.3. *Let f be a square integrable function on a probability space $(X, \mathcal{F}, \mu)$, centred at the expectation value. Assume that*

$$\sum_{n=0}^{\infty} n\left|\int f \cdot (f \circ T^n)\, d\mu\right| < \infty. \qquad (2.11.12)$$

Then the following three conditions are equivalent:

(a) $\sigma^2(f) = 0$;

(b) All the sums $S_n = S_n f = \sum_{j=0}^{n-1} f \circ T^j$ *have the norm in* L^2 *(the space square integrable functions) bounded by the same constant;*

(c) f *is co-homologous to 0 in the space* $\mathcal{H} = L^2$.

Proof. The implication (c)⇒(a) follows immediately from (2.11.1) after substituting $f = h \circ T - h$. Let us prove (a)⇒(b). Write C_j for the correlations $\int f \cdot (f \circ T^j)\, d\mu, j = 0, 1, \ldots$. Then

$$\int |S_n|^2\, d\mu = nC_0^2 + 2\sum_{j=1}^{n}(n-j)C_j$$

$$= n\Big(C_0^2 + 2\sum_{j=1}^{\infty} C_j\Big) - 2n \sum_{j=n+1}^{\infty} C_j - 2\sum_{j=1}^{n} j \cdot C_j = n\sigma^2 - I_n - II_n.$$

Since $I_n \to 0$ and II_n stays bounded as $n \to \infty$ and $\sigma^2 = 0$, we deduce that all the sums S_n are uniformly bounded in L^2.

(b)⇒(c): $f = h \circ T - h$ for any h, a limit in the weak topology, of the sequence $\frac{1}{n} S_n$ bounded in $L^2(\mu)$. We leave this easy computation to the reader. (This computation will be given in detail in the similar situation of the Bogolyubov–Krylov Theorem, in Remark 3.1.14.) ♣

Now Theorem 2.11.1 for $\sigma^2 = 0$ follows from (c), which gives $\sum_{j=0}^{n-1} f \circ T^j = h \circ T^n - h$, with the use of the Borel–Cantelli lemma.

Remark. Theorem 2.11.1 in the two-sided case: where g depends on $\mathcal{G}_j = T^j(\mathcal{G})$ for $j = \ldots, -1, 0, 1, \ldots$ for an automorphism T, also holds. In (2.11.8) one should replace $\mathcal{G}_0^n$ by $\mathcal{G}_{-n}^n$.

Given two finite partitions $\mathcal{A}$ and $\mathcal{B}$ of a probability space and $\varepsilon \geq 0$, we say that $\mathcal{B}$ is *ε-independent* of $\mathcal{A}$ if there is a subfamily $\mathcal{A}' \subset \mathcal{A}$ such that $\mu(\bigcup \mathcal{A}') > 1 - \varepsilon$, and for every $A \in \mathcal{A}'$

$$\sum_{B \in \mathcal{B}} \left| \frac{\mu(A \cap B)}{\mu(A)} - \mu(B) \right| \leq \varepsilon. \qquad (2.11.13)$$

Given an ergodic measure-preserving endomorphism $T : X \to X$ of a Lebesgue space, a finite partition $\mathcal{A}$ is called *weakly Bernoulli* (abbr. WB) if for every $\varepsilon > 0$ there is an $N = N(\varepsilon)$ such that the partition $\bigvee_{j=n}^{s} T^{-j}(\mathcal{A})$ is ε-independent of the partition $\bigvee_{j=0}^{m} T^{-j}(\mathcal{A})$ for every $0 \leq m \leq n \leq s$ such that $n - m \geq N$.

Of course, in the definition of ε-independence we can consider any measurable (possibly uncountable) partition $\mathcal{A}$ and write conditional measures $\mu_A(B)$ in (2.11.13). Then for T an automorphism we can replace in the definition of WB $\bigvee_{j=n}^{s} T^{-j}(\mathcal{A})$ by $\bigvee_{j=0}^{s-n} T^{-j}(\mathcal{A})$ and $\bigvee_{j=0}^{m} T^{-j}(\mathcal{A})$ by $\bigvee_{j=-n}^{m-n} T^{-j}(\mathcal{A})$, and set $n = \infty, n - m \geq N$. WB in this formulation becomes one more version of weak dependence of the present (and future) from the remote past.

If $\varepsilon = 0$ and $N = 1$, then all partitions $T^{-j}(\mathcal{A})$ are mutually independent (recall that $\mathcal{A}, \mathcal{B}$ are said to be independent if $\mu(A \cap B) = \mu(A)\mu(B)$ for every $A \in \mathcal{A}, B \in \mathcal{B}$.). We then say that $\mathcal{A}$ is Bernoulli. If $\mathcal{A}$ is a one-sided generator (two-sided generator), then clearly T on $(X, \mathcal{F}, \mu)$ is isomorphic to a one-sided (two-sided) Bernoulli shift of $\sharp\mathcal{A}$ symbols: see Chapter 1, Example 1.9. The following famous theorem of Friedman and Ornstein holds.

Theorem 2.11.4. *If $\mathcal{A}$ is a finite, weakly Bernoulli, two-sided generating partition of X for an automorphism T, then T is isomorphic to a two-sided Bernoulli shift.*

Of course, the standard Bernoulli partition (and in particular the number of its states) in the above Bernoulli shift can be different from the image under the isomorphism of the WB partition.

The Bernoulli shift above is unique in the sense that all two-sided Bernoulli shifts of the same entropy are isomorphic [Ornstein 1970].

Note that ϕ-mixing in the sense of (2.11.10), with $\phi(n) \to 0$, for $\mathcal{G}$ associated to a finite partition $\mathcal{A}$, implies a weak Bernoulli property.

The Central Limit Theorem is a much weaker property than LIL. We end this section with a useful abstract theorem that allows us to deduce CLT for g without specifying $\mathcal{G}$. This theorem, similarly to Theorem 2.11.1, can be proved with the use of an approximation by a martingale difference sequence.

Theorem 2.11.5. *Let $(X, \mathcal{F}, \mu)$ be a probability space and $T : X \to X$ an automorphism-preserving μ. Let $\mathcal{F}_0 \subset \mathcal{F}$ be a σ-algebra such that $T^{-1}(\mathcal{F}_0) \subset F_0$. Denote $\mathcal{F}_n = T^{-n}(\mathcal{F}_0)$ for all integers $n = \ldots, -1, 0, 1, \ldots$ Let g be a real-valued square integrable function. If*

$$\sum_{n \geq 0} \|E(g|\mathcal{F}_n) - Eg\|_2 + \|g - E(g|\mathcal{F}_{-n})\|_2 < \infty, \tag{2.11.14}$$

then g satisfies CLT.

Exercises

Ergodic theorems, ergodicity

2.1. Prove that for any two σ-algebras $\mathcal{F} \supset \mathcal{F}'$ and ϕ an $\mathcal{F}$-measurable function, the conditional expectation value operator $L^p(X, \mathcal{F}, \mu) \ni \phi \to E(\phi|\mathcal{F}')$ has norm 1 in L^p, for every $1 \leq p \leq \infty$.

Hint: Prove that $E((\vartheta \circ |\phi|)|\mathcal{F}') \geq \vartheta \circ E((|\phi|)|\mathcal{F}')$ for convex ϑ, and in particular for $t \mapsto t^p$.

2.2. Prove that if $S : X \to X'$ is a measure-preserving surjective map for measure spaces $(X, \mathcal{F}, \mu)$ and $(X', \mathcal{F}', \mu')$, and there are measure-preserving endomorphisms $T : X \to X$ and $T' : X' \to X'$ satisfying $S \circ T = T' \circ S$, then T ergodic implies T' is ergodic, but not *vice versa*. Prove that if $(X, \mathcal{F}, \mu)$ is Rokhlin's natural extension of $(X', \mathcal{F}', \mu')$, then $(X', \mathcal{F}', \mu')$ implies $(X, \mathcal{F}, \mu)$ is ergodic.

2.3. (a) Prove the Maximal Ergodic Theorem: Let $f \in L^1(\mu)$ for T a measure-preserving endomorphism of a probability space $(X, \mathcal{F}, \mu)$. Then for $A := \{x : \sup_{n \geq 0} \sum_{i=0}^{n} f(T^i(x)) > 0\}$ it holds that $\int_A f \geq 0$.
(b) Note that this implies the Maximal Inequality for the so-called maximal function $f^* := \sup_{n \geq 1} \frac{1}{n} \sum_{i=0}^{n-1} f(T^i(x))$:

$$\mu(\{f^* > \alpha\}) \leq \frac{1}{\alpha} \int_{\{f^* > \alpha\}} f \, d\mu, \quad \text{for every real } \alpha.$$

(c) Deduce Birkhoff's Ergodic Theorem.

Hint: One can proceed directly. Another way is to first prove the a.e. convergence on a set D of functions dense in L^1. Decomposed functions in L^2 in sums $g = h_1 + h_2$, where h_1 is T invariant in the case where T is an automorphism (i.e. $h_1 = h_1 \circ T$) and $h_2 = -g \circ T + g$. Consider only $g \in L^\infty$. If T is not an automorphism, $h_1 = U^*(h_1)$ for U^* being conjugate to the Koopman operator $U(f) = f \circ T$: compare Sections 5.2 and 5.7. To pass to the closure of D use the Maximal Inequality. One can also just refer to the Banach Principle below. Its assumption, $\sup_n |T_n f| < \infty$ a.e. for $T_n f = \frac{0}{n-1} \sum_{k=0}^{n-1} f \circ T^k$, follows from the Maximal Inequality. See [Petersen 1983].

2.4. Prove the Banach Principle: Let $1 \leq p < \infty$ and let $\{T_n\}$ be a sequence of bounded linear operators on L^p. If $\sup_n |T_n f| < \infty$ a.e. for each $f \in L^p$, then the set of f for which $T_n f$ converges a.e. is closed in L^p.

2.5. Let $(X, \mathcal{F}, \mu)$ be a probability space, and let $\mathcal{F}_1 \subset \mathcal{F}_2 \subset \cdots \subset \mathcal{F}$ be an increasing to $\mathcal{F}$ sequence of σ-algebras and $\phi \in L^p$, $1 \leq p \leq \infty$. Prove the Martingale Convergence Theorem in the version of Theorem 2.1.4, saying that $\phi_n := E(\phi|\mathcal{F}_n) \to E(\phi|\mathcal{F})$ a.e. and in L^p.

Steps:

(a) Prove: $\mu\{\max_{i \leq n} \phi_i > \alpha\} \leq \frac{1}{\alpha} \int_{\{\max_{i \leq n} \phi_i > \alpha\}} \phi_n \, d\mu$. (Hint: decompose $X = \bigcup_{k=1}^{n} A_k$, where $A_k := \{\max_{i<k} \phi_i \leq \alpha < \phi_k\}$, and use the Tchebyshev inequality on each A_k. Compare Lemma 2.5.1.)

(b) Use the Banach Principle, first checking the convergence a.e. on the set of indicator functions on each $\mathcal{F}_n$.

2.6. For a Lebesgue integrable function $f : \mathbb{R} \to \mathbb{R}$ the Hardy–Littlewod maximal function is

$$Mf(t) = \sup_{\varepsilon > 0} \frac{1}{2\varepsilon} \int_{-\varepsilon}^{\varepsilon} |f(t+s)| \, ds.$$

(a) Prove the Maximal Inequality of F. Riesz, $m(\{x \in \mathbb{R} : Mf(x) > \alpha\}) \leq \frac{2}{\alpha}\|f\|_1$, for every $\alpha > 0$, where m is the Lebesgue measure.
(b) Prove the Lebesgue Differentiation Theorem: For a.e. t

$$\lim_{\varepsilon \to 0} \frac{1}{2\varepsilon} \int_{-\varepsilon}^{\varepsilon} |f(t+s)|\, ds = f(t).$$

Hint: Use the Banach Principle (Exercise 2.4), using the fact that the above equality holds on the set of differentiable functions, which is dense in L^1.
(c) Generalize this theory to $f : \mathbb{R}^d \to R$, $d > 1$; the constant 2 is then replaced by another constant resulting from the Besicovitch Covering Theorem: see Chapter 8.

Lebesgue spaces, measurable partitions

2.7. Let T be an ergodic automorphism of a probability non-atomic measure space, and $\mathcal{A}$ its partition into orbits $\{T^n(x), n = \ldots, -1, 0, 1 \ldots\}$. Prove that $\mathcal{A}$ is not measurable.

Suppose we do not assume ergodicity of T. What is the largest measurable partition, smaller than the partition into orbits? (Hint: Theorem 2.8.11.)

2.8. Prove that the following partitions of measure spaces are not measurable:

(a) Let $T : S^1 \to S^1$ be a mapping of the unit circle with Haar (length) measure defined by $T(z) = e^{2\pi i \alpha} z$ for an irrational α. $\mathcal{P}$ is the partition into orbits.

(b) T is the automorphism of the two-dimensional torus $\mathbb{R}^2/\mathbb{Z}^2$, given by a hyperbolic integer matrix of determinant 1. Let $\mathcal{P}$ be the partition into stable, or unstable, lines (i.e. straight lines parallel to an eigenvector of the matrix).

(c) Let $T : S^1 \to S^1$ be defined by $T(z) = z^2$. Let $\mathcal{P}$ be the partition into *grand orbits*, i.e. equivalence classes of the relation $x \sim y$ iff $\exists m, n \geq 0$ such that $T^m(x) = T^n(y)$.

2.9. Prove that every Lebesgue space is isomorphic to the unit interval equipped with the Lebesgue measure together with countably many atoms.

2.10. Prove that every separable complete metrisable (Polish) space with a measure on the σ-algebra containing all open sets, minimal among complete measures, is a Lebesgue space.

Hint: [Rokhlin, 1949, 2.7].

2.11. Let $(X, \mathcal{F}, \mu)$ be a Lebesgue space. Then $Y \subset X, \mu_e(Y) > 0$ is measurable iff $(Y, \mathcal{F}_Y, \mu_Y)$ is a Lebesgue space, where μ_e is the outer measure, $\mathcal{F}_Y = \{A \cap Y : A \in \mathcal{F}\}$ and $\mu_Y(A) = \frac{\mu_e(A \cap Y)}{\mu_e(Y)}$.

Hint: If $\mathbf{B}=(B_n)$ is a basis for $(X, \mathcal{F}, \mu)$, then $(B'_n) = (B_n \cap Y)$ is a basis for $(Y, \mathcal{F}_Y, \mu_Y)$. Add to Y one point for each sequence $(B'^{\varepsilon_n}_n)$ whose intersection is missing in Y, and in the space $\tilde{Y}$ obtained in such a way generate the complete measure space $(\tilde{Y}, \tilde{\mathcal{F}}, \tilde{\mu})$ from the extension $\tilde{\mathbf{B}}$ of the basis (B'_n). Borel sets with respect to $\mathbf{B}$ in X correspond to Borel sets with respect to $\tilde{\mathbf{B}}$, and sets of μ

measure 0 correspond to sets of $\tilde{\mu}$ measure 0. So measurability of Y implies $\tilde{\mu}(\tilde{Y} \setminus Y) = 0$.

2.12. Prove Theorem 2.6.6.

Hint: In the case where both spaces are unit intervals with standard Lebesgue measure, consider all intervals J' with rational end points. Each $J = T^{-1}(J')$ is contained in a Borel set B_J with $\mu(B_J \setminus J) = 0$. Remove from X a Borel set of measure 0 containing $\bigcup_J (B_J \setminus J)$. Then T becomes a Borel map: hence it is a Baire function, and hence, owing to the injectivity, it maps Borel sets to Borel sets.

2.13. (a) Consider the unit square $[0,1] \times [0,1]$ equipped with a Lebesgue measure. For each $x \in [0,1]$ let $\mathcal{A}_x$ be the partition into points (x', y) for $x' \neq x$ and the interval $\{x\} \times [0,1]$. What is $\bigwedge_x \mathcal{A}_x$? Let $\mathcal{B}_x$ be the partition into the intervals $\{x'\} \times [0,1]$ for $x' \neq x$ and the points $\{(x,y) : y \in [0,1]\}$. What is $\bigvee_x \mathcal{B}_x$?

(b) Find two measurable partitions $\mathcal{A}, \mathcal{A}'$ of a Lebesgue space such that their set-theoretic intersection (i.e. the largest partition such that $\mathcal{A}, \mathcal{A}'$ are finer than this partition) is not measurable.

2.14. Prove that if $F : X \to X$ is an ergodic endomorphism of a Lebesgue space then its natural extension is also ergodic.

Hint: See [Kornfeld, Fomin & Sinai, 1982, Sec. 10.4].

2.15. Find an example of $T : X \to X$ an endomorphism of a probability space $(X, \mathcal{F}, \mu)$, injective and onto, such that for the system $\cdots \xrightarrow{T} X \xrightarrow{T} X$, natural extension does not exist.

Hint: Set X to be the unit circle and T irrational rotation. Let A be a set consisting of exactly one point in each T-orbit. Set $B = \bigcup_{j \geq 0} T^j(A)$. Notice that B is not Lebesgue measurable, and that the outer measure of B is 1 (use *unique ergodicity* of T, i.e. that (2.2.2) holds for every x).

Let $\mathcal{F}$ be the σ-algebra consisting of all the sets $C = B \cap D$ for D Lebesgue measurable; set $\mu(C) = \mathrm{Leb}(D)$, and for $C \subset X \setminus B$, set $\mu(C) = 0$. Note that $\bigcap_{n \geq 0} T^n(B) = \emptyset$, and in the set-theoretic inverse limit the set $\pi_{-n}^{-1}(B) = \pi_0^{-1}(T^n(B))$ would be of measure 1 for every $n \geq 0$.

Entropy, generators, mixing

2.16. Prove Theorem 2.4.7 provided $(X, \mathcal{F}, \rho)$ is a Lebesgue space, using Theorem 2.8.11 (ergodic decomposition theorem) for ρ.

2.17. (a) Prove that in a Lebesgue space $d(\mathcal{A}, \mathcal{B}) := H(\mathcal{A}|\mathcal{B}) + H(\mathcal{B}|\mathcal{A})$ is a metric in the space Z of countable partitions (mod 0) of finite entropy. Prove that the metric space (Z, d) is separable and complete.

(b) Prove that if T is an endomorphism of the Lebesgue space, then the function $\mathcal{A} \to \mathrm{h}(T, \mathcal{A})$ is continuous for $\mathcal{A} \in Z$ with respect to the above metric d.

Hint: $|\,\mathrm{h}(T, \mathcal{A}) - \mathrm{h}(T, \mathcal{B})| \leq \max\{\mathrm{H}(\mathcal{A}|\mathcal{B}), \mathrm{H}(\mathcal{B}|\mathcal{A})\}$. Compare the proof of Theorem 2.4.5.

2.18. (a) Let $d_0(\mathcal{A},\mathcal{B}) := \sum_i \mu(A_i \div B_i)$ for partitions of a probability space into r measurable sets $\mathcal{A} = \{A_i, i = 1,\dots,r\}$ and $\mathcal{B} = \{B_i, i = 1,\dots,r\}$. Prove that for every r and every $d > 0$ there exists $d_0 > 0$ such that if $\mathcal{A},\mathcal{B}$ are partitions into r sets and $d_0(\mathcal{A},\mathcal{B}) < d_0$, then $d(\mathcal{A},\mathcal{B}) < d$

(b) Using (a) give a simple proof of Corollary 2.8.10. (Hint: Given an arbitrary finite $\mathcal{A}$ construct $\mathcal{B} \le \mathcal{B}_m$ so that $d_0(\mathcal{A},\mathcal{B})$ is small for m large. Next use (a) and Theorem 2.4.4(d).)

2.19. Prove that there exists a finite one-sided generator for every T, a continuous positively expansive map of a compact metric space (see the definition of *positively expansive* in Chapter 3, Section 3.5).

2.20. Compute the entropy $\mathrm{h}(T)$ for Markov chains: see Chapter 1.

2.21. Prove that the entropy $\mathrm{h}(T)$ defined as supremum of $\mathrm{h}(T,\mathcal{A})$ over finite partitions, or over countable partitions of finite entropy, or as $\sup \mathrm{H}(\xi|\xi^-)$ over all measurable partitions ξ that are forward invariant (i.e. $T^{-1}(\xi) \le \xi$), is the same.

2.22. Let T be an endomorphism of the two-dimensional torus $\mathbb{R}^2/\mathbb{Z}^2$, given by an integer matrix of determinant larger than 1 and with eigenvalues λ_1, λ_2 such that $|\lambda_1| < 1$ and $|\lambda_2| > 1$. Let S be the endomorphism of $\mathbb{R}^2/\mathbb{Z}^2$, being the Cartesian product of $S_1(x) = 2x \pmod 1$ on the circle $\mathbb{R}/\mathbb{Z}$ and of $S_2(y) = y + \alpha \pmod 1$, the rotation by an irrational angle α. Which of the maps T, S is exact? Which has a countable one-sided generator of finite entropy?

Answer: T does not have the generator, but it is exact. The latter holds because for each small parallelepiped P spanned by the eigendirections there exists n such that $T^n(P)$ covers the torus (that is, T is topologically exact; see Definition 4.3.3) with multiplicity bounded by a constant not depending on P. This follows from the fact that λ_j are algebraic numbers, and from Roth's theorem about Diophantine approximation. S is not exact, but it is ergodic and has a generator.

2.23. (a) Prove that ergodicity of an endomorphism $T: X \to X$ for a probability space $(X,\mathcal{F},\mu)$ is equivalent to the non-existence of a non-constant measurable function ϕ such that $U_T(\phi) = \phi$, where U_T is the Koopman operator: see 2.2.1 and the notes following it.

(b) Prove that for an automorphism T, weak mixing is equivalent to the non-existence of a non-constant eigenfunction for U_T acting on $L^2(X,\mathcal{F},\mu)$.

(c) Prove that if T is a K-mixing automorphism then $L^2 \ominus$constant functions decomposes in a countable product of pairwise orthogonal U_T-invariant subspaces H_i, each of which contains h_i such that for each i all $U_T^j(h_i), j \in \mathbb{Z}$ are pairwise orthogonal and span H_i. (This property is called a countable Lebesgue spectrum.)

Hint: Use condition (d) in 2.10.1.

2.24. Prove that if the definition of partition $\mathcal{A}$ ε-independent of partition $\mathcal{B}$ is replaced by $\sum_{A\in\mathcal{A},B\in\mathcal{B}} |\mu(A\cap B) - \mu(A)\mu(B)|$, then the definition of weakly Bernoulli is equivalent to the old one. (Note that now the expression is symmetric with respect to $\mathcal{A},\mathcal{B}$.)

Bibliographical notes

For the Martingale Convergence Theorem see for example [Doob 1953], [Billingsley 1979], [Petersen 1983] or [Stroock 1993]. Its standard proofs go via a *maximal function*: see Exercise 2.5. We borrowed the idea of relying on the Banach Principle in Exercise 2.4 from [Petersen 1983]. We followed the way to use a maximal inequality in the proof of the Shannon, McMillan, Breiman Theorem in Section 2.5, Lemma 2.5.1, where we relied on [Petersen 1983] and [Parry 1969]. Remark 2.1.5 is taken from [Neveu, 1964, Ch. 4.3]: see for example [Hoover 1991] for a more advanced theory. In Exercises 2.3–2.6 we again took the idea of relying on the Banach Principle from [Petersen 1983].

Standard proofs of Birkhoff's Ergodic Theorem also use the idea of maximal function. This concerns in particular the extremely simple proof in Section 2.2, which has been taken from [Katok & Hasselblatt 1995]. It uses Garsia's celebrated proof of the Maximal Ergodic Theorem.

Koopman's operator was introduced by Koopman in the L^2 setting in [Koopman 1931].

For the material of Section 2.6 and related exercises see [Rokhlin 1949]. It is also written in an elegant and very concise way in [Cornfeld, Fomin & Sinai 1982].

The consideration in Section 2.7 leading to the extension of the compatible family $\tilde{\mu}_{\Pi,n}$ to $\tilde{\mu}_{\Pi}$ is known as the Kolmogorov Extension Theorem (or the Kolmogorov Theorem on the existence of stochastic process). First, one verifies the σ-additivity of a measure on an algebra; next one uses the Extension Theorem 2.7.2. Our proof of σ-additivity of $\tilde{\mu}$ on $\tilde{X}$ via the Lusin Theorem is also a variant of Kolmogorov's proof. The proofs of σ-additivity on algebras depend, unfortunately, on topological concepts. Halmos wrote [Halmos 1950, p. 212]: 'This peculiar and somewhat undesirable circumstance appears to be unavoidable.' Indeed, the σ-additivity may be not true: see [Halmos 1950, p. 214]. Our example of the non-existence of natural extension (Exercise 2.15) is in the spirit of Halmos's example. There might even be trouble with extending a measure from cylinders in the product of two measure spaces: see [Marczewski & Ryll-Nardzewski 1953] for a counter-example. On the other hand, product measures extend to generated σ-algebras without any additional assumptions [Halmos 1950], [Billingsley 1979].

For Theorem 2.9.1, the existence of a countable partition into cross-sections, see [Rokhlin 1949]; for bounds of its entropy, see for example [Rokhlin 1967, Theorem 10.2], or [Parry 1969]. The simple proof of Theorem 2.8.6 via convergence in measure has been taken from [Rokhlin 1967] and [Walters 1982]. The proof of Theorem 2.8.11(b) is taken from [Rokhlin 1967, sec. 8.10-11 and 9.8].

For Theorem 2.9.6 see [Parry 1969, L. 10.5]; our proof is different. For the construction of a one-sided generator and a two-sided generator see again [Rokhlin 1967], [Parry 1969] or [Cornfeld, Fomin & Sinai 1982]. The same are references to the theory of measurable invariant partitions (exhausting and extremal), and to the Pinsker partition, which we omitted because we do not need these notions further in the book, but which are fundamental to a deeper understanding of the measure-theoretic entropy theory. Finally we encourage the reader to become acquainted with the spectral theory of dynamical systems, and in particular in relation to mixing properties: for an introduction see for example [Cornfeld, Fomin & Sinai 1982], [Parry 1981] (in particular Appendix), and [Walters 1982].

Theorem 2.11.1 can be found in [Philipp & Stout 1975]. See also [Przytycki, Urbański & Zdunik 1989]. For (2.11.9) see [Ibragimov 1962, 1.1.2] or [Billingsley 1968]. For Theorem 2.11.3 see [Leonov 1961], [Ibragimov 1962, 1.5.2] or [Przytycki, Urbański & Zdunik 1989, Lemma 1]. Theorem 2.11.5 can be found in [Gordin 1969]. We owe the idea of the proof of the exactness via Roth's theorem in 2.22 to Wieslaw Szlenk. Generalizations to higher dimensions lead to Wolfgang Schmidt's Diophantine Approximation Theorem.

3

Ergodic theory on compact metric spaces

In the previous chapter a measure preserved by a measurable map T was given *a priori*. Here a continuous mapping T of a topological compact space is given, and we look for various measures preserved by T. Given a real continuous function ϕ on X we try to maximize the functional *measure theoretical entropy* +*integral*, i.e. $\mathrm{h}_\mu(T) + \int \phi \, d\mu$. Supremum over all probability measures on the Borel σ-algebra turns out to be *topological pressure*, similar to P in the Finite Variational Principle or $P(\alpha)$ for ϕ_α on the Cantor set, discussed in the Introduction. We discuss equilibria, that is, measures on which supremum is attained. This chapter provides an introduction to the theory called *thermodynamical formalism*, which will be the main technical tool in this book. We shall continue to develop the thermodynamical formalism in more specific situations in Chapter 5.

3.1 Invariant measures for continuous mappings

We recall in this section some basic facts from functional analysis needed to study the space of measures and invariant measures. We recall the Riesz Representation Theorem, weak* topology, and the Schauder Fixed Point Theorem. We also recall the Krein–Milman Theorem on extremal points, and its stronger form, the Choquet Representation Theorem. This gives a variant of the Ergodic Decomposition Theorem from Chapter 2.

Let X be a topological space. The *Borel* σ-algebra $\mathcal{B}$ of subsets of X is defined to be generated by open subsets of X. We call every probability measure on the Borel σ-algebra of subsets of X a *Borel probability measure* on X. We denote the set of all such measures by $M(X)$.

Denote by $C(X)$ the Banach space of real-valued continuous functions on X with the supremum norm: $\sup|\phi| := \sup_{x\in X} |\phi(x)|$. Sometimes we shall use the notation $||\phi||_\infty$, introduced in Section 2.1 in $L^\infty(\mu)$, although it is compatible only if μ is positive on open sets, even in the absence of μ.

Note that each Borel probability measure μ on X induces a bounded linear functional F_μ on $C(X)$ defined by the formula

$$F_\mu(\phi) = \int \phi \, d\mu. \tag{3.1.1}$$

One can extend the notion of measure and consider σ-additive set functions, known as *signed measures.* Just as in the definition of measure from Section 2.1 consider $\mu : \mathcal{F} \to [-\infty, \infty)$ or $\mu : \mathcal{F} \to (-\infty, \infty]$ and keep the notation $(X, \mathcal{F}, \mu)$ from Chapter 2. The set of signed measures is a linear space. On the set of finite signed measures – that is, with the range $\mathbb{R}$ – one can introduce the following total variation norm:

$$v(\mu) := \sup\left\{ \sum_{i=1}^{n} |\mu(A_i)| \right\},$$

where the supremum is taken over all finite sequences (A_i) of disjoint sets in $\mathcal{F}$.

It is easy to prove that every finite signed measure is bounded, and that it has finite total variation. It is also not difficult to prove the following theorem.

Theorem 3.1.1 (Hahn–Jordan decomposition). *For every signed measure μ on a σ-algebra $\mathcal{F}$ there exist $A_\mu \in \mathcal{F}$ and two measures μ^+ and μ^- such that $\mu = \mu^+ - \mu^-$, μ^- is zero on all measurable subsets of A_μ, and μ^- is zero on all measurable subsets of $X \setminus A_\mu$.*

Note that $v(\mu) = \mu^+(X) + \mu^-(X)$.

A measure (or signed measure) is called *regular* if for every $A \in \mathcal{F}$ and $\varepsilon > 0$ there exist $E_1, E_2 \in \mathcal{F}$ such that $\overline{E}_1 \subset A \subset \operatorname{Int} E_2$, and for every $C \in \mathcal{F}$ with $C \subset E_2 \setminus E_1$ we have $|\mu(C)| < \varepsilon$.

If X is a topological space, denote the space of all regular finite signed measures with the total variation norm by $\operatorname{rca}(X)$. The abbreviation 'rca' replaces *regular countably additive.*

If $\mathcal{F} = \mathcal{B}$, the Borel σ-algebra, and X is metrizable, regularity holds for every finite signed measure. This can be proved by Carathéodory's outer measure argument: compare the proof of Corollary 2.8.10.

Denote by $C(X)^*$ the space of all bounded linear functionals on $C(X)$. This is called the *dual space* (or *conjugate space*). *Bounded* means here bounded on the unit ball in $C(X)$, which is equivalent to *continuous.* The space $C(X)^*$ is equipped with the norm $||F|| = \sup\{F(\phi) : \phi \in C(X), |\phi| \leq 1\}$, which makes it a Banach space.

There is a natural order in $\operatorname{rca}(X)$: $\nu_1 \leq \nu_2$ if and only if $\nu_2 - \nu_1$ is a measure.

Also in the space $C(X)^*$ one can distinguish positive functionals, similar to measures amongst signed measures, as those that are non-negative on the set of functions $C^+(X) := \{\phi \in C(X) : \phi(x) \geq 0 \text{ for every } x \in X\}$. This gives the order: $F \leq G$ for $F, G \in C(X)^*$ if and only if $G - F$ is positive.

Remark that $F \in C(X)^*$ is positive if and only if $||F|| = F(\mathbb{1})$, where $\mathbb{1}$ is the function on X identically equal to 1. Also, for every bounded linear operator

$F : C(X) \to C(X)$ that is positive, namely $F(C^+(X)) \subset C^+(X)$, we have $||F|| = |F(\mathbb{1})|$.

Note that (3.1.1) transforms measures to positive linear functionals.

The following fundamental theorem of F. Riesz says more about the transformation $\mu \mapsto F_\mu$ in (3.1.1) (see [Dunford & Schwartz, 1958, pp. 373, 380] for the history of this theorem).

Theorem 3.1.2 (Riesz Representation Theorem). *If X is a compact Hausdorff space, the transformation $\mu \mapsto F_\mu$ defined by (3.1.1) is an isometric isomorphism between the Banach space $C(X)^*$ and $rca(X)$. Furthermore, this isomorphism preserves order.*

In the sequel we shall often write μ instead of F_μ and *vice versa*, and $\mu(\phi)$ or $\mu\phi$ instead of $F_\mu(\phi)$ or $\int \phi \, d\mu$.

Note that in Theorem 3.1.2 the hard part is the existence: that is, that for every $F \in C(X)^*$ there exists $\mu \in \mathrm{rca}(X)$ such that $F = F_\mu$. The uniqueness is just the following.

Lemma 3.1.3. *If μ and ν are finite regular Borel signed measures on a compact Hausdorff space X, such that $\int \phi \, d\mu = \int \phi \, d\nu$ for each $\phi \in C(X)$, then $\mu = \nu$.*

Proof. This is an exercise in the use of the regularity of μ and ν. Let $\eta := \mu - \nu = \eta^+ - \eta^-$ in the Hahn–Jordan decomposition. Suppose that $\mu \neq \nu$. Then η^+ (or η^-) is non-zero, say $\eta^+(X) = \eta^+(A_\eta) = \varepsilon > 0$, where A_η is the set defined in Theorem 3.1.1. Let E_1 be a closed set and E_2 be an open set, such that $E_1 \subset A_\eta \subset E_2$, $\eta^-(E_2 \setminus A_\eta) < \varepsilon/2$ and $\eta^+(A_\eta \setminus E_1) < \varepsilon/2$. There exists $\phi \in C(X)$ with values in $[0,1]$ identically equal 1 on E_1 and 0 on $X \setminus E_2$. Then

$$\begin{aligned}\int \phi \, d\eta &= \int_{E_1} \phi \, d\eta + \int_{A_\eta \setminus E_1} \phi \, d\eta + \int_{E_2 \setminus A_\eta} \phi \, d\eta + \int_{X \setminus E_2} \phi \, d\eta \\ &= \int_{E_1} \phi \, d\eta^+ + \int_{A_\eta \setminus E_1} \phi \, d\eta^+ - \int_{E_2 \setminus A_\eta} \phi \, d\eta^- \\ &\geq \eta^+(E_1) - \int_{E_2 \setminus A_\eta} \phi \, d\eta^- \\ &\geq \varepsilon - \varepsilon/2 > 0. \end{aligned} \tag{3.1.2}$$

♣

The space $C(X)^*$ can be also equipped with the weak* topology. In the case where X is *metrizable* – that is, if there exists a metric on X such that the topology induced by this metric is the original topology on X – weak* topology is characterized by the property that a sequence $\{F_n : n = 1, 2, \ldots\}$ of functionals in $C(X)^*$ converges to a functional $F \in C(X)^*$ if and only if

$$\lim_{n \to \infty} F_n(\phi) = F(\phi) \tag{3.1.3}$$

for every function $\phi \in C(X)$.

If we do not assume X to be metrizable, weak* topology is defined as the smallest topology in which all elements of $C(X)$ are continuous on $C(X)^*$ (recall that $\phi \in C(X)$ acts on $F \in C(X)^*$ by $F(\phi)$). One says *weak** to distinguish this topology from the *weak topology*, where one considers all continuous functionals on $C(X)^*$ and not only those represented by $f \in C(X)$. This discussion of topologies of course concerns every Banach space B and its dual B^*.

Using the bijection established by the Riesz Representation Theorem we can move the weak* topology from $C(X)^*$ to $\mathrm{rca}(X)$ and restrict it to $M(X)$. The topology on $M(X)$ obtained in this way is usually called the *weak* topology* on the space of probability measures (sometimes one omits * to simplify the language and notation, but one still has in mind weak*, unless stated otherwise). In view of (3.1.3), if X is metrizable, this topology is characterized by the property that a sequence $\{\mu_n : n = 1, 2, \ldots\}$ of measures in $M(X)$ converges to a measure $\mu \in M(X)$ if and only if

$$\lim_{n\to\infty} \mu_n(\phi) = \mu(\phi) \tag{3.1.4}$$

for every function $\phi \in C(X)$. Such a convergence of measures will be called a *weak* convergence* or *weak convergence*, and can be also characterized as follows.

Theorem 3.1.4. *Suppose that X is metrizable (we do not assume compactness here). A sequence $\{\mu_n : n = 1, 2, \ldots\}$, of Borel probability measures on X converges weakly to a measure μ if and only if $\lim_{n\to\infty} \mu_n(A) = \mu(A)$ for every Borel set A such that $\mu(\partial A) = 0$.*

Proof. Suppose that $\mu_n \to \mu$ and $\mu(\partial A) = 0$. Then there exist sets $E_1 \subset \mathrm{Int} A$ and $E_2 \supset \overline{A}$ such that $\mu(E_2 \setminus E_1) = \varepsilon$ is arbitrarily small. Indeed, metrizability of X implies that every open set, and in particular $\mathrm{int} A$, is the union of a sequence of closed sets, and every closed set is the intersection of a sequence of open sets. For example, $\mathrm{Int} A = \bigcup_{n=1}^{\infty} \{x \in X : \inf_{z \notin \mathrm{int} A} \rho(x, z) \geq 1/n\}$ for a metric ρ.

Next, there exist $f, g \in C(X)$ with range in the unit interval $[0, 1]$ such that f is identically 1 on E_1, 0 on $X \setminus \mathrm{int} A$, g identically 1 on $\mathrm{cl}\, A$ and 0 on $X \setminus E_2$. Then $\mu_n(f) \to \mu(f)$ and $\mu_n(g) \to \mu(g)$. As $\mu(E_1) \leq \mu(f) \leq \mu(g) \leq \mu(E_2)$ and $\mu_n(f) \leq \mu_n(A) \leq \mu_n(g)$, we obtain

$$\begin{aligned}\mu(E_1) \leq \mu(f) = \lim_{n\to\infty} \mu_n(f) &\leq \liminf_{n\to\infty} \mu_n(A) \\ &\leq \limsup_{n\to\infty} \mu_n(A) \leq \lim_{n\to\infty} \mu_n(g) = \mu(g) \leq \mu(E_2).\end{aligned}$$

As also $\mu(E_1) \leq \mu(A) \leq \mu(E_2)$, letting $\varepsilon \to 0$ we obtain $\lim_{n\to\infty} \mu_n(A) = \mu(A)$.

Proof in the opposite direction follows from the definition of an integral: approximate an arbitrary continuous function f uniformly by simple functions $\sum_{i=1}^{k} \varepsilon_i \mathbb{1}_{E_i}$, where $E_i = \{x \in X : \varepsilon_i \leq f(x) < \varepsilon_{i+1}\}$, for an increasing sequence $\varepsilon_i, i = 1, \ldots, k$ such that $\varepsilon_i - \varepsilon_{i-1} < \varepsilon$ and $\mu(f^{-1}(\{\varepsilon_i\})) = 0$, with $\varepsilon \to 0$. It is possible to find such numbers ε_i because only countably many sets $f^{-1}(a)$ for $a \in \mathbb{R}$ can have non-zero measure. ♣

Example 3.1.5. The assumption $\mu(\partial A) = 0$ is substantial. Let X be the interval $[0,1]$. Denote by δ_x the delta Dirac measure concentrated at the point x, which is defined by the formula

$$\delta_x(A) = \begin{cases} 1, & \text{if } x \in A \\ 0, & \text{if } x \notin A \end{cases}$$

for all sets $A \in \mathcal{B}$.

Consider non-atomic probability measures μ_n supported respectively on the ball $B(x, \frac{1}{n})$. The sequence μ_n converges weakly to δ_x but does not converge on $\{x\}$.

Of particular importance is the following theorem.

Theorem 3.1.6. *The space $M(X)$ is compact in the weak* topology.*

This theorem follows immediately from compactness in the weak* topology of any subset of $C(X)^*$ closed in weak* topology, which is bounded in the standard norm of the dual space $C(X)^*$ (compare for example [Dunford & Schwartz, 1958, V.4.3], where this result is proved for all spaces dual to Banach spaces). $M(X)$ is weak*-closed, since it is closed in the dual space norm, and convex by the Hahn–Banach Theorem. (Caution: convexity is a substantial assumption. Indeed, the unit sphere in an infinite dimensional Banach space, for example, is never weak*-closed, as 0 is in its closure.)

It turns out (see [Dunford & Schwartz, 1958, V.5.1]) that if X is compact metrizable, then every weak*-compact subset of the space $C(X)^*$ with weak* topology is metrizable: hence, in particular, $M(X)$ is metrizable. (Caution: $C(X)^*$ itself is not metrizable for infinite X. The reason is for example that it does not have a countable basis of topology at 0.)

Let now $T : X \to X$ be a continuous transformation of X. The mapping T is measurable with respect to the Borel σ-algebra. At the very begining of Section 2.2 we defined T-invariant measures μ to satisfy the condition $\mu = \mu \circ T^{-1}$. This means that Borel probability T-invariant meaures are exactly the fixed points of the transformation $T_* : M(X) \to M(X)$ defined by the formula $T_*(\mu) = \mu \circ T^{-1}$.

We denote the set of all T-invariant measures in $M(X)$ by $M(X,T)$. This notation is consistent with the notation from Section 2.2. We omit here the σ-algebra $\mathcal{F}$ because it is always the Borel σ-algebra $\mathcal{B}$.

Noting that $\int \phi \, d(\mu \circ T^{-1}) = \int \phi \circ T \, d\mu$ for any $\mu \in M(X)$ and any integrable function ϕ (Proposition 2.2.1), it follows from Lemma 3.1.3 that a Borel probability measure μ is T-invariant if and only if for every continuous function $\phi : X \to \mathbb{R}$

$$\int \phi \, d\mu = \int \phi \circ T \, d\mu. \tag{3.1.5}$$

In order to look for fixed points of T_* one can apply the following very general result, whose proof (and the definition of locally convex topological vector spaces, abbreviation: LCTVS) can be found for example in [Dunford & Schwartz 1958] or [Edwards 1995].

Theorem 3.1.7 (Schauder–Tychonoff Theorem [Dunford & Schwartz, 1958, V.10.5]). *If K is a non-empty compact convex subset of an LCTVS, then any continuous transformation $H : K \to K$ has a fixed point.*

Assume from now on that X is compact and metrizable. In order to apply the Schauder–Tychonoff Theorem consider the LCTVS $C(X)^*$ with weak* topology and $K \subset C(X)^*$, being the image of $M(X)$ under the identification between measures and functionals, given by the Riesz Representation Theorem. With this identification we can consider T_* acting on K. Note that T_* is continuous on $M(X)$ (or K) in the weak* topology. Indeed, if $\mu_n \to \mu$ weakly*, then for every continuous function $\phi : X \to \mathbb{R}$, since $\phi \circ T$ is continuous, we get $\mu_n(\phi \circ T) \to \mu(\phi \circ T)$, i.e. $T_*(\mu_n)(\phi) \to T_*(\mu)(\phi)$, hence $T_*(\mu_n) \to T_*(\mu)$ weakly*.
We obtain

Theorem 3.1.8 (Bogolyubov–Krylov Theorem [Walters 1982, 6.9.1]). *If $T : X \to X$ is a continuous mapping of a compact metric space X, then there exists on X a Borel probability measure μ invariant under T.*

Thus our space $M(X,T)$ is non-empty. It is also weak* compact, since it is closed as the set of fixed points for a continuous transformation.

As an immediate consequence of this theorem and Theorem 2.8.11 (the Ergodic Decomposition Theorem), we get the following:

Corollary 3.1.9. *If $T : X \to X$ is a continuous mapping of a compact metric space X, then there exists a Borel ergodic probability measure μ invariant under T.*

We shall use the notation $M_e(X,T)$ for the set of all ergodic measures in $M(X,T)$. Write also $\mathcal{E}(M(X,T))$ for the set of extreme points in $M(X,T)$.

Thus, because of Theorem 2.2.8 and Corollary 3.1.9, we know that $M_e(X,T) = \mathcal{E}(M(X,T)) \neq \emptyset$.

In fact, Corollary 3.1.9 can be obtained in a more elementary way without using Theorem 2.8.11: it now follows immediately from Theorem 2.2.8 and the following theorem.

Theorem 3.1.10 (Krein–Milman theorem on extremal points [Dunford & Schwartz, 1958, V.8.4]). *If K is a non-empty compact convex subset of an LCTVS, then the set $\mathcal{E}(K)$ of extreme points of K is non-empty; moreover K is the closure of the convex hull of $\mathcal{E}(K)$.*

Below we state Choquet's Representation Theorem, which is stronger than the Krein–Milman theorem. It corresponds to the Ergodic Decomposition Theorem (Theorem 2.8.11). We formulate it in $C(X)^*$ with weak* topology as in [Walters, 1982, p. 153]. The reader can find a general LCTVS version in [Phelps 1966]. For example, it is sufficient to add to the assumptions of the Krein–Milman theorem the metrizability of K.

We rely here also on [Ruelle 1978a, Appendix A.5], where the reader can find further references.

Theorem 3.1.11 (Choquet Representation Theorem). *Let K be a non-empty compact convex set in $M(X)$ with weak* topology, for X a compact metric space. Then for every $\mu \in K$ there exists a 'mass distribution', i.e. a measure $\alpha_\mu \in M(\mathcal{E}(K))$, such that*

$$\mu = \int m \, d\alpha_\mu(m).$$

This integral converges in weak topology, which means that for every $f \in C(X)$*

$$\mu(f) = \int m(f) \, d\alpha_\mu(m). \tag{3.1.6}$$

Note that we have already had a formula analogous to (3.1.6) in Remark 2.6.10.

Note that the Krein–Milman Theorem follows from the Choquet Representation Theorem because one can weakly approximate α_μ by measures on $\mathcal{E}(K)$ with finite support (finite linear combinations of Dirac measures).

Exercise. Prove that if we allow α_μ to be supported on the closure of $\mathcal{E}(K)$, then the existence of such α_μ follows from the Krein–Milman Theorem.

Example 3.1.12. For $K = M(X)$ we have $\mathcal{E}(K) = \{\text{Dirac measures on } X\}$. Then $\alpha_\mu\{\delta_x : x \in A\} = \mu(A)$ for every $A \in \mathcal{B}$ defines a Choquet representation for every $\mu \in M(X)$. (Exercise)

Choquet's Theorem asserts the existence of α_μ satisfying (3.1.6) but does not claim uniqueness, which is usually not true. A compact closed set K with the uniqueness of α_μ satisfying (3.1.6) for every $\mu \in M(K)$ is called *simplex* (or *Choquet simplex*).

Theorem 3.1.13. *The set $K = M(X)$ or $K = M(X,T)$ for every continuous $T : X \to X$ is a simplex.*

A proof in the case of $K = M(X)$ is very easy: see Example 3.1.12. A proof for $K = M(X,T)$ is not hard either. The reader can look in [Ruelle 1978a, A.5.5]. The proof there relies on the fact that two different measures $\mu_1, \mu_2 \in \mathcal{E}(M(X,T))$ are singular (see Theorem 2.2.6). Observe that $||\mu_1 - \mu_2|| = 2$. One proves in fact that for every $\mu_1, \mu_2 \in M(X,T)$, $||\alpha_{\mu_1} - \alpha_{\mu_2}|| = ||\mu_1 - \mu_2||$.

Let us go back to the Schauder–Tychonoff Theorem (Theorem 3.1.7). We shall use it in this book later, in Section 5.2, for maps different from T_*. The Bogolyubov–Krylov Theorem proved above with the help of Theorem 3.1.7 has a different, more elementary proof owing to the fact that T_* is affine. A general theorem on the existence of a fixed point for a family of commuting continuous affine maps on K is called the Markov–Kakutani Theorem [Dunford & Schwartz, 1958, V.10.6], [Walters, 1982, 6.9]).

Remark 3.1.14. An alternative proof of Theorem 3.1.8. Take an arbitrary $\nu \in M(X)$ and consider the sequence

$$\mu_n = \mu_n(\nu) = \frac{1}{n} \sum_{j=0}^{n-1} T_*^j(\nu).$$

In view of Theorem 3.1.4, it has a weakly convergent sub-sequence, say $\{\mu_{n_k} : k = 1, 2, \ldots\}$. Denote its limit by μ. We shall show that μ is T-invariant.

We have

$$T_*(\mu_{n_k}) = T_* \left(\frac{1}{n_k} \sum_{j=0}^{n_k-1} T_*^j(\nu) \right) = \left(\frac{1}{n_k} \sum_{j=0}^{n_k-1} T_*^{j+1}(\nu) \right).$$

So for every $\phi \in C(X)$ we have

$$|\mu(\phi) - T_*(\mu(\phi))| = \left| \lim_{k\to\infty} \left(\mu_{n_k}(\phi) - T_*(\mu_{n_k})(\phi) \right) \right|$$

$$\le \lim_{k\to\infty} \frac{1}{n_k} |\nu(\phi) - T_*^{n_k}(\nu)(\phi)| \le \lim_{k\to\infty} \frac{2}{n_k} ||\phi||_\infty = 0.$$

This, in view of Lemma 3.1.3, finishes the proof.

Remark 3.1.15. If in the above proof we consider $\nu = \delta_x$, a Dirac measure, then $T_*^j(\delta_x) = \delta_{T^j(x)}$ and $\mu_n(\phi) = \frac{1}{n} \sum_{j=0}^{n-1} \phi(T^j(x))$. If we have *a priori* $\mu \in M(X,T)$ then

$$\mu_n(\delta_x) = \frac{1}{n} \sum_{j=0}^{n-1} \delta_{T^j(x)}$$

is weakly convergent for μ-a.e. $x \in X$ by Birkhoff's Ergodic Theorem.

Remark 3.1.16. Recall that in Birkhoff's Ergodic Theorem (Chapter 2), for $\mu \in M(X,T)$ for every integrable function $\phi : X \to \mathbb{R}$ one considers $\lim_{n\to\infty} \frac{1}{n} \sum_{j=0}^{n-1} \phi(T^j(x))$ for a.e. x. This 'almost every' depends on ϕ. If X is compact, as is the case in this chapter, one can reverse the order of quantifiers for continuous functions.

That is, there exists $\Lambda \in \mathcal{B}$ such that $\mu(\Lambda) = 1$ and for every $\phi \in C(X)$ and $x \in \Lambda$ the limit $\lim_{n\to\infty} \frac{1}{n} \sum_{j=0}^{n-1} \phi(T^j(x))$ exists.

Remark 3.1.17. We could take in Remark 3.1.14 an arbitrary sequence $\nu_n \in M(X)$ and take $\mu_n := \mu_n(\nu_n)$. This gives a general method of constructing measures in the space $M(X,T)$: see for example the proof of the Variational Principle in Section 3.5. This point of view is taken from [Walters 1982].

We end this section with the following lemma, useful in the sequel.

Lemma 3.1.18. *For every finite partition* P *of the space* $(X, \mathcal{B}, \mu)$, *with* X *a compact metric space,* $\mathcal{B}$ *the Borel* σ*-algebra and* $\mu \in M(X,T)$, *if* $\sum_{A\in\mathrm{P}} \mu(\partial A) = 0$, *then the entropy* $\mathrm{H}_\nu(\mathrm{P})$ *is a continuous function of* $\nu \in M(X,T)$ *at* μ. *The entropy* $\mathrm{h}_\nu(T,\mathrm{P})$ *is upper semi-continuous at* μ.

Proof. The continuity of $\mathrm{H}_\nu(\mathrm{P})$ follows immediately from Theorem 3.1.4. This fact, applied to the partitions $\bigvee_{i=1}^{n-1} T^{-i}\mathrm{P}$, gives the upper semi-continuity of $\mathrm{h}_\nu(T,\mathrm{P})$ being the limit of the decreasing sequence of continuous functions $\frac{1}{n}\mathrm{H}_\nu(\bigvee_{i=1}^{n-1} T^{-i}\mathrm{P})$. See Lemma 2.4.3. ♣

3.2 Topological pressure and topological entropy

This section is of topological character, and no measure is involved. We introduce and examine here some basic topological invariants coming from the thermodynamic formalism of statistical physics.

Let $\mathcal{U} = \{A_i\}_{i\in I}$ and $\mathcal{V} = \{B_j\}_{j\in J}$ be two covers of a compact metric space X, that is $\bigcup\mathcal{U} = \bigcup\mathcal{V} = X$. We define the new cover $\mathcal{U}\vee\mathcal{V}$, putting

$$\mathcal{U}\vee\mathcal{V} = \{A_i \cap B_j : i \in I, j \in J\} \tag{3.2.1}$$

and we write

$$\mathcal{U} \prec \mathcal{V} \iff \forall_{j\in J}\exists_{i\in I} B_j \subset A_i. \tag{3.2.2}$$

Let, as in the previous section, $T : X \to X$ be a continuous transformation of X. Let $\phi : X \to \mathbb{R}$ be a continuous function. In the context of this book such a function is often called a *potential*. Let $\mathcal{U}$ be a finite, open cover of X. For every integer $n \geq 1$, we set

$$\mathcal{U}^n = \mathcal{U} \vee T^{-1}(\mathcal{U}) \vee \ldots \vee T^{-(n-1)}(\mathcal{U}),$$

for every set $Y \subset X$,

$$S_n\phi(Y) = \sup\left\{\sum_{k=0}^{n-1} \phi\circ T^k(x) : x \in Y\right\},$$

and for every $n \geq 1$,

$$Z_n(\phi,\mathcal{U}) = \inf_{\mathcal{V}}\left\{\sum_{U\in\mathcal{V}} \exp S_n\phi(U)\right\}, \tag{3.2.3}$$

where $\mathcal{V}$ ranges over all covers of X contained (in the sense of inclusion) in $\mathcal{U}^n$. The quantity $Z_n(\phi,\mathcal{U})$ is sometimes called the *partition function.*

Lemma 3.2.1. *The limit* $\mathrm{P}(\phi,\mathcal{U}) = \lim_{n\to\infty} \frac{1}{n}\log Z_n(\phi,\mathcal{U})$ *exists, and moreover it is finite. In addition,* $\mathrm{P}(\phi,\mathcal{U}) \geq -||\phi||_\infty$.

Proof. Fix $m, n \geq 1$ and consider arbitrary covers $\mathcal{V} \subset \mathcal{U}^m$, $\mathcal{G} \subset \mathcal{U}^n$ of X. If $U \in \mathcal{V}$ and $V \in \mathcal{G}$ then

$$S_{m+n}\phi(U \cap T^{-m}(V)) \leq S_m\phi(U) + S_n\phi(V)$$

and thus

$$\exp\bigl(S_{m+n}\phi(U\cap T^{-m}(V))\bigr) \leq \exp S_m\phi(U)\exp S_n\phi(V).$$

Since $U \cap T^{-m}(V) \in \mathcal{V}\vee T^{-m}(\mathcal{G}) \subset \mathcal{U}^m \vee T^{-m}(\mathcal{U}^n) = \mathcal{U}^{m+n}$, we therefore obtain

$$\begin{aligned} Z_{m+n}(\phi,\mathcal{U}) &\leq \sum_{U\in\mathcal{V}}\sum_{V\in\mathcal{G}} \exp\bigl(S_{m+n}f(U\cap T^{-m}(V))\bigr) \\ &\leq \sum_{U\in\mathcal{V}}\sum_{V\in\mathcal{G}} \exp S_m\phi(U)\exp S_n\phi(V) \\ &= \sum_{U\in\mathcal{V}} \exp S_m\phi(U) \times \sum_{V\in\mathcal{G}} \exp S_n\phi(V). \end{aligned} \tag{3.2.4}$$

Ranging now over all $\mathcal{V}$ and $\mathcal{G}$ as specified in (3.2.3), we get $Z_{m+n}(\phi,\mathcal{U}) \le Z_m(\phi,\mathcal{U}) \cdot Z_n(\phi,\mathcal{U})$. This implies that

$$\log Z_{m+n}(\phi,\mathcal{U}) \le \log Z_m(\phi,\mathcal{U}) + \log Z_n(\phi,\mathcal{U}).$$

Moreover, $Z_n(\phi,\mathcal{U}) \ge \exp(-n||\phi||_\infty)$. So $\log Z_n(\phi,\mathcal{U}) \ge -n||\phi||_\infty$. Now to complete the proof we apply Lemma 2.4.3. ♣

Notice that although, in the notation $\mathrm{P}(\phi,\mathcal{U})$, the transformation T does not directly appear, this quantity obviously also depends on T. If we want to indicate this dependence we write $\mathrm{P}(T,\phi,\mathcal{U})$, and similarly $Z_n(T,\phi,\mathcal{U})$ for $Z_n(\phi,\mathcal{U})$. Given an open cover $\mathcal{V}$ of X let

$$\mathrm{osc}(\phi,\mathcal{V}) = \sup_{V\in\mathcal{V}} \big(\sup\{|\phi(x)-\phi(y)| : x,y \in V\}\big).$$

Lemma 3.2.2. *If $\mathcal{U}$ and $\mathcal{V}$ are finite open covers of X such that $\mathcal{U} \succ \mathcal{V}$, then* $\mathrm{P}(\phi,\mathcal{U}) \ge \mathrm{P}(\phi,\mathcal{V}) - \mathrm{osc}(\phi,\mathcal{V})$.

Proof. Take $U \in \mathcal{U}^n$. Then there exists $V = i(U) \in \mathcal{V}^n$ such that $U \subset V$. For every $x,y \in V$ we have $|S_n\phi(x) - S_n\phi(y)| \le \mathrm{osc}(\phi,\mathcal{V})n$, and therefore

$$S_n\phi(U) \ge S_n\phi(V) - \mathrm{osc}(\phi,\mathcal{V})n. \tag{3.2.5}$$

Let now $\mathcal{G} \subset \mathcal{U}^n$ be a cover of X and let $\tilde{\mathcal{G}} = \{i(U) : U \in \mathcal{U}^n\}$. The family $\tilde{\mathcal{G}}$ is also an open finite cover of X and $\tilde{\mathcal{G}} \subset \mathcal{V}^n$. In view of (3.2.5) and (3.2.3) we get

$$\sum_{U\in\mathcal{G}} \exp S_n\phi(U) \ge \sum_{V\in\tilde{\mathcal{G}}} \exp S_n\phi(V) e^{-\mathrm{osc}(\phi,\mathcal{V})n} \ge e^{-\mathrm{osc}(\phi,\mathcal{V})n} Z_n(\phi,\mathcal{V}).$$

Therefore, applying (3.2.3) again, we get $Z_n(\phi,\mathcal{U}) \ge \exp(-\mathrm{osc}(\phi,\mathcal{V})n) Z_n(\phi,\mathcal{V})$. Hence $\mathrm{P}(\phi,\mathcal{U}) \ge \mathrm{P}(\phi,\mathcal{V}) - \mathrm{osc}(\phi,\mathcal{V})$. ♣

Definition 3.2.3 (topological pressure). Consider now the family of all sequences $(\mathcal{V}_n)_{n=1}^\infty$ of open finite covers of X such that

$$\lim_{n\to\infty} \mathrm{diam}(\mathcal{V}_n) = 0, \tag{3.2.6}$$

and define the *topological pressure* $\mathrm{P}(T,\phi)$ as the supremum of upper limits

$$\limsup_{n\to\infty} \mathrm{P}(\phi,\mathcal{V}_n),$$

taken over all such sequences. Note that, by Lemma 3.2.1, $\mathrm{P}(T,\phi) \ge -||\phi||_\infty$.

The following lemma gives us a simpler way to calculate topological pressure, showing that in fact we do not have to take the supremum in its definition.

Lemma 3.2.4. *If $(\mathcal{U}_n)_{n=1}^\infty$ is a sequence of open finite covers of X such that $\lim_{n\to\infty} \mathrm{diam}(\mathcal{U}_n) = 0$, then the limit $\lim_{n\to\infty} \mathrm{P}(\phi,\mathcal{U}_n)$ exists and is equal to* $\mathrm{P}(T,\phi)$.

Proof. Assume first that $\mathrm{P}(T,\phi)$ is finite, and fix $\varepsilon > 0$. By the definition of pressure and uniform continuity of ϕ there exists $\mathcal{W}$, an open cover of X, such that

$$\operatorname{osc}(\phi,\mathcal{W}) \le \frac{\varepsilon}{2} \text{ and } \mathrm{P}(\phi,\mathcal{W}) \ge \mathrm{P}(T,\phi) - \frac{\varepsilon}{2}. \tag{3.2.7}$$

Fix now $q \ge 1$ so large that, for all $n \ge q$, $\operatorname{diam}(\mathcal{U}_n)$ does not exceed a Lebesgue number of the cover $\mathcal{W}$. Take $n \ge q$. Then $\mathcal{U}_n \succ \mathcal{W}$, and applying (3.2.7) and Lemma 3.2.2 we get

$$\mathrm{P}(\phi,\mathcal{U}_n) \ge \mathrm{P}(\phi,\mathcal{W}) - \frac{\varepsilon}{2} \ge \mathrm{P}(T,\phi) - \frac{\varepsilon}{2} - \frac{\varepsilon}{2} = \mathrm{P}(T,\phi) - \varepsilon. \tag{3.2.8}$$

Hence letting $\varepsilon \to 0$, $\liminf_{n\to\infty} \mathrm{P}(\phi,\mathcal{U}_n) \ge \mathrm{P}(T,\phi)$. This completes the proof in the case of finite pressure $\mathrm{P}(T,\phi)$. Note also that the same proof actually goes through in the infinite case. ♣

Since in the definition of numbers $\mathrm{P}(\phi,\mathcal{U})$ no metric was involved, they do not depend on a compatible metric under consideration. And since also the convergence to zero of diameters of a sequence of subsets of X does not depend on a compatible metric, we come to the conclusion that the topological pressure $\mathrm{P}(T,\phi)$ is independent of any compatible metric (this of course depends on the topology).

Readers familiar with directed sets will notice easily that the family of all finite open covers $\mathcal{U}$ of X equipped with the relation '$\prec$' is a directed set, and topological pressure $\mathrm{P}(T,\phi)$ is the limit of the generalized sequence $\mathrm{P}(\phi,\mathcal{U})$. However, we can assure them that this remark will not be used anywhere in this book.

Definition 3.2.5 (topological entropy)**.** If the function ϕ is identically zero, the pressure $\mathrm{P}(T,\phi)$ is usually called the *topological entropy* of the map T, and is denoted by $\mathrm{h}_{\mathrm{top}}(T)$. Thus we can define

$$\mathrm{h}_{\mathrm{top}}(T) := \sup_{\mathcal{U}} \limsup_{n\to\infty} \frac{1}{n} \log\left(\inf_{\mathcal{U}^n \prec \mathcal{V}} \#\mathcal{V}\right).$$

Note that, because $\phi \equiv 0$, we could replace $\lim_{\operatorname{diam}(\mathcal{U})\to 0} \mathrm{P}(\phi,\mathcal{U})$ in the definition of topological pressure by $\sup_{\mathcal{U}}$ here, and $\mathcal{V}$ being a subset of $\mathcal{U}^n$ by $\mathcal{U}^n \prec \mathcal{V}$.

In the rest of this section we establish some basic elementary properties of pressure and provide its more effective characterizations. Applying Lemma 3.2.2, we obtain the following.

Corollary 3.2.6. *If $\mathcal{U}$ is a finite, open cover of X, then* $\mathrm{P}(T,\phi) \ge \mathrm{P}(\phi,\mathcal{U}) - \operatorname{osc}(\phi,\mathcal{U})$.

Lemma 3.2.7. $\mathrm{P}(T^n, S_n\phi) = n\,\mathrm{P}(T,\phi)$ *for every* $n \ge 1$. *In particular,* $\mathrm{h}_{\mathrm{top}}(T^n) = n\mathrm{h}_{\mathrm{top}}(T)$.

Proof. Put $g = S_n\phi$. Take $\mathcal{U}$, a finite open cover of X. Let $\overline{\mathcal{U}} = \mathcal{U} \vee T^{-1}(\mathcal{U}) \vee \ldots \vee T^{-(n-1)}(\mathcal{U})$. Since now we are actually dealing with two transformations T and

T^n, we do not use the symbol $\mathcal{U}^n$, in order to avoid possible misunderstandings. For any $m \geq 1$ consider an open set $U \in \mathcal{U} \vee T^{-1}(\mathcal{U}) \vee \ldots \vee T^{-(nm-1)}(\mathcal{U}) = \overline{\mathcal{U}} \vee T^{-n}(\overline{\mathcal{U}}) \vee \ldots \vee T^{-n(m-1)}(\overline{\mathcal{U}})$. Then for every $x \in U$ we have

$$\sum_{k=0}^{mn-1} \phi \circ T^k(x) = \sum_{k=0}^{m-1} g \circ T^{nk}(x),$$

and therefore $S_{mn}\phi(U) = S_m g(U)$, where the symbol S_m is considered here with respect to the map T^n. Hence $Z_{mn}(T,\phi,\mathcal{U}) = Z_m(T^n,g,\overline{\mathcal{U}})$, and this implies that $\mathrm{P}(T^n,g,\overline{\mathcal{U}}) = n\,\mathrm{P}(T,\phi,\mathcal{U})$. Since, given a sequence $(\mathcal{U}_k)_{k=1}^{\infty}$ of open covers of X whose diameters converge to zero, the diameters of the sequence of its refinements $\overline{\mathcal{U}_k}$ also converge to zero, applying Lemma 3.2.4 now completes the proof. ♣

Lemma 3.2.8. *If $T : X \to X$ and $S : Y \to Y$ are continuous mappings of compact metric spaces, and $\pi : X \to Y$ is a continuous surjection such that $S \circ \pi = \pi \circ T$, then for every continuous function $\phi : Y \to \mathbb{R}$ we have* $\mathrm{P}(S,\phi) \leq \mathrm{P}(T,\phi \circ \pi)$.

Proof. For every finite, open cover $\mathcal{U}$ of Y we get

$$\mathrm{P}(S,\phi,\mathcal{U}) = \mathrm{P}(T,\phi \circ \pi,\pi^{-1}(\mathcal{U})). \tag{3.2.9}$$

In view of Corollary 3.2.6 we have

$$\begin{aligned}\mathrm{P}(T,\phi \circ \pi) &\geq \mathrm{P}(T,\phi \circ \pi,\pi^{-1}(\mathcal{U})) - \mathrm{osc}(\phi \circ \pi,\pi^{-1}(\mathcal{U})) \\ &= \mathrm{P}(T,\phi \circ \pi,\pi^{-1}(\mathcal{U})) - \mathrm{osc}(\phi,\mathcal{U}).\end{aligned} \tag{3.2.10}$$

Let $(\mathcal{U}_n)_{n=1}^{\infty}$ be a sequence of open finite covers of Y whose diameters converge to 0. Then also $\lim_{n\to\infty} \mathrm{osc}(\phi,\mathcal{U}_n)) = 0$ and therefore, using Lemma 3.2.4, (3.2.9) and (3.2.10) we obtain

$$\mathrm{P}(S,\phi) = \lim_{n\to\infty} \mathrm{P}(S,\phi,\mathcal{U}_n) = \lim_{n\to\infty} \mathrm{P}(T,\phi \circ \pi,\pi^{-1}(\mathcal{U}_n)) \leq \mathrm{P}(T,\phi \circ \pi).$$

The proof is complete. ♣

In the sequel we shall need the following technical result.

Lemma 3.2.9. *If $\mathcal{U}$ is a finite open cover of X, then* $\mathrm{P}(\phi,\mathcal{U}^k) = \mathrm{P}(\phi,\mathcal{U})$ *for every $k \geq 1$.*

Proof. Fix $k \geq 1$ and let $\gamma = \sup\{|S_{k-1}\phi(x)| : x \in X\}$. Since $S_{k+n-1}\phi(x) = S_n\phi(x) + S_{k-1}\phi(T^n(x))$, for every $n \geq 1$ and $x \in X$ we get

$$S_n\phi(x) - \gamma \leq S_{k+n-1}\phi(x) \leq S_n\phi(x) + \gamma.$$

Therefore, for every $n \geq 1$ and every $U \in \mathcal{U}^{k+n-1}$,

$$S_n\phi(U) - \gamma \leq S_{k+n-1}\phi(U) \leq S_n\phi(U) + \gamma.$$

Since $(\mathcal{U}^k)^n = \mathcal{U}^{k+n-1}$, these inequalities imply that

$$e^{-\gamma} Z_n(\phi,\mathcal{U}^k) \leq Z_{n+k-1}(\phi,\mathcal{U}) \leq e^{\gamma} Z_n(\phi,\mathcal{U}^k).$$

Letting now $n \to \infty$, the required result follows. ♣

3.3 Pressure on compact metric spaces

Let ρ be a metric on X. For every $n \geq 1$ we define the new metric ρ_n on X by putting

$$\rho_n(x,y) = \max\{\rho(T^j(x), T^j(y)) : j = 0, 1, \ldots, n-1\}.$$

Given $r > 0$ and $x \in X$, by $B_n(x,r)$ we denote the open ball in the metric ρ_n centred at x and of radius r. Let $\varepsilon > 0$ and let $n \geq 1$ be an integer. A set $F \subset X$ is said to be (n,ε)*-spanning* if and only if the family of balls $\{B_n(x,\varepsilon) : x \in F\}$ covers the space X. A set $S \subset X$ is said to be (n,ε)*-separated* if and only if $\rho_n(x,y) \geq \varepsilon$ for any pair x, y of different points in S. The following fact is obvious.

Lemma 3.3.1. *Every maximal, in the sense of inclusion, (n,ε)-separated set forms an (n,ε)-spanning set.*

We should like to emphasize here that the word 'maximal' refering to separated sets will in this book always be understood in the sense of inclusion and not in the sense of cardinality. We finish this section with the following characterization of pressure.

Theorem 3.3.2. *For every $\varepsilon > 0$ and every $n \geq 1$ let $F_n(\varepsilon)$ be a maximal (n,ε)-separated set in X. Then*

$$\begin{aligned} \mathrm{P}(T,\phi) &= \lim_{\varepsilon\to 0} \limsup_{n\to\infty} \frac{1}{n} \log \sum_{x\in F_n(\varepsilon)} \exp S_n\phi(x) \\ &= \lim_{\varepsilon\to 0} \liminf_{n\to\infty} \frac{1}{n} \log \sum_{x\in F_n(\varepsilon)} \exp S_n\phi(x). \end{aligned}$$

Proof. Fix $\varepsilon > 0$, and let $\mathcal{U}(\varepsilon)$ be a finite cover of X by open balls of radii $\varepsilon/2$. For any $n \geq 1$ consider $\underline{\mathcal{U}}$, a subcover of $\mathcal{U}(\varepsilon)^n$ such that

$$Z_n(\phi, \mathcal{U}(\varepsilon)) = \sum_{U\in\underline{\mathcal{U}}} \exp S_n\phi(U),$$

where $Z_n(\phi, \mathcal{U}(\varepsilon))$ was defined by formula (3.2.3). For every $x \in F_n(\varepsilon)$, let $U(x)$ be an element of $\underline{\mathcal{U}}$ containing x. Since $F_n(\varepsilon)$ is an (n,ε)-separated set, we deduce that the function $x \mapsto U(x)$ is injective. Therefore

$$Z_n(\phi, \mathcal{U}(\varepsilon)) = \sum_{U\in\underline{\mathcal{U}}} \exp S_n\phi(U) \geq \sum_{x\in F_n(\varepsilon)} \exp S_n\phi(U(x)) \geq \sum_{x\in F_n(\varepsilon)} \exp S_n\phi(x).$$

Thus, by Lemma 3.2.1,

$$\mathrm{P}(\phi, \mathcal{U}(\varepsilon)) \geq \limsup_{n\to\infty} \frac{1}{n} \log \sum_{x\in F_n(\varepsilon)} \exp S_n\phi(x).$$

Hence, letting $\varepsilon \to 0$ and applying Lemma 3.2.4, we get

$$P(T,\phi) \geq \limsup_{\varepsilon\to 0} \limsup_{n\to\infty} \frac{1}{n} \log \sum_{x\in F_n(\varepsilon)} \exp S_n\phi(x). \tag{3.3.1}$$

Now let $\mathcal{V}$ be an arbitrary finite open cover of X, and let $\delta > 0$ be a Lebesgue number of $\mathcal{V}$. Take $\varepsilon < \delta/2$. Since for any $k = 0, 1, \ldots, n-1$ and for every $x \in F_n(\varepsilon)$,

$$\operatorname{diam}\big(T^k(B_n(x,\varepsilon))\big) \leq 2\varepsilon < \delta,$$

we conclude that, for some $U_k(x) \in \mathcal{V}$,

$$T^k(B_n(x,\varepsilon)) \subset U_k(x).$$

Since the family $\{B_n(x,\varepsilon) : x \in F_n(\varepsilon)\}$ covers X (by Lemma 3.3.1), this implies that the family $\{U(x) : x \in F_n(\varepsilon)\} \subset \mathcal{V}^n$ also covers X, where $U(x) = U_0(x) \cap T^{-1}(U_1(x)) \cap \ldots \cap T^{-(n-1)}(U_{n-1}(x))$. Therefore

$$Z_n(\phi,\mathcal{V}) \leq \sum_{x\in F_n(\varepsilon)} \exp S_n\phi(U(x)) \leq \exp\big(\operatorname{osc}(\phi,\mathcal{V})n\big) \sum_{x\in F_n(\varepsilon)} \exp S_n\phi(x).$$

Hence

$$P(\phi,\mathcal{V}) \leq \operatorname{osc}(\phi,\mathcal{V}) + \liminf_{n\to\infty} \frac{1}{n} \log \sum_{x\in F_n(\varepsilon)} \exp S_n\phi(x),$$

and consequently

$$P(\phi,\mathcal{V}) - \operatorname{osc}(\phi,\mathcal{V}) \leq \liminf_{\varepsilon\to 0} \liminf_{n\to\infty} \frac{1}{n} \log \sum_{x\in F_n(\varepsilon)} \exp S_n\phi(x).$$

Letting $\operatorname{diam}(\mathcal{V}) \to 0$, we get

$$P(T,\phi) \leq \liminf_{\varepsilon\to 0} \liminf_{n\to\infty} \frac{1}{n} \log \sum_{x\in F_n(\varepsilon)} \exp S_n\phi(x).$$

Combining this and (3.3.1) completes the proof. ♣

Frequently we shall use the notation

$$\overline{P}(T,\phi,\varepsilon) := \limsup_{n\to\infty} \frac{1}{n} \log \sum_{x\in F_n(\varepsilon)} \exp S_n\phi(x)$$

and

$$\underline{P}(T,\phi,\varepsilon) := \liminf_{n\to\infty} \frac{1}{n} \log \sum_{x\in F_n(\varepsilon)} \exp S_n\phi(x).$$

These limits also depend on the sequence $(F_n(\varepsilon))_{n=1}^{\infty}$ of maximal (n,ε)-separated sets under consideration. However, it will be always clear from the context which sequence is being considered.

3.4 Variational Principle

In this section we shall prove the theorem called the *Variational Principle.* It has a long history, and establishes a useful relationship between measure-theoretic dynamics and topological dynamics.

Theorem 3.4.1 (Variational Principle). *If $T : X \to X$ is a continuous transformation of a compact metric space X, and $\phi : X \to \mathbb{R}$ is a continuous function, then*

$$\mathrm{P}(T,\phi) = \sup\left\{\mathrm{h}_\mu(T) + \int \phi\, d\mu : \mu \in M(T)\right\},$$

where $M(T)$ denotes the set of all Borel probability T-invariant measures on X. In particular, for $\phi \equiv 0$,

$$\mathrm{h}_{\mathrm{top}}(T) = \sup\{h_\mu(T) : \mu \in M(T)\}.$$

The proof of this theorem consists of two parts. In Part I we show that $\mathrm{h}_\mu(T) + \int \phi\, d\mu \le \mathrm{P}(T,\phi)$ for every measure $\mu \in M(T)$, and Part II is devoted to proving the inequality $\sup\{\mathrm{h}_\mu(T) + \int \phi\, d\mu : \mu \in M(T)\} \ge \mathrm{P}(T,\phi)$.

Proof of Part I. Let $\mu \in M(T)$. Fix $\varepsilon > 0$ and consider a finite partition $\mathcal{U} = \{A_1, \dots, A_s\}$ of X into Borel sets. One can find compact sets $B_i \subset A_i$, $i = 1, 2, \dots, s$, such that for the partition $\mathcal{V} = \{B_1, \dots, B_s, X \setminus (B_1 \cup \ldots \cup B_s)\}$ we have

$$\mathrm{H}_\mu(\mathcal{U}|\mathcal{V}) \le \varepsilon,$$

where the conditional entropy $\mathrm{H}_\mu(\mathcal{U}|\mathcal{V})$ has been defined in (2.3.3). Therefore, as in the proof of Theorem 2.4.4(d), we get for every $n \ge 1$ that

$$\mathrm{H}_\mu(\mathcal{U}^n) \le \mathrm{H}_\mu(\mathcal{V}^n) + n\varepsilon. \tag{3.4.1}$$

Our first aim is to estimate from the above the number $\mathrm{H}_\mu(\mathcal{V}^n) + \int S_n\phi\, d\mu$. Putting $b_n = \sum_{B \in \mathcal{V}^n} \exp S_n\phi(B)$, keeping the notation $k(x) = -x\log x$, and using concavity of the logarithmic function, we obtain by Jensen inequality

$$\begin{aligned}
\mathrm{H}_\mu(\mathcal{V}^n) + \int S_n\phi\, d\mu &\le \sum_{B \in \mathcal{V}^n} \mu(B)\big(S_n\phi(B) - \log\mu(B)\big) \\
&= \sum_{B \in \mathcal{V}^n} \mu(B)\log\big(e^{S_n\phi(B)}/\mu(B)\big) \\
&\le \log\left(\sum_{B \in \mathcal{V}^n} e^{S_n\phi(B)}\right).
\end{aligned} \tag{3.4.2}$$

(Compare the Finite Variational Principle in the Introduction).

Now take $0 < \delta < \frac{1}{2}\inf\{\rho(B_i, B_j) : 1 \le i \ne j \le s\} > 0$ so small that

$$|\phi(x) - \phi(y)| < \varepsilon \tag{3.4.3}$$

whenever $\rho(x,y) < \delta$. Consider an arbitrary maximal (n,δ)-separated set $E_n(\delta)$. Fix $B \in \mathcal{V}^n$. Then, by Lemma 3.3.1, for every $x \in B$ there exists $y \in E_n(\delta)$ such that $x \in B_n(y,\delta)$, whence $|S_n\phi(x) - S_n\phi(y)| \le \varepsilon n$ by (3.4.3). Therefore, using the finiteness of the set $E_n(\delta)$, we see that there exists $y(B) \in E_n(\delta)$ such that

$$S_n\phi(B) \le S_n\phi(y(B)) + \varepsilon n \tag{3.4.4}$$

and

$$B \cap B_n(y(B),\delta) \ne \emptyset.$$

The definitions of δ and of the partition $\mathcal{V}$ imply that, for every $z \in X$,

$$\#\{B \in \mathcal{V} : B \cap B_1(z,\delta) \ne \emptyset\} \le 2.$$

Thus

$$\#\{B \in \mathcal{V}^n : B \cap B_n(z,\delta) \ne \emptyset\} \le 2^n.$$

Therefore the function $\mathcal{V}^n \ni B \mapsto y(B) \in E_n(\delta)$ is at most 2^n to 1. Hence, using (3.4.4),

$$2^n \sum_{y \in E_n(\delta)} \exp S_n\phi(y) \ge \sum_{B \in \mathcal{V}^n} \exp\big(S_n\phi(B) - \varepsilon n\big) = e^{-\varepsilon n} \sum_{B \in \mathcal{V}^n} \exp S_n\phi(B).$$

Taking now the logarithms of both sides of this inequality, dividing them by n and applying (3.4.2), we get

$$\begin{aligned} \log 2 + \frac{1}{n}\log\left(\sum_{y \in E_n(\delta)} \exp S_n\phi(y)\right) &\ge -\varepsilon + \frac{1}{n}\log\left(\sum_{B \in \mathcal{V}^n} \exp S_n\phi(B)\right) \\ &\ge \frac{1}{n}\mathrm{H}_\mu(\mathcal{V}^n) + \frac{1}{n}\int S_n\phi\,d\mu - \varepsilon. \end{aligned}$$

So, by (3.4.1),

$$\frac{1}{n}\log\left(\sum_{y \in E_n(\delta)} \exp S_n\phi(y)\right) \ge \frac{1}{n}\mathrm{H}_\mu(\mathcal{U}^n) + \int \phi\,d\mu - (2\varepsilon + \log 2).$$

In view of the definition of entropy $\mathrm{h}_\mu(T,\mathcal{U})$, presented just after Lemma 2.4.2, by letting $n \to \infty$ we get

$$\underline{\mathrm{P}}(T,\phi,\delta) \ge \mathrm{h}_\mu(T,\mathcal{U}) + \int \phi\,d\mu - (2\varepsilon + \log 2).$$

Applying now Theorem 3.3.2 with $\delta \to 0$ and next letting $\varepsilon \to 0$, and finally taking supremum over all Borel partitions $\mathcal{U}$, leads us to

$$\mathrm{P}(T,\phi) \ge \mathrm{h}_\mu(T) + \int \phi\,d\mu - \log 2.$$

And applying with every $n \ge 1$ this estimate to the transformation T^n and to the function $S_n\phi$, we obtain

$$\mathrm{P}(T^n, S_n\phi) \ge \mathrm{h}_\mu(T^n) + \int S_n\phi\,d\mu - \log 2$$

or equivalently, by Lemma 3.2.7 and Theorem 2.4.6(a),

$$n\,\mathrm{P}(T,\phi) \geq n\,\mathrm{h}_\mu(T) + n\int \phi\,d\mu - \log 2.$$

Dividing both sides of this inequality by n and then letting $n \to \infty$, the proof of Part I follows. ♣

In the proof of Part II we shall need the following two lemmas.

Lemma 3.4.2. *If μ is a Borel probability measure on X, then for every $\varepsilon > 0$ there exists a finite partition $\mathcal{A}$ such that* $\mathrm{diam}(\mathcal{A}) \leq \varepsilon$ *and $\mu(\partial A) = 0$ for every $A \in \mathcal{A}$.*

Proof. Let $E = \{x_1, \dots, x_s\}$ be an $\varepsilon/4$-spanning set (that is, with respect to the metric $\rho = \rho_1$) of X. Since for every $i \in \{1, \dots, s\}$ the sets $\{x : \rho(x, x_i) = r\}$, $\varepsilon/4 < r < \varepsilon/2$, are closed and mutually disjoint, only countably many of them can have positive measure μ. Hence there exists $\varepsilon/4 < t < \varepsilon/2$ such that, for every $i \in \{1, \dots, s\}$,

$$\mu(\{x : \rho(x, x_i) = t\}) = 0. \tag{3.4.5}$$

Define inductively the sets $A_1, A_2, \dots, A_s$, putting $A_1 = \{x : \rho(x, x_1) \leq t\}$ and for every $i = 2, 3, \dots, s$

$$A_i = \{x : \rho(x, x_i) \leq t\} \setminus (A_1 \cup A_2 \cup \ldots \cup A_{i-1}).$$

The family $\mathcal{U} = \{A_1, \dots, A_s\}$ is a partition of X with diameter not exceeding ε. Using (3.4.5) and noting that generally $\partial(A \setminus B) \subset \partial A \cup \partial B$, we conclude by induction that $\mu(\partial A_i) = 0$ for every $i = 1, 2, \dots, s$. ♣

Proof of Part II. Fix $\varepsilon > 0$ and let $E_n(\varepsilon)$, $n = 1, 2, \dots$, be a sequence of maximal (n, ε)-separated sets in X. For every $n \geq 1$ define measures

$$\mu_n = \frac{\sum_{x \in E_n(\varepsilon)} \delta_x \exp S_n\phi(x)}{\sum_{x \in E_n(\varepsilon)} \exp S_n\phi(x)} \text{ and } m_n = \frac{1}{n}\sum_{k=0}^{n-1} \mu_n \circ T^{-k},$$

where δ_x denotes the Dirac measure concentrated at the point x (see Example 3.1.5). Let $(n_i)_{i=1}^\infty$ be an increasing sequence such that m_{n_i} converges weakly, say to m, and

$$\lim_{i\to\infty} \frac{1}{n_i} \log \sum_{x \in E_{n_i}(\varepsilon)} \exp S_n\phi(x) = \limsup_{n\to\infty} \frac{1}{n} \log \sum_{x \in E_n(\varepsilon)} \exp S_n\phi(x). \tag{3.4.6}$$

Clearly $m \in M(T)$. In view of Lemma 3.4.2 there exists a finite partition γ such that $\mathrm{diam}(\gamma) \leq \varepsilon$ and $\mu(\partial G) = 0$ for every $G \in \gamma$. For any $n \geq 1$ put $g_n = \sum_{x \in E_n(\varepsilon)} \exp S_n\phi(x)$. Since $\#(G \cap E_n(\varepsilon)) \leq 1$ for every $G \in \gamma^n$, we obtain

$$\begin{aligned}
\mathrm{H}_{\mu_n}(\gamma^n) + \int S_n\phi\, d\mu_n &= \sum_{x\in E_n(\varepsilon)} (-\log \mu_n(x) + S_n\phi(x))\mu_n(x) \\
&= \sum_{x\in E_n(\varepsilon)} \frac{\exp S_n\phi(x)}{g_n}\left(S_n\phi(x) - \log\left(\frac{\exp S_n\phi(x)}{g_n}\right)\right) \\
&= g_n^{-1} \sum_{x\in E_n(\varepsilon)} \exp S_n\phi(x)\big(S_n\phi(x) - S_n\phi(x) + \log g_n\big) \\
&= \log g_n. \qquad (3.4.7)
\end{aligned}$$

Now fix $M \in \mathbb{N}$ and $n \geq 2M$. For $j = 0, 1, \ldots, M-1$, let $s(j) = \mathrm{E}(\frac{n-j}{M}) - 1$, where $\mathrm{E}(x)$ denotes the integer part of x. Note that

$$\begin{aligned}
\bigvee_{k=0}^{s(j)} T^{-(kM+j)}\gamma^M &= T^{-j}\gamma \vee \ldots \vee T^{-(s(j)M+j)-(M-1)}\gamma \\
&= T^{-j}\gamma \vee \ldots \vee T^{-((s(j)+1)M+j-1)}\gamma
\end{aligned}$$

and

$$(s(j)+1)M + j - 1 \leq n - j + j - 1 = n - 1.$$

Therefore, setting $R_j = \{0, 1, \ldots, j-1, (s(j)+1)M + j, \ldots, n-1\}$, we can write

$$\gamma^n = \bigvee_{k=0}^{s(j)} T^{-(kM+j)}\gamma^M \vee \bigvee_{i\in R_j} T^{-i}\gamma.$$

Hence

$$\begin{aligned}
\mathrm{H}_{\mu_n}(\gamma^n) &\leq \sum_{k=0}^{s(j)} \mathrm{H}_{\mu_n}\big(T^{-(kM+j)}\gamma^M\big) + \mathrm{H}_{\mu_n}\left(\bigvee_{i\in R_j} T^{-i}\gamma\right) \\
&\leq \sum_{k=0}^{s(j)} \mathrm{H}_{\mu_n\circ T^{-(kM+j)}}(\gamma^M) + \log\left(\#\left(\bigvee_{i\in R_j} T^{-i}\gamma\right)\right).
\end{aligned}$$

Summing now over all $j = 0, 1, \ldots, M-1$, we get

$$\begin{aligned}
M\,\mathrm{H}_{\mu_n}(\gamma^n) &\leq \sum_{j=0}^{M-1}\sum_{k=0}^{s(j)} \mathrm{H}_{\mu_n\circ T^{-(kM+j)}}(\gamma^M) + \sum_{j=0}^{M-1} \log(\#\gamma^{\#R_j}) \\
&\leq \sum_{l=0}^{n-1} \mathrm{H}_{\mu_n\circ T^{-l}}(\gamma^M) + 2M^2\log\#\gamma \\
&\leq n\,\mathrm{H}_{\frac{1}{n}\sum_{l=0}^{n-1}\mu_n\circ T^{-l}}(\gamma^M) + 2M^2\log\#\gamma.
\end{aligned}$$

And applying (3.4.7) we obtain

$$M\log\left(\sum_{x\in E_n(\varepsilon)} \exp S_n\phi(x)\right) \leq n\,\mathrm{H}_{m_n}(\gamma^M) + M\int S_n\phi\, d\mu_n + 2M^2\log\#\gamma.$$

Dividing both sides of this inequality by Mn, we get

$$\frac{1}{n}\log\left(\sum_{x\in E_n(\varepsilon)}\exp S_n\phi(x)\right) \le \frac{1}{M}\mathrm{H}_{m_n}(\gamma^M) + \int \phi\, dm_n + 2\frac{M}{n}\log\#\gamma.$$

Since $\partial T^{-1}(A) \subset T^{-1}(\partial A)$ for every set $A \subset X$, the measure m of the boundaries of the partition γ^M is equal to 0. Therefore, letting $n \to \infty$ along the sub-sequence $\{n_i\}$, we conclude from this inequality, Lemma 3.1.18 and Theorem 3.1.4 that

$$\overline{\mathrm{P}}(T,\phi,\varepsilon) \le \frac{1}{M}\mathrm{H}_m(\gamma^M) + \int \phi\, dm.$$

Now letting $M \to \infty$, we get

$$\overline{\mathrm{P}}(T,\phi,\varepsilon) \le \mathrm{h}_m(T,\gamma) + \int \phi\, dm \le \sup\left\{h_\mu(T) + \int \phi\, d\mu : \mu \in M(T)\right\}.$$

Finally, applying Theorem 3.3.2 and letting $\varepsilon \searrow 0$, we get the desired inequality. ♣

Corollary 3.4.3. *Under the assumptions of Theorem 3.4.1,*

$$\mathrm{P}(T,\phi) = \sup\{\mathrm{h}_\mu(T) + \int \phi\, d\mu : \mu \in M_e(T)\},$$

where $M_e(T)$ denotes the set of all Borel ergodic probability T-invariant measures on X.

Proof. Let $\mu \in M(T)$, and let $\{\mu_x : x \in X\}$ be the ergodic decomposition of μ. Then $\mathrm{h}_\mu = \int \mathrm{h}_{\mu_x}\, d\mu(x)$ and $\int \phi\, d\mu = \int(\int \phi\, d\mu_x)\, d\mu(x)$. Therefore

$$\mathrm{h}_\mu + \int \phi\, d\mu = \int \left(\mathrm{h}_{\mu_x} + \int \phi\, d\mu_x\right) d\mu(x),$$

and consequently there exists $x \in X$ such that $\mathrm{h}_{\mu_x} + \int \phi\, d\mu_x \ge \mathrm{h}_\mu + \int \phi\, d\mu$, which completes the proof. ♣

Corollary 3.4.4. *If $T : X \to X$ is a continuous transformation of a compact metric space X, $\phi : X \to \mathbb{R}$ is a continuous function and Y is a forward invariant subset of X (i.e. $T(Y) \subset Y$), then $\mathrm{P}(T|_Y, \phi|_Y) \le \mathrm{P}(T,\phi)$.*

Proof. The proof follows immediatly from Theorem 3.4.1 by the remark that each $T|_Y$-invariant measure on Y can be treated as a measure on X, and it is then T-invariant. ♣

3.5 Equilibrium states and expansive maps

We keep in this section the notation from the previous one. A measure $\mu \in M(T)$ is called an *equilibrium state* for the transformation T and function ϕ if

$$\mathrm{P}(T,\phi) = \mathrm{h}_\mu(T) + \int \phi\, d\mu.$$

The set of all these measures will be denoted by $E(\phi)$. In the case $\phi = 0$ the equilibrium states are also called *maximal measures.* Similarly to Corollary 3.4.4 (and in fact more easily) one can prove the following.

Proposition 3.5.1. *If $E(\phi) \neq \emptyset$, then $E(\phi)$ contains ergodic measures.*

As the following example shows, there exist transformations and functions that admit no equilibrium states.

Example 3.5.2. Let $\{T_n : X_n \to X_n\}_{n\geq 1}$ be a sequence of continuous mappings of compact metric spaces X_n, such that for every $n \geq 1$

$$\mathrm{h_{top}}(T_n) < \mathrm{h_{top}}(T_{n+1}) \text{ and } \sup_n \mathrm{h_{top}}(T_n) < \infty. \tag{3.5.1}$$

The disjoint union $\oplus_{n=1}^{\infty} X_n$ of the spaces X_n is a locally compact space, and let $X = \{\omega\} \cup \oplus_{n=1}^{\infty} X_n$ be its Alexandrov one-point compactification. Define the map $T : X \to X$ by $T|_{X_n} = T_n$ and $T(\omega) = \omega$. The reader can check easily that T is continuous. By Corollary 3.4.4, $\mathrm{h_{top}}(T_n) \leq \mathrm{h_{top}}(T)$ for all $n \geq 1$. Suppose that μ is an ergodic maximal measure for T. Then $\mu(X_n) = 1$ for some $n \geq 1$, and therefore

$$\mathrm{h_{top}}(T) = \mathrm{h}_\mu(T_n) \leq \mathrm{h_{top}}(T_n) < \mathrm{h_{top}}(T_{n+1}) \leq \mathrm{h_{top}}(T),$$

which is a contradiction. In view of Proposition 3.5.1, this shows that T has no maximal measure.

A more difficult problem is to find a transitive and smooth example without maximal measure (see for instance [Misiurewicz 1973]).

The remaining part of this section is devoted to providing sufficient conditions for the existence of equilibrium states. We start with the following simple general criterion, which will provide the basis for obtaining all others.

Proposition 3.5.3. *If the function $M(T) \ni \mu \to \mathrm{h}_\mu(T)$ is upper semi-continuous, then each continuous function $\phi : X \to \mathbb{R}$ has an equilibrium state.*

Proof. By the definition of weak* topology the function $M(T) \ni \mu \to \int \phi \, d\mu$ is continuous. Therefore the lemma follows from the assumption, the weak*-compactness of the set $M(T)$, and Theorem 3.4.1 (the Variational Principle). ♣

As an immediate consequence of Theorem 3.4.1 we obtain the following.

Corollary 3.5.4. *If $\mathrm{h_{top}}(T) = 0$, then each continuous function on X has an equilibrium state.*

A continuous transformation $T : X \to X$ of a compact metric space X equipped with a metric ρ is said to be *(positively) expansive* if and only if

$$\exists \delta > 0 \text{ such that } (\rho(T^n(x), T^n(y)) \leq \delta \; \forall n \geq 0 \,) \Longrightarrow x = y.$$

The number δ that appears in this definition is called an *expansive constant* for $T : X \to X$.

Although, at the end of this section, we shall introduce a related but different notion of expansiveness of homeomorphisms, we shall frequently omit the word 'positively'. Note that the property of being expansive does not depend on the choice of a metric compatible with the topology. From now on in this chapter the transformation T will be assumed to be positively expansive, unless stated otherwise. The following lemma is an immediate consequence of expansiveness.

Lemma 3.5.5. *If $\mathcal{A}$ is a finite Borel partition of X with diameter not exceeding an expansive constant, then $\mathcal{A}$ is a generator for every Borel probability T-invariant measure μ on X.*

The main result concerning expansive maps is as follows.

Theorem 3.5.6. *If $T : X \to X$ is expansive, then the function $M(T) \ni \mu \to \mathrm{h}_\mu(T)$ is upper semi-continuous, and consequently (by Proposition 3.5.3) each continuous function on X has an equilibrium state.*

Proof. Let $\delta > 0$ be an expansive constant of T, and let $\mu \in M(T)$. By Lemma 3.4.2 there exists a finite partition $\mathcal{A}$ of X such that $\mathrm{diam}(\mathcal{A}) \le \delta$ and $\mu(\partial A) = 0$ for every $A \in \mathcal{A}$.

Consider now a sequence $(\mu_n)_{n=1}^\infty$ of invariant measures converging weakly to μ. In view of Lemma 3.5.5 and Theorem 2.8.7(b), we have

$$\mathrm{h}_\nu(T) = \mathrm{h}_\nu(T, \mathcal{A})$$

for every $\nu \in M(T)$, and in particular for $\nu = \mu$ and $\nu = \mu_n$ with $n = 1, 2, \ldots$. Hence, because of Lemma 3.1.18,

$$\mathrm{h}_\mu(T) = \mathrm{h}_\mu(T, \mathcal{A}) \ge \limsup_{n\to\infty} \mathrm{h}_{\mu_n}(T, \mathcal{A}) = \limsup_{n\to\infty} \mathrm{h}_{\mu_n}(T).$$

The proof is complete. ♣

Below we prove three additional interesting results about expansive maps.

Lemma 3.5.7. *If $\mathcal{U}$ is a finite open cover of X with diameter not exceeding an expansive constant of an expansive map $T : X \to X$, then $\lim_{n\to\infty} \mathrm{diam}(\mathcal{U}^n) = 0$.*

Proof. Let $\mathcal{U} = \{U_1, U_2, \ldots, U_s\}$. By expansiveness for every sequence $(a_n)_{n=0}^\infty$ of elements of the set $\{1, 2, \ldots, s\}$

$$\#\left(\bigcap_{n=0}^{\infty} T^{-n}(\overline{U_{a_n}})\right) \le 1$$

and hence

$$\lim_{k\to\infty} \mathrm{diam}\left(\bigcap_{n=0}^{k} T^{-n}(\overline{U_{a_n}})\right) = 0.$$

Therefore, given a fixed $\varepsilon > 0$, there exists a minimal finite $k = k(\{a_n\})$ such that

$$\operatorname{diam}\left(\bigcap_{n=0}^{k} T^{-n}(\overline{U_{a_n}})\right) < \varepsilon.$$

Note now that the function $\{1,2,\dots,s\}^{\mathbb{N}} \ni \{a_n\} \mapsto k(\{a_n\})$ is continuous: what is more, it is locally constant. Thus, by compactness of the space $\{1,2,\dots,s\}^{\mathbb{N}}$, this function is bounded, say by t, and therefore

$$\operatorname{diam}(\mathcal{U}^n) < \varepsilon$$

for every $n \geq t$. The proof is complete. ♣

Combining now Lemma 3.2.4, Lemma 3.5.7 and Lemma 3.2.9, we get the following fact corresponding to Theorem 2.8.7 (b).

Proposition 3.5.8. *If $\mathcal{U}$ is a finite open cover of X with diameter not exceeding an expansive constant, then* $\mathrm{P}(T,\phi) = \mathrm{P}(T,\phi,\mathcal{U})$.

As the last result of this section we shall prove the following.

Proposition 3.5.9. *There exists a constant $\eta > 0$ such that $\forall\, \varepsilon > 0 \exists\, n(\varepsilon) \geq 1$, such that*

$$\rho(x,y) \geq \varepsilon \Longrightarrow \rho_{n(\varepsilon)}(x,y) > \eta.$$

Proof. Let $\mathcal{U} = \{U_1, U_2, \dots, U_s\}$ be a finite open cover of X with diameter not exceeding an expansive constant δ, and let η be a Lebesgue number of $\mathcal{U}$. Fix $\varepsilon > 0$. In view of Lemma 3.5.7 there exists an $n(\varepsilon) \geq 1$ such that

$$\operatorname{diam}(\mathcal{U}^{n(\varepsilon)}) < \varepsilon. \tag{3.5.2}$$

Let $\rho(x,y) \geq \varepsilon$, and suppose that $\rho_{n(\varepsilon)}(x,y) \leq \eta$. Then

$$\forall\, (0 \leq j \leq n(\varepsilon) - 1) \;\; \exists\, (U_{i_j} \in \mathcal{U}) \;\text{ such that } T^j(x), T^j(y) \in U_{i_j}$$

and therefore

$$x, y \in \bigcap_{j=0}^{n(\varepsilon)-1} T^{-j}(U_{i_j}) \in \mathcal{U}^{n(\varepsilon)}.$$

Hence $\operatorname{diam}(\mathcal{U}^{n(\varepsilon)}) \geq \rho(x,y) \geq \varepsilon$, which contradicts (3.5.2). The proof is complete. ♣

As we mentioned at the begining of this section, there is a notion related to positive expansiveness that makes sense only for homeomorphisms. We say that a homeomorphism $T: X \to X$ is *expansive* if and only if

$$\exists \delta > 0 \text{ such that } (\rho(T^n(x), T^n(y)) \leq \delta \;\; \forall n \in \mathbb{Z}) \Longrightarrow x = y.$$

We shall not explore this notion in this book; we want only to emphasize that for expansive homeomorphisms analogous results (with obvious modifications) can be proved (in the same way) as for positively expansive mappings. Of course, each positively expansive homeomorphism is expansive. However, if there exists a positively expansive homeomorhism $T: X \to X$ for X a compact metric space, then X is finite. See for example [Coven & Keane 2006].

3.6 Topological pressure as a function on the Banach space of continuous functions; the issue of uniqueness of equilibrium states

Let $T : X \to X$ be a continuous mapping of a compact topological space X. We shall discuss here the topological pressure function $\mathrm{P} : C(X) \to \mathbb{R}$, $\mathrm{P}(\phi) = \mathrm{P}(T, \phi)$. Assume that the topological entropy is finite, $\mathrm{h_{top}}(T) < \infty$. Hence the pressure P is also finite, because for example

$$\mathrm{P}(\phi) \leq \mathrm{h_{top}}(T) + \sup \phi. \tag{3.6.1}$$

This estimate follows directly from the definitions: see Section 2.2. It is also an immediate consequence of Theorem 3.4.1 (the Variational Principle) in the case where X is metrizable.

Let us start with the following easy theorem.

Theorem 3.6.1. *The pressure function* P *is Lipschitz continuous with the Lipschitz constant 1.*

Proof. Let $\phi \in C(X)$. Recall from Section 3.2 that in the definition of pressure we have considered the following partition function:

$$Z_n(\phi, \mathcal{U}) = \inf_{\mathcal{V}} \left\{ \sum_{U \in \mathcal{V}} \exp S_n\phi(U) \right\},$$

where $\mathcal{V}$ ranges over all covers of X contained in $\mathcal{U}^n$. Now, if also $\psi \in C(X)$, then we obtain for every open cover $\mathcal{U}$ and positive integer n that

$$Z_n(\psi, \mathcal{U}) e^{-||\phi - \psi||_\infty n} \leq Z_n(\phi, \mathcal{U}) \leq Z_n(\psi, \mathcal{U}) e^{||\phi - \psi||_\infty n}.$$

Taking limits, if $n \nearrow \infty$ we get $\mathrm{P}(\psi) - ||\phi - \psi||_\infty \leq \mathrm{P}(\phi) \leq \mathrm{P}(\psi) + ||\phi - \psi||_\infty$: hence $|\mathrm{P}(\psi) - \mathrm{P}(\phi)| \leq ||\psi - \phi||_\infty$. ♣

Theorem 3.6.2. *If X is a compact metric space, then the topological pressure function* $\mathrm{P} : C(X) \to \mathbb{R}$ *is convex.*

We provide two different proofs of this important theorem: one elementary, the other relying on the Variational Principle (Theorem 3.4.1).

Proof 1. By Hölder inequality applied with the exponents $a = 1/\alpha, b = 1/(1-\alpha)$, so that $1/a + 1/b = \alpha + 1 - \alpha = 1$, we obtain for an arbitrary finite set $E \subset X$

$$\begin{aligned}
\frac{1}{n} \log \sum_E e^{S_n(\alpha\phi) + S_n(1-\alpha)\psi)} &= \frac{1}{n} \log \sum_E e^{\alpha S_n(\phi)} e^{(1-\alpha) S_n(\psi)} \\
&\leq \frac{1}{n} \log \left(\sum_E e^{S_n(\phi)} \right)^\alpha \left(\sum_E e^{S_n(\psi)} \right)^{1-\alpha} \\
&\leq \alpha \frac{1}{n} \log \left(\sum_E e^{S_n(\phi)} \right) + (1-\alpha) \frac{1}{n} \log \left(\sum_E e^{S_n(\psi)} \right).
\end{aligned}$$

To conclude the proof, now apply the definition of pressure with $E = F_n(\varepsilon)$ that are (n, ε)-separated sets: see Theorem 3.3.2. ♣

Proof 2. It is sufficient to prove that the function

$$\hat{\mathrm{P}} := \sup_{\mu \in M(X,T)} L_\mu \phi \text{ where } L_\mu \phi := \mathrm{h}_\mu(T) + \mu\phi$$

(where $\mu\phi$ abbreviates $\int \phi \, d\mu$: see Section 3.1) is convex, because by the Variational Principle $\hat{\mathrm{P}}(\phi) = \mathrm{P}(\phi)$.

That is, we need to prove that the set

$$A := \{(\phi, y) \in C(X) \times \mathbb{R} : y \geq \hat{\mathrm{P}}(\phi)\}$$

is convex. Observe, however, that by its definition $A = \bigcap_{\mu \in M(X,T)} L_\mu^+$, where by L_μ^+ we denote the upper half-space $\{(\phi, y) : y \geq L_\mu \phi\}$. Since all the half-spaces L_μ^+ are convex, the set A is convex as their intersection. ♣

Remark 3.6.3. We can write $L_\mu \phi = \mu\phi - (-h_\mu(T))$. The function $\hat{\mathrm{P}}(\phi) = \sup_{\mu \in M(T)} L_\mu \phi$ defined on the space $C(X)$ is called the *Legendre–Fenchel transform* of the convex function $\mu \mapsto -h_\mu(T)$ on the weakly*-compact convex set $M(T)$. We shall abbreviate the name Legendre–Fenchel transform to LF-transform. Observe that this transform generalizes the standard Legendre transform of a strictly convex function h on a finite dimensional linear space, say $\mathbb{R}^n$,

$$y \mapsto \sup_{x \in \mathbb{R}^n} \{\langle x, y \rangle - h(x)\},$$

where $\langle x, y \rangle$ is the scalar (inner) product of x and y.

Note that $-h_\mu(T)$ is not strictly convex (unless $M(X,T)$ is a one-element space), because it is affine: see Theorem 2.4.7.

Proof 2 just repeats the standard proof that the Legendre transform is convex.

In the sequel we shall need the so-called geometric form of the Hahn–Banach Theorem (see [Bourbaki, 1981, Theorem 1, Chapter 2.5], or Chapter 1.7 of [Edwards 1995]).

Theorem 3.6.4 (Hahn–Banach). *Let A be an open convex non-empty subset of a real topological vector space V, and let M be a non-empty affine subset of V (linear subspace moved by a vector) that does not meet A. Then there exists a codimension 1 closed affine subset H that contains M and does not meet A.*

Suppose now that $P : V \to \mathbb{R}$ is an arbitrary convex continuous function on a real topological vector space V. We call a continuous linear functional $F : V \to \mathbb{R}$ *tangent* to P at $x \in V$ if

$$F(y) \leq P(x + y) - P(x) \tag{3.6.2}$$

for every $y \in V$. We denote the set of all such functionals by $V_{x,P}^*$. (Sometimes the term *supporting functional* is used in the literature.)

Applying Theorem 3.6.4 we easily prove that for every x the set $V^*_{x,P}$ is non-empty. Indeed, we can consider the open convex set $A = \{(x,y) \in V \times \mathbb{R}\} : y > P(x)\}$ in the vector space $V \times \mathbb{R}$ with the product topology and the one-point set $M = \{x, P(x)\}$, and define a supporting functional we look for as having the graph $H - \{x, P(x)\}$ in $V \times \mathbb{R}$.

We would also like to bring to the reader's attention another general fact from functional analysis, as follows

Theorem 3.6.5. *Let V be a separable Banach space and $P : V \to \mathbb{R}$ be a convex continuous function. Then for every $x \in V$ the function P is differentiable at x in every direction (Gateaux differentiable), or in a dense (in the weak topology) set of directions, if and only if $V^*_{x,P}$ is a singleton.*

Proof. Suppose first that P is not differentiable at some point x and direction y. Choose an arbitrary $F \in V^*_{x,P}$. Non-differentiability in the direction $y \in V$ implies that there exist $\varepsilon > 0$ and a sequence $\{t_n\}_{n\geq 1}$ converging to 0 such that

$$P(x + t_n y) - P(x) \geq t_n F(y) + \varepsilon |t_n|. \tag{3.6.3}$$

In fact, we can assume that all t_n, $n \geq 1$, are positive by passing to a sub-sequence and replacing y by $-y$ if necessary. We shall prove that (3.6.3) implies the existence of $\hat{F} \in V^*_{x,P} \setminus \{F\}$. Indeed, take $F_n \in V^*_{x+t_n y,P}$. Then, by (3.6.2) applied for F_n at $x + t_n y$ and $-t_n y$ in place of x and y, we have

$$P(x) - P(x + t_n y) \geq F_n(-t_n y). \tag{3.6.4}$$

The inequalities (3.6.3) and (3.6.4) give

$$t_n F(y) + \varepsilon t_n \leq t_n F_n(y).$$

Hence

$$(F_n - F)(y) \geq \varepsilon. \tag{3.6.5}$$

In the case when P is Lipschitz continuous, and this is the case for topological pressure (see Theorem 3.6.1), which we are mostly interested in, all F_n's, $n \geq 1$, are uniformly bounded. Indeed, let L be a Lipschitz constant of P. Then, for every $z \in V$ and every $n \geq 1$,

$$F_n(z) \leq P(x + t_n y + z) - P(x + t_n y) \leq L||z||.$$

So $||F_n|| \leq L$ for every $n \geq 1$. Thus there exists $\hat{F} = \lim_{n\to\infty} F_n$, a weak*-limit of a sequence $\{F_n\}_{n\geq 1}$ (sub-sequence of the previous sequence). We used here the fact that a bounded set is metrizable in weak*-topology (compare section 3.1).

By (3.6.5) $(\hat{F} - F)(y) \geq \varepsilon$. Hence $\hat{F} \neq F$. Since

$$P(x + t_n y + v) - P(x + t_n y) \geq F_n(v) \quad \text{for all } n \text{ and } v \in V$$

passing with n to ∞ and using continuity of P, we conclude that $\hat{F} \in V^*_{x,P}$.

If we do not assume that P is Lipschitz continuous, we restrict F_n to the one-dimensional space spanned by y: that is, we consider $F_n|_{\mathbb{R}y}$. In view of (3.6.5), for every $n \geq 1$ there exists $0 \leq s_n \leq 1$ such that $F_n(s_n y) - F(s_n y) = \varepsilon$. Passing to a sub-sequence, we may assume that $\lim_{n\to\infty} s_n = s$ for some $s \in [0,1]$. Define

$$f_n = s_n F_n|_{\mathbb{R}y} + (1-s_n)F|_{\mathbb{R}y}.$$

Then $f_n(y) - F(y) = \varepsilon$: hence $||f_n - F|_{\mathbb{R}y}|| = \frac{\varepsilon}{||y||}$ for every $n \geq 1$. Thus the sequence $\{f_n\}_{n\geq 1}$ is uniformly bounded and, consequently, it has a weak-* limit $\hat{f} : \mathbb{R}y \to \mathbb{R}$. Now we use Theorem 3.6.4 (Hahn–Banach) for the affine set M being the graph of $\hat{f}$ translated by $(x, P(x))$ in $V \times \mathbb{R}$. We extend M to H and find the linear functional $\hat{F} \in V^*_{x,P}$ whose graph is $H - (x, P(x))$, continuous since H is closed. Since $\hat{F}(y) - F(y) = \hat{f}(y) - F(y) = \varepsilon$, $\hat{F} \neq F$.

Suppose now that $V^*_{x,P}$ contains at least two distinct linear functionals, say F and $\hat{F}$. So $F(y) - \hat{F}(y) > 0$ for some $y \in V$. Suppose on the contrary that P is differentiable in every direction at the point x. In particular, P is differentiable in the direction y. Hence

$$\lim_{t\to 0} \frac{P(x+ty) - P(x)}{t} = \lim_{t\to 0} \frac{P(x-ty) - P(x)}{-t}$$

and consequently

$$\lim_{t\to 0} \frac{P(x+ty) + P(x-ty) - 2P(x)}{t} = 0.$$

On the other hand, for every $t > 0$, we have $P(x+ty) - P(x) \geq F(t) = tF(y)$ and $P(x-ty) - P(x) \geq \hat{F}(-ty) = -t\hat{F}(y)$: hence

$$\liminf_{t\to 0} \frac{P(x+ty) + P(x-ty) - 2P(x)}{t} \geq F(y) - \hat{F}(y) > 0,$$

which is a contradiction.

In fact $F(y) - \hat{F}(y) = \varepsilon > 0$ implies $F(y') - \hat{F}(y') \geq \varepsilon/2 > 0$ for all y' in the neighbourhood of y in the weak topology defined just by $\{y' : (F - \hat{F})(y - y') < \varepsilon/2\}$. Hence P is not differentiable in a weak*-open set of directions. ♣

Let us go back now to our special situation:

Proposition 3.6.6. *If $\mu \in M(T)$ is an equilibrium state for $\phi \in C(X)$, then the linear functional represented by μ is tangent to* P *at ϕ.*

Proof. We have

$$\mu(\phi) + \mathrm{h}_\mu = \mathrm{P}(\phi)$$

and for every $\psi \in C(X)$

$$\mu(\phi + \psi) + \mathrm{h}_\mu \leq \mathrm{P}(\phi + \psi).$$

Subtracting the sides of the equality from the respective sides of the latter inequality we obtain $\mu(\psi) \leq \mathrm{P}(\phi+\psi) - \mathrm{P}(\phi)$, which is just the inequality defining tangent functionals. ♣

As an immediate consequence of Proposition 3.6.6 and Theorem 3.6.5 we get the following.

Corollary 3.6.7. *If the pressure function* P *is differentiable at* ϕ *in every direction, or at least in a dense (in the weak topology) set of directions, then there is at most one equilibrium state for* ϕ.

Because of this Corollary, in future (see Chapter 5), in order to prove uniqueness it will be sufficient to prove differentiability of the pressure function in a weak*-dense set of directions.

The next part of this section will be devoted to a sort of reversal of Proposition 3.6.6 and Corollary 3.6.7, and to a better understanding of the mutual Legendre–Fenchel transforms $-\mathrm{h}$ and P. This is a beautiful topic, but will not have applications in the rest of this book. Let us start with a characterization of T-invariant measures in the space of all signed measures $C(X)^*$ formulated by means of the pressure function P.

Theorem 3.6.8. *For every* $F \in C(X)^*$ *the following three conditions are equivalent:*

(i) *For every* $\phi \in C(X)$ *it holds that* $F(\phi) \leq \mathrm{P}(\phi)$.

(ii) *There exists* $C \in \mathbb{R}$ *such that for every* $\phi \in C(X)$ *it holds that* $F(\phi) \leq \mathrm{P}(\phi) + C$.

(iii) F *is represented by a probability-invariant measure* $\mu \in M(X,T)$.

Proof. (iii) $\Rightarrow$ (i) follows immediately from the Variational Principle:

$$F(\phi) \leq F(\phi) + \mathrm{h}_\mu(T) \leq \mathrm{P}(\phi) \text{ for every } \phi \in C(X).$$

(i) $\Rightarrow$ (ii) is obvious. Let us prove that (ii) $\Rightarrow$ (iii). Take an arbitrary non-negative $\phi \in C(X)$: that is, such that for every $x \in X, \phi(x) \geq 0$. For every real $t < 0$ we have

$$F(t\phi) \leq \mathrm{P}(t\phi) + C.$$

Since $t\phi \leq 0$, it immediately follows from (3.6.1) that $\mathrm{P}(t\phi) \leq \mathrm{P}(0)$. Hence $F(t\phi) \leq P(0) + C$. So

$$|t|F(\phi) \geq -(C + \mathrm{P}(0)), \text{ hence } F(\phi) \geq \frac{-(C + \mathrm{P}(0))}{|t|}.$$

Letting $t \to -\infty$, we obtain $F(\phi) \geq 0$. We estimate the value of F on constant functions t. For every $t > 0$ we have $F(t) \leq \mathrm{P}(t) + C \leq \mathrm{P}(0) + t + C$. Hence $F(1) \leq 1 + \frac{\mathrm{P}(0)+C}{t}$. Similarly $F(-t) \leq \mathrm{P}(-t) + C = \mathrm{P}(0) - t + C$, and therefore $F(1) \geq 1 - \frac{\mathrm{P}(0)+C}{t}$. Letting $t \to \infty$ we thus obtain $F(1) = 1$. Therefore, by Theorem 3.1.1 (the Riesz Representation Theorem), the functional F is represented by a probability measure $\mu \in M(X)$. Let us finally prove that μ is T-invariant. For every $\phi \in C(X)$ and every $t \in \mathbb{R}$ we have by (i) that

$$F(t(\phi \circ T - \phi)) \leq \mathrm{P}(t(\phi \circ T - \phi)) + C.$$

It immediately follows from Theorem 3.4.1 (the Variational Principle) that $\mathrm{P}(t(\phi\circ T-\phi))=\mathrm{P}(0)$. Hence

$$|F(\phi\circ T)-F(\phi)|\le\left|\frac{\mathrm{P}(0)+C}{t}\right|.$$

Thus, letting $|t|\to\infty$, we obtain $F(\phi\circ T)=F(\phi)$, i.e T-invariance of μ. ♣

We shall prove the following.

Corollary 3.6.9. *Every functional F tangent to* P *at $\phi\in C(X)$, that is, $F\in C(X)^*_{\phi,\mathrm{P}}$, is represented by a probability T-invariant measure $\mu\in M(X,T)$.*

Proof. Using Theorem 3.6.1, we get for every $\psi\in C(X)$ that

$$F(\psi)\le\mathrm{P}(\phi+\psi)-\mathrm{P}(\phi)\le\mathrm{P}(\psi)+|\mathrm{P}(\phi+\psi)-\mathrm{P}(\psi)|-\mathrm{P}(\phi)\le\mathrm{P}(\psi)+||\phi||_\infty-\mathrm{P}(\phi).$$

So condition (ii) of Theorem 3.6.8 holds: hence (iii) holds, which means that F is represented by $\mu\in M(X,T)$. ♣

We can now almost reverse Proposition 3.6.6. That is, being a functional tangent to P at ϕ implies being an 'almost' equilibrium state for ϕ.

Theorem 3.6.10. *It holds that $F\in C(X)^*_{\phi,\mathrm{P}}$ if and only if F, or actually the measure $\mu=\mu_F\in M(X,T)$ representing F, is a weak*-limit of measures $\mu_n\in M(X,T)$ such that*

$$\mu_n\phi+\mathrm{h}_{\mu_n}(T)\to\mathrm{P}(\phi).$$

Proof. In one way the proof is simple. Assume that $\mu=\lim_{n\to\infty}\mu_n$ in the weak* topology and $\mu_n\phi+\mathrm{h}_{\mu_n}(T)\to\mathrm{P}(\phi)$. We proceed as in Proof of Proposition 3.6.6. In view of Theorem 3.4.1 (the Variational Principle) $\mu_n(\phi+\psi)+\mathrm{h}_{\mu_n}(T)\le\mathrm{P}(\phi+\psi)$ which means that $\mu_n(\psi)\le\mathrm{P}(\phi+\psi)-(\mu_n\phi+\mathrm{h}_{\mu_n}(T))$. Thus, letting $n\to\infty$, we get $\mu(\psi)\le\mathrm{P}(\phi+\psi)-\mathrm{P}(\phi)$. This means that $\mu\in C(X)^*_{\phi,\mathrm{P}}$.

Now, let us prove our theorem in the other direction. Recall again that the function $\mu\mapsto\mathrm{h}_\mu(T)$ on $M(X,T)$ is affine (Theorem 2.4.7), and hence concave. Set $\overline{\mathrm{h}}_\mu(T)=\limsup_{\nu\to\mu}\mathrm{h}_\nu(T)$, with $\nu\to\mu$ in weak*-topology. The function $\mu\mapsto\overline{\mathrm{h}}_\mu(T)$ is also concave and upper semi-continuous on $M(T)=M(X,T)$. In the sequel we shall prefer to consider the function $\mu\mapsto-\overline{\mathrm{h}}_\mu(T)$, which is lower semi-continuous and convex.

We need the following.

Lemma 3.6.11 (On composing two LF-transformations.). *For every $\mu\in M(T)$*

$$\sup_{\vartheta\in C(X)}\left(\mu\vartheta-\sup_{\nu\in M(T)}(\nu\vartheta-(-\mathrm{h}_\nu(T)))\right)=-\overline{\mathrm{h}}_\mu(T)),\qquad(3.6.6)$$

which, because of the Variational Principle, takes the form

$$\sup_{\vartheta\in C(X)}\Big(\mu\vartheta-\mathrm{P}(\vartheta)\Big)=-\overline{\mathrm{h}}_\mu(T)).\qquad(3.6.7)$$

Proof. To prove (3.6.6), observe first that for every $\vartheta \in C(X)$,

$$\mu\vartheta - \sup_{\nu\in M(T)} (\nu\vartheta - (-\mathrm{h}_\nu(T))) \le \mu\vartheta - (\mu\vartheta - (-\overline{\mathrm{h}}_\mu(T))) = -\overline{\mathrm{h}}_\mu(T)).$$

Note that we obtained above $-\overline{\mathrm{h}}_\nu(T))$ rather than merely $-\mathrm{h}_\nu(T))$, by taking all sequences $\mu_n \to \mu$, writing on the right-hand side of the above inequality the expression $\mu\vartheta - (\mu_n\vartheta - (-\mathrm{h}_{\mu_n}\vartheta(T)))$, and letting $n \to \infty$. So

$$\sup_{\vartheta\in C(X)} \left(\mu\vartheta - \sup_{\nu\in M(T)} (\nu\vartheta - (-\mathrm{h}_\nu(T)))\right) \le -\overline{\mathrm{h}}_\mu(T). \tag{3.6.8}$$

This says that the LF-transform of the LF-transform of $-\mathrm{h}_\mu(T)$ is less than or equal to $-\overline{\mathrm{h}}_\mu(T)$. The preceding LF-transform was discussed in Remark 3.6.3. The following LF-transformation, leading from $\vartheta \to \mathrm{P}(\vartheta)$ to $\nu \to -(-\mathrm{h}_\nu(T))$, is defined by $\sup_{\vartheta\in C(X)}\Big(\mu\vartheta - \mathrm{P}(\vartheta)\Big)$.

Let us now prove the opposite inequality. We refer to the following consequence of the geometric form of the Hahn–Banach Theorem [Bourbaki, 1981, Chapter II.§5. Prop. 5].

Let M be a closed convex set in a locally convex vector space V. Then every lower semi-continuous convex function f defined on M is the supremum of a family of functions bounded above by f, which are restrictions to M of continuous affine functions on V.

We shall apply this theorem to $V = C^*(X)$ endowed with the weak*-topology, to $f(\nu) = \mathrm{h}_\nu(T)$. We use the fact that every linear functional continuous with respect to this topology is represented by an element belonging to $C(X)$. (This is a general fact concerning dual pairs of vector spaces [Bourbaki, 1981, Ch. II.§6. Prop. 3].) Thus, for every $\varepsilon > 0$, there exists $\psi \in C(X)$ such that for every $\nu \in M(T)$

$$(\nu - \mu)(\psi) \le -\mathrm{h}_\nu(T) - (-\overline{\mathrm{h}}_\mu(T)) + \varepsilon. \tag{3.6.9}$$

So

$$\mu\psi - \sup_{\nu\in M(T)} (\nu\psi - (-\mathrm{h}_\nu(T))) \ge -\overline{\mathrm{h}}_\mu(T) - \varepsilon.$$

Letting $\varepsilon \to 0$, we obtain

$$\sup_{\vartheta\in C(X)} \left(\mu\vartheta - \sup_{\nu\in M(T)} (\nu\vartheta - (-\mathrm{h}_\nu(T)))\right) \ge -\overline{\mathrm{h}}_\mu(T).$$

♣

Continuation of Proof of Theorem 3.6.10. Fix $\mu = \mu_F \in C(X)^*_{\phi,\mathrm{P}}$. From $\mu\psi \le \mathrm{P}(\phi+\psi) - \mathrm{P}(\phi)$ we obtain

$$\mathrm{P}(\phi+\psi) - \mu(\phi+\psi) \ge \mathrm{P}(\phi) - \mu\phi \text{ for all } \psi \in C(X).$$

So

$$\inf_{\psi\in C(X)} \{\mathrm{P}(\psi) - \mu\psi\} \ge \mathrm{P}(\phi) - \mu\phi. \tag{3.6.10}$$

This expresses the fact that the supremum (– infimum above) in the definition of the LF-transform of P at F is attained at ϕ at which F is tangent to P.

By Lemma 3.6.11 and (3.6.10) we obtain

$$\overline{\mathrm{h}}_\mu \geq \mathrm{P}(\phi) - \mu\phi. \tag{3.6.11}$$

So, by the definition of $\overline{\mathrm{h}}_\mu$, there exists a sequence of measures $\mu_n \in M(T)$ such that $\lim_{n\to\infty} \mu_n = \mu$ and $\lim_{n\to\infty} h_{\mu_n} \geq \mathrm{P}(\phi) - \mu\phi$. The proof is complete. ♣

Remark 3.6.12. In Lemma 3.6.11 we considered as $\mu = \mu_F$ an arbitrary $\mu \in M(T)$; we did not assume that μ_F is tangent to P, i.e. that $F \in C(X)^*_{\phi,\mathrm{P}}$. Then considering $\varepsilon > 0$ in (3.6.9) was necessary; without $\varepsilon > 0$ this formula might be false: see Example 3.6.15.

In the proof of Theorem 3.6.10, for $\mu \in C(X)^*_{\phi,\mathrm{P}}$, we obtain from (3.6.11) and the inequality $\mathrm{h}_\nu(T) \leq \mathrm{P}(\phi) - \nu\phi$ for every $\nu \in M(T)$ that

$$h_\nu(T) - \overline{\mathrm{h}}_\mu(T) \leq (\mu - \nu)\phi, \tag{3.6.12}$$

which is just (3.6.9) with $\varepsilon = 0$.

The meaning of this is that if $\mu = \mu_F$ is tangent to P at ϕ, then ϕ is tangent to $-\overline{\mathrm{h}}$, the LF-transform of P, at μ.

Conversely, if ψ satisfies (3.6.12), i.e. ψ is tangent to $-\overline{\mathrm{h}}$ at $\mu \in M(T)$, then, as in the second part of the proof of Theorem 3.6.10 we can prove the inequality analogous to (3.6.10), namely that

$$\sup_{\nu \in M(T)} \nu\psi - (-\overline{\mathrm{h}}_\mu(T)) = \mathrm{P}(\psi) \leq \mu\psi - (-\overline{\mathrm{h}}_\mu(T)).$$

Hence μ is tangent to P at ψ.

Assume now the upper semi-continuity of the entropy $\mathrm{h}_\mu(T)$ as a function of μ. Then, as an immediate consequence of Theorem 3.6.10, we obtain the following corollary.

Corollary 3.6.13. *If the entropy is upper semi-continuous, then a functional $F \in C(X)^*$ is tangent to* P *at $\phi \in C(X)$ if and only if it is represented by a measure that is an equilibrium state for ϕ.*

Recall that the upper semi-continuity of entropy implies the existence of at least one equilibrium state for every continuous $\phi : X \to \mathbb{R}$, already by Proposition 3.5.3.

Now we can complete Corollary 3.6.7.

Corollary 3.6.14. *If the entropy is upper semi-continuous, then the pressure function P is differentiable at $\phi \in C(X)$ in every direction, or in a set of directions dense in the weak topology, if and only if there is at most one equilibrium state for ϕ.*

Proof. This corollary follows directly from Corollary 3.6.13 and Theorem 3.6.5. ♣

After discussing functionals tangent to P, and proving that they coincide with the set of equilibrium states for maps for which the entropy is upper semi-continuous as the function on $M(T)$, the question arises of whether all measures in $M(T)$ are equilibrium states of some continuous functions. The answer given below is no.

Example 3.6.15. We shall construct a measure $m \in M(T)$ that is not an equilibrium state for any $\phi \in C(X)$. Here X is the one-sided shift space Σ^2 with the left-side shift map σ. Since this map is obviously expansive, it follows from Theorem 3.5.6 that the entropy function is upper semi-continuous. Let $m_n \in M(\sigma)$ be the measure equidistributed on the set Per_n of points of period n: that is,

$$m_n = \sum_{x \in \mathrm{Per}_n} \frac{1}{\mathrm{Card}\,\mathrm{Per}_n} \delta_x,$$

where δ_x is the Dirac measure supported by x. m_n converge weakly* to $\mu_{\max}$, the measure of maximal entropy: $\log 2$. (Check that this follows, for example, from Part II of the proof of the Variational Principle.) Let $t_n, n = 0, 1, 2, \ldots$ be a sequence of positive real numbers such that $\sum_{n=0}^{\infty} t_n = 1$. Finally, define

$$m = \sum_{n=0}^{\infty} t_n m_n.$$

Let us prove that there is no $\phi \in C(X)$ tangent to h at m. Let $\mu_n = R_n \mu_{\max} + \sum_{j=0}^{n-1} t_j m_j$, where $R_n = \sum_{j=n}^{\infty} t_j$. We have of course $\mathrm{h}_{m_n}(\sigma) = 0$, $n = 1, 2, \ldots$. Therefore $\mathrm{h}_m(\sigma) = 0$. This follows for example from Theorem 2.8.11 (the Ergodic Decomposition Theorem), or just from the fact that h is affine on $M(\sigma)$, Theorem 2.4.7.

Thus, since h is affine,

$$\mathrm{h}_{\mu_n}(\sigma) - \mathrm{h}_m(\sigma) = R_n \, \mathrm{h}_{\mu_{\max}}(\sigma) = R_n \log 2, \tag{3.6.13}$$

and for an arbitrary $\phi \in C(\Sigma^2)$

$$(\mu_n - m)\phi = \left(R_n \mu_{\max} - \sum_{j=n}^{\infty} t_j m_j \right) \phi \le R_n \varepsilon_n, \tag{3.6.14}$$

where $\varepsilon_n \to 0$ as $n \to \infty$ because $m_j \to \mu_{\max}$. The inequalities (3.6.13) and (3.6.14) prove that ϕ is not 'tangent' to h at m. More precisely, we obtain $\mathrm{h}_{\mu_n}(\sigma) - \mathrm{h}_m(\sigma) > (\mu_n - m)\phi$ for n large: that is,

$$-\mathrm{h}_{\mu_n}(\sigma) - (-\mathrm{h}_m(\sigma)) > (\mu_n - m)\psi$$

for $\psi = -\phi$, opposite to the tangency inequality (3.6.12). So, by Remark 3.6.12, m is not tangent to any ϕ for the pressure function P.

In fact it is easy to see that m is not an equilibrium state for any $\phi \in C(\Sigma^2)$ directly. For an arbitrary $\phi \in C(\Sigma^2)$ we have $\mu_{\max}\phi < \mathrm{P}(\phi)$, because

$h_{\mu_{\max}}(\sigma)>0$. So $m_n\phi < P(\phi)$ for all n large enough as $m_n \to \mu_{\max}$. Also, $m_n\phi \le P(\phi)$ for all n's. So for m being the average of m_n's we have $m\phi = m\phi + h_m(\sigma) < P(\phi)$. So ϕ is not an equilibrium state.

The measure m in this example is very non-ergodic: this is necessary, as will follow from Exercise 3.15.

Exercises

Topological entropy

3.1. Let $T : X \to X$ and $S : Y \to Y$ be two continuous maps of compact metric spaces X and Y respectively. Show that $h_{top}(T \times S) = h_{top}(T) + h_{top}(S)$.

3.2. Prove that if $T : X \to X$ is an isometry of a compact metric space X, then $h_{top}(T) = 0$.

3.3. Show that if $T : X \to X$ is a local homeomorphism of a compact connected metric space and $d = \#T^{-1}(x)$ (note that it is independent of $x \in X$), then $h_{top}(T) \ge \log d$.

3.4. Prove that if $f : M \to M$ is a C^1 endomorphism of a compact differentiable manifold M, then $h_{top}(f) \ge \log \deg(f)$, where $\deg(f)$ means degree of f.

Hint: Look for (n,ϵ)-separated points in $f^{-n}(x)$ for 'good' x.
See [Misiurewicz & Przytycki 1977] or [Katok & Hasselblatt 1995].

3.5. Let $S^1 = \{z \in \mathbb{C} : |z| = 1\}$ be the unit circle, and let $f_d : S^1 \to S^1$ be the map defined by the formula $f_d(z) = z^d$. Show that $h_{top}(f_d) = \log d$.

3.6. Let $\sigma_A : \Sigma_A \to \Sigma_A$ be the shift map generated by the incidence matrix A. Prove that $h_{top}(\sigma_A)$ is equal to the logarithm of the spectral radius of A.

3.7. Show that for every continuous potential ϕ, $P(\phi) \le h_{top}(T) + \sup(\phi)$ (see (3.6.1)).

3.8. Provide an example of a topologically transitive diffeomorphism without measures of maximal entropy.

3.9. Provide an example of a topologically transitive diffeomorphism with at least two measures of maximal entropy.

3.10. Find a sequence of continuous maps $T_n : X_n \to X_n$ such that $h_{top}(T_{n+1}) > h_{top}(T_n)$ and $\lim_{n\to\infty} h_{top}(T_n) < \infty$.

Topological pressure: functional analysis approach

3.11. Prove that for an arbitrary convex continuous function $P : V \to \mathbb{R}$ on a real Banach space V the set of tangent functionals: $\bigcup_{x\in V} V^*_{x,P}$ is dense in the norm topology in the set of so-called P-bounded functionals:

$$\{F \in V^* : (\exists C \in \mathbb{R}) \text{ such that } (\forall\, x \in V),\ \ F(x) \le P(x) + C\}.$$

Remark. The conclusion is that for P being the pressure function on $C(X)$, tangent measures are dense in $M(X,T)$: see Theorem 3.5.6. Hint: This follows from the Bishop-Phelps Theorem (see [Bishop & Phelps 1963] or [Israel 1979, pp. 112–115]), which can be stated as follows:

For every P-bounded functional F_0, for every $x_0 \in V$ and for every $\varepsilon > 0$ there exists $x \in V$ and $F \in V^*$ tangent to P at x such that

$$\|F - F_0\| \leq \varepsilon \text{ and } \|x - x_0\| \leq \frac{1}{\varepsilon}\Big(P(x_0) - F_0(x_0) + s(F_0)\Big),$$

where $s(F_0) := \sup_{x' \in V}\{F_0 x' - P(x')\}$ (this is $-\overline{\mathrm{h}}$, the LF-transform of P).

The reader can imagine F_0 as asymptotic to P and estimate how far the tangency point x of a functional F is close to F_0.

3.12. Prove that in the situation from Exercise 3.11, for every $x \in V$, the set $V^*_{x,P}$ is convex and weak*-compact.

3.13. Let E_ϕ denote the set of all equilibrium states for $\phi \in C(X)$.

(i) Prove that E_ϕ is convex.

(ii) Find an example that E_ϕ is not weak*-compact.

(iii) Prove that extremal points of E_ϕ are extremal points of $M(X,T)$.

(iv) Prove that almost all measures in the ergodic decomposition of an arbitrary $\mu \in E_\phi$ belong also to E_ϕ. (One says that every equilibrium state has a unique decomposition into pure, i.e. ergodic, equilibrium states.)

Hints: In (ii) consider a sequence of Smale horseshoes of topological entropies $\log 2$ converging to a point fixed for T. To prove (iii) and (iv) use the fact that entropy is an affine function of measure.

3.14. Find an example showing that part (iii) of Exercise 3.13 is false if we consider $C(X)^*_{\phi,P}$ rather than E_ϕ.

Hint: An idea is to have two fixed points p, q and two trajectories $(x_n), (y_n)$ such that $x_n \to p, y_n \to q$ for $n \to \infty$ and $x_n \to q, y_n \to p$ for $n \to -\infty$. Now take a sequence of periodic orbits γ_k approaching $\{p,q\} \cup \{x_n\} \cup \{y_n\}$ with periods tending to ∞. Take their Cartesian products with corresponding invariant subsets A_k of small horseshoes of topological entropies less than $\log 2$ but tending to $\log 2$, with the diameters of the horseshoes shrinking to 0 as $k \to \infty$. Then, for $\phi \equiv 0$, the set $C(X)^*_{\phi,P}$ consists of exactly one measure: $\frac{1}{2}(\delta_p + \delta_q)$. (One cannot repeat the proof in Exercise 3.13(iii) with the function $\overline{\mathrm{h}}_\mu$ instead of the entropy function h_μ, because $\overline{\mathrm{h}}_\mu$ is no longer affine!)

This is Peter Walters' example: for details see [Walters 1992].

3.15. Suppose that the entropy function h_μ is upper semi-continuous (then for each $\phi \in C(x)$ $C(X)^*_{\phi,\mathrm{P}} = E_\phi$: see Corollary 3.6.13). Prove that:

(i) Every $\mu \in M(T)$ that is a finite combination of ergodic masures $\mu = \sum t_j m_j$, $m_j \in M(T)$, is tangent to P; more precisely, there exists $\phi \in C(X)$ such that $\mu, m_j \in C(X)^*_{\phi,\mathrm{P}}$, and moreover they are equilibrium states for ϕ.

(ii) If $\mu = \int_{M_e(T)} m \, d\alpha(m)$, where $M_e(X,T)$ consists of ergodic measures in $M(X,T)$ and α is a probability non-atomic measure on $M_e(X,T)$, then there exists $\phi \in C(X)$, which has uncountably many ergodic equilibria in the support of α.

(iii) The set of elements of $C(X)$ with uncountably many ergodic equilibria is dense in $C(X)$.

Hint: By the Bishop–Phelps Theorem (see Remark in Exercise 3.11) there exists $\nu \in E_\phi$ arbitrarily close to μ. Then in its ergodic decomposition there are all the measures μ_j, because all ergodic measures are far apart from each other (in the norm in $C(X)^*$). These measures, by Exercise 3.13, belong to the same E_ϕ, which proves (i). For more details and proofs of (ii) and (iii) see [Israel 1979, Theorem V.2.2] or [Ruelle 1978a, 3.17, 6.15].

Remark. In statistical physics the occurence of more than one equilibrium for $\phi \in C(X)$ is called 'phase transition'. Part (iii) says that the set of functions with 'very rich' phase transition is dense. For further discussion see also [Israel, 1979, V.2].

3.16. Prove the following. Let $P : V \to \mathbb{R}$ be a continuous convex function on a real Banach space V with norm $\|\cdot\|_V$. Suppose P is differentiable at $x \in V$ in every direction. Let $W \subset V$ be an arbitrary linear subspace with norm $\|\cdot\|_W$, such that the embedding $W \subset V$ is continuous and the unit ball in $(W, \|\cdot\|_W)$ is compact in $(V, \|\cdot\|_V)$. Then $P|_W$ is differentiable in the sense that there exists a functional $F \in V^*$ such that for $y \in W$ it holds that

$$|P(x+y) - P(x) - F(y)| = o(\|y\|_W).$$

Remark. In Chapter 4 we shall discuss W being the space of Hölder continuous functions with an arbitrary exponent $\alpha < 1$, and the entropy function will be upper semi-continuous. So the conclusion will be that uniqueness of the equilibrium state at an arbitrary $\phi \in C(X)$ is equivalent to differentiability in the direction of this space of Hölder functions.

3.17. (Walters) Prove that the pressure function P is Frechet differentiable at $\phi \in C(X)$ if and only if P is affine in a neighbourhood of ϕ. Prove also the following conclusion: P is Frechet differentiable at every $\phi \in C(X)$ if and only if T is uniquely ergodic, that is, if $M(X,T)$ consists of one element.

3.18. Prove S. Mazur's Theorem: If $P : V \to \mathbb{R}$ is a continuous convex function on a real separable Banach space V, then the set of points at which there exists a unique functional tangent to P is dense G_δ.

Remark. In the case of the pressure function on $C(X)$, this says that for a dense G_δ set of functions there exists at most one equilibrium state. Mazur's Theorem contrasts with the theorem from Exercise 3.15(iii).

Bibliographical notes

The concept of topological pressure in the dynamical context was introduced by D. Ruelle in [Ruelle 1973], and since then it has been studied in many papers and books. We mention here only [Bowen 1975], [Walters 1976], [Walters 1982] and [Ruelle 1978a]. Topological entropy was introduced earlier in [Adler, Konheim & McAndrew 1965]. The Variational Principle (Theorem 3.4.1) has been proved for some maps in [Ruelle 1973]. The first proofs of this principle in its full generality can be found in [Walters 1976] and [Bowen 1975]. The simplest proof presented in this chapter is taken from [Misiurewicz 1976]. In the case of topological entropy (potential $\phi = 0$) the corresponding results have been obtained earlier: Goodwyn in [Goodwyn 1969] proved the first part of the Variational Principle; Dinaburg in [Dinaburg 1971] proved its full version, assuming that the space X has finite covering topological dimension; and finally Goodman in [Goodman 1971] proved the Variational Principle for topological entropy without any additional assumptions. The concept of equilibrium states and expansive maps in mathematical setting was introduced in [Ruelle 1973], where the first existence and uniqueness type results appeared. Since then these concepts have been explored by many authors, in particular in [Bowen 1975] and [Ruelle 1978a]. The material of Section 3.6 is taken mostly from [Ruelle 1978a], [Israel 1979] and [Ellis 1985]. See also [Walters 1992].

4

Distance-expanding maps

We devote this chapter to a study in detail of the topological properties of distance-expanding maps. Often, however, weaker assumptions will be sufficient. We always assume that the maps are continuous on a compact metric space X, and we usually assume that the maps are open, which means that open sets have open images. This is equivalent to saying that if $f(x) = y$ and $y_n \to y$, then there exist $x_n \to x$ such that $f(x_n) = y_n$ for n large enough.

In view of Section 4.6, in theorems with assertions of topological character, only the assumption that a map is expansive leads to the same conclusions as if we assumed that the map is expanding. We shall prove in Section 4.6 that for every expansive map there exists a metric compatible with the topology on X given by an original metric, such that the map is distance-expanding with respect to this new metric.

Recall that for (X, ρ), a compact metric space, a continuous mapping $T : X \to X$ is said to be *distance-expanding* (with respect to the metric ρ) if there exist constants $\lambda > 1$, $\eta > 0$ and $n \geq 0$, such that for all $x, y \in X$,

$$\rho(x,y) \leq 2\eta \implies \rho(T^n(x), T^n(y)) \geq \lambda\rho(x,y). \tag{4.0.1}$$

We say that T is *distance-expanding at a set* $Y \subset X$ if the above holds for all $z \in Y$ and for every $x, y \in B(z, \eta)$.

In the sequel we shall always assume that $n = 1$: that is, that

$$\rho(x,y) \leq 2\eta \implies \rho(T(x), T(y)) \geq \lambda\rho(x,y), \tag{4.0.2}$$

unless otherwise stated. One can achieve this in two ways:

(1) If T is Lipschitz continuous (say with constant $L > 1$), replace the metric $\rho(x,y)$ by $\sum_{j=0}^{n-1} \rho(T^j(x), T^j(y))$. Of course, then λ and η change. As an exercise the reader can check that the number $1 + (\lambda - 1)(\frac{L-1}{L^n-1})$ can play the role of λ in (4.0.2).

Note that the ratio of both metrics is bounded; in particular, they yield the same topologies.

For another improvement of ρ, working without assuming Lipschitz continuity of T, see Lemma 4.6.3.

(2) Work with T^n instead of T.

Sometimes, in order to simplify notation, we shall write *expanding*, instead of *distance-expanding*.

4.1 Distance-expanding open maps: basic properties

Let us first make a simple observation relating the properties of being *expanding* and being *expansive*.

Theorem 4.1.1. *The distance-expanding property implies forward expansive property.*

Proof. By the definition of 'expanding' above, if $0 < \rho(x,y) \leq 2\eta$, then $\rho(T(x),T(y)) \geq \lambda\rho(x,y) \ldots \rho(T^n(x),T^n(y)) \geq \lambda^n\rho(x,y)$, until for the first time n it happens that $\rho(T^n(x),T^n(y)) > 2\eta$. Such n exists, since $\lambda > 1$. Therefore T is forward expansive, with expansivness constant $\delta = 2\eta$. ♣

Let us prove now a lemma where we assume only $T : X \to X$ to be a continuous open map of a compact metric space X. We do not need to assume in this lemma that T is distance-expanding.

Lemma 4.1.2. *If $T : X \to X$ is a continuous open map, then for every $\eta > 0$ there exists $\xi > 0$ such that $T(B(x,\eta)) \supset B(T(x),\xi)$ for every $x \in X$.*

Proof. For every $x \in X$ let

$$\xi(x) = \sup\{r > 0 : T(B(x,\eta)) \supset B(T(x),r)\}.$$

Since T is open, $\xi(x) > 0$. Since $T(B(x,\eta)) \supset B(T(x),\xi(x))$, it suffices to show that $\xi = \inf\{\xi(x) : x \in X\} > 0$. Suppose conversely that $\xi = 0$. Then there exists a sequence of points $x_n \in X$ such that

$$\xi(x_n) \to 0 \quad \text{as} \quad n \to \infty, \tag{4.1.1}$$

and, as X is compact, we can assume that $x_n \to y$ for some $y \in X$. Hence $B(x_n,\eta) \supset B(y,\frac{1}{2}\eta)$ for all n large enough. Therefore

$$T(B(x_n,\eta)) \supset T\left(B\left(y,\frac{1}{2}\eta\right)\right) \supset B(T(y),\varepsilon) \supset B\left(T(x_n),\frac{1}{2}\varepsilon\right)$$

for some $\varepsilon > 0$ and again for every n large enough. The existence of ε such that the second inclusion holds follows from the openness of T. Consequently $\xi(x_n) \geq \frac{1}{2}\varepsilon$ for these n, which contradicts (4.1.1). ♣

Definition 4.1.3. If $T : X \to X$ is an expanding map, then by (4.0.1), for all $x \in X$, the restriction $T|_{B(x,\eta)}$ is injective, and therefore it has the inverse map on $T(B(x,\eta))$. (The same holds for expanding at a set Y for all $x \in Y$.)

If additionally $T : X \to X$ is an open map, then, in view of Lemma 4.1.2, the domain of the inverse map contains the ball $B(T(x), \xi)$. So it makes sense to define the restriction of the inverse map,

$$T_x^{-1} : B(T(x), \xi) \to B(x, \eta). \tag{4.1.2}$$

Observe that for every $y \in X$ and every $A \subset B(y, \xi)$,

$$T^{-1}(A) = \bigcup_{x \in T^{-1}(y)} T_x^{-1}(A). \tag{4.1.3}$$

Indeed, the inclusion $\supset$ is obvious. So suppose that $x' \in T^{-1}(A)$. Then $y' = T(x') \in B(y, \xi)$. Hence $y \in B(y', \xi)$. Let $x = T_{x'}^{-1}(y)$. As T_x^{-1} and $T_{x'}^{-1}$ coincide on y, they coincide on y', because they map y' into $B(x, \eta)$, and T is injective on $B(x, \eta)$. Thus $x' = T_x^{-1}(y')$.

The formula (4.1.3) for all $A = B(y, \xi)$ implies that T is a so-called *covering map*.

(This property is in fact a standard definition of a covering map except that, for general covering maps, on non-compact spaces ξ may depend on y. We have proved in fact that a local homeomorphism of a compact space is a covering map.)

From now on throughout this section, wherever the notation T^{-1} appears, we assume also the expanding property, i.e. (4.0.2). We then get the following.

Lemma 4.1.4. *If $x \in X$ and $y, z \in B(T(x), \xi)$ then*

$$\rho(T_x^{-1}(y), T_x^{-1}(z)) \le \lambda^{-1} \rho(y, z).$$

In particular, $T_x^{-1}(B(T(x), \xi)) \subset B(x, \lambda^{-1}\xi) \subset B(x, \xi)$ and

$$T(B(x, \lambda^{-1}\xi)) \supset B(T(x), \xi) \tag{4.1.4}$$

for all $\xi > 0$ small enough (which specifies the inclusion in Lemma 4.1.2).

Definition 4.1.5. For every $x \in X$, every $n \ge 1$ and every $j = 0, 1, \ldots, n-1$ write $x_j = T^j(x)$. In view of Lemma 4.1.4, the composition

$$T_{x_0}^{-1} \circ T_{x_1}^{-1} \circ \ldots \circ T_{x_{n-1}}^{-1} : B(T^n(x), \xi) \to X$$

is well defined, and will be denoted by T_x^{-n}.

Below we collect the basic elementary properties of maps T_x^{-n}. They follow immediately from (4.1.3) and Lemma 4.1.4. For every $y \in X$

$$T^{-n}(A) = \bigcup_{x \in T^{-n}(y)} T_x^{-n}(A) \tag{4.1.5}$$

for all sets $A \subset B(y, \xi)$;

$$\rho(T_x^{-n}(y), T_x^{-n}(z)) \le \lambda^{-n} \rho(y, z) \text{ for all } y, z \in B(T^n(x), \xi); \tag{4.1.6}$$

$$T_x^{-n}(B(T^n(x), r)) \subset B(x, \min\{\eta, \lambda^{-n} r\}) \text{ for every } r \le \xi. \tag{4.1.7}$$

Remark. All these properties also hold, and the notation makes sense, for open maps $T : X \to X$ expanding at $Y \subset X$, provided $x, T(x), \ldots, T^n(x) \in Y$.

4.2 Shadowing of pseudo-orbits

We keep the notation of Section 4.1. We consider an open, distance-expanding map $T : X \to X$ with the constants η, λ, ξ.

Let n be a non-negative integer or $+\infty$. Given $\alpha \geq 0$, a sequence $(x_i)_0^n$ is said to be an *α-pseudo-orbit* (alternatively called: *α-orbit, α-trajectory, α-T-trajectory*) for $T : X \to X$ of length $n+1$ if, for every $i = 0, \dots, n-1$,

$$\rho(T(x_i), x_{i+1}) \leq \alpha. \tag{4.2.1}$$

Of course, every (genuine) orbit $(x, T(x), \dots, T^n(x))$, $x \in X$, is an α-pseudo-orbit for every $\alpha \geq 0$. We shall prove a kind of converse fact, that in the case of open, distance-expanding maps, each 'sufficiently good' pseudo-orbit can be approximated (shadowed) by an orbit. To make this precise we proceed as follows. Let $\beta > 0$. We say that an orbit of $x \in X$, *β-shadows* the pseudo-orbit $(x_i)_0^n$ if and only if for every $i = 0, \dots, n$

$$\rho(T^i(x), x_i) \leq \beta. \tag{4.2.2}$$

Definition 4.2.1. We say that a continuous map $T : X \to X$ has the *shadowing property* if for every $\beta > 0$ there exists $\alpha > 0$ such that every α-pseudo-orbit of finite or infinite length can be β-shadowed by an orbit.

Note that, owing to the compactness of X, the shadowing property for all finite n implies shadowing with $n = \infty$.

Here is a simple observation yielding the uniqueness of the shadowing. Assume only that T is expansive (cf. Section 3.2).

Proposition 4.2.2. *If 2β is less than an expansiveness constant of T (we do not need to assume here that T is expanding with respect to the metric ρ), and $(x_i)_0^\infty$ is an arbitrary sequence of points in X, then there exists at most one point x whose orbit β-shadows the sequence $(x_i)_{i=0}^\infty$.*

Proof. Suppose the forward orbits of x and y β-shadow (x_i). Then for every $n \geq 0$ we have $\rho(T^n(x), T^n(y)) \leq 2\beta$. Then since 2β is the expansiveness constant for T, we get $x = y$. ♣

We shall now prove some less trivial results, concerning the existence of β-shadowing orbits.

Lemma 4.2.3. *Let $T : X \to X$ be an open, distance-expanding map. Let $0 < \beta < \xi$, $0 < \alpha \leq \min\{(\lambda - 1)\beta, \xi\}$. If $(x_i)_0^\infty$ is an α-pseudo-orbit, then the points $x_i' = T_{x_i}^{-1}(x_{i+1})$ are well defined, and*

(a) For all $i = 0, 1, 2, \dots, n-1$,

$$T_{x_i'}^{-1}(\overline{B(x_{i+1}, \beta)}) \subset \overline{B(x_i, \beta)}$$

and consequently, for all $i = 0, 1, \dots, n$, the compositions

$$S_i := T_{x_0'}^{-1} \circ T_{x_1'}^{-1} \circ \dots \circ T_{x_{i-1}'}^{-1} : \overline{B(x_i, \beta)} \to X$$

are well defined.

(b) The sequence of closed sets $S_i(\overline{B(x_i,\beta)})$, $i=0,1,\dots,n$, is descending.

(c) The intersection

$$\bigcap_{i=0}^{n} S_i(\overline{B(x_i,\beta)}$$

is non-empty, and the forward orbits (for times $0,1,\dots,n$) of all the points of this intersection β-shadow the pseudo-orbit $(x_i)_0^n$.

Proof. x_i' are well defined by $\alpha \le \xi$. In order to prove (a), observe that by (4.1.7) we have

$$T_{x_i'}^{-1}(\overline{B(x_{i+1},\beta)}) \subset \overline{B(x_i',\lambda^{-1}\beta)} \subset \overline{B(x_i,\lambda^{-1}\beta+\lambda^{-1}\alpha)}$$

and $\lambda^{-1}\beta+\lambda^{-1}\alpha \le \beta$ as $\alpha \le (\lambda-1)\beta$. Statement (b) follows immediately from (a). The first part of (c) follows immediately from (b) and the compactness of the space X. To prove the second part denote the intersection that appears in (c) by A. Then $T^i(A) \subset \overline{B(x_i,\beta)}$ for all $i=0,1,\dots,n$. Thus the forward orbit of every point in A, β shadows $(x_i)_0^n$. The proof is complete. ♣

As an immediate consequence of Lemma 4.2.3 we get the following.

Corollary 4.2.4 (Shadowing lemma). *Every open, distance-expanding map satisfies the shadowing property. More precisely, for all $\beta>0$ and $\alpha>0$ as in Lemma 4.2.3 every α-pseudo-orbit $(x_i)_0^n$ can be β-shadowed by an orbit in X.*

As a consequence of Corollary 4.2.4 we shall prove the following.

Corollary 4.2.5 (Closing lemma). *Let $T: X \to X$ be an expansive map, satisfying the shadowing property. Then for every $\beta>0$ there exists $\alpha>0$ such that if $x \in X$ and $\rho(x,T^l(x)) \le \alpha$ for some $l \ge 1$, then there exists a periodic point of period l whose orbit β-shadows the pseudo-orbit $(x,T(x),\dots,T^{l-1}(x))$. The choice of α to β is the same as in the definition of shadowing.*

In particular, the above holds for every $T: X \to X$ an open, distance-expanding map.

Proof. We can assume without loss of generality that 2β is less than the expansiveness constant for T. Since $\rho(x,T^l(x)) \le \alpha$, the sequence made up as the infinite concatenation of the sequence $(x,T(x),\dots,T^{l-1}(x))$ is an α-pseudo-orbit. Hence, by shadowing with $n=\infty$, there is a point $y \in X$ whose orbit β-shadows this pseudo-orbit. But note that then the orbit of the point $T^l(y)$ also does this, and therefore, by Proposition 4.2.2, $T^l(y)=y$. The proof is complete. ♣

Note that the assumption T is expansive is substantial. The adding machine map (see Example 1.4) satisfies the shadowing property, whereas it has no periodic orbits at all. In fact the same proof yields the following *periodic shadowing.*

Definition 4.2.6. We say that a continuous map $T : X \to X$ satisfies the *periodic shadowing* property if for every $\beta > 0$ there exists $\alpha > 0$ such that for every finite n and every *periodic α-pseudo-orbit* $x_0, \dots, x_{n-1}$ – that is, a sequence of points $x_0, \dots, x_{n-1}$ such that $\rho(T(x_i), x_{(i+1)(\mathrm{mod}\, n)}) \le \alpha$ – there exists a point $y \in X$ of period n such that for all $0 \le i < n$, $\rho(T^i(y), x_i) \le \beta$.

Note that shadowing and periodic shadowing can hold for maps that are not expansive. One can just add artificially the missing periodic orbits, of periods 2^n, to the adding machine space. This example in fact appears as the non-wandering set for any Feigenbaum-like map of the interval: see Section 7.6.

4.3 Spectral decomposition; mixing properties

Let us start with general observations concerning iterations of continuous mappings.

Definition 4.3.1. Let X be a compact metric space. We call a continuous mapping $T : X \to X$ *topologically transitive* if for all non-empty open sets $U, V \subset X$ there exists $n \ge 0$ such that $T^n(U) \cap V \ne \emptyset$. By compactness of X, topological transitivity implies that T maps X onto X.

Example 4.3.2. Consider a topological Markov chain Σ_A, or $\tilde{\Sigma}_A$ in a one-sided or two-sided shift space of d states: see Example 1.3. Observe that the left-shift map s on the topological Markov chain is topologically transitive if and only if the matrix A is *irreducible*: that is, for each i, j there exists an $n > 0$ such that the i, j-th entry $A^n_{i,j}$ of the n-th composition matrix A^n is non-zero.

One can consider a directed graph consisting of d vertices such that there is an edge from a vertex v_i to v_j iff $A_{i,j} \ne 0$; then one can identify elements of the topological Markov chain with infinite paths in the graph (that is, sequences of edges indexed by all integers or non-negative integers, depending on whether we consider the two-sided or the one-sided case, such that each edge begins at the vertex, where the preceding edge ends). Then it is easy to see that A is irreducible if and only if for every two vertices v_1, v_2 there exists a finite path from v_i to v_j.

A notion stronger than topological transitivity, which makes non-trivial sense only for non-invertible maps T, is the following.

Definition 4.3.3. A continuous mapping $T : X \to X$ for a compact metric space X is called *topologically exact* (or *locally eventually onto*) if for every open set $U \subset X$ there exists $n > 0$ such that $T^n(U) = X$.

In Example 4.3.2, in the one-sided shift space case, topological exactness is equivalent to the property that there exists $n > 0$ such that the matrix A^n has all entries positive. Such a matrix is called *aperiodic*.

In the two-sided case, aperiodicity of the matrix is equivalent to *topological mixing* of the shift map. We say a continuous map is *topologically mixing* if for every non-empty open set $U, V \subset X$ there exists $N > 0$ such that for every $n \ge N$ we have $T^n(U) \cap V \ne \emptyset$.

Proposition 4.3.4. *The following three conditions are equivalent:*

(1) $T : X \to X$ is topologically transitive.

(2) For all non-empty open sets $U, V \subset X$ and every $N \geq 0$ there exists $n \geq N$ such that $T^n(U) \cap V \neq \emptyset$.

(3) There exists $x \in X$ such that every $y \in X$ is its ω-limit point: that is, for every $N \geq 0$ the set $\{T^n(x)\}_{n=N}^{\infty}$ is dense in X.

Proof. Let us prove first the implication (1)⇒(3). So, suppose $T : X \to X$ is topologically transitive. Then for every open non-empty set $V \subset X$, the set

$$K(V) := \{x \in X : \text{there exists } n \geq 0 \text{ such that } T^n(x) \in V\} = \bigcup_{n \geq 0} T^{-n}(V)$$

is open and dense in X. Let $\{V_k\}_{k \geq 1}$ be a countable basis of topology of X. By Baire's Category Theorem, the intersection

$$K := \bigcap_{k \geq 1} \bigcap_{N \geq 0} K(T^{-N}(V_k))$$

is a dense G_δ subset of X. In particular, K is non-empty, and by its definition the trajectory $(T^n(x))_{n=N}^{\infty}$ is dense in X for every $x \in K$. Thus (1) implies (3).

Let us now prove that (3)⇒(2). Indeed, if $(T^n(x))_0^{\infty}$ is a trajectory satisfying the condition (3), then for all non-empty open sets $U, V \subset X$ and $N \geq 0$, there exist $n \geq m > 0, n - m \geq N$ such that $T^m(x) \in U$ and $T^n(x) \in V$. Hence $T^{n-m}(U) \cap V \neq \emptyset$. Thus (3) implies (2). Since (2) implies (1) trivially, the proof is complete. ♣

Definition 4.3.5. A point $x \in X$ is called *wandering* if there exists an open neighhbourhood V of x such that $V \cap T^n(V) = \emptyset$ for all $n \geq 1$. Otherwise x is called *non-wandering*. We denote the set of all non-wandering points for T by Ω or $\Omega(T)$.

Proposition 4.3.6. *For $T : X \to X$ satisfying the periodic shadowing property, the set of periodic points is dense in the set Ω of non-wandering points.*

Proof. Given $\beta > 0$, let $\alpha > 0$ come from the definition of shadowing. Take any $x \in \Omega(T)$. Then by the definition of $\Omega(T)$ there exists $y \in B(x, \alpha/2)$ and $n > 0$ such that $T^n(y) \in B(x, \alpha/2)$. So $\rho(y, T^n(y)) \leq \alpha$. Therefore $(y, T(y), \ldots, T^n(y))$ can be β-shadowed by a periodic orbit. Since we can take β arbitrarily small, we obtain the density of periodic points in $\Omega(T)$. ♣

Remark 4.3.7. It is not true that for every open, distance-expanding map $T : X \to X$ we have $\overline{\text{Per}} = X$. Here is an example. Let $X = \{(1/2)^n : n = 0, 1, 2, \ldots\} \cup \{0\}$. Let $T((1/2)^n) = (1/2)^{n-1}$ for $n > 0$, $T(0) = 0, T(1) = 1$. Let the metric be the restriction to X of the standard metric on the real line. Then $T : X \to X$ is distance-expanding, but $\Omega(T) = \text{Per}(T) = \{0\} \cup \{1\}$. See also Exercise 4.3.

Here is the main theorem of this section. Its assertion holds under the assumption that $T : X \to X$ is open, and distance-expanding, and even under weaker assumptions below.

Theorem 4.3.8 (on the existence of spectral decomposition). *Suppose that $T : X \to X$ is an open map that also satisfies the periodic shadowing property, and is expanding at the set of non-wandering points $\Omega(T)$ (equal here to $\overline{\mathrm{Per}(T)}$, the closure of the set of periodic points, by Proposition 4.3.6). Then $\Omega(T)$ is the union of finitely many disjoint compact sets $\Omega_j, j = 1, \ldots, J$, with*

$$(T|_{\Omega(T)})^{-1}(\Omega_j) = \Omega_j$$

and each $T|_{\Omega_j}$ is topologically transitive.

Each Ω_j is the union of $k(j)$ disjoint compact sets Ω_j^k, which are cyclically permuted by T and such that $T^{k(j)}|_{\Omega_j^k}$ is topologically exact.

Proof of Theorem 4.3.8. Let us start by defining an equivalence relation $\sim$ on $\mathrm{Per}(T)$. For $x, y \in \mathrm{Per}(T)$ we write $x \sim_{\to} y$ if for every $\varepsilon > 0$ there exist $x' \in X$ and positive integer m such that $\rho(x, x') < \varepsilon$ and $T^m(x') = T^m(y)$. We write $x \sim y$ if $x \sim_{\to} y$ and $y \sim_{\to} x$. Of course, for every $x \in \mathrm{Per}(T)$, $x \sim x$, so the relation is symmetric.

Now we shall prove it is transitive. Suppose that $x \sim y$ and $y \sim z$. Let k_y, k_z denote periods of y, z respectively.

Let x' be close to x and $T^n(x') = T^n(y) = y$; an integer n satisfiying the latter equality exists, since we can take an integer so that the first equality holds, and then take any larger integer divisible by k_y. Choose n divisible by $k_y k_z$. Next, since T is open, for y' close enough to y, with $T^m(y') = T^m(z) = z$ for m divisible by k_z, there exists x'' close to x' such that $T^n(x'') = y'$. Hence $T^{n+m}(x'') = T^m(y') = z = T^{n+m}(z)$, since both m and n are divisible by k_z. Thus $x \sim z$. We have thus shown that $\sim$ is an equivalence relation. This proof is illustrated in Figure 4.1.

(Figure 4.2 illustrates transitivity for hyperbolic sets $\overline{\mathrm{Per}(T)}$ – see Exercises or [Katok & Hasselblatt 1995] – where $x \sim y$ if the unstable manifold of x

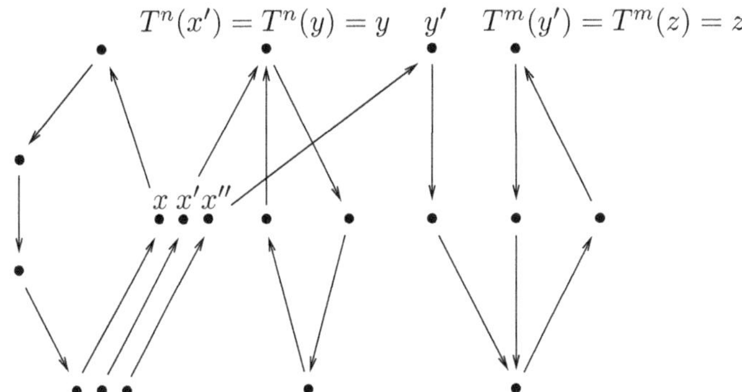

Figure 4.1 Transitivity: the expanding case.

intersects the stable manifold of y transversally. In our expanding case the role of transversality is played by the openness of T.)

So far we have not used the expanding assumption.

Observe now that for all $x, y \in \mathrm{Per}(T)$, $\rho(x,y) \leq \xi$ implies $x \sim y$. Indeed, we can take $x' = T_x^{-nk_xk_y}(y)$ for n arbitrarily large. Then x' is arbitrarily close to x, and $T^{nk_xk_y}(x') = y = T^{nk_xk_y}(y)$. Hence the number of equivalence classes of $\sim$ is finite. Denote them by $P_1, \ldots, P_N$. Moreover, the sets $\overline{P_1}, \ldots, \overline{P_N}$ are pairwise disjoint, and the distances between them are at least ξ. We have $T(\mathrm{Per}(T)) = \mathrm{Per}(T)$, and if $x \sim y$ then $T(x) \sim T(y)$. The latter follows directly from the definition of $\sim$. So T permutes the sets P_i. This permutation decomposes into the cyclic permutations we were looking for. More precisely: consider the partition of $\overline{\mathrm{Per}(T)}$ into sets of the form

$$\bigcup_{n=0}^{\infty} T^n(\overline{P_i}), \; i = 1, \ldots, N.$$

The unions are in fact formed over finite families. It does not matter in which place the closure is placed in these unions, because X is compact, so for every $A \subset X$ we have $T(\overline{A}) = \overline{T(A)}$. We consider this partition as a partition into the Ω_j's we were looking for. The Ω_j^k's are the summands $T^n(\overline{P_i})$ in the unions.

Observe now that T is topologically transitive on each Ω_j.

Indeed, if periodic x, y belong to the same Ω_j, there exist $x' \in B(x,\xi)$ and $y' \in B(y,\xi)$ such that $T^n(x') = T^{n_0}(y)$ and $T^m(y') = T^{m_0}(x)$ for some natural numbers n, m and $n_0 \leq k_y, m_0 \leq k_x$. For an arbitrary $\beta > 0$ choose $\alpha > 0$ from the definition of periodic shadowing, and consider x'', y'' such that $\rho(x'', x) \leq \alpha$, $\rho(y'', y) \leq \alpha$ and $T^{n_1}(x'') = x'$, $T^{m_1}(y'') = y'$ for some natural numbers n_1, m_1, existing by the expanding property at $\overline{\mathrm{Per}(T)}$. Then the sequence of points $T(x''), \ldots, T^{n_1+n+k_y-n_0}(x''), T(y''), \ldots, T^{m_1+m+k_x-m_0}(y'')$ is a periodic α-pseudo-orbit, of period $n_1 + n + k_y - n_0 + m_1 + m + k_x - m_0$, so it can be β-shadowed by a periodic orbit. Thus there exists $z \in \mathrm{Per}(T)$ such that $\rho(z, x) \leq \beta$ and $\rho(T^N(z), y) \leq \beta$ for an integer $N > 0$. Now take arbitrary open sets U and V in X intersecting Ω_j and consider periodic points $x \in \Omega_j \cap U$ and $y \in \Omega_j \cap V$.

Take β such that $B(x,\beta) \subset U$ and $B(y,\beta) \subset V$. We find a periodic point z as above. Note that, provided $\beta \leq \xi$, $z \sim x$ and $T^N(z) \sim y$. We obtain $T^N(z) \in T^N(U \cap \Omega_j) \cap (V \cap \Omega_j)$, so this set is non-empty. This proves the topological transitivity of $T|_{\Omega_j}$.

Note that we proved that the orbits (their finite parts) $x'', \ldots, T^{n_1}(x'') = x', \ldots, T^n(x')$ and $y'', \ldots, T^{m_1}(y'') = y', \ldots, T^m(y')$, with n_1, n and m_1, m arbitrarily large, can be arbitrarily well shadowed by parts of periodic orbits. This corresponds to the approximation of a transversal homoclinic orbit or of cycles of transversal heteroclinic orbits by periodic ones, in the hyperbolic theory for diffeomorphisms (see also Exercise 4.6).

This analogy justifies the name *heteroclinic cycle points* for the points x' and y', or *heteroclinic cycle orbits* for their orbits as discussed above. Thus we have proved:

Lemma 4.3.9. *Under the assumptions of Theorem 4.3.8, every heteroclinic cycle point is a limit of periodic points.*

The following is interesting in itself:

Lemma 4.3.10. $T|_{\overline{\mathrm{Per}(T)}}$ *is an open map.*

Proof. Fix $x, y \in \overline{\mathrm{Per}(T)}$ and $\rho(T(x), y) \le \varepsilon \le \xi/3$. Since T is open, by Lemma 4.1.2, and because of the expanding property at $\overline{\mathrm{Per}(T)}$, there exists $\hat{y} = T_x^{-1}(y) \in B(x, \lambda^{-1}\xi/3)$. We want to prove that $\hat{y} \in \overline{\mathrm{Per}(T)}$.

There exist $z_1, z_2 \in \mathrm{Per}(T)$ such that $\rho(z_1, x) \le \lambda^{-1}\xi/3$ and $\rho(z_2, y) \le \xi/3$. Hence $\rho(T(z_1), z_2) \le \xi$, hence $T(z_1) \sim z_2$, and hence z_1 and z_2 belong to the same Ω_j. Then $T_x^{-1}(z_2)$ is a heteroclinic cycle point, so by Lemma 4.3.9 $T_x^{-1}(z_2)$, and hence $\hat{y}$, are limits of periodic points. ♣

Continuation of the proof of Theorem 4.3.8. We can now prove the topological exactness of $T^{k(j)}|_{\Omega_j^k}$. Fix $\Omega_j^k = \overline{P_i}$ with $T^{k(j)}(P_i) = P_i$. Let $\{x_s\}, s = 1, \dots, S$ be a $\xi'/2$-spanning set in P_i, where ξ' is a constant having the properties of ξ for the map $T|_{\overline{\mathrm{Per}}}$, existing by the openness of $T|_{\overline{\mathrm{Per}(T)}}$ (Lemmas 4.1.2 and 4.3.10). Write $k(P_i) = \prod_{s=1}^{S} k_{x_s}$. Take an arbitrary open set $U \subset \overline{P_i}$. It contains a periodic point x.

Note that for every ball $B = B(y, r)$ in $\overline{\mathrm{Per}(T)}$ with the origin at $y \in \mathrm{Per}(T)$ and radius r less than η and $\lambda^{-k_y}\xi'$, we have $T^{k_y}(B) \supset B(y, \lambda^{k_y} r)$. Repeating this step by step, we obtain $T^{nk(y)}(B) \supset B(y, \xi')$: see (4.1.7).

Let us go back to U and consider $B_x = B(x, r) \subset U$ with $r \le \lambda^{-k(P_i)}\xi'$. Then $T^{nk(P_i)}(B_x)$ is an increasing family of sets for $n = 0, 1, 2, \dots$.

By the definition of $\sim$, the set $\bigcup_{n \ge 0} T^{nk(P_i)}(B_x)$ contains $\{x_s : s = 1, \dots, S\}$, because the points x_s are in the relation $\sim$ with x. This uses the fact proved above (see Lemma 4.3.9) that x' in the definition of $\sim$, such that $T^m(x') = T^m(x_s)$, belongs to $\overline{\mathrm{Per}(T)}$. It belongs even to $\overline{P_i}$, since for $z \in \mathrm{Per}(T)$ close to x' we have $z \sim x_s$, with the use of the same x' as that of heteroclinic cycle points. Hence, by the observation above, $\bigcup_{n \ge 0} T^{nk(P_i)}(B_x)$ contains the ball $B(x_s, \xi')$ for each s. So it contains $\overline{P_i}$. Since $T^{nk(P_i)}(B_x)$ is an increasing family of open sets in $\overline{\mathrm{Per}(T)}$ that is compact, just one of these sets covers $\overline{\mathrm{Per}(T)}$. The topological exactness and therefore Theorem 4.3.8 is proved. ♣

Remark 4.3.11. In Theorem 4.3.8 one can replace the assumption of periodic shadowing by just $\overline{\mathrm{Per}\, T} = \Omega(T)$. (By analogy to Axiom A diffeomorphisms we can call an open map $T : X \to X$ expanding on $\Omega(T)$, and such that $\overline{\mathrm{Per}(T)} = \Omega(T)$, an *Axiom A Ω-distance-expanding map*.)

Indeed, in the proof of Theorem 4.3.8 we used shadowing only to approximate heteroclinic cycle points by periodic ones. It is sufficient, however, to notice that heteroclinic cycle points are non-wandering, by the openness of T.

(In particular, periodic shadowing is not needed in Lemma 4.3.9 to conclude the non-wandering.)

This yields topological transitivity of each $T|_{\Omega_j}$ with the proof as before. We find the periodic point z by $\overline{\mathrm{Per}(T)} = \Omega(T)$.

We do not know whether expanding on $\Omega(T)$ implies $\Omega(T) = \overline{\mathrm{Per}(T)}$; for diffeomorphisms hyperbolic on Ω it does not.

As a corollary, we obtain the following two theorems.

Theorem 4.3.12. *Let $T : X \to X$ be a continuous mappping for X a compact metric space. Assume that T is open distance-expanding, or at least expanding at the set $\overline{\mathrm{Per}(T)}$ satisfying the periodic shadowing property. Then, if T is topologically transitive, or is surjective and its spectral decomposition consists of just one set $\Omega_1 = \bigcup_{k=1}^{k(1)} \Omega_1^k$, the following properties hold:*

(1) The set of periodic points is dense in X, which is thus equal to Ω_1.

(2) For every open $U \subset X$ there exists $N = N(U)$ such that $\bigcup_{j=0}^{N} T^j(U) = X$.

(3) $(\forall r > 0)(\exists N)(\forall x \in X) \bigcup_{j=0}^{N} T^j(B(x,r)) = X$.

(4) The following specification property holds: for every $\beta > 0$ there exists a positive integer N such that for every $n \geq 0$ and every T-orbit $(x_0, \dots x_n)$ there exists a periodic point y of period not larger than $n+N$ whose orbit for the times $0, \dots, n$ β-shadows $(x_0, \dots x_n)$.

Proof. By topological transitivity, for every open set U there exists $n \geq 1$ such that $T^n(U) \cap U \neq \emptyset$ (use condition (2) in Proposition 4.3.4 for $N = 1$). Hence for the set Ω of non-wandering points we have $\Omega = X$. This gives the density of $\mathrm{Per}(T)$ by Proposition 4.3.6.

If we assume only that there is one $\Omega_1 (= \Omega = \overline{\mathrm{Per}(T)})$ in the spectral decomposition, then for an arbitrary $z \in X$ we find by the surjectivity of T an infinite backward orbit z_{-n} of z. Note that $z_{-n} \to \Omega$ and $T^n(z) \to \Omega$, which follows easily from the definition of Ω. So for every $\alpha > 0$ there exist $w_1, w_2 \in \mathrm{Per}(T)$ and natural numbers k, n such that $T^k(w_2) \sim w_1$, $\rho(w_1, z_{-n}) \leq \alpha$, and $\rho(w_2, T^n(z)) \leq \alpha$. This allows us to find a periodic point in $B(z, \beta)$, where $\beta > 0$ is arbitrarily small and α is chosen for β from the periodic shadowing property.

We conclude that $X = \bigcup_{j=1}^{J} \Omega_j$, each Ω_j is T-invariant, closed, and also open since Ω_j's are at least ξ-distant from each other. So $J = 1$. Otherwise, by topological transitivity, for $j \neq i$ there exists n such that $T^n(\Omega_j) \cap \Omega_i \neq \emptyset$, which would contradict the T-invariance of Ω_j.

Thus $X = \bigcup_{k=1}^{k(1)} (\Omega_1^k)$, and assertion (2) follows immediately from the topological exactness of $T^{k(1)}$ on each set Ω_1^k, $k = 1, \dots, k(1)$.

Property (3) follows from (2), where given r we choose $N = \max\{N(U)\}$ associated to a finite cover of X by sets U of diameter not exceeding $r/2$. Indeed, then for every $B(x,r)$ the set U containing x is a subset of $B(x,r)$.

Now let us prove the specification property. By property (3), for every $\alpha > 0$ there exists $N = N(\alpha)$ such that for every $v, w \in X$ there exists $m \leq N$ and $z \in B(v, \alpha)$ such that $T^m(z) \in B(w, \alpha)$.

Consider any T-orbit $x_0, \ldots x_n$. Then consider an α-pseudo-orbit $x_0, \ldots x_{n-1}, z, \ldots, T^{m-1}(z)$ with $m \leq N$ and $z \in B(x_n, \alpha, T^m(z)) \in B(x_0, \alpha)$. By Corollary 4.2.5 we can β-shadow it by a periodic orbit of period $n + m \leq n + N$. ♣

The same proof yields this.

Theorem 4.3.13. *Let T satisfy the assumptions of Theorem 4.3.12, and let it be also topologically mixing: that is, $k(1) = 1$. Then*

(1) T is topologically exact: that is, for every open $U \subset X$ there exists $N = N(U)$ such that $T^N(U) = X$.

(2) $(\forall r > 0)(\exists N)(\forall x \in X)\ \ T^N(B(x, r)) = X$.

4.4 Hölder continuous functions

For distance-expanding maps, Hölder continuous functions play a special role. Recall that a function $\phi : X \to \mathbb{C}$ (or $\mathbb{R}$) is said to be *Hölder continuous* with an exponent $0 < \alpha \leq 1$ if and only if there exists $C > 0$ such that

$$|\phi(y) - \phi(x)| \leq C\rho(y, x)^\alpha$$

for all $x, y \in X$. All Hölder continuous functions are continuous; if $\alpha = 1$ they are usually called *Lipschitz continuous.*

Let $C(X)$ denote, as in previous chapters, the space of all continuous, real or complex-valued functions defined on a compact metric space X, and for $\psi : X \to \mathbb{C}$ we write $||\psi||_\infty := \sup\{|\psi(x)| : x \in X\}$ for its supremum norm. For any $\alpha > 0$ let $\mathcal{H}_\alpha(X)$ denote the space of all Hölder continuous functions with exponent $\alpha > 0$. If $\psi \in \mathcal{H}_\alpha(X)$, let

$$\vartheta_{\alpha,\xi}(\psi) = \sup\left\{\frac{|\psi(y) - \psi(x)|}{\rho(y, x)^\alpha} : x, y \in X,\ x \neq y,\ \rho(x, y) \leq \xi\right\}$$

and

$$\vartheta_\alpha(\psi) = \sup\left\{\frac{|\psi(y) - \psi(x)|}{\rho(y, x)^\alpha} : x, y \in X, x \neq y\right\}.$$

Note that

$$\vartheta_\alpha(\psi) \leq \max\left\{\frac{2||\psi||_\infty}{\xi^\alpha}, \vartheta_{\alpha,\xi}(\psi)\right\}.$$

The reader can check easily that $\mathcal{H}_\alpha(X)$ becomes a Banach space when equipped with the norm

$$||\psi||_{\mathcal{H}_\alpha} = \vartheta_\alpha(\psi) + ||\psi||_\infty.$$

Thus, in future, to estimate $||\psi||_{\mathcal{H}_\alpha}$ it is sufficient to estimate $\vartheta_{\alpha,\xi}(\psi)$ and $||\psi||_\infty$.

The following result is a straightforward consequence of the Arzela–Ascoli Theorem.

Theorem 4.4.1. *Any bounded subset of the Banach space $\mathcal{H}_\alpha(X)$ with the norm $\|\cdot\|_{\mathcal{H}_\alpha}$ is relatively compact as a subset of the Banach space $C(X)$ with the supremum norm $\|\cdot\|_\infty$. Moreover, if $\{\psi_n : n = 1, 2, \ldots\}$ is a sequence of continuous functions in $\mathcal{H}_\alpha(X)$ such that $\|x_n\|_{\mathcal{H}_\alpha} \le C$ for all $n \ge 1$ and some constant C, and if $\lim_{n\to\infty} \|\psi_n - \psi\|_\infty = 0$ for some $\psi \in C(X)$, then $\psi \in \mathcal{H}_\alpha(X)$ and $\|\psi\|_{\mathcal{H}_\alpha} \le C$.*

Now let us formulate a simple but very basic lemma that demonstrates coherence of the expanding property of T and the Hölder continuity property of a function.

Lemma 4.4.2 (Pre-Bounded Distortion Lemma for Iteration)**.** *Let $T : X \to X$ be a distance-expanding map and $\phi : X \to \mathbb{C}$ be a Hölder continuous function with the exponent α. Then for every positive integer n and all $x, y \in X$ such that*

$$\rho(T^j(x), T^j(y)) < 2\eta \quad \textit{for all} \quad j = 0, 1, \ldots, n-1, \tag{4.4.1}$$

we have, with $C(T,\phi) := \left(\frac{\vartheta_\alpha(\phi)}{1-\lambda^{-\alpha}}\right)$,

$$|S_n\phi(x) - S_n\phi(y)| \le C(T,\phi)\rho(T^n(x), T^n(y))^\alpha, \tag{4.4.2}$$

where $S_n\phi(z) := \sum_{j=0}^{n-1} \phi \circ T^j(x)$.

If T is open we can assume $x, y \in T_z^{-n}(B(T^n(z), \xi)$ for a point $z \in X$, instead of (4.4.1). Then in (4.4.2) we can replace ϑ_α by $\vartheta_{\alpha,\xi}$.

The point of (4.4.2) is that the coefficient $C(T,\phi) = \frac{\vartheta_\alpha(\phi)}{1-\lambda^{-\alpha}}$ does not depend on x, y or on n.

Proof. By (4.0.2) we have $\rho(T^j(x), T^j(y)) \le \lambda^{-(n-j)}\rho(T^n(y), T^n(z))$ for every $0 \le j \le n$. Hence

$$|\phi(T^j(y)) - \phi(T^j(z))| \le \vartheta_\alpha(\phi)\lambda^{-(n-j)\alpha}\rho(T^n(y), T^n(z))^\alpha.$$

Thus

$$\begin{aligned}
|S_n\phi(y) - S_n\phi(z)| &\le \vartheta_\alpha(\phi)\rho(T^n(y), T^n(z))^\alpha \sum_{j=0}^{n-1} \lambda^{-(n-j)\alpha} \\
&\le \vartheta_\alpha(\phi)\rho(T^n(y), T^n(z))^\alpha \sum_{j=0}^{\infty} \lambda^{-j\alpha} \\
&= \frac{\vartheta_\alpha(\phi)}{1-\lambda^{-\alpha}}\rho(T^n(y), T^n(z))^\alpha.
\end{aligned}$$

The proof is complete. ♣

For an open distance-expanding topologically transitive map we can replace topological pressure, defined in Chapter 3, by a corresponding notion related to a ‘tree’ of pre-images of an arbitrary point.

Proposition 4.4.3. *If $T : X \to X$ is a topologically transitive open distance-expanding map, then for every Hölder continuous potential $\phi : X \to \mathbb{R}$ and for every $x \in X$ there exists the limit*

$$\mathrm{P}_x(T,\phi) := \lim_{n\to\infty} \frac{1}{n} \log \sum_{\overline{x}\in T^{-n}(x)} \exp S_n\phi(\overline{x}),$$

and it is equal to the topological pressure $\mathrm{P}(T,\phi)$. In addition, there exists a constant C such that for all $x, y \in X$ and every positive integer n

$$\frac{\sum_{\overline{x}\in T^{-n}(x)} \exp S_n\phi(\overline{x})}{\sum_{\overline{y}\in T^{-n}(y)} \exp S_n\phi(\overline{y})} < C. \tag{4.4.3}$$

Proof. If $\rho(x,y) < \xi$ then (4.4.3) follows immediately from Lemma 4.4.2 with the constant, $C = C_1 := \exp(C(T,\phi)\xi^\alpha)$, since this is the bound for the ratio of corresponding summands for each backward trajectory, by Lemma 4.4.2. Now observe that by the topological transitivity of T there exists N (depending on ξ) such that for all $x, y \in X$ there exists $0 \le m < N$ such that $T^m(B(x,\xi)) \cap B(y,\xi) \ne \emptyset$. Indeed, by condition (3) in Proposition 4.3.4 we can find two blocks of a trajectory of z with dense ω-limit set, say $T^k(z), \dots, T^{k'}(z)$ and $T^l(z), \dots, T^{l'}(z)$ with $l > k'$, each ξ-dense in X. Then we set $N = l' - k$. We can find t between k and k' and s between l and l' so that $T^t(z) \in B(x,\xi)$ and $T^s(z) \in B(y,\xi)$. We have $m := s - t \le N$.

Now fix arbitrary $x, y \in X$. There therefore exists a point $y' \in T^{-m}(B(y,\xi)) \cap B(x,\xi)$. We then have

$$\begin{aligned}
\sum_{\overline{x}\in T^{-n}(x)} \exp S_n\phi(\overline{x}) &\le C_1 \sum_{\overline{y'}\in T^{-n}(y')} \exp S_n\phi(\overline{y'}) \\
&= C_1 \exp(-S_m\phi(T^n(y'))) \sum_{\overline{y'}\in T^{-n}(y')} \exp S_{n+m}\phi(\overline{y'}) \\
&\le C_1 \exp(-m \inf \phi) \sum_{\overline{y'}\in T^{-(n+m)}(T^m(y'))} \exp S_{n+m}\phi(\overline{y'}) \\
&\le C_1 \exp(-m \inf \phi) \sum_{\overline{y'}\in T^{-(n+m)}(T^m(y'))} \exp S_n\phi(T^m(\overline{y'})) \exp S_m\phi(\overline{y'}) \\
&\le C_1 \exp(m \sup \phi - m \inf \phi) \sum_{\overline{y'}\in T^{-(n+m)}(T^m(y'))} \exp S_n\phi(T^m(\overline{y'})) \\
&\le C_1 \exp(2N||\phi||_\infty) D^N \sum_{\overline{y'}\in T^{-n}(T^m(y'))} \exp S_n\phi(\overline{y'}) \\
&\le C_1^2 \exp(2N||\phi||) D^N \sum_{\overline{y}\in T^{-n}(y)} \exp S_n\phi(\overline{y}),
\end{aligned}$$

where $D = \sup\{\#(T^{-1}(z)) : z \in X\} < \infty$. This proves (4.4.3).

Observe that each set $T^{-n}(x)$ is $(n, 2\eta)$-separated, whence

$$\limsup_{n\to\infty} \frac{1}{n} \log \sum_{\overline{x}\in T^{-n}(x)} \exp S_n\phi(\overline{x}) \le \mathrm{P}(T,\phi),$$

by the characterization of pressure given in Theorem 3.3.2.

In order to prove the opposite inequality, fix $\varepsilon < 2\xi$, and for every $n \ge 1$ an (n,ε)-separated set F_n. Cover X by finitely many balls

$$B(z_1,\varepsilon/2), B(z_2,\varepsilon/2), \ldots, B(z_k,\varepsilon/2).$$

Then $F_n = F_n \cap \left(\bigcup_{j=1}^k T^{-n}(B(z_j,\varepsilon/2))\right)$, and therefore

$$\sum_{z\in F_n} \exp(S_n\phi(z)) \le \sum_{j=1}^k \sum_{F_n\cap T^{-n}(B(z_j,\varepsilon/2))} \exp(S_n\phi(z)).$$

Consider an arbitrary j and $y \in F_n \cap T^{-n}(B(z_j,\varepsilon/2))$. Let $\overline{z_{j,y}} \in T^{-n}(z_j)$ be defined by $y \in T^{-n}_{\overline{z_{j,y}}}(B(z_j,\varepsilon/2)$. We shall show that the function $y \mapsto \overline{z_{j,y}}$ is injective. Indeed, suppose that $\overline{z_j} = \overline{z_{j,a}} = \overline{z_{j,b}}$ for some $a, b \in F_n \cap T^{-n}(B(z_j,\varepsilon/2))$. Then

$$\rho(T^l(a), T^l(b)) \le \rho(T^l(a), T^l(\overline{z_j})) + \rho(T^l(\overline{z_j}), T^l(b)) \le \frac{\varepsilon}{2} + \frac{\varepsilon}{2} = \varepsilon$$

for every $0 \le l \le n$. So $a = b$, since F_n is (n,ε)-separated.

Hence, using Lemma 4.4.2 (compare (4.4.3)), we obtain

$$\sum_{z\in F_n} \exp(S_n\phi(z)) \le \sum_{j=1}^k C \sum_{\overline{z_j}} \exp(S_n\phi(\overline{z_j})) \le kC^2 \sum_{\overline{x}\in T^{-n}(x)} \exp(S_n\phi(\overline{x})).$$

Letting $n \nearrow \infty$, next $\varepsilon \to 0$, and then applying Theorem 3.3.2, we therefore get

$$\mathrm{P}(T,\phi) \le \liminf_{n\to\infty} \frac{1}{n} \log \sum_{\overline{x}\in T^{-n}(x)} \exp S_n\phi(\overline{x}).$$

Thus

$$\liminf_{n\to\infty} \frac{1}{n} \log \sum_{\overline{x}\in T^{-n}(x)} \exp S_n\phi(\overline{x}) \ge \mathrm{P}(T,\phi) \ge \limsup_{n\to\infty} \frac{1}{n} \log \sum_{\overline{x}\in T^{-n}(x)} \exp S_n\phi(\overline{x}).$$

So lim inf = lim sup above, the limit exists and is equal to $\mathrm{P}(T,\phi)$. ♣

Remark 4.4.4. It follows from Proposition 4.4.3, the proof of the Variational Principle Part II (see Section 3.4), and the expansiveness of T that for every $x \in X$ every weak limit of the measures $\frac{1}{n}\sum_{k=0}^{n-1} \mu_n \circ T^{-k}$, for

$$\mu_n = \frac{\sum_{\overline{x}\in T^{-n}(x)} \delta_x \exp S_n\phi(\overline{x})}{\sum_{\overline{x}\in T^{-n}(x)} \exp S_n\phi(\overline{x})},$$

and δ_x denoting the Dirac measure concentrated at the point x, is an equilibrium state for ϕ. In fact, in our very special situation we can say a lot more about the measures involved. Chapter 5 will be devoted to this end.

Let us finish this section with one more very useful fact (compare Theorem 2.11.3.)

Proposition 4.4.5. *Let $T : X \to X$ be an open, distance-expanding, topologically transitive map. If $\phi, \psi \in \mathcal{H}_\alpha(X)$, then the following conditions are equivalent.*

(1) If $x \in X$ is a periodic point of T, and if n denotes its period, then $S_n\phi(x) - S_n\psi(x) = 0$.

(2) There exists a constant $C > 0$ such that for every $x \in X$ and integer $n \geq 0$, we have $|S_n\phi(x) - S_n\psi(x)| \leq C$.

(3) There exists a function $u \in \mathcal{H}_a$ such that $\phi - \psi = u \circ T - u$.

Proof. The implications (3) $\Longrightarrow$ (2) $\Longrightarrow$ (1) are very easy. The first is obtained by summing up the equation in (3) along the orbit $x, T(x), \ldots, T^{n-1}(x)$, which gives $C = 2\sup|u|$. The second holds because otherwise, if $S_n\phi(x) - S_n\psi(x) = K \neq 0$ for x of period n, then we have $S_{jn}\phi(x) - S_{jn}\psi(x) = jK$, which contradicts (2) for j large enough. Now let us prove (1) $\Longrightarrow$ (3). Let $x \in X$ be a point such that for every $N \geq 0$ the orbit $(x_n)_N^\infty$ is dense in X. Such x exists by topological transitivity of T: see Proposition 4.3.4. Write $\eta = \phi - \psi$. Define u on the forward orbit of x, the set $A = \{T^n(x)\}_0^\infty$ by $u(x_n) = S_n\eta(x)$. If x is periodic then X is just the orbit of x, and the function u is well defined owing to the equality in (1). So, suppose that x is not periodic. Set $x_n = T^n(x)$. Then $x_n \neq x_m$ for $m \neq n$: hence u is well defined on A. We shall show that it extends in a Hölder continuous manner to $\overline{A} = X$. Indeed, if we take points $x_m, x_n \in A$ such that $m < n$ and $\rho(x_m, x_n) < \varepsilon$ for ε small enough, then $x_m, \ldots, x_{n-1}$ can be β-shadowed by a periodic orbit $y, \ldots, T^{n-m-1}(y)$ of period $n - m$ by Corollary 4.2.5, where ε is related to β in the same way as α is related to β in that corollary. Then by Lemma 4.4.2,

$$\begin{aligned} |u(x_n) - u(x_m)| &= |S_n\eta(x) - S_m\eta(x)| = |S_{n-m}\eta(x_m)| \\ &= |S_{n-m}\eta(x_m) - S_{n-m}\eta(y)| \leq \vartheta(\phi)_\alpha \varepsilon^\alpha. \end{aligned}$$

In particular, we proved that u is uniformly continuous on A, which allows us to extend u continuously to $\overline{A}$. By taking limits we see that this extension satisfies the same Hölder estimate on $\overline{A}$ as on A. Also, the equality in (3), true on A, extends to $\overline{A}$ by the definition of u and by the continuity of η and u . The proof is complete. ♣

Remark 4.4.6. The equality in (3) is called the *cohomology equation*, and u is a solution of this equation: compare Section 2.11. Here the cohomology equation is solvable in the space $\mathcal{K} = \mathcal{H}_\alpha$. Note that in proving (3) $\Longrightarrow$ (2) we used only the assumption that u is bounded. So, going through (2) $\Longrightarrow$ (1) $\Longrightarrow$ (3) we prove that if the cohomology equation is solvable with u bounded, then automatically $u \in \mathcal{H}_\alpha$. The reader will see later that frequently, even under assumptions of T weaker than expanding, to prove that u is a 'good' function it suffices to assume u to be measurable and finite almost everywhere, for some probability T-invariant measure with support X. Often u is forced to be as regular as ϕ and ψ are. These types of theorem are called *Livšic-type theorems.*

4.5 Markov partitions and symbolic representation

We shall prove in this section that the topological Markov chains (Chapter 1, Example 1.3) describe quite precisely the dynamics of general open expanding maps.

This can be done through so-called Markov partitions of X. The sets of a partition will play the role of 'cylinders' $\{i_0 = \text{Const}\}$ in the symbolic space Σ_A.

Definition 4.5.1. A finite cover $\mathfrak{R} = \{R_1, \ldots, R_n\}$ of X is said to be a *Markov partition* of the space X for the mapping T if $\operatorname{diam}(\mathfrak{R}) < \min\{\eta, \xi\}$ and the following conditions are satisfied.

(a) $R_i = \overline{\operatorname{Int} R_i}$ for all $i = 1, 2, \ldots, d$.

(b) $\operatorname{Int} R_i \cap \operatorname{Int} R_j = \emptyset$ for all $i \neq j$.

(c) $\operatorname{Int} R_j \cap T(\operatorname{Int} R_i) \neq \emptyset \implies R_j \subset T(R_i)$ for all $i, j = 1, 2, \ldots, d$.

Theorem 4.5.2. *For an open, distance-expanding map $T : X \to X$ there exist Markov partitions of arbitrarily small diameters.*

Proof. Fix $\beta < \min\{\eta/4, \xi\}$ and let α be the number associated to β as in Lemma 4.2.3. Choose $0 < \gamma \leq \min\{\beta/2, \alpha/2\}$ so small that

$$\rho(x, y) \leq \gamma \implies \rho(T(x), T(y)) \leq \alpha/2. \tag{4.5.1}$$

Let $E = \{z_1, \ldots, z_r\}$ be a γ-spanning set of X. Define the space Ω by putting

$$\Omega = \{q = (q_i) \in E^{\mathbb{Z}^+} : \rho(T(q_i), q_{i+1}) \leq \alpha \text{ for all } i \geq 0\}.$$

By definition, all elements of the space Ω are α-pseudo-orbits, and therefore, in view of Corollary 4.2.4 and Lemma 4.2.3, for every sequence $q \in \Omega$ there exists a unique point whose orbit β-shadows q. Denote this point by $\Theta(q)$. In this way we have defined a map $\Theta : \Omega \to X$. We shall need some of its properties.

Let us show first that Θ is surjective. Indeed, since E is a γ-spanning set, for every $x \in X$ and every $i \geq 0$ there exists $q_i \in E$ such that

$$\rho(T^i(x), q_i) < \gamma.$$

Therefore, using also (4.5.1),

$$\rho(T(q_i), q_{i+1}) \leq \rho(T(q_i), T(T^i(x))) + \rho(T^{i+1}(x), q_{i+1}) < \alpha/2 + \gamma \leq \alpha/2 + \alpha/2 = \alpha$$

for all $i \geq 0$. Thus $q = (q_i)_{i=0}^{\infty} \in \Omega$ and (as $\gamma < \beta$) $x = \Theta(q)$. The surjectivity of Θ is proved.

Now we shall show that Θ is continuous. For this aim we shall need the following notation. If $q \in \Omega$ then we put

$$q(n) = \{p \in \Omega : p_i = q_i \text{ for every } i = 0, 1, \ldots, n\}. \tag{4.5.2}$$

To prove continuity suppose that $p, q \in \Omega$, $p(n) = q(n)$ with some $n \geq 0$, and put $x = \Theta(q)$, $y = \Theta(p)$. Then for all $i = 0, 1, \ldots, n$,

$$\rho(T^i(x), T^i(y)) \leq \rho(T^i(x), q_i) + \rho(p_i, T^i(y)) \leq \beta + \beta = 2\beta.$$

As $\beta < \eta$, we therefore obtain by (4.0.2) that

$$\rho(T^{i+1}(x), T^{i+1}(y)) \geq \lambda \rho(T^i(x), T^i(y))$$

for $i = 0, 1, \ldots, n-1$, (see (4.1.6)), and consequently $\rho(x, y) \leq \lambda^{-n} 2\beta$. The continuity of Θ is proved.

Now for every $k = 1, \ldots, r$ define the sets

$$P_k = \Theta(\{q \in \Omega : q_0 = z_k\}).$$

Since Θ is continuous, Ω is a compact space, and the sets $\{q \in \Omega : q_0 = z_k\}$ are closed in Ω, all sets P_k are closed in X.

Denote

$$W(k) = \{l : \rho(T(z_k), z_l) \leq \alpha\}.$$

The following basic property is satisfied:

$$T(P_k) = \bigcup_{l \in W(k)} P_l. \tag{4.5.3}$$

Indeed, if $x \in P_k$ then $x = \Theta(q)$ for $q \in \Omega$ with $q_0 = z_k$. By the definition of Ω we have $q_1 = z_l$ for some $l \in W(k)$. We obtain $T(x) \in P_l$.

Conversely, let $x \in P_l$ for $l \in W(k)$. This means that $x = \Theta(q)$ for some $q \in \Omega$ with $q_0 = z_l$. By the definition of $W(k)$ the concatenation $z_k q$ belongs to Ω, and therefore the point $T(\Theta(z_k q))$ β-shadows q. Thus $T(\Theta(z_k q)) = \Theta(q) = x$, and hence $x \in T(P_k)$.

Let now

$$Z = X \setminus \bigcup_{n=0}^{\infty} T^{-n}\Big(\bigcup_{k=1}^{r} \partial P_k\Big).$$

Note that the boundary set $\partial P_k := \overline{P_k} \setminus \operatorname{Int} P_k$ is closed, by definition. It is also nowhere dense, since P_k itself is closed. Indeed, by the definition of interior, each point in ∂P_k is a limit of a sequence of points belonging to $X \setminus P_k$, and hence belonging to $X \setminus \overline{P_k}$, but not belonging to ∂P_k. Since T is open, also all the sets $T^{-n}(\partial P_k)$ are nowhere dense. They are closed by the continuity of T. We conclude, referring to the Baire Theorem, that Z is dense in X; its complement is of the first Baire category.

For any $x \in Z$ denote

$$P(x) = \{k \in \{1, \ldots, r\} : x \in P_k\},$$

$$Q(x) = \Big\{l \notin P(x) : P_l \cap \big(\bigcup_{k \in P(x)} P_k\big) \neq \emptyset\Big\},$$

and

$$S(x) = \bigcap_{k \in P(x)} \operatorname{Int} P_k \setminus \left(\bigcup_{k \in Q(x)} P_k \right) = \bigcap_{k \in P(x)} \operatorname{Int} P_k \setminus \left(\bigcup_{k \notin P(x)} P_k) \right).$$

We shall show that the family $\{S(x) : x \in Z\}$ is in fact finite and, moreover, that the family $\{\overline{S(x)} : x \in Z\}$ is a Markov partition of diameter not exceeding 2β.

Indeed, since $\operatorname{diam}(P_k) \le 2\beta$ for every $k = 1, \ldots, r$ we have

$$\operatorname{diam}(S(x)) \le 2\beta. \tag{4.5.4}$$

As the sets $S(x)$ are open, we have

$$\overline{\operatorname{Int} \overline{S(x)}} = \overline{S(x)} \tag{4.5.5}$$

for all $x \in Z$. This proves the property (a) in Definition 4.5.1.

We shall now show that for every $x \in Z$

$$T(S(x)) \supset S(Tx). \tag{4.5.6}$$

Note first that for $K(x) := \bigcup_{k \in P(x)} P_k \cup \bigcup_{l \in Q(x)} P_l$ we have $\operatorname{diam}(K(x)) \le 8\beta$ and therefore, by the assumption $\beta < \eta/4$, the map T restricted to $K(x)$ (and even to its neighbourhood U) is injective.

Consider $k \in P(x)$. Then there exists $l \in W(k)$ such that $T(x) \in P_l$ (see (4.5.3)), and using the definition of Z we get $T(x) \in \operatorname{Int}(P_l)$. Using the injectivity of $T|_U$ and the continuity of T, and then (4.5.3), we obtain $\operatorname{Int} P_k \supset T|_U^{-1}(\operatorname{Int}(T(P_k))$: hence

$$T(\operatorname{Int} P_k) \supset \operatorname{Int}(T(P_k)) \supset \operatorname{Int} P_l \supset S(T(x)),$$

and therefore

$$T\left(\bigcap_{k \in P(x)} \operatorname{Int} P_k \right) \supset S(T(x)). \tag{4.5.7}$$

Now consider $k \in Q(x)$. Observe that by the injectivity of $T|_{K(x)}$ the assumption $x \notin P_k$ implies $T(x) \notin P_l$, $l \in W(k)$.

Thus, using (4.5.3), we obtain

$$T(P_k) \subset \bigcup_{l \notin P(T(x))} P_l.$$

Hence

$$T\left(\bigcup_{l\in Q(x)} P_l\right)\cap S(T(x))=\emptyset.$$

Combining this and (4.5.7) gives

$$T\left(\bigcap_{k\in P(x)} \operatorname{Int} P_k \setminus \left(\bigcup_{k\in Q(x)} P_k\right)\right)\supset S(T(x)),$$

which means that formula (4.5.6) is satisfied, and therefore

$$T(\overline{S(x)})\supset \overline{S(Tx)}. \tag{4.5.8}$$

We shall now prove the following claim.

Claim. If $x,y\in Z$ then either $S(x)=S(y)$ or $S(x)\cap S(y)=\emptyset$.

Indeed, if $P(x)=P(y)$ then also $Q(x)=Q(y)$ and consequently $S(x)=S(y)$. If $P(x)\neq P(y)$ then there exists $k\in P(x)\div P(y)$, say $k\in P(x)\setminus P(y)$. Hence $S(x)\subset \operatorname{Int} P_k$ and $S(y)\subset X\setminus P_k$. Therefore $S(x)\cap S(y)=\emptyset$ and the claim is proved.

(One can write the family $S(x)$ as $\bigvee_{k=1,\dots,r}\{\operatorname{Int} P_k, X\setminus P_k\}$: compare notation in Chapter 2. Then the assertion of the claim is immediate.)

Since the family $\{P(x):x\in Z\}$ is finite, so is the family $\{S(x):x\in Z\}$. Note that $S(x)\cap S(y)=\emptyset$ implies $\operatorname{Int}\overline{S(x)}\cap \operatorname{Int}\overline{S(y)}=\emptyset$. This is a general property of pairs of open sets: $U\cap V=\emptyset$ implies $\overline{U}\cap V=\emptyset$ implies $\operatorname{Int}\overline{U}\cap V=\emptyset$ implies $\operatorname{Int}\overline{U}\cap\overline{V}=\emptyset$ implies $\operatorname{Int}\overline{U}\cap\operatorname{Int}\overline{V}=\emptyset$.

Since $\bigcup_{x\in Z} S(x)\supset Z$ and Z is dense in X, we thus have $\bigcup_{x\in Z}\overline{S(x)}=X$. That the family $\{\overline{S(x)}:x\in Z\}$ is a Markov partition for T of diameter not exceeding 2β now follows from (4.5.5), (4.5.6), (4.5.4) and from the claim. The proof is complete. ♣

Remark 4.5.3. If in Theorem 4.5.2 we omit the assumption that T is an open map, but assume that $X\subset W$ and T extends to an open map in W, then the assertion about the existence of the Markov partition holds for $\tilde{X}$, an arbitrarily small T invariant extension of X.

The proof is the same. One finds $\tilde{X}:=\Theta(\Omega)\supset X$; it need not be equal to X. The only difficulty is to verify that the sets $T^{-j}(\partial P_k)$ for all $j\geq 1$ are nowhere dense. We can prove it by assuming that $\lambda>2$, where it follows immediately from the following lemma.

Lemma 4.5.4. *For each cylinder $[q_0,\dots,q_n]$ its Θ-image contains an open set in $\tilde{X}$.*

Proof. Let $L:=\sup|T'|$. Assume $\gamma<<\alpha$. Choose an arbitrary $q_{n+1},\dots$ such that for all $j\geq n$ we have $\rho(T(q_j),q_{j+1})\leq\gamma$. Let $x=\Theta((q_0,\dots))$. We prove that every $y\in\tilde{X}$ close enough to x, $\rho(x,y)\leq\epsilon$, belongs to $\Theta([q_1,\dots,q_n])$.

Since $y \in \tilde{X}$, by forward invariance of $\tilde{X}$ we get $T^{n+1}(y) \in \tilde{X}$. Hence there exists a sequence of points $z_0, \dots \in E$ such that $T^{n+1}(y) = \Theta((z_0, \dots))$. As a consequence the sequence $s = (q_0, \dots, q_n, z_0, z_1, \dots)$ of points in E satisfies the following

$$\rho(T(q_j), q_{j+1}) \leq \alpha \ \text{ for } j = 0, 1, \dots, n-1.$$

$$\rho(T(q_n), z_0) \leq \rho(T(q_n), q_{n+1}) + \rho(q_{n+1}, T^{n+1}(x)) + \rho(T^{n+1}(x), T^{n+1}(y))$$
$$+\rho(T^{n+1}(y), z_0) \leq \gamma + \frac{1/\lambda}{1-1/\lambda}\gamma + L^{n+1}\epsilon + \frac{1/\lambda}{1-1/\lambda}\alpha \leq \alpha,$$

as we can assume $\lambda > 2$ and

$$\rho(T(z_j), z_{j+1}) \leq \alpha \ \text{ for } j = 0, 1, \dots.$$

Therefore $\Theta(s) = y$ and $s \in [q_0, \dots, q_n]$. ♣

Therefore there is an arbitrary small extension of X to a compact set $\tilde{X}$ that is $F = T^n$ invariant for an integer $n > 0$ and has Markov partition $\{R_i\}$ for F. Then take $\hat{X} = \bigcup_{j \geq 0} T^j(\tilde{X})$. It is easy to check that the family of the closures of the intersections of the sets $T^{-j}_{T^{n-j}(x)}(\text{Int } R_i)$, for $x \in \text{Int } R_k$ and interiors in $\tilde{X}$ constitutes a Markov partition of $\hat{X}$ for T.

Example 4.5.5. It is not true that in the situation of Remark 4.5.3 one can always extend X to a T-invariant set $\tilde{X}$, in an arbitrarily small neighbourhood of X, on which T is open (i.e. $(\tilde{X}, T)$ is a repeller: see Section 6.1). Indeed, consider in the plane the set X, being the union of a circle together with its diameter interval. It is easy to find a mapping T defined on a neighbourhood of X, preserving X, smooth and expanding. Then at least one of the pre-images of one of two triple points (end points of the diameter) is not a triple point. Denote it by A. T restricted to X is not open at A. Adding a short arc γ starting at A, disjoint from X (except A), a pre-image of an arc in X does not make T open. Indeed, it is not open at the second end of γ. (It is not open either at T-pre-images of A, but we can cope with this trouble by adding pre-images of γ under iteration of T.)

On the other hand, (X, T) can be extended to a repeller if X is a Cantor set. This fact will be applied in Section 11.6.

Proposition 4.5.6. *Let $T : W \to W$ be an open continuous map of a compact metric space (W, ρ). Let $X \subset W$ be a T-invariant Cantor set, such that T is expanding in a neighbourhood U of X: that is, (4.0.1) holds for $x, y \in U$. Then, in an arbitrarily small neighbourhood of X in W, there exists a Cantor set $\tilde{X}$ containing X such that T is open on it.*

Proof. One can change the metric ρ on W to a metric ρ' giving the same topology, such that (X, T) is distance-expanding on U in ρ' in the sense of (4.0.2): see Section 4.1 for T Lipschitz or the formula defining ρ' in Lemma 4.6.3 in the general case.

First we prove that there exist arbitrarily small $r > 0$ such that $B(X,r) := \{z \in W : \rho'(z,X) < r\} \subset U$, consists of a finite number of open domains $U_k(r) \subset W$, with pairwise disjoint closures in W.

For any $z, z' \in B(X,r)$ define $z \sim_r z'$ if there exists a sequence $x_1, ..., x_n \in X$ such that $z \in B(x_1,r), z' \in B(x_n,r)$ and for all $k = 1, ..., n-1$, $B(x_k,r) \cap B(x_{k+1},r) \neq \emptyset$, the balls in (W, ρ'). This is an equivalence relation: each equivalence class contains a point in X, and for $x \in V_r \cap X, x' \in V_r' \cap X'$ for two different equivalence classes V_r, V_r', we have $\rho(x,x') \geq r$. So by compactness of X there is at most a finite number of the equivalence classes. Denote their number by $N(r)$. Clearly, for every $r < r'$ for every V_r there exists $V_{r'}$ such that $V_r \subset V_{r'}$ and every $V_{r'}$ contains some V_r. Hence the function $r \mapsto N(r)$ is monotone decreasing. Let $r_1 > r_2 > ... > r_n > ... \searrow 0$ be the sequence of consecutive points of its discontinuity. Take any $r > 0$ not belonging to this sequence. Let $r_j < r < r_{j-1}$. Denote $\varepsilon = (r_{j-1} - r)/2$. Consider two different sets V_r and V_r'. Suppose there is $z_0 \in \overline{V}_r \cap \overline{V}'_r$. Then there are points $z \in V_r$ and $z' \in V_r'$ such that $\rho'(z,z') < \varepsilon$. Then $z \sim_{r+\varepsilon} z'$. So both V_r and V_r' are contained in the same equivalence class of $\sim_{r+\varepsilon}$. So $N(r) > N(r+\varepsilon)$, which contradicts the definition of ε.

Observe that $\sup_k \operatorname{diam} U_k(r) \to 0$ as $r \to 0$ since X is a Cantor set. Indeed, for every $\delta > 0$ there is a covering of X by pairwise disjoint closed sets A_j of diameter $< \delta$. Then for $r < \inf_{j \neq j'} \operatorname{dist}(A_j, A_{j'})/2$ each two distinct $A_j, A_{j'}$ belong to different $\sim_r$ equivalence classes.

Thus we can assume, for $U_k = U_k(r)$, that $\operatorname{diam} U_k < \xi$. So we can consider the branches of $g = T_x^{-1}$ on U_k's for all $x \in X$: see Lemma 4.1.2. Then each g maps $\bar{U}_k$ into some $U_{k'}$ because it is a contraction (by the factor λ^{-1}). Then denote g by $g_{k',k}$. Finally define

$$\tilde{X} = \bigcap_{n=0}^{\infty} \bigcup_{k_1,\dots,k_n} g_{k_1,k_2} \circ g_{k_2,k_3} \circ \dots \circ g_{k_{n-1},k_n}(U_{k_n}), \tag{4.5.9}$$

the union over all $k_1, ..., k_n$ such that $g_{k_j,k_{j+1}}$ exist for all $j = 1, ..., n-1$. It follows that for r small enough the family of sets $\{U_k(r) \cap \tilde{X}\}$ is a Markov partition of $\tilde{X}$ with pairwise disjoint 'cylinders', and $(\tilde{X}, T)$ is topologically conjugate to a topological Markov chain: see Example 1.3. Hence T is open on $\tilde{X}$ (see more details below). ♣

Each Markov partition gives rise to a coding (symbolic representation) of $T : X \to X$ as follows (an example was provided in Proposition 4.5.6 above).

Theorem 4.5.7. *Let $T : X \to X$ be an open, distance-expanding map. Let $\{R_1, \dots, R_d\}$ be a Markov partition. Let $A = (a_{i,j})$ be a $d \times d$ matrix with $a_{i,j} = 0$ or 1 according to whether the intersection $T(\operatorname{Int} R_i) \cap \operatorname{Int} R_j$ is empty or not. Then consider the corresponding one-sided topological Markov chain Σ_A with the left shift map $\sigma : \Sigma_A \to \Sigma_A$: see Example 1.3. Define the map $\pi : \Sigma_A \to X$ by*

$$\pi((i_0, i_1, \dots)) = \bigcap_{n=0}^{\infty} T^{-n}(R_{i_n}).$$

Then π is a well-defined Hölder continuous mapping onto X, and $T \circ \pi = \pi \circ \sigma$. Moreover, $\pi|_{\pi^{-1}(X \setminus \bigcup_{n=0}^{\infty} T^{-n}(\bigcup_i \partial R_i))}$ is injective.

Proof. For an arbitrary sequence $(i_0, i_1, \dots) \in \Sigma_A$, $a_{i,j} = 1$ implies $T(R_{i_n}) \supset R_{i_{n+1}}$. Since $\operatorname{diam} R_{i_n} < 2\eta$, T is injective on R_{i_n}: hence there exists an inverse branch $T^{-1}_{R_{i_n}}$ on $R_{i_{n+1}}$ The subscript R_{i_n} indicates that we take the branch leading to R_{i_n}: compare notation from Section 4.1. Thus $T^{-1}_{R_{i_n}}(R_{i_{n+1}}) \subset R_{i_n}$. Hence

$$T^{-1}_{R_{i_0}} T^{-1}_{R_{i_1}} \dots T^{-1}_{R_{i_n}}(R_{i_{n+1}}) \subset T^{-1}_{R_{i_0}} T^{-1}_{R_{i_1}} \dots T^{-1}_{R_{i_{n-1}}}(R_{i_n}).$$

So $\bigcap_{n \geq 0} T^{-n}(R_{i_n}) \neq \emptyset$, as the intersection of a descending family of compact sets. We have used here that

$$\begin{aligned} T^{-1}_{R_{i_0}} \dots T^{-1}_{R_{i_{n-1}}}(R_{i_n}) &= T^{-1}_{R_{i_0}} \dots T^{-1}_{R_{i_{n-2}}}(T^{-1}(R_{i_n}) \cap R_{i_{n-1}}) \\ &= T^{-1}_{R_{i_0}} \dots T^{-1}_{R_{i_{n-3}}}(T^{-2}(R_{i_n}) \cap T^{-1} R_{i_{n-1}} \cap R_{i_{n-2}}) \\ &= \dots \\ &= \bigcap_{k=0}^{n} T^{-k}(R_{i_k}), \end{aligned}$$

following from $T^{-1}_{R_{i_k}}(A) = T^{-1}(A) \cap R_{i_k}$ for every $A \subset R_{i_{k+1}}$, $k = 0, \dots, n-1$.

Our infinite intersection consists of only one point, since $\operatorname{diam}(R_i)$ are all less than an expansivness constant.

Let us prove now that π is Hölder continuous. Indeed, $\rho'((i_n), (i'_n)) \leq \lambda_1^{-N}$ implies $i_n = i'_n$ for all $n = 0, \dots, N-1$, where the metric ρ' comes from Example 1.3, with the factor $\lambda = \lambda(\rho') > 1$. Then, for $x = \pi((i_n))$, $y = \pi((i'_n))$ and every $n : 0 \leq n < N$ we have $T^n(x), T^n(y) \in R_{i_n}$: hence $\rho(T^n(x), T^n(y)) \leq \operatorname{diam} R_{i_n} \leq \xi$, and hence $\rho(x, y) \leq \lambda^{-(N-1)}\xi$. Therefore π is Hölder continuous, with exponent $\min\{1, \log \lambda / \log \lambda(\rho')\}$.

Let us deal now with the injectivity. If $x = \pi((i_n))$ and $T^n(x) \in \operatorname{Int} R_{i_n}$ for all $n = 0, 1, \dots$, then $T^n(x) \notin R_j$ for all $j \neq i_n$. So, if $x \in \bigcap_n T^{-n}(R_{i'_n})$, then all $i'_n = i_n$.

Finally, π maps Σ_A onto X. Indeed, by definition, $\pi(\Sigma_A)$ contains $X \setminus \bigcup_{n=0}^{\infty} T^{-n}(\bigcup_i \partial R_i)$ which is dense in X. Since $\pi(\Sigma_A)$ is compact, it is therefore equal to X. ♣

Remark. One should not think that π is always injective on the whole Σ_A. Consider for example the mapping of the unit interval $T(x) = 2x (\operatorname{mod} 1)$: compare Example 1.5. Then the dyadic expansion of x is not unique for $x \in \bigcup_{n=0}^{\infty} T^{-n}(\{\frac{1}{2}\})$. Dyadic expansion is the inverse, π^{-1}, of the coding obtained from the Markov partition $[0,1] = \{[0, \frac{1}{2}], [\frac{1}{2}, 1]\}$.

Recall finally that $\sigma : \Sigma_A \to \Sigma_A$ is an open, distance-expanding map. The partition into the cylinders $C_i := \{(i_n) : i_0 = i\}$ for $i = 1, \dots, d$, is a Markov partition into closed-open sets. The corresponding coding π is just the identity.

Another fact concerning a similarity between (Σ_A, σ) and (X, T) is the following theorem.

Theorem 4.5.8. *For every Hölder continuous function $\phi : X \to \mathbb{R}$ the function $\phi\circ\pi$ is Hölder continuous on Σ_A and the pressures coincide, $\mathrm{P}(T, \phi) = \mathrm{P}(\sigma, \phi\circ\pi)$.*

Proof. The function $\phi \circ \pi$ is Hölder continuous as a composition of Hölder continuous functions. Consider next an arbitrary point $x \in X \setminus \bigcup_{n=0}^{\infty} T^{-n}(\bigcup_i \partial R_i)$. Then, using Proposition 4.4.3 for T and σ, we obtain

$$\mathrm{P}(T, \phi) = \mathrm{P}_x(T, \phi) = \mathrm{P}_{\pi^{-1}(x)}(\sigma, \phi \circ \pi) = \mathrm{P}(\sigma, \phi \circ \pi).$$

The middle equality follows directly from the definitions. ♣

Finally we shall prove that π is injective in the measure-theoretic sense.

Theorem 4.5.9. *For every ergodic Borel probability measure μ on Σ_A, invariant under the left-shift map σ, positive on open sets, the mapping π yields an isomorphism between the probability spaces $(\Sigma_A, \mathcal{F}_{\Sigma_A}, \mu)$ and $(X, \mathcal{F}_X, \mu \circ \pi^{-1})$, for $\mathcal{F}$ respective (completed) Borel σ-algebras, conjugating the shift map σ to the transformation $T : X \to X$ (i.e. $\pi \circ \sigma = T \circ \pi$).*

Proof. The set $\partial = \bigcup_{i=1}^{d} \partial(R_i)$, and hence $\pi^{-1}(\partial)$, have non-empty open complements in Σ_A. Since $T(\partial) \subset \partial$, we have $\sigma(\pi^{-1}(\partial)) \subset \pi^{-1}(\partial)$: hence $\pi^{-1}(\partial)) \subset \sigma^{-1}(\pi^{-1}(\partial))$. Since μ is σ-invariant, we conclude by ergodicity of μ that $\mu(\pi^{-1}(\partial))$ is equal either to 0 or to 1. But the complement of $\pi^{-1}(\partial)$, as a non-empty open set, has positive measure μ. Hence $\mu(\pi^{-1}(\partial)) = 0$. Hence $\mu(E) = 0$ for $E := \bigcup_{n=0}^{\infty} \sigma^{-n}(\pi^{-1}(\partial))$, and by Theorem 4.5.7 π is injective on $\Sigma_A \setminus E$. This proves that π is the required isomorphism. ♣

4.6 Expansive maps are expanding in some metric

Theorem 4.1.1 says that distance-expanding maps are expansive. In this section we prove the following much more difficult result, which can be considered as a sort of converse statement, and which provides an additional strong justification for exploring expanding maps.

Theorem 4.6.1. *If a continuous map $T : X \to X$ of a compact metric space X is (positively) expansive, then there exists a metric on X, compatible with the topology, such that the mapping T is distance-expanding with respect to this metric.*

The proof of Theorem 4.6.1 given here relies heavily on the topological result of Frink (see [Frink 1937], comp. [Kelley, 1955, p. 185]), which we state below without proof.

Lemma 4.6.2 (The Metrization Lemma of Frink). *Let $\{U_n\}_{n=0}^\infty\}$ be a sequence of open neighborhoods of the diagonal $\Delta \subset X \times X$ such that $U_0 = X \times X$,*

$$\bigcap_{n=1}^{\infty} U_n = \Delta, \tag{4.6.1}$$

and for every $n \geq 1$

$$U_n \circ U_n \circ U_n \subset U_{n-1}. \tag{4.6.2}$$

Then there exists a metric ρ, compatible with the topology on X, such that for every $n \geq 1$,

$$U_n \subset \{(x,y) : \rho(x,y) < 2^{-n}\} \subset U_{n-1}. \tag{4.6.3}$$

We shall also need the following, almost obvious, result.

Lemma 4.6.3. *If $T : X \to X$ is a continuous map of a compact metric space X, and T is distance-expanding for some $n \geq 1$, then T is distance-expanding for $n = 1$ with respect to some metric compatible with the topology on X.*

Proof. Let ρ be a compatible metric with respect to which T^n is distance-expanding, and let $\lambda > 1$ and $\eta > 0$ be constants such that

$$\rho(T^n(x), T^n(y)) \geq \lambda \rho(x,y)$$

whenever $\rho(x,y) < 2\eta$. Put $\xi = \lambda^{\frac{1}{n}}$ and define the new metric ρ' by setting

$$\rho'(x,y) = \rho(x,y) + \frac{1}{\xi}\rho(T(x),T(y)) + \ldots + \frac{1}{\xi^{n-1}}\rho(T^{n-1}(x),T^{n-1}(y)).$$

Then ρ' is a metric on X compatible with the topology and $\rho'(T(x),T(y)) \geq \xi\rho'(x,y)$ whenever $\rho'(x,y) < 2\eta$. ♣

Now we can pass to the proof of Theorem 4.6.1.

Proof of Theorem 4.6.1. Let d be a metric on X compatible with the topology, and let $3\theta > 0$ be an expansive constant associated to T which does not exceed the constant η claimed in Proposition 3.5.9. For any $n \geq 1$ and $\gamma > 0$ let

$$V_n(\gamma) = \{(x,y) \in (X \times X) : d(T^j(x), T^j(y)) < \gamma \quad \text{for every } j = 0, \ldots, n\}.$$

Then in view of Proposition 3.5.9 there exists $M \geq 1$ such that

$$V_M(3\theta) \subset \{(x,y) : d(x,y) < \theta\}. \tag{4.6.4}$$

Define $U_0 = X \times X$ and $U_n = V_{Mn}(\theta)$ for every $n \geq 1$. We shall check that the sequence $\{U_n\}_{n=0}^\infty$ satisfies the assumptions of Lemma 4.6.2. Indeed, (4.6.1) follows immediately from expansiveness of T. Now we shall prove condition (4.6.2). We shall proceed by induction. For $n = 1$ nothing has to be proved. Suppose that (4.6.2) holds for some $n \geq 1$. Let $(x,u),\ (u,v),\ (v,y) \in U_{n+1}$. Then by the triangle inequality

$$d(T^j(y), T^j(x)) < 3\theta \qquad \text{for every } j = 0, \ldots, (n+1)M.$$

Therefore, using (4.6.4), we conclude that

$$d(T^j(y), T^j(x)) < \theta \qquad \text{for every } j = 0, \ldots, Mn.$$

Equivalently $(x, y) \in V_{Mn}(\theta) = U_n$, which completes the proof of (4.6.2).

So, we have shown that the assumptions of Lemma 4.6.2 are satisfied, and therefore we obtain a compatible metric ρ on X satisfying (4.6.3). In view of Lemma 4.6.3 it suffices to show that T^{3M} is expanding with respect to the metric ρ. So suppose that $0 < \rho(x, y) < \frac{1}{16}$. Then by (4.6.1) there exists an $n \geq 0$ such that

$$(x, y) \in U_n \setminus U_{n+1}. \tag{4.6.5}$$

As $0 < \rho(x, y) < \frac{1}{16}$, this and (4.6.3) imply that $n \geq 3$. It follows from (4.6.5) and the definitions of U_n and $V_{Mn}(\theta)$ that there exists $Mn < j \leq (n+1)M$ such that $d(T^j(y), T^j(x)) \geq \theta$. Since $3 \leq n$ we conclude that $d(T^i(T^{3M}(x)), T^i(T^{3M}(y))) \geq \theta$ for some $0 \leq i \leq (n-2)M$, and therefore $(T^{3M}(x), T^{3M}(y)) \notin U_{n-2}$. Consequently, by (4.6.3) and (4.6.5) we obtain that

$$\rho(T^{3M}(x), T^{3M}(y)) \geq 2^{-(n-1)} = 2 \cdot 2^{-n} > 2\rho(x, y).$$

The proof is complete. ♣

Exercises

4.1. Prove the following *Shadowing Theorem* generalizing Corollary 4.2.4 (Shadowing lemma) and Corollary 4.2.5 (Closing lemma):

Let $T : X \to X$ be an open map, expanding at a compact set $Y \subset X$. Then for every $\beta > 0$ there exists $\alpha > 0$ such that for every map $\Gamma : Z \to Z$ for a set Z and a map $\Phi : Z \to B(Y, \alpha)$ satisfying $\rho(T\Phi(z), \Phi\Gamma(z)) \leq \alpha$ for every $z \in Z$, there exists a map $\Psi : Z \to X$ satisfying $T\Phi = \Phi\Gamma$ (hence $T(Y') \subset Y'$ for $Y' = \Psi(Z)$) and such that for every $z \in Z$, $\rho(\Psi(z), \Phi(z)) \leq \beta$. If Z is a metric space and Γ, Φ are continuous, then Ψ is continuous. If $T(Y) \subset Y$ and the map $T|_Y : Y \to Y$ be open, then $Y' \subset Y$.

(Hint: See Section 6.1.)

4.2. Prove the following *Structural Stability Theorem.*

Let $T : X \to X$ be an open map with a compact set $Y \subset X$ such that $T(Y) \subset Y$. Then for every $\lambda > 1$ and $\beta > 0$ there exists $\alpha > 0$ such that if $S : X \to X$ is distance-expanding at Y with the expansion factor λ and for all $y \in Y$ $\rho(S(y), T(y)) \leq \alpha$ then there exists a continuous mapping $h : Y \to X$ such that $Sh|_Y = hT|_Y$; in particular, $S(Y') \subset Y'$ for $Y' = h(Y)$, and $\rho(h(z), z) \leq \beta$.

Hint: Apply the previous exercise for $Z = Y, \Gamma = T|_Y, \Phi = \text{id}$, $T = S$ and $Y = Y$. Compare also Section 6.1.

4.3. Prove that if $T : X \to X$ is an open, distance-expanding map and X is compact connected, then $T : X \to X$ is topologically exact.

4.4. Prove that for $T : X \to X$ a continuous map on a compact metric space X the topological entropy is attained on the set of non-wandering points: that is, $\mathrm{h_{top}}(T) = \mathrm{h_{top}}(T|_{\Omega}(T)$.

Hint: Use the Variational Principle (Theorem 3.4.1).

4.5. Prove Lemma 4.3.9 and hence Theorem 4.3.8 (Spectral Decomposition) without the assumption of periodic shadowing, assuming that T is a branched covering of the Riemann sphere.

4.6. Prove the existence of stable and unstable manifolds for hyperbolic sets and *Smale's Spectral Decomposition Theorem for Axiom A diffeomorphisms.*

An invariant set Λ for a diffeomorphism $T : X \to X$ of a compact manifold X, is called *hyperbolic* if there exist constants $\lambda > 1$ and $C > 0$ such that the tangent bundle on X restricted to Λ, $T_\Lambda X$ decomposes into DT-invariant subbundles $T_\Lambda X = T^u_\Lambda X \oplus T^s_\Lambda X$ such that $||DT^n(v)|| \geq C\lambda^n$ for all $v \in T^u_\Lambda X$ and $n \geq 0$ and $||DT^n(v)|| \geq C\lambda^n$ for all $v \in T^s_\Lambda X$ and $n \leq 0$.

Prove that for every $x \in \Lambda$ the sets $W^u(x) = \{y \in X : \rho(T^n(x), T^n(y)) \to 0 \text{ as } n \to -\infty\}$, and $W^s(x) = \{y \in X : \rho(T^n(x), T^n(y)) \to 0 \text{ as } n \to \infty\}$ are immersed manifolds. (They are called *unstable* and *stable manifolds.*)

Assume next that a diffeomorphism $T : X \to X$ satisfies *Smale's Axiom A condition*: that is, the set of non-wandering points Ω is hyperbolic and $\Omega = \overline{\text{Per}}$.

Then the relation between periodic points is as follows. $x \sim y$ if there are points $z \in W^u(x) \cap W^s(y)$ and $z' \in W^u(y) \cap W^s(x)$ where $W^u(x)$ and $W^s(y)$, and $W^u(y)$ and $W^s(x)$ respectively, intersect transversally: that is, the tangent spaces to these manifolds at z and z' span the whole spaces tangent to X.

Prove that this relation yields spectral decomposition, as in Theorem 4.3.8, with the topological transitivity assertion rather than topological exactness, of course (Figure 4.2).

As one of the steps prove a lemma corresponding to Lemma 4.3.9 about approximation of transversal heteroclinic cycle points by periodic ones. That is,

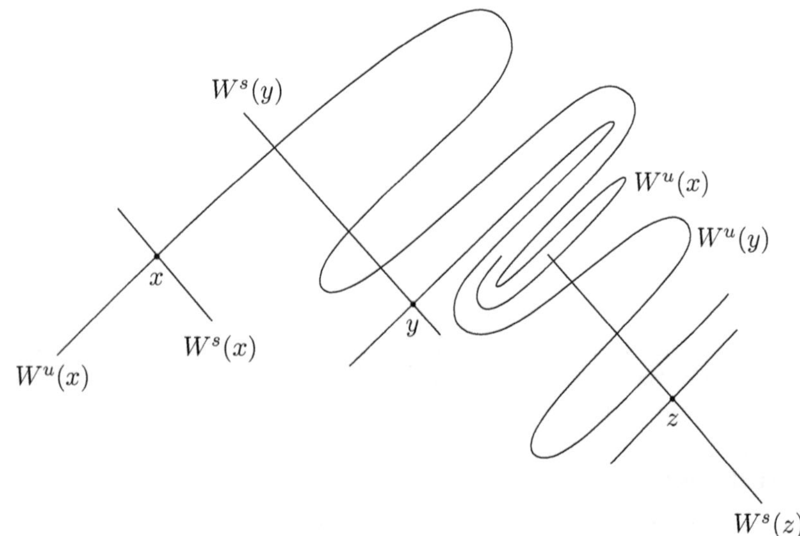

Figure 4.2 Transitivity for diffeomorphisms.

assume that $x_1, x_2, \ldots, x_n$ are hyperbolic periodic points (i.e. their orbits are hyperbolic sets) for a diffeomorphism, and $W^u_{x_i}$ has a point p_i of transversal intersection with $W^s_{x_{(i+1)\mathrm{mod}\, n}}$ for each $i = 1, \ldots, n$. Then $p_i \in \overline{\mathrm{Per}}$.

(For the theory of hyperbolic sets for diffeomorphisms see for example [Katok & Hasselblatt 1995].)

4.7. Prove directly that 1) $\Longrightarrow$ 2) in Proposition 4.4.5, using the specification property, Theorem 4.3.12.

4.8. Suppose $T : X \to X$ is a distance-expanding map on a closed surface. Prove that there exists a Markov partition for an iterate T^N compatible with a cell complex structure. That is, elements R_i of the partitions are topological discs, the one-dimensional 'skeleton' $\bigcup_i \partial R_i$ is a graph consisting of a finite number of continuous curves ('edges') intersecting one another only at end points, called 'vertices'. The intersection of each two R_i is empty or one vertex or one edge; each vertex is contained in two or three edges.

(Hint: Start with any cellular partition, with R_i being nice topological discs and correct it by adding or subtracting components of $T^{-N}(R_i), T^{-2N}(R_i)$, etc. See [Farrell & Jones 1979] for details.)

4.9. Prove that if T is an expanding map of the two-dimensional torus $\mathbb{R}^2/\mathbb{Z}^2$, a factor map of a linear map of $\mathbb{R}^2$ given by an integer matrix with two irrational eigenvalues of different moduli (for example $\left(\begin{smallmatrix} 0 & 11 \\ -1 & 7 \end{smallmatrix}\right)$ but not $\left(\begin{smallmatrix} 2 & 0 \\ 0 & 3 \end{smallmatrix}\right)$), then ∂R_i cannot be differentiable.

(Hint: Smooth curves $T^n(\partial R_i)$ become more and more dense in $\mathbb{R}^2/\mathbb{Z}^2$ as $n \to \infty$, stretching in the direction of the eigenspace corresponding to the eigenvalue with a larger modulus. So they cannot omit $\mathrm{Int}\, R_i$.

The same argument, looking backwards, says that the components of $T^{-n}(\mathrm{Int}\, R_i)$ are dense and very distorted, since the eigenvalues have different moduli. The curve ∂R_i must manoeuvre between them, so it is 'fractal'. See [Przytycki & Urbański 1989] for more details.)

Bibliographical notes

The Shadowing Lemma in the hyperbolic setting has appeared in [Anosov 1970], [Bowen 1979] and [Kushnirenko 1972]. See [Katok & Hasselblatt 1995] for the variant as in Exercise 4.1. In the context of C^1-differentiable (distance) expanding maps on smooth compact manifolds, the shadowing property was proved in [Shub 1969], where structural stability was also extablished. D. Sullivan introduced in [Sullivan 1982] the notion of a *telescope* for the sequence $T^{-1}_{x'_i}(\overline{B(x_{i+1}, \beta)}) \subset \overline{B(x_i, \beta)}$ to capture a shadowing orbit, and hence to prove the stability of expanding repellers: compare Section 6.1 in the context of hyperbolic rational functions. This stability was also proved in [Przytycki 1977]. Later a comprehensive monograph on shadowing by S. Yu. Pilyugin [Pilyugin 1999] appeared.

The existence of spectral decomposition in the sense of Theorem 4.3.8 (see Exercise 4.6) was first proved by S. Smale [Smale 1967] for diffeomorphisms, called by him *Axiom A*, defined by the property that the set of non-wandering points Ω is hyperbolic and $\Omega = \overline{\mathrm{Per}}$: see also [Katok & Hasselblatt 1995] and further historical information therein. In a topological setting an analogous theory was founded by [Bowen 1970], for (introduced by him) *Axiom A^** homeomorphisms. *Axiom A endomorphisms* were studied in [Przytycki 1977], comprising the diffeomorphisms and expanding (smooth) cases. For open, distance-expanding maps $\Omega = \overline{\mathrm{Per}}$ (Proposition 4.3.6) corresponds to the analogous fact for Anosov diffeomorphisms. It is not known whether $\Omega = X$ for all Anosov diffeomorphisms. It is not true for some distance-expanding endomorphisms (Remark 4.3.7), but it is true for T smooth and X a connected smooth manifold (Exercise 4.3): see [Shub 1969].

The construction of the Markov partition in Section 4.5 is similar to the construction for basic sets of Axiom A diffeomorphisms in [Bowen 1975]. The case of X not locally maximal for T has been studied recently in the case of diffeomorphisms [Crovisier 2002], [Fisher 2006]. The non-invertible case is considered here for the first time, to our knowledge.

For a general theory of cellular Markov partitions, including Exercise 4.8, see [Farrell & Jones 1993]. The fact that the Hausdorff dimension of the boundaries of two-dimensional cells is greater than 1, and in particular their non-differentiability, Exercise 4.9, follows from [Przytycki & Urbański 1989].

5

Thermodynamical formalism

In Chapter 3 (Theorem 3.5.6) we proved that for every positively expansive map $T : X \to X$ of a compact metric space and an arbitrary continuous function $\phi : X \to \mathbb{R}$ there exists an equilibrium state. In Remark 4.4.4 we provided a specific construction for T an open, distance-expanding topologically transitive map and a Hölder continuous function ϕ. Here we shall construct this equilibrium measure with greater care and study its amazing regularity with respect to the 'potential' function ϕ, its 'mixing' properties and uniqueness. So, for the entire chapter we fix an open, continuous, distance-expanding, topologically transitive map $T : X \to X$ of a compact metric space (X, ρ), with constants η, λ, ξ introduced in Chapter 4.

5.1 Gibbs measures: introductory remarks

A probability measure μ on X and the Borel σ-algebra of sets is said to be a *Gibbs state (measure)* for the potential $\phi : X \to \mathbb{R}$ if there exist $P \in \mathbb{R}$ and $C \geq 1$ such that for all $x \in X$ and all $n \geq 1$,

$$C^{-1} \leq \frac{\mu(T_x^{-n}(B(T^n(x), \xi)))}{\exp(S_n\phi(x) - Pn)} \leq C. \tag{5.1.1}$$

If in addition μ is T-invariant, we call μ an *invariant Gibbs state* (or *measure*).

We denote the set of all Gibbs states of ϕ by G_ϕ. It is obvious that if μ is a Gibbs state of ϕ, and ν is equivalent to μ with Radon–Nikodym derivatives uniformly bounded from above and below, then ν is also a Gibbs state. The following proposition shows that the converse is also true, and it identifies the constant P appearing in the definition of Gibbs states as the topological pressure of ϕ.

Proposition 5.1.1. *If μ and ν are Gibbs states associated to the map T and a Hölder continuous function ϕ, and the corresponding constants are denoted respectively by P, C and Q, D, then $P = Q = \mathrm{P}(T, \phi)$, and the measures μ and ν are equivalent with mutual Radon–Nikodym derivatives uniformly bounded.*

Proof. Since X is a compact space, there exist finitely many points $x_1, \ldots, x_l \in X$ such that $B(x_1, \xi) \cup \ldots \cup B(x_l, \xi) = X$. We claim that for every compact set $A \subset X$, every $\delta > 0$, and for all $n \geq 1$ large enough,

$$\mu(A) \leq CDl \exp((Q - P)n)(\nu(A) + \delta). \tag{5.1.2}$$

By the compactness of A, and by the regularity of the measure ν, there exists $\varepsilon > 0$ such that $\nu(B(A, \varepsilon)) \leq \nu(A) + \delta$. Fix an integer $n \geq 1$ so large that $\xi \lambda^{-n} < \frac{\varepsilon}{2}$, and for every $1 \leq i \leq l$ let

$$X(i) = \{x \in T^{-n}(x_i) : A \cap T_x^{-n}(B(x_i, \xi)) \neq \emptyset\}.$$

Then

$$A \subset \bigcup_{i=1}^{l} \bigcup_{x \in X(i)} T_x^{-n}(B(x_i, \xi)) \subset B(A, \varepsilon),$$

and, since for any fixed $1 \leq i \leq l$ the sets $T_x^{-n}(B(x_i, \xi))$ for $x \in T^{-n}(x_i)$ are mutually disjoint, it follows from (5.1.1) that

$$\begin{aligned}
\mu(A) &\leq \mu\left(\bigcup_{i=1}^{l} \bigcup_{x \in X(i)} T_x^{-n}(B(x_i, \xi))\right) \leq \sum_{i=1}^{l} \sum_{x \in X(i)} \mu(T_x^{-n}(B(x_i, \xi))) \\
&\leq C \sum_{i=1}^{l} \sum_{x \in X(i)} \exp(S_n\phi(x) - Pn) \\
&= C \exp((Q - P)n) \sum_{i=1}^{l} \sum_{x \in X(i)} \exp(S_n\phi(x) - Qn) \\
&\leq CD \exp((Q - P)n) \sum_{i=1}^{l} \sum_{x \in X(i)} \nu(T_x^{-n}(B(x_i, \xi))) \\
&\leq CD \exp((Q - P)n) l \nu(B(A, \varepsilon)) \\
&\leq CDl \exp((Q - P)n)(\nu(A) + \delta).
\end{aligned}$$

Exchanging the roles of μ and ν we also obtain

$$\nu(A) \leq CDl \exp((P - Q)n)(\mu(A) + \delta) \tag{5.1.3}$$

for all $n \geq 1$ large enough. So, if $P \neq Q$, say $P < Q$, then it follows from (5.1.3) applied to the compact set X that $\nu(X) = 0$. Hence $P = Q$, and as, by regularity of μ and ν, (5.1.2) and (5.1.3) continue to be true for all Borel subsets of X, we conclude that μ and ν are equivalent, with the Radon–Nikodym derivative $d\mu/d\nu$ bounded from above by CDl and from below by $(CDl)^{-1}$ (letting $\delta \to 0$).

It is left to show that $P = \mathrm{P}(T, \phi)$. Looking at the expression after the third inequality sign in our estimates of $\mu(A)$ with $A = X$, we get

$$0 = \log \mu(X) \leq \log C + \log \left(\sum_{i=1}^{l} \sum_{x \in X(i)} \exp(S_n\phi(x)) \right) - Pn.$$

Since for every i, $X(i)$ is an (η, n)-separated set, taking into account division by n in the definition of pressure, we can replace $\sum_i$ here by a largest summand for each n. We get $P \leq P(T, \phi)$.

On the other hand, for an arbitrary $x \in X$,

$$\sum_{y \in T^{-n}(x)} \exp(S_n\phi(y) - Pn) \leq C \sum_{y \in T^{-n}(x)} \mu(T_y^{-n}(B(x, \xi))) \leq C\mu(X) = C$$

gives $\mathrm{P}(T, \phi) = \mathrm{P}_x(T, \phi) \leq P$, for P_x defined in 4.4.3 applicable owing to the topological transitivity of T. The proof is complete. ♣

Remark 5.1.2. In order to prove Proposition 5.1.1, except the part identifying P as $\mathrm{P}(T, \phi)$, we used only the inequalities

$$C^{-1} \leq \frac{\mu(T_x^{-n}(B(T^n(x), \xi)) \exp Pn}{\nu(T_x^{-n}(B(T^n(x), \xi)) \exp Qn} \leq C.$$

We needed the function ϕ in (5.1.1) and its Hölder continuity only to prove that $P = Q = P(T, \phi)$. Hölder continuity also allows us to replace x in $S_n\phi(x)$ by an arbitrary point contained in $T_x^{-n}(B(T^n(x), \xi))$.

Remark 5.1.3. For $\mathcal{R} = \{R_1, \ldots, R_d\}$, a Markov partition of diameter smaller than ξ, (5.1.1) produces a constant C depending on $\mathcal{R}$ (see Exercise 5.1) such that

$$C^{-1} \leq \frac{\mu(R_{j_0,\ldots,j_{n-1}})}{\exp(S_n\phi(x) - Pn)} \leq C \tag{5.1.4}$$

for every admissible sequence $j_0, j_1, \ldots, j_{n-1}$ and every $x \in R_{j_0,\ldots,j_{n-1}}$. In particular, (5.1.4) holds for the shift map of a one-sided topological Markov chain.

The following completes Proposition 5.1.1.

Proposition 5.1.4. *If ϕ and ψ are two arbitrary Hölder continuous functions on X, then the following conditions are equivalent:*

(1) *$\phi - \psi$ is co-homologous to a constant in the space of bounded functions (see Definition 2.11.2).*

(2) *$G_\phi = G_\psi$.*

(3) *$G_\phi \cap G_\psi \neq \emptyset$.*

Proof. Of course, (2) implies (3). That (1) implies (2) is also obvious. If (3) is satisfied, that is if there exists $\mu \in G_\phi \cap G_\psi$, then it follows from (5.1.1) that

$$D^{-1} \le \exp(S_n(\phi)(x) - S_n(\psi)(x) - n\,\mathrm{P}(\phi) + n\,\mathrm{P}(\psi)) \le D$$

for some constant D, all $x \in X$ and $n \in \mathbb{N}$. Applying logarithms we see that condition (2) in Proposition 4.4.5 is satisfied with ϕ and ψ replaced by $\phi - \mathrm{P}(\phi)$ and $\psi - \mathrm{P}(\psi)$ respectively. Hence, by this proposition, $\phi - P(\phi)$ and $\psi - \mathrm{P}(\psi)$ are co-homologous, which completes the proof. ♣

We shall prove later that the class of Gibbs states associated to T and ϕ is not empty (Section 5.3) and contains exactly one Gibbs state, which is T-invariant (Corollary 5.2.14). Actually we shall prove a stronger uniqueness theorem. We shall prove that any invariant Gibbs state is an equilibrium state for T and ϕ, and prove (Section 5.6) uniqueness of the equilibrium states for open expanding topologically transitive maps T and Hölder continuous functions $\phi : X \to \mathbb{R}$.

Proposition 5.1.5. *A probability T-invariant Gibbs state μ is an equilibrium state for T and ϕ.*

Proof. Consider an arbitrary finite partition $\mathcal{P}$ into Borel sets of diameter less than $\min(\eta, \xi)$. Then for every $x \in X$ we have $T_x^{-n}(B(T^n(x), \xi)) \supset \mathcal{P}^n(x)$, where $\mathcal{P}^n(x)$ is the element of the partition $\mathcal{P}^n = \bigvee_{j=0}^n \mathcal{P}$ that contains x. Hence $\mu(T_x^{-n}(B(T^n(x), \xi))) \ge \mu(\mathcal{P}^n(x))$. Therefore by the Shannon–McMillan–Breiman Theorem and (5.1.1) we obtain

$$\mathrm{h}_\mu(T) \ge \mathrm{h}_\mu(T, \mathcal{P}) \ge \int \left(\limsup_{n\to\infty} \frac{1}{n}(n\,\mathrm{P}(T, \phi)) - S_n\phi(x) \right) d\mu = \mathrm{P}(T, \phi) - \int \phi \, d\mu$$

or, in other words, $\mathrm{h}_\mu(T) + \int \phi \, d\mu \ge \mathrm{P}(T, \phi)$ which just means that μ is an equilibrium state. ♣

5.2 Transfer operator and its conjugate; measures with prescribed Jacobians

Suppose first that we are in the situation of Chapter 2: that is, T is a measurable map. Suppose that m is *backward quasi-invariant* with respect to T: that is,

$$T_*(m) = m \circ T^{-1} \prec m. \tag{5.2.1}$$

(Sometimes this property is called *non-singular.*) Then by the Radon–Nikodym Theorem there exists an m-integrable function $\Phi : X \to [0, \infty)$ such that for every measurable set $A \subset X$ we have

$$m(T^{-1}(A)) = \int_A \Phi dm.$$

One writes $d(m \circ T^{-1})/dm = \Phi$. In the situation of this chapter, where T is a local homeomorphism (it does not need expanding yet), if T^{-1} has d branches on a ball $B(x,\xi)$ mapping the ball onto $U_1, \dots, U_d$ respectively, then

$$\Phi = \sum_{j=1}^{d} \Phi_j \quad \text{where} \quad \Phi_j := d(m \circ (T|_{U_j})^{-1})/dm.$$

If we consider measures absolutely continuous with respect to a backward quasi-invariant 'reference measure' m, then the transformation $\mu \mapsto T_*(\mu)$ can be rewritten in the language of densities with respect to m as follows:

$$d\mu/dm \mapsto d(T_*\mu)/dm = \sum_{j=1}^{d} \big((d\mu/dm) \circ (T|_{U_j})^{-1}\big)\Phi_j. \tag{5.2.2}$$

It is convenient to define

$$\Psi(z) = \frac{d(m \circ (T|_{U_j})^{-1})}{dm}(T(z)), \tag{5.2.3}$$

that is, $\Psi = \Phi_j \circ T$ for $z \in U_j$. Note that Ψ is defined on a set whose T-image has full measure (which is possibly larger than just a set of full measure, in the case where a set of measure zero is mapped by T to a set of positive measure). See Section 5.6 for further discussion.

The transformation in (5.2.2) can be considered as a linear operator $\mathcal{L}_m : L^1(m) \to L^1(m)$, called the *transfer operator*,

$$\mathcal{L}_m(u)(x) = \sum_{\overline{x} \in T^{-1}(x)} u(\overline{x})\Psi(\overline{x}).$$

This definition makes sense, because if we change u on a set A of measure 0, then even if $m(T(A)) > 0$, we have $\Phi_j|_{T(A)\cap B(x,\xi)} = 0$ m-a.e.: hence $\mathcal{L}_m(u)$ does not depend on the values of u on $T(A)$. We have the convention that if u is not defined (on a set of measure 0) and $\Psi = 0$, then $u\Psi = 0$.

The transformation $\mathcal{L}_m$ in fact makes sense in a more general situation, where $T : X \to X$ is a measurable map of a probability space $(X, \mathcal{F}, m)$, backward quasi-invariant (non-singular), finite (or countable) to one. Instead of U_j we write $X = \bigcup X_j$, where X_j are measurable, pairwise disjoint, and for each j the map $T|_{X_j} \to T(X_j)$ is a measurable isomorphism.

Proposition 5.2.1.

$$\int \mathcal{L}_m(u)\, dm = \int u\, dm \quad \textit{for all} \quad u \in L^1(m). \tag{5.2.4}$$

Conversely, if (5.2.4) holds where in the definition of $\mathcal{L}_m$ we put an arbitrary m integrable function Ψ, then Ψ satisfies (5.2.3).

Proof. It is sufficient to consider $u = \mathbb{1}_A$ the indicator function for an arbitrary measurable $A \subset X_j$. We have

$$\int \mathcal{L}_m(\mathbb{1}_A)\,dm = \int_{T(A)} \Psi \circ (T|_{X_j})^{-1}\,dm = \int_{T(A)} \Phi_j\,dm = m(A),$$

the latter true by change of coordinates if and only if Φ_j is Jacobian as above. Compare Lemma 5.2.5. ♣

It follows from (5.2.4) that $\mathcal{L}_m$ restricted to non-negative functions is an isometry in the $L^1(m)$ norm. The transfer operator $\mathcal{L}_m : L^1(m) \to L^1(m)$ is an example of a Markov operator: see Exercise 5.4.

By (5.2.2) we obtain the following characterization of probability T-invariant measures absolutely continuous with respect to m.

Proposition 5.2.2. *The probability measure $\mu = hm$ for $h \in L^1(m)$, $h \geq 0$, is T-invariant if and only if*

$$\mathcal{L}_m(h) = h.$$

Remark 5.2.3. For the operator $\mathcal{L}_m$ we have the identity

$$\mathcal{L}_m\big(f \cdot (g \circ T)\big) = \mathcal{L}_m(f) \cdot g. \tag{5.2.5}$$

making sense for any measurable functions $f, g : X \to \mathbb{R}$. Hence, using (5.2.4), for all $f \in L^\infty(\mu)$ and $g \in L^1(\mu)$, we get

$$\int f \cdot (g \circ T)\,dm = \int \mathcal{L}_m(f \cdot (g \circ T))\,dm = \int \mathcal{L}_m(f) \cdot g\,dm, \tag{5.2.6}$$

and, iterating this equality, we get

$$\int f \cdot (g \circ T^n)\,dm = \int \mathcal{L}_m^n(f) \cdot g\,dm \tag{5.2.7}$$

for all $n = 1, 2, \ldots$.

Remark 5.2.4. Since $\mathcal{L}_m$ acts on $L^1(m)$, we can consider the *conjugate* (another name: *adjoint*) operator $\mathcal{L}_m^* : L^\infty(m) \to L^\infty(m)$. Notice that

$$\int \mathcal{L}_m^*(f) \cdot g\,dm = \int f \cdot \mathcal{L}_m(g)\,dm = \int \mathcal{L}_m((f \circ T) \cdot g)\,dm = \int (f \circ T) \cdot g\,dm,$$

by definition and (5.2.4). Hence $\mathcal{L}_m^*(f) = f \circ T$.

Recall from Section 2.2 that $h \to h \circ T$ is called the Koopman operator, here acting on $L^\infty(m)$. So the operator conjugate to $\mathcal{L}_m$ is this Koopman operator. If one considers both operators acting on $L^2(m)$, which is the case for m being T invariant (see Exercise 5.3), then these operators are mutually conjugate.

Continuous case

After this introduction, the appearance of the following linear operator, called the Perron–Frobenius–Ruelle or Ruelle or Araki or also *transfer operator*, is not surprising:

$$\mathcal{L}_\phi(u)(x) = \sum_{\overline{x}\in T^{-1}(x)} u(\overline{x})\exp(\phi(\overline{x})). \tag{5.2.8}$$

If the function ϕ is fixed, we sometimes omit the subscript ϕ at $\mathcal{L}$. The function ϕ is often called a *potential* function. This term is compatible with the term used for ϕ in Section 5.1 for $P=0$. It will become clear later on. The transfer's conjugate operator will be our tool to find a backward quasi-invariant measure m such that Ψ will be a scalar multiple of $\exp\phi$: hence $\mathcal{L}_m$ will be a scalar multiple of $\mathcal{L}_\phi$. Then in turn we shall look for fixed points of $\mathcal{L}_m$ to find invariant measures. Restricting our attention to $\exp\phi$, we restrict considerations to Ψ strictly positive defined everywhere. One sometimes allows ϕ to have the value $-\infty$, but we do not consider this case in our book. See, for example, [Keller 1998].

Let us now be more specific. Let $\phi : X \to \mathbb{R}$ be a continuous function. Consider $\mathcal{L}_\phi$ acting on the Banach space of continuous functions $\mathcal{L}_\phi : C(X) \to C(X)$. It is a continuous linear operator, and its norm is equal to $\sup_x \sum_{\overline{x}\in T^{-1}(x)} \exp(\phi(\overline{x})) = \sup \mathcal{L}_\phi(1\!\!1)$, as this is a positive operator: that is, it maps real non-negative functions to real non-negative functions (see Section 3.1). Consider the conjugate operator $\mathcal{L}_\phi^* : C^*(X) \to C^*(X)$. Note that as conjugate to a positive operator it is also positive, that is, it transforms measures into measures.

Lemma 5.2.5. *For every $\mu \in C^*(X)$ and every Borel set $A \subset X$ on which T is injective,*

$$\mathcal{L}_\phi^*(\mu)(A) = \int_{T(A)} \exp(\phi\circ(T|_A)^{-1})d\mu. \tag{5.2.9}$$

Proof. It is sufficient to prove (5.2.9) for $A \subset B(x,r)$ with any $x \in X$ and $r > 0$ such that T is injective on $B(x,2r)$ (say $r=\eta$). Now approximate in pointwise convergence the indicator function χ_A by uniformly bounded continuous functions with support in $B = B(x,2r)$. We have, for any such function f,

$$\mathcal{L}_\phi^*(\mu)(f) = \mu(\mathcal{L}_\phi(f)) = \int_{T(B)} (f\exp(\phi))\circ(T|_B)^{-1}d\mu.$$

We used here the fact that the only branch of T^{-1} mapping $T(B)$ to the support of f is the one leading $T(B)$ to B. Passing with f to the limit χ_A on both sides (Dominated Convergence Theorem, Section 2.1) gives (5.2.9). ♣

Observe that whereas $\mathcal{L}_\phi$ transports a measure from the past (more precisely, transports a density: see (5.2.2)), $\mathcal{L}_\phi^*$ pulls a measure back from the future with Jacobian $\exp\phi\circ T^{-1}$. This is the right operator to use, to look for the missing 'reference measure' m.

Definition 5.2.6. J is called the *weak Jacobian* if $J : X \to [0,\infty)$ and there exists a Borel set $E \subset X$ such that $\mu(E)=0$ and for every Borel set $A \subset X$ on which T is injective, $\mu(T(A\setminus E)) = \int_A J d\mu$.

Recall from Chapter 2 (Definition 2.9.4) that a measurable function $J : X \to [0,\infty)$ is called the *Jacobian* or the *strong Jacobian* of a map $T : X \to X$ with respect to a measure μ if for every Borel set $A \subset X$ on which T is injective $\mu(T(A)) = \int_A J d\mu$. In particular, μ is forward quasi-invariant (cf. Lemma 2.9.3 and Definition 2.9.4).

Notice that if μ is backward quasi-invariant then the condition that J is the weak Jacobian translates to $\mu(A) = \int_{T(A)} \frac{1}{J\circ(T|_A)^{-1}}\, d\mu$.

Corollary 5.2.7. *If a probability measure μ satisfies $\mathcal{L}^*_\phi(\mu) = c\mu$ (i.e. μ is an eigen-measure of $\mathcal{L}^*_\phi$ corresponding to a positive eigenvalue c), then $c\exp(-\phi)$ is the Jacobian of T with respect to μ.*

Proof. Substitute $c\mu$ in place of $\mathcal{L}^*(\mu)$ in (5.2.9). It then follows that μ is backward quasi-invariant, and $c\exp(-\phi)$ is the weak Jacobian of T with respect to μ. Since $\frac{1}{\exp(-\phi)} = \exp\phi$ is positive everywhere, $c\exp(-\phi)$ is the strong Jacobian of T. ♣

Theorem 5.2.8. *Let $T : X \to X$ be a local homeomorphism of a compact metric space X, and let $\phi : X \to \mathbb{R}$ be a continuous function. Then there exists a Borel probability measure $m = m_\phi$ and a constant $c > 0$, such that $\mathcal{L}^*_\phi(m) = cm$. The function $c\exp(-\phi)$ is the strong Jacobian for T with respect to the measure m.*

Proof. Consider the map $l(\mu) := \frac{\mathcal{L}^*(\mu)}{\mathcal{L}^*(\mu)(1\!\!1)}$ on the convex set of probability measures on X, that is, on $M(X)$, endowed with the weak* topology (Section 3.1). The transformation l is continuous in this topology, since $\mu_n \to \mu$ weak* implies for every $u \in C(X)$ that $\mathcal{L}^*(\mu_n)(u) = \mu_n(\mathcal{L}(u)) \to \mu(\mathcal{L}(u)) = \mathcal{L}^*(\mu)(u)$. As $M(X)$ is weak* compact (see Theorem 3.1.6) we can use Theorem 3.1.7 (the Schauder-Tychonoff Fixed Point Theorem) to find $m \in M(X)$ such that $l(m) = m$. Hence $\mathcal{L}^*(m) = cm$ for $c = \mathcal{L}^*(m)(1\!\!1)$. Thus T has the Jacobian equal to $c\exp(-\phi)$, by Corollary 5.2.7. ♣

Note again that we write $\exp\phi$ in order to guarantee that it never vanishes, so that there exists the Jacobian for T with respect to m. To find an eigen-measure m for $\mathcal{L}^*$ (i.e. with a weak Jacobian being a multiple of $\exp(-\phi)$) we could perfectly well allow $\exp\phi = 0$.

We have the following complementary fact in the case when Jacobian J exists.

Proposition 5.2.9. *If $T : X \to X$ is a local homeomorphism of a compact metric space X, and a Jacobian J with respect to a probability measure m exists, then for every Borel set A*

$$\frac{1}{d}\int_A J\, dm \le m(T(A)) \le \int_A J\, dm,$$

where d is the degree of T ($d := \sup_{x\in X} \sharp T^{-1}(\{x\})$). In particular, if $m(A) = 0$, then $m(T(A)) = 0$.

Proof. Let us partition A into finitely many Borel sets, say $A_1, A_2, \ldots, A_n$, of diameters so small that T restricted to each of them is injective. Then, on the one hand,

$$m(T(A)) = m\left(\bigcup_{i=1}^{n} T(A_i)\right) \le \sum_{i=1}^{n} m(T(A_i)) = \sum_{i=1}^{n} \int_{A_i} J\,dm = \int_A J\,dm,$$

and on the other hand, since the multiplicity of the family $\{T(A_i) : 1 \le i \le n\}$ does not exceed d,

$$m(T(A)) = m\left(\bigcup_{i=1}^{n} T(A_i)\right) \ge \frac{1}{d}\sum_{i=1}^{n} m(T(A_i)) = \frac{1}{d}\sum_{i=1}^{n} \int_{A_i} J\,dm = \frac{1}{d}\int_A J\,dm.$$

The proof is complete. ♣

Let us go back to T, a distance-expanding topologically transitive open map.

Proposition 5.2.10. *The measure m produced in Theorem 5.2.8 is positive on non-empty open sets. Moreover, for every $r > 0$ there exists $\alpha = \alpha(r) > 0$ such that for every $x \in X$, $m(B(x,r)) \ge \alpha$.*

Proof. For every open $U \subset X$ there exists $n \ge 0$ such that $\bigcup_{j=0}^{n} T^j(U) = X$ (Theorem 4.3.12). So, by Proposition 5.2.9, $m(U) = 0$ would imply that $1 = m(X) \le \sum_{j=0}^{n} m(T^j(U)) = 0$, a contradiction.

Passing to the second part of the proof, let $x_1, \ldots, x_m$ be an $r/2$-net in X and $\alpha := \min_{1\le j\le m}\{m(B(x_j, r/2))\}$. Since for every $x \in X$ there exists j such that $\rho(x, x_j) \le r/2$, we have $B(x,r) \supset B(x_j, r/2)$, and so $m(B(x,r)) \ge m(B(x_j, r/2))$. Thus it is enough to set $\alpha(r) := \alpha$. ♣

Proposition 5.2.11. *The measure m is a Gibbs state of ϕ and* $\log c = \mathrm{P}(T,\phi)$.

Proof. We have for every $x \in X$ and every integer $n \ge 0$,

$$m(B(T^n(x), \xi)) = \int_{T_x^{-n}(B(T^n(x),\xi))} c^n \exp(-S_n\phi)\,dm.$$

Since, by Lemma 4.4.2, the ratio of the supremum and infimum of the integrand of the above integral is bounded from above by a constant $C > 0$ and is bounded from below by C^{-1}, we obtain

$$1 \ge m(B(T^n(x), \xi)) \ge C^{-1} c^n \exp(-S_n\phi(x)) m\big(T_x^{-n}(B(T^n(x), \xi))\big)$$

and

$$\alpha(\xi) \le m(B(T^n(x), \xi)) \le C c^n \exp(-S_n\phi(x)) m(T_x^{-n}(B(T^n(x), \xi))).$$

Hence

$$\alpha(\xi) C^{-1} \le \frac{m(T_x^{-n}(B(T^n(x), \xi)))}{\exp(S_n\phi(x) - n\log c)} \le C,$$

and therefore m is a Gibbs state. That $\log c = \mathrm{P}(T,\phi)$ now follows from Proposition 5.1.1. ♣

We now give a simple direct proof of the equality $\log c = \mathrm{P}(T,\phi)$. First note that by the definition of $\mathcal{L}_\phi$ and a simple inductive argument, for every integer $n \geq 0$,

$$\mathcal{L}_\phi^n(u)(x) = \sum_{\overline{x}\in T^{-n}(x)} u(\overline{x}) \exp(S_n\phi(\overline{x})). \tag{5.2.10}$$

The estimate (4.4.3) can be rewritten as

$$C^{-1} \leq \mathcal{L}^n(1\!\!1)(x)/\mathcal{L}^n(1\!\!1)(y) \leq C \quad \text{for every} \quad x, y \in X. \tag{5.2.11}$$

Now $c^n = c^n m(1\!\!1) = (\mathcal{L}^*)^n(m)(1\!\!1) = m(\mathcal{L}^n(1\!\!1))$, and hence

$$\log c = \lim_{n\to\infty} \frac{1}{n} \log m(\mathcal{L}^n(1\!\!1)) = \mathrm{P}(T,\phi),$$

where the last equality follows from (5.2.11) and Proposition 4.4.3.

Note that in the last equality above we used the property that m is a measure, or more precisely that the linear functional corresponding to m is positive. For m a signed eigen-measure and c a complex eigenvalue for $\mathcal{L}^*$ we would obtain only $\log|c| \leq P(T,\phi)$ (one should consider a function u such that $\sup|u| = 1$ and $m(u) = 1$ rather than the function $1\!\!1$), and indeed the point spectrum of $\mathcal{L}^*$ is usually large: see for example [Baladi, 2000, Theorem 2.5].

We are now in a position to prove some ergodic properties of Gibbs states.

Theorem 5.2.12. *If $T : X \to X$ is an open, topologically exact, distance-expanding map, then the system (T, m) is exact in the measure-theoretic sense: that is, for every A of positive measure $m(T^n(A)) \to 1$ as $n \to \infty$ (see Definition 2.10.2 and the exercise following it).*

Proof. Let E be an arbitrary Borel subset of X with $m(E) > 0$. By regularity of the measure m we can find a compact set $A \subset E$ such that $m(A) > 0$. Fix an arbitrary $\varepsilon > 0$. As in the proof of Proposition 5.1.1, we find for every n large enough, a cover of A by sets D_ν of the form $T_x^{-n}(B(x_i,\xi)), x \in X(i), i = 1,\dots,l$ such that $m(\bigcup_\nu D_\nu) \leq m(A) + \varepsilon$. Hence $m(\bigcup_\nu(D_\nu \setminus A)) \leq \varepsilon$. Since the multiplicity of this cover is at most l, we have

$$\sum_\nu m(D_\nu \setminus A) \leq l\varepsilon.$$

Hence

$$\frac{\sum_\nu m(D_\nu \setminus A)}{\sum_\nu m(D_\nu)} \leq \frac{l\varepsilon}{m(A)}.$$

Therefore for all n large enough there exists $D = D_\nu = T_x^{-n}(B)$, with some $B = B(x_i,\xi))$, $1 \leq i \leq l$, such that

$$\frac{m(D \setminus A)}{m(D)} \leq \frac{l\varepsilon}{m(A)}.$$

Hence, as $B \setminus T^n(A) \subset T^n(D \setminus A)$,

$$\frac{m(B \setminus T^n(A))}{m(B)} \leq \frac{\int_{D\setminus A} c^n \exp(-S_n\phi)dm}{\int_D c^n \exp(-S_n\phi)dm}$$
$$\leq C^2 \frac{m(D \setminus A)}{m(D)} \leq C^2 \frac{l\varepsilon}{m(A)}.$$

By the topological exactness of T, there exists $N \geq 0$ such that for every i we have $T^N(B(x_i, \xi)) = X$. In particular, $T^N(B) = X$. So, using Proposition 5.2.9, we get

$$m(X \setminus T^N(T^n(A))) \leq m(T^N(B \setminus T^n(A))) \leq c^N (\inf \exp\phi)^{-N} \frac{Cl\varepsilon}{m(A)}.$$

Letting $\varepsilon \to 0$, we obtain $m(X \setminus T^N(T^n(A))) \to 0$ as $n \to \infty$. Hence $m(T^{N+n}(A)) \to 1$. ♣

We have considered here a special Gibbs measure $m = m_\phi$. Notice, however, that by Proposition 5.1.1 the assertion of Theorem 5.2.12 holds for every Gibbs measure associated to T and ϕ.

Corollary 5.2.13. *If $T : X \to X$ is an open, topologically transitive, distance-expanding map, then for every Hölder potential $\phi : X \to \mathbb{R}$, every Gibbs measure for ϕ is ergodic.*

Proof. By Theorem 4.3.8 and Theorem 4.3.12 there exists a positive integer N such that T^N is topologically mixing on a T^N-invariant closed-open set $Y \subset X$, where all $T^j(Y)$ are pairwise disjoint and $\bigcup_{j=0,\dots,N-1} T^j(Y) = X$. So our $T^N|_Y$, being also an open expanding map, is topologically exact by Theorem 4.3.8, and hence exact in the measure-theoretic sense by Theorem 5.2.12. Let $m(E) > 0$. Then there is $k \geq 0$ such that $m(E \cap T^k(Y)) > 0$. Then for every $j = 0, \dots, N-1$ we have $m(T^{Nn}T^j(E \cap T^k(Y))) \to m(T^j(T^k(Y)))$: hence $m(\bigcup_{n\geq 0} T^n(E)) \to 1$. For E being T-invariant this yields $m(E) = 1$. This implies ergodicity. ♣

With the use of Proposition 2.2.7 we get the following fact, promised in Section 5.1.

Corollary 5.2.14. *If $T : X \to X$ is an open, topologically transitive, distance-expanding map, then for every Hölder continuous potential $\phi : X \to \mathbb{R}$, there is at most one invariant Gibbs measure for ϕ.*

5.3 Iteration of the transfer operator; existence of invariant Gibbs measures

It is convenient to consider the normalized operator $\mathcal{L}_{\overline{\phi}}$ with $\overline{\phi} = \phi - \mathrm{P}(T, \phi)$. We have $\mathcal{L}_{\overline{\phi}} = \mathrm{e}^{-\mathrm{P}(T,\phi)} \mathcal{L}_\phi$ (recall that $\mathrm{P}(T, \phi) = \log c$). Then for the reference measure $m = m_\phi$ satisfying $\mathcal{L}^*_\phi(m) = \mathrm{e}^{\mathrm{P}(\phi)} m$ we have $\mathcal{L}^*_{\overline{\phi}}(m) = m$: that is,

$$\int u dm = \int \mathcal{L}_{\overline{\phi}}(u) dm \quad \text{for every } \ u \in C(X). \tag{5.3.1}$$

For a fixed potential ϕ we often denote $\mathcal{L}_{\overline{\phi}}$ by $\mathcal{L}_0$. By (5.2.11), for all $x, y \in X$, and all non-negative integers n,

$$\mathcal{L}_0^n(1\!\!1)(x)/\mathcal{L}_0^n(1\!\!1)(y) \leq C. \tag{5.3.2}$$

Multiplying this inequality by $\mathcal{L}_0^n(1\!\!1)(y)$ and then integrating with respect to the variables x and y we get respectively the first and the third of the following inequalities:

$$C^{-1} \leq \inf \mathcal{L}_0^n(1\!\!1) \leq \sup \mathcal{L}_0^n(1\!\!1) \leq C. \tag{5.3.3}$$

By (4.4.2), for every $x, y \in X$ such that $x \in B(y, \xi)$ we have an inequality that is more refined than (4.4.3):

$$\begin{aligned} \frac{\mathcal{L}_\phi^n(1\!\!1)(x)}{\mathcal{L}_\phi^n(1\!\!1)(y)} &= \frac{\sum_{\overline{x} \in T^{-n}(x)} \exp S_n\phi(\overline{x})}{\sum_{\overline{y} \in T^{-n}(y)} \exp S_n\phi(\overline{y})} \\ &\leq \sup_{\overline{x} \in T^{-n}(x)} \frac{\exp S_n\phi(\overline{x})}{\exp S_n\phi(y_n(\overline{x}))} \leq \exp(C_1 \rho(x,y)^\alpha), \end{aligned} \tag{5.3.4}$$

where $C_1 = \frac{\vartheta_\alpha(\phi)}{1-\lambda^{-\alpha}}$ and $y_n(\overline{x}) := T_{\overline{x}}^{-n}(y)$. By this estimate and by (5.3.3) we get for all $n \geq 1$ and all $x, y \in X$ such that $x \in B(y, \xi)$, the following:

$$\begin{aligned} \mathcal{L}_0^n(1\!\!1)(x) - \mathcal{L}_0^n(1\!\!1)(y) &= \Big(\frac{\mathcal{L}_0^n(1\!\!1)(x)}{\mathcal{L}_0^n(1\!\!1)(y)} - 1 \Big) \mathcal{L}_0^n(1\!\!1)(y) \\ &\leq C |\exp(C_1\rho(x,y)^\alpha) - 1| \leq C_2 \rho(x,y)^\alpha \end{aligned} \tag{5.3.5}$$

with C_2 depending on C, C_1 and ξ.

Proposition 5.3.1. *There exists a positive function $u_\phi \in \mathcal{H}_\alpha(X)$ such that $\mathcal{L}_0(u_\phi) = u_\phi$ and $\int u_\phi \, dm = 1$.*

Proof. By (5.3.5) and (5.3.3) the functions $\mathcal{L}_0^n(1\!\!1)$ have uniformly bounded norms in the space $\mathcal{H}_\alpha(X)$ of all Hölder continuous functions: see Section 4.4. Hence by the Arzela–Ascoli Theorem there exists a limit $u_\phi \in C(X)$ for a sub-sequence of $u_n = \frac{1}{n} \sum_{j=0}^{n-1} \mathcal{L}_0^j(1\!\!1)$, $n = 1, \ldots$. Of course, $u_\phi \in \mathcal{H}_\alpha(X)$, $C^{-1} \leq u_\phi \leq C$, and using (5.3.3), a straightforward computation shows that $\mathcal{L}_0(u_\phi) = u_\phi$ (compare 3.1.14). Also, $\int u_\phi \, dm = \lim_{n\to\infty} \int u_n \, dm = \int 1\!\!1 \, dm = 1$. The proof is complete. ♣

Combining this proposition, Proposition 5.2.2, Proposition 5.2.11 and Corollary 5.2.14, we get the following.

Theorem 5.3.2. *For every Hölder continuous function $\phi : X \to \mathbb{R}$ there exists a unique invariant Gibbs state associated to T and ϕ, namely $\mu_\phi = u_\phi m_\phi$.*

In the rest of this section we study in detail the iteration of $\mathcal{L}_0$ on the real or complex Banach spaces $C(X)$ and an $\mathcal{H}_\alpha$.

Definition 5.3.3. We call a bounded linear operator $Q : B \to B$ on a Banach space B *almost periodic* if for every $b \in B$ the family $\{Q^n(b)\}_{n=0}^{\infty}$ is relatively compact: that is, its closure in B is compact in the norm topology.

Proposition 5.3.4. *The operators $\mathcal{L}_0^n$ acting on $C(X)$ have the norms uniformly bounded for all $n = 1, 2, \ldots$.*

Proof. By the definition of $\mathcal{L}$, by (5.3.3) and by $\int \mathcal{L}_0^n(1\!\!1)\, d\mu_\phi = 1$, for every $u \in C(X)$ we get

$$\sup |\mathcal{L}_0^n(u)| \le \sup |u| \sup \mathcal{L}_0^n(1\!\!1) \le C \sup |u|. \tag{5.3.6}$$

♣

Remark that in the proof above, instead of referring to the form of $\mathcal{L}$, one can refer only to the fact that $\mathcal{L}$ is a positive operator: hence its norm is attained at $1\!\!1$.

Consider an arbitrary function $h : [0, \infty) \to [0, \infty)$ such that $h(0) = 0$, continuous at 0 and monotone increasing. We call such a function an *abstract modulus of continuity.* If $u : X \to \mathbb{C}$ is a function such that there is $\xi > 0$ such that for all $x, y \in X$ with $\rho(x, y) \le \xi$

$$|u(x) - u(y)| \le h(\rho(x, y)), \tag{5.3.7}$$

we say that h is a modulus of continuity of u. Given also $b \ge 0$, we denote by $C_h^b(X)$ the set of all functions $u \in C(X)$ such that $\|u\|_\infty \le b$ and h is a modulus of continuity of u with fixed $\xi > 0$. By the Arzela–Ascoli Theorem each $C_h^b(X)$ is a compact subset of $C(X)$.

Theorem 5.3.5. *The operator $\mathcal{L}_0 : C(X) \to C(X)$ is almost periodic. Moreover, if $b \ge 0$, h is an abstract modulus of continuity, $\theta \ge 0$, and ξ as in Lemma 4.1.2, then for all $\phi \in \mathcal{H}_\alpha$ with $\vartheta_\alpha(\phi) \le \theta$ there exist $\hat{b}$ and $\hat{C}$ depending only on b and θ such that for the abstract modulus of continuity $\hat{h}(t) = \hat{C}(t^\alpha + h(t))$*

$$\{\mathcal{L}_0^n(u) : u \in C_h^b(X), n \ge 0\} \subset C_{\hat{h}}^{\hat{b}}(X). \tag{5.3.8}$$

Proof. It follows from (5.3.6) that we can set $\hat{b} = Cb$. For every $x \in X$ and $n \ge 0$ denote $\exp(S_n\overline{\phi}(x))$ by $E_n(x)$. Consider arbitrary points $x \in X$ and $y \in B(x, \xi)$. Use the notation $y_n(\overline{x}) := T_{\overline{x}}^{-n}(y)$, the same as in (5.3.4). Fix $u \in C_h^b$ By (5.3.5) and (5.3.3) we have for every $u \in C(X)$

$$\begin{aligned}
|\mathcal{L}_0^n(u)(x) - \mathcal{L}_0^n(u)(y)| &= \Big| \sum_{\overline{x} \in T^{-n}(x)} u(\overline{x}) E_n(\overline{x}) - u(y_n(\overline{x})) E_n(y_n(\overline{x})\Big| \\
&\le \Big| \sum_{\overline{x} \in T^{-n}(x)} u(\overline{x})(E_n(\overline{x}) - E_n(y_n(\overline{x})))\Big| \\
&+ \Big| \sum_{\overline{x} \in T^{-n}(x)} E_n(y_n(\overline{x}))(u(\overline{x}) - u(y_n(\overline{x}))\Big|
\end{aligned}$$

$$\begin{aligned}
&\le \|u\|_\infty C_2\rho(x,y)^\alpha + Ch\left(\sup_{\overline{x}\in T^{-n}(x)} |u(\overline{x}) - u(y_n(\overline{x}))|\right)\\
&\le bC_2\rho(x,y)^\alpha + Ch(\lambda^{-n}\rho(x,y))\\
&\le bC_2\rho(x,y)^\alpha + Ch(\rho(x,y)). \qquad (5.3.9)
\end{aligned}$$

Therefore we are done by setting $\hat{C} := \max(bC_2, C)$. ♣

For $u \in \mathcal{H}_\alpha$ we obtain the fundamental estimate (5.3.10).

Theorem 5.3.6. *There exist constants $C_3, C_4 > 0$ such that for every $u \in \mathcal{H}_\alpha$, all $n = 1, 2, \ldots$ and $\lambda > 1$ from the expanding property of T,*

$$\vartheta_\alpha(\mathcal{L}_0^n(u)) \le C_3\lambda^{-n\alpha}\vartheta_\alpha(u) + C_4\|u\|_\infty. \qquad (5.3.10)$$

Proof. Continuing the last line of (5.3.9) and using $\rho(\overline{x}, y_n(\overline{x})) \le \lambda^{-n}\rho(\overline{x}, \overline{y})$, we obtain

$$|\mathcal{L}_0^n(u)(x) - \mathcal{L}_0^n(u)(y)| \le \|u\|_\infty C_2\rho(x,y)^\alpha + C\vartheta_{\alpha,\xi}(u)\lambda^{-n\alpha}\rho(x,y)^\alpha.$$

This proves (5.3.10), provisionally with $\vartheta_{\alpha,\xi}$ rather than ϑ_α, with $C_3 = C$ from (4.4.3) and (5.3.3) and with $C_4 = C_2$ (recall that the latter constant is of order CC_1 where C_1 appeared in (5.3.4)). To get a bound on ϑ_α replace C_4 by $\max\{C_4, 2C/\xi^\alpha\}$, see (5.3.6) and Section 4.4. ♣

Corollary 5.3.7. *There exist an integer $N > 0$ and real numbers $0 < \tau < 1, C_5 > 0$ such that for every $u \in \mathcal{H}_\alpha$,*

$$\|\mathcal{L}_0^N(u)\|_{\mathcal{H}_\alpha} \le \tau\|u\|_{\mathcal{H}_\alpha} + C_5\|u\|_\infty. \qquad (5.3.11)$$

Proof. This Corollary immediately follows from (5.3.10) and Proposition 5.3.4. ♣

In fact a reverse implication, yielding (5.3.10) for iterates of $\mathcal{L}^N$, holds:

Proposition 5.3.8. *(5.3.11) together with (5.3.6) imply*

$$\exists C_6 > 0\ \forall n = 1, 2, \ldots \qquad \|\mathcal{L}_0^{nN}(u)\|_{\mathcal{H}_\alpha} \le \tau^n\|(u)\|_{\mathcal{H}_\alpha} + C_6\|u\|_\infty. \qquad (5.3.12)$$

Proof. Substitute in (5.3.11) $\mathcal{L}_0^N(u)$ in place of u etc. n times using $\|\mathcal{L}_0^j(u)\|_\infty \le C\|u\|_\infty$. We obtain (5.3.12) with $C_6 = CC_5/(1-\tau)$. ♣

5.4 Convergence of $\mathcal{L}^n$; mixing properties of Gibbs measures

Recall that by Proposition 5.3.1 there exists a positive function $u_\phi \in \mathcal{H}_\alpha(X)$ such that $\mathcal{L}_0(u_\phi) = u_\phi$ and $\int u_\phi dm_\phi = 1$.

It is convenient to replace the operator $\mathcal{L}_0 = \mathcal{L}_{\overline{\phi}}$ by the operator $\hat{\mathcal{L}} = \hat{\mathcal{L}}_\phi$, defined by

$$\hat{\mathcal{L}}(u) = \frac{1}{u_\phi}\mathcal{L}_0(uu_\phi).$$

If we denote the operator of multiplication by a function w by the same symbol w, then we can write

$$\hat{\mathcal{L}}(u) = u_\phi^{-1} \circ \mathcal{L}_0 \circ u_\phi.$$

Since $\hat{\mathcal{L}}$ and $\mathcal{L}_0$ are conjugate by the operator u_ϕ, their spectra are the same. In addition, as this operator u_ϕ is positive, non-negative functions go to non-negative functions. Hence measures are mapped to measures by the conjugate operator.

Proposition 5.4.1. *$\hat{\mathcal{L}} = \mathcal{L}_\psi$ where $\psi = \overline{\phi} + \log u_\phi - \log u_\phi \circ T = \phi - \mathrm{P}(T,\phi) + \log u_\phi - \log u_\phi \circ T$.*

Proof.

$$\begin{aligned}\hat{\mathcal{L}}(u)(x) &= \frac{1}{u_\phi(x)} \sum_{T(\overline{x})=x} u(\overline{x})u_\phi(\overline{x}) \exp\phi(\overline{x}) \\ &= \sum_{T(\overline{x})=x} u(\overline{x}) \exp(\phi(\overline{x}) + \log u_\phi(\overline{x}) - \log u_\phi(x)).\end{aligned}$$

♣

Note that the eigenfunction u_ϕ for $\mathcal{L}_0$ has changed to the eigenfunction $1\!\!1$ for $\hat{\mathcal{L}}$. In other words, we have the following.

Proposition 5.4.2. $\hat{\mathcal{L}}(1\!\!1) = 1\!\!1$*: that is, for every $x \in X$*

$$\sum_{\overline{x}\in T^{-1}(x)} \exp\psi(\overline{x}) = 1. \tag{5.4.1}$$

♣

Note that the Jacobian of T with respect to the Gibbs measure $\mu = u_\phi m$ (see Theorem 5.3.2) is $(u_\phi \circ T)(\exp(-\overline{\phi}))u_\phi^{-1} = \exp(-\psi)$. So for ψ the reference measure (with Jacobian $\exp(-\psi)$) and the invariant Gibbs measure coincide.

Note that passing from $\mathcal{L}_\phi$, through $\mathcal{L}_{\overline{\phi}}$, to $\mathcal{L}_\psi$ we have been replacing ϕ by co-homological (up to a constant) functions. By Proposition 5.1.4 this does not change the set of Gibbs states.

One can think of the transformation $u \mapsto u/u_\phi$ as new coordinates on $C(X)$ or $\mathcal{H}_\alpha(X)$ (real or complex-valued functions). $\mathcal{L}_0$ changes in these coordinates to $\mathcal{L}_\psi$, and the functional $m(u)$ changes to $m(u_\phi u)$. The latter (denote it by m_ψ) is the eigen-measure for $\mathcal{L}_\psi^*$ with eigenvalue 1. It is positive because the operator u_ϕ is positive (see the comment above). So $\exp(-\psi)$ is the Jacobian for m_ψ by Corollary 5.2.7. Hence, by (5.4.1), m_ψ is T-invariant. This is our invariant Gibbs measure μ.

Proposition 5.3.4 applied to $\hat{\mathcal{L}}$ takes the following form.

Proposition 5.4.3. $\|\hat{\mathcal{L}}\|_\infty = 1$.

Proof. $\sup|\hat{\mathcal{L}}(u)| \le \sup|u|$ because $\hat{\mathcal{L}}$ is an operator of 'taking an average' of u from the past (by Proposition 5.4.2). The equality follows from $\hat{\mathcal{L}}(1\!\!1) = 1\!\!1$. ♣

The topological exactness of T gives a stronger result, as follows.

Lemma 5.4.4. *Let $T : X \to X$ be a continuous, topologically exact, distance-expanding open map. Suppose that $g : [0,\infty) \to [0,\infty)$ is an abstract modulus of continuity. Then for every $K > 0$ and $\delta_1 > 0$ there exist $\delta_2 > 0$ and $n > 0$ such that*

- *for all $\phi \in \mathcal{H}_\alpha$ with $\|\phi\|_{\mathcal{H}_\alpha} \le K$ and*
- *for all $u \in C(X,\mathbb{R})$ with g being its modulus of continuity and such that $\int u\,d\mu = 0$ and $\|u\|_\infty \ge \delta_1$,*

we have for $\hat{\mathcal{L}} = \hat{\mathcal{L}}_\phi$

$$\|\hat{\mathcal{L}}^n(u)\|_\infty \le \|u\|_\infty - \delta_2.$$

Proof. Fix $\varepsilon > 0$ so small that $g(\varepsilon) < \delta_1/2$. Let n be ascribed to ε according to Theorem 4.3.13(2): that is, $\forall x\ T^n(B(x,\varepsilon)) = X$. Since $\int u\,d\mu = 0$, there exist $y_1, y_2 \in X$ such that $u(y_1) \le 0$ and $u(y_2) \ge 0$. For an arbitrary $x \in X$ choose $x' \in B(y_1,\varepsilon) \cap T^{-n}(x)$ (it exists by the definition of n). We have $u(x') \le u(y_1) + g(\varepsilon) \le \delta_1/2 \le \|u\|_\infty - \delta_1/2$. So:

$$\begin{aligned}
\hat{\mathcal{L}}^n(u)(x) &= u(x')\exp S_n\psi(x') + \sum_{\overline{x}\in T^{-n}(x)\setminus\{x'\}} u(\overline{x})\exp S_n\psi(\overline{x}) \\
&\le (\|u\|_\infty - \delta_1/2)\exp S_n\psi(x') + \|u\|_\infty \sum_{\overline{x}\in T^{-n}(x)\setminus\{x'\}} \exp S_n\psi(\overline{x}) \\
&\le \|u\|_\infty \left(\sum_{\overline{x}\in T^{-n}(x)} \exp S_n\psi(\overline{x}) \right) - (\delta_1/2)\exp S_n\psi(x') \\
&= \|u\|_\infty - (\delta_1/2)\exp S_n\psi(x').
\end{aligned}$$

Similarly for $x'' \in B(y_2,\varepsilon) \cap T^{-n}(x)$:

$$\hat{\mathcal{L}}^n(u)(x) \ge -\|u\|_\infty + (\delta_1/2)\exp S_n\psi(x'').$$

Thus we have proved our lemma with $\delta_2 := (\delta_1/2)\inf_{x\in X}\exp S_n\psi(x)$. To complete the proof we need to relate δ_2 to ϕ rather than to ψ. To this end, note that for every $x \in X$ we have $\psi(x) \ge \phi(x) - 2\log\|(u_\phi)\|_\infty - P(T,\phi) \ge -3\log\|(u_\phi)\|_\infty - \mathrm{h_{top}}(T)$, and $\|u_\phi)\|_\infty \le C$, where C depends on K: see (5.3.3), (4.4.2), (4.4.3). ♣

We shall prove now a theorem that completes Proposition 5.3.4 and Theorem 5.3.5.

Theorem 5.4.5. *For every $u \in C(X, \mathbb{C})$, and T a continuous, topologically exact, distance-expanding open map, we have for $c = e^{P(T,\phi)}$*

$$\lim_{n\to\infty} \|c^{-n}\mathcal{L}_\phi^n(u) - m_\phi(u)u_\phi\|_\infty = 0. \tag{5.4.2}$$

In particular, if $\int u\, d\mu = 0$, then

$$\lim_{n\to\infty} \|\hat{\mathcal{L}}^n(u)\|_\infty = 0. \tag{5.4.3}$$

Moreover, the convergences in (5.4.2) and (5.4.3) are uniform for $u \in C_h^b$ and ϕ in an arbitrary bounded subset H of in $\mathcal{H}_\alpha(X)$.

Proof. For real-valued u, with $\int u d\mu = 0$, the sequence $a_n(u) := \|\hat{\mathcal{L}}^n(u)\|_\infty$ is monotone decreasing, by Proposition 5.4.3. Suppose that $\lim_{n\to\infty} a_n = a > 0$. By Theorem 5.3.5 all the iterates $\hat{\mathcal{L}}^n(u)$ have a common modulus of continuity g. So applying Lemma 5.4.4 with this g and $\delta_1 = a$ we find n_0 and $\delta_2 > 0$ such that $\|\hat{\mathcal{L}}^{n_0}(\hat{\mathcal{L}}^n(u))\|_\infty \le \|\hat{\mathcal{L}}^n(u)\|_\infty - \delta_2$ for every $n \ge 0$. So, for n such that $\|\hat{\mathcal{L}}^n(u)\|_\infty < a + \delta_2$, we obtain $\|\hat{\mathcal{L}}^{n+n_0}(u)\|_\infty < a$, which contradicts the definition of a. This proves (5.4.3) for u real-valued. For u complex-valued with $\int u d\mu = 0$, decompose u into the real and complex parts.

To prove (5.4.2), note first that, for an arbitrary $u \in C(X, \mathbb{C})$ the convergence in (5.4.3) yields, owing to $\hat{\mathcal{L}}(1\!\!1) = 1\!\!1$,

$$||\hat{\mathcal{L}}^n(u) - \mu(u)1\!\!1||_\infty = ||\hat{\mathcal{L}}^n(u - \mu(u)1\!\!1)||_\infty \to 0.$$

Now change coordinates on $C(X)$ to go back to $\mathcal{L}_0$ and then replace it by $c^{-1}\mathcal{L}_\phi$. One obtains (5.4.2).

For the last part of the theorem, set

$$a_n := \sup\{||\hat{\mathcal{L}}_\phi^n(u) : \phi \in H, u \in C_h^b; u \ge 0\}$$

and proceed in the same way as above, with the help of the full power of Lemma 5.4.4. ♣

Note that (5.4.2) means weak*-convergence of measures

$$\lim_{n\to\infty} \sum_{\overline{x}\in T^{-n}(x)} c^{-n} \exp(S_n\phi(\overline{x}))\delta_{\overline{x}} \to u_\phi(x)m_\phi$$

for every $x \in X$. Using (5.4.2) also for $u = 1\!\!1$, we obtain

$$\lim_{n\to\infty} \mathcal{L}_\phi^n(1\!\!1)(x))^{-1} \sum_{\overline{x}\in T^{-n}(x)} (\exp(S_n\phi(\overline{x}))\delta_{\overline{x}} \to m_\phi. \tag{5.4.4}$$

In the sequel one can consider either $C(X, \mathbb{R})$ or $C(X, \mathbb{C})$. Let us choose $C(X, \mathbb{C})$.

Note that by $\mathcal{L}_\phi^*(m_\phi) = cm_\phi$, we have the $\mathcal{L}$-invariant decomposition

$$C(X) = \text{span}(u_\phi) \oplus \ker(m_\phi). \tag{5.4.5}$$

For $u \in \text{span}(u_\phi)$ we have $\mathcal{L}_\phi(u) = cu$. On $\ker(m_\phi)$, by Theorem 5.4.5, c^{-n} $\mathcal{L}_\phi^n \to 0$ in strong topology. Denote $(\mathcal{L}_\phi)|_{\ker(m_\phi)}$ by $\mathcal{L}_{\ker,\phi}$. For $\mathcal{L}_{\ker,\phi}$ restricted to $\mathcal{H}_\alpha$ we can say more about the above convergence.

Theorem 5.4.6. *There exists an integer $n > 0$ such that for $c = e^{P(T,\phi)}$*

$$\|c^{-n}\mathcal{L}_{\ker,\phi}^n\|_{\mathcal{H}_\alpha} < 1.$$

Proof. Again, it is sufficient to consider a real-valued function u with $\mu(u) = 0$ and the operator $\hat{\mathcal{L}}$. Set $\delta = \min\{1/8C_4, 1/4\}$, with C_4 taken from (5.3.10). By Theorem 5.3.6 for u such that $\|u\|_{\mathcal{H}_\alpha} \le 1$, all functions $\hat{\mathcal{L}}^n(u)$ have the same modulus of continuity $g(\varepsilon) = C_7\varepsilon^\alpha$ with $C_7 = C_3 + C_4 > 0$. Hence from Theorem 5.4.5 we conclude that $(\exists n_1)(\forall n \ge n_1)(\forall u : \|u\|_{\mathcal{H}_\alpha} \le 1)$

$$\|\hat{\mathcal{L}}^n(u)\|_\infty \le \delta. \tag{5.4.6}$$

Next, for n_2 satisfying $C_3\lambda^{-n_2\alpha}C_7 + C_4\delta \le 1/4$, again by Theorem 5.3.6, we obtain

$$\vartheta_\alpha(\hat{\mathcal{L}}^{n_2}(\hat{\mathcal{L}}^{n_1}(u)) \le 1/4.$$

Hence $\|\hat{\mathcal{L}}^{n_1+n_2}(u)\|_{\mathcal{H}_\alpha} \le 1/2$. The theorem has thus been proved with $n = n_1 + n_2$. ♣

Note that Theorem 5.4.5 could be deduced from Theorem 5.4.6 by approximation of continuous functions uniformly by Hölder ones, and using Proposition 5.3.4.

Corollary 5.4.7. *The convergences in Theorem 5.4.5 for $u \in \mathcal{H}_\alpha$ are exponential: that is, there exist $0 < \tau < 1$ and $C \ge 0$ such that for every function $u \in \mathcal{H}_\alpha$*

$$\begin{aligned} \|c^{-n}\mathcal{L}_\phi^n(u) - m_\phi(u)u_\phi\|_\infty &\le \|c^{-n}\mathcal{L}_\phi^n(u) - m_\phi(u)u_\phi\|_{\mathcal{H}_\alpha} \\ &\le C\|u - m_\phi(u)u_\phi\|_{\mathcal{H}_\alpha}\tau^n. \end{aligned} \tag{5.4.7}$$

In particular, if $\int u\,d\mu = 0$, then

$$\|\hat{\mathcal{L}}^n(u)\|_\infty \le \|\hat{\mathcal{L}}^n(u)\|_{\mathcal{H}_\alpha} \le C\|u\|_{\mathcal{H}_\alpha}\tau^n. \tag{5.4.8}$$

Remark 5.4.8. Theorem 5.4.6 along with (5.4.5) and the fact that $c^{-1}\mathcal{L}\phi(u_\phi) = u_\phi$ implies that the spectrum of the operator $\mathcal{L}_\phi : \mathcal{H}_\alpha \to \mathcal{H}_\alpha$ consists of two parts: the number $c = e^{P(T,\phi)}$, which is its simple and isolated eigenvalue; and the rest, contained in a disc centred at 0 with radius $< c$. There thus exists a 'spectral gap'. An isolated eigenvalue moves analytically for an analytic family of transfer operators induced by analytic families of maps T and potential functions ϕ, yielding the analyticity of $P(T, \phi)$. See Section 6.4 and the notes at the end of this chapter.

Now we can study the 'mixing' properties of the dynamical system (T, μ) for our invariant Gibbs measure μ. Roughly speaking, the speed of mixing is related to the speed of convergence of $\mathcal{L}_{\ker,\phi}^n$ to 0.

The first dynamical (mixing) consequence of Theorem 5.4.6 is the following result, known in the literature as the exponential decay of correlations: see the definition in Section 2.11.

Theorem 5.4.9. *There exist $C \geq 1$ and $\rho < 1$ such that for all $f \in \mathcal{H}_\alpha$ and all $g \in L^1(\mu)$,*

$$C_n(f,g) \leq C\rho^n \|f - Ef\|_{\mathcal{H}_\alpha} \|g - Eg\|_1.$$

Proof. Set $F = f - Ef$, $G = g - Eg$, and consider $\hat{\mathcal{L}}$ acting on $C(X)$, as a restriction of $\mathcal{L}_\mu$ acting on $L^1(\mu)$. By (5.2.7) and (5.4.8) we obtain

$$|C_n(f,g)| = \left| \int F \cdot (G \circ T^n)\, d\mu \right| = \left| \int \hat{\mathcal{L}}^n(F) \cdot G\, d\mu \right| \leq C\tau^n \|F\|_{\mathcal{H}_\alpha} \|G\|_1.$$

♣

Exercise. Prove that for all μ square-integrable functions f, g one has $\int f \cdot (g \circ T^n)\, d\mu \to Ef \cdot Eg$. (Hint: Approximate f and g by Hölder functions. Of course, the information on the speed of convergence would become lost.)

The convergence in the exercise is one of several equivalent definitions of the *mixing* property: see Section 2.10. However, we proved earlier the stronger property, measure-theoretical exactness (Theorem 5.2.12).

We can make better use of the exponential convergence in Theorem 5.4.9 for T being the shift on the one-sided shift space:

Theorem 5.4.10. *Let $\sigma : \Sigma_A \to \Sigma_A$ be a topologically mixing topological one-sided Markov chain with the alphabet $\{1, \dots, d\}$ and σ the left shift (see Chapter 1). Let $\mathcal{F}$ be the σ-algebra generated by the partition $\mathcal{A}$ into 0-cylinders, that is, sets with fixed 0-th symbol. For every $0 \leq k \leq l$ denote by $\mathcal{F}_k^l$ the σ-algebra generated by $\mathcal{A}_k^l = \bigvee_{j=k}^l T^{-j}(\mathcal{A})$, that is, by the sets (cylinders) with fixed $k, k+1, \dots, l$'th symbols. Let $\phi : \Sigma_A \to \mathbb{R}$ be a Hölder continuous function.*

Then there exist $0 < \rho < 1$ and $C > 0$ such that for every $n \geq k \geq 0$, every function $f : \Sigma_A \to \mathbb{R}$ measurable with respect to $\mathcal{F}_0^k$ and every μ_ϕ-integrable function $g : \Sigma_A \to \mathbb{R}$

$$\left| \int f \cdot (g \circ T^n)\, d\mu_\phi - Ef \cdot Eg \right| \leq C\rho^{n-k} \|f - Ef\|_1 \|g - Eg\|_1. \tag{5.4.9}$$

Proof. Assume $Ef = Eg = 0$. By Theorem 5.4.9,

$$\left| \int f \cdot (g \circ T^n)\, d\mu \right| = \left| \int g \cdot \hat{\mathcal{L}}^{n-k}(\hat{\mathcal{L}}^k(f))\, d\mu \right| \leq \|g\|_1 C\rho^{n-k} \|\hat{\mathcal{L}}^k(f)\|_{\mathcal{H}_\alpha}. \tag{5.4.10}$$

Decompose f into real and imaginary parts, and represent each one by the difference of nowhere-negative functions. This allows us, in the estimates to follow, to assume that $f \geq 0$.

Notice that for every cylinder $A \in \mathcal{A}$ and $x \in A$, in the expression

$$\hat{\mathcal{L}}^k(f)(x) = \sum_{T^k(y)=x} f(y) \exp S_k\psi(y)$$

there is no dependence of $f(y)$ on $x \in A$, because f is constant on cylinders of $\mathcal{A}_0^k$. So

$$\frac{\sup_A \hat{\mathcal{L}}^k(f)}{\inf_A \hat{\mathcal{L}}^k(f} \leq \sup_{B\in\mathcal{A}_0^k} \sup_{y,y'\in B} \exp\big(S_k\psi(y) - S_k\psi(y')\big) \leq C,$$

with the constant C resulting from Section 4.4. So

$$\sup_A \hat{\mathcal{L}}^k(f) \leq \frac{C}{\mu(A)} \int \hat{\mathcal{L}}^k(f)\, d\mu = \frac{C}{\mu(A)} \|f\|_1 \leq \left(\frac{C}{\inf_{A\in\mathcal{A}} \mu(A)}\right) \|f\|_1 = C'\|f\|_1,$$

where the last equality defines C'.

It is still left to estimate the pseudonorms $\vartheta_{\alpha,\xi}$ and ϑ_α of $\hat{\mathcal{L}}^k(f)$: cf. Section 4.4. We assume that ξ is less than the minimal distance between the cylinders in $\mathcal{A}$. We have, similarly to (5.3.5), for x, y belonging to the same cylinder $A \in \mathcal{A}$,

$$\begin{aligned} |\hat{\mathcal{L}}^k(f)(x) - \hat{\mathcal{L}}^k(f)(y)| &= \left|\left(\frac{\hat{\mathcal{L}}^k(f)(x)}{\hat{\mathcal{L}}^k(f)(y)} - 1\right)\right| |\hat{\mathcal{L}}^k(f)(y)| \\ &\leq (\exp C_1\rho(x,y)^\alpha - 1)\|C'\|f\|_1 \leq C''\rho(x,y)^\alpha \|f\|_1. \end{aligned}$$

for a constant C''.

Hence $\vartheta_{\alpha,\xi}(\hat{\mathcal{L}}^k(f)) \leq \|f\|_1 C''$ and, passing to ϑ_α as in Section 4.4, we get

$$\vartheta_\alpha(\hat{\mathcal{L}}^k(f)) \leq \|f\|_1 \max\{C'', 2C'\xi^{-\alpha}\}.$$

Thus, continuing (5.4.10), we obtain for a constant C that

$$C_n(f,g) \leq \|f\|_1 \|g\|_1 C\rho^{n-k}.$$

♣

An immediate corollary from Theorem 5.4.10 is that for every $B_1 \in \mathcal{F}_0^k$ and a Borel B_2 (i.e. $B_2 \in \mathcal{F}_0^\infty$),

$$|\mu(B_1 \cap T^{-n}(B_2)) - \mu(B_1)\mu(B_2)| \leq C\rho^{n-k}\mu(B_1)\mu(B_2). \tag{5.4.11}$$

Compare this with (2.11.10). Therefore, for any non-negative integer t and every $A \in \mathcal{F}_0^k$,

$$\sum_{B\in\mathcal{A}_0^t} |\mu(T^{-n}(B)|A) - \mu(B)| \leq C\rho^{n-k},$$

for the conditional measures $\mu(\cdot|A)$, with respect to A.

This means that $\mathcal{A}$ satisfies the weak Bernoulli property: hence the natural extension $(\tilde{X}, \tilde{T}, \tilde{\mu})$ is measure-theoretically isomorphic to a two-sided Bernoulli shift (see Section 2.11).

Corollary 5.4.11. *Every continuous, topologically exact, distance-expanding open map $T : X \to X$, with invariant Gibbs measure $\mu = \mu_\phi$ for a Hölder continuous function $\phi : X \to \mathbb{R}$, has the natural extension $(\tilde{X}, \tilde{T}, \tilde{\mu})$ measure-theoretically isomorphic to a two-sided Bernoulli shift.*

Proof. Let $\pi : \Sigma_A \to X$ be the coding map from a one-sided topological Markov chain, due to a Markov partition: see Theorem 4.5.7. Since the map π is Hölder continuous, the function $\phi \circ \pi : \Sigma_A \to \mathbb{R}$ is also Hölder continuous: hence we have the invariant Gibbs measure $\mu_{\phi\circ\pi}$. For this measure we can apply Theorem 5.4.10 and its corollaries. Recall also that by Theorem 4.5.9 π yields a measure-theoretical isomorphism between $\mu_{\phi\circ\pi}$ and $\mu_{\phi\circ\pi} \circ \pi^{-1}$. Therefore, to complete the proof, it is sufficient to prove the following.

Lemma 5.4.12. *The measures μ_ϕ and $\mu_{\phi\circ\pi} \circ \pi^{-1}$ coincide.*

Proof. The function $\exp(-\phi \circ \pi + P - h)$ for $h := \log u_{\phi\circ\pi} + \log u_{\phi\circ\pi} \circ \sigma)$, is the strong Jacobian for the shift map σ and the measure $\mu_{\phi\circ\pi}$, where P is the topological pressure for both $(\sigma, \phi \circ \pi)$ and (T, ϕ): see Theorem 4.5.8. Since π yields a measure-theoretical isomorphism between $\mu_{\phi\circ\pi}$ and $\mu_{\phi\circ\pi} \circ \pi^{-1}$, the measure $\mu_{\phi\circ\pi} \circ \pi^{-1}$ is forward quasi-invariant under T with the strong Jacobian $\exp(-\phi + P - h \circ \pi^{-1})$. T with respect to μ_ϕ has a strong Jacobian of the same form, possibly with a priori different h co-homologous to 0 in bounded functions. Therefore the two measures are equivalent: hence, as they are ergodic, they coincide. ♣

5.5 More on almost periodic operators

In this section we show how to deduce Theorem 5.4.5 (on convergence) and Theorem 5.4.6 and Corollary 5.4.7 (exponential convergence) from general functional analysis theorems. We do not need this later in this book, but the theorems are useful in other important situations.

Recall (Definition 5.3.3) that $Q : F \to F$ a bounded linear operator of a Banach space is called almost periodic if for every $b \in F$ the sequence $Q^n(b)$ is relatively compact. By the Banach–Steinhaus Theorem there is a constant $C \geq 0$ such that $\|Q^n\| \leq C$ for every $n \geq 0$.

Theorem 5.5.1. *If $Q : F \to F$ is an almost periodic operator on a complex Banach space F, then*

$$F = F_0 \oplus F_{\mathrm{u}}, \tag{5.5.1}$$

where $F_0 = \{x \in F : \lim_{n\to\infty} A^n(x) = 0\}$, and F_{u} is the closure of the linear subspace of F generated by all eigenfunctions of eigenvalues of modulus 1.

Adding additional assumptions one gains additional information on the above decomposition.

Definition 5.5.2. Let $F = C(X)$, and suppose $Q : F \to F$ is *positive*: that is, $f \geq 0$ implies $Q(f) \geq 0$. Then Q is called *primitive* if for every $f \in C(X)$, $f \geq 0, f \not\equiv 0$ there exists $n \geq 0$ such that for every $x \in X$ it holds that $Q^n(f)(x) > 0$. If we change the order of the quantifiers to 'for every x there exists n', then we call Q *non-decomposable.*

Theorem 5.5.3. *For $Q : C(X) \to C(X)$, a (real or complex) linear almost periodic positive primitive operator of spectral radius equal to 1, we have*

(1) *dim*$(C(X)_{\mathrm{u}}) = 1$ *in the decomposition (5.5.1)*

(2) *the eigenvalue corresponding to* $C(X)_{\mathrm{u}}$ *is equal to 1, and the respective eigenfunction* u_Q *is positive (everywhere* > 0*).*

(3) *In addition there exists a probability measure* m_Q *on* X *invariant under the conjugate operator* Q^**, such that for every* $u \in C(X)$ *we have the strong convergence*

$$Q^n(u) \to u_Q \int u\, dm_Q.$$

Proof. This is just a repetition of considerations of Sections 5.2–5.4. Find first a probability measure m such that $Q^*(m) = m$ as in Theorem 5.2.8 (we leave to the reader the proof that the eigenvalue is equal to 1). Next observe that by the almost periodicity of Q the sequence of averages $a_n := \frac{1}{n}\sum_{j=0}^{n-1} Q^j(\mathbb{1})$ is relatively compact (an exercise). Let u_Q be any function in the limit. Then $u_Q \geq 0$ is an eigenfunction for the eigenvalue 1. It is not identically 0, since $\int a_n dm = 1$ for all $n \geq 0$. We have $u_Q = Q(u_Q) > 0$, because Q is non-decomposable. Finally, for $\hat{Q}(u) := Q(uu_Q)u_Q^{-1}$ we have $\hat{Q}(\mathbb{1}) = \mathbb{1}$ and we repeat the proof of Theorem 5.4.5, replacing the property of topological exactness by primitivity. ♣

Notice that this yields Theorem 5.4.5 because of the following proposition.

Proposition 5.5.4. *If an open expanding map* T *is topologically exact, then for every continuous function* ϕ *the transfer operator* $Q = \mathcal{L}_{\overline{\phi}}$ *is primitive.*

The proof is easy: it is in fact contained in the proof of Lemma 5.4.4.

Assume now only that T is topologically transitive. Let Ω^k denote the sets from spectral decomposition $X = \Omega = \bigcup_{k=1}^{n} \Omega^k$, as in Theorem 4.3.8. Write $u_Q \in C(X)$ for an eigenfunction of the operator Q as before. Now note (exercise!) that the space F_{u} for the operator $Q = \mathcal{L}_{\overline{\phi}}$ is spanned by n eigenfunctions $v_t = \sum_{k=1}^{n} \chi_{\Omega^k} \lambda^{-tk} u_Q, \quad t = 1, \ldots, n$, where χ means indicator functions, with $\lambda = \varepsilon^{2\pi i/n}$. Each v_t corresponds to the eigenvalue λ^t. Thus the set of these eigenvalues is a cyclic group.

It is also an easy exercise to describe F_{u} if $X = \Omega = \bigcup_{j=1}^{J} \bigcup_{k=1}^{k(j)} \Omega_j^k$. The set of eigenvalues is the union of J cyclic groups. It is harder to understand F_{u} and the corresponding set of eigenvalues for T open expanding, without assuming $\Omega = X$.

A general theorem related to Theorem 5.4.6 and Corollary 5.4.7 is the following.

Theorem 5.5.5 (Ionescu Tulcea and Marinescu). *Let* $(F, |\cdot|)$ *be a Banach space equipped with a norm* $|\cdot|$*, and let* $E \subset F$ *be its dense linear subspace.* E *is assumed to be a Banach space with respect to a norm* $\|\cdot\|$ *defined on it. Let* $Q : F \to F$ *be a bounded linear operator that preserves* E*, whose restriction to* E *is also bounded with respect to the norm* $\|\cdot\|$*.*

Suppose the following conditions are satisfied.

(1) *If* $(x_n : n = 1, 2, \ldots)$ *is a sequence of points in* E *such that* $\|x_n\| \le K_1$ *for all* $n \ge 1$ *and some constant* K_1, *and if* $\lim_{n\to\infty} |x_n - x| = 0$ *for some* $x \in F$, *then* $x \in E$ *and* $\|x\| \le K_1$.

(2) *There exists a constant* K *such that* $|Q^n| \le K$ *for all* $n = 1, 2, \ldots$.

(3) $\exists N \ge 1 \quad \exists \tau < 1 \quad \exists K_2 > 0 \quad \|Q^N(x)\| \le \tau\|x\| + K_2|x|$ *for all* $x \in E$.

(4) *For any bounded subset* A *of the Banach space* E *with norm* $\|\cdot\|$, *the set* $Q^N(A)$ *is relatively compact as a subset of the Banach space* F *with norm* $|\cdot|$.

Then

(5) *There exist at most finitely many eigenvalues of* $Q : F \to F$ *of modulus* 1, *say* $\gamma_1, \ldots, \gamma_p$.

(6) *Let* $F_i = \{x \in F : Q(x) = \gamma_i x\}$, $i = 1, \ldots, p$. *Then* $F_i \subset E$ *and* $\dim(F_i) < \infty$.

(7) *The operator* $Q : F \to F$ *can be represented as*

$$Q = \sum_{i=1}^{p} \gamma_i Q_i + S,$$

where Q_i *and* S *are bounded,* $Q_i(F) = F_i$, $\sup_{n\ge1} |S^n| < \infty$ *and*

$$Q_i^2 = Q_i, \quad Q_iQ_j = 0\ (i \ne j), \quad Q_iS = SQ_i = 0$$

Moreover,

(8) $S(E) \subset E$ *and* $S|_E$ *considered as a linear operator on* $(E, \|\cdot\|)$ *is bounded, and there exist constants* $K_3 > 0$ *and* $0 < \tilde\tau < 1$ *such that*

$$\|S^n|_E\| \le K_3\tilde\tau^n$$

for all $n \ge 1$.

The proof of this theorem can be found in [Ionescu Tulcea & Marinescu 1950].
Now, in view of Theorem 4.4.1 and Corollary 5.3.7, Theorem 5.5.5 applies to the operator $Q = \mathcal{L}_{\overline{\phi}} : C(X) \to C(X)$ if one substitutes $F = C(X)$, $E = \mathcal{H}_\alpha(X)$. If T is topologically exact and in consequence Q is primitive on $C(X)$, then $\dim(\oplus F_i) = 1$ and the corresponding eigenvalue is equal to 1, as in Theorem 5.5.3.

5.6 Uniqueness of equilibrium states

We have already proved the existence (Theorem 5.3.2) and uniqueness (Corollary 5.2.14) of invariant Gibbs states, and proved that invariant Gibbs states are equilibrium states (Proposition 5.1.5). Here we shall give three different proofs of the uniqueness of equilibrium states.

Let ν be a T-invariant measure, and let a finite real function J_ν be the corresponding Jacobian in the weak sense; J_ν is defined ν-a.e. By the invariance of ν we have $\nu(E) = 0 \Rightarrow \nu(T^{-1}(E)) = \nu(E) = 0$: that is, ν is backward quasi-invariant. At the beginning of Section 5.2 we defined in this situation $\Psi = \Phi_x \circ T$ with $\Phi_x = \frac{d\nu \circ T_x^{-1}}{d\nu}$ defined for ν-a.e. point in the domain of a branch T_x^{-1}. (In Section 5.2 we used the notation Φ_j for Φ_x.) By definition, Φ_x is strong Jacobian for T_x^{-1}.

Notice that for ν-a.e. z

$$(J_\nu \circ T_x^{-1}) \cdot \Phi_x(z) = \begin{cases} 1, & \text{if } \Phi_x(z) \neq 0; \\ 0, & \text{if } \Phi_x(z) = 0. \end{cases} \tag{5.6.1}$$

Indeed, after removal of $\{z : \Phi_x(z) = 0\}$, the measures ν and $\nu \circ T^{-1}$ are equivalent: hence Jacobians of T and T_x^{-1} are mutual reciprocals. We can fix J_ν on the set $T^{-1}(\{z : \Phi_x(z) = 0\})$ arbitrarily, since this set has measure ν equal to 0.

Recall that we have defined $\mathcal{L}_\nu : L^1(\nu) \to L^1(\nu)$, the transfer operator associated with the measure ν, as follows:

$$\mathcal{L}_\nu(g)(x) = \sum_{y \in T^{-1}(x)} g(y)\Psi(y).$$

Recall that if T maps a set A of measure 0 to a set of positive measure, then Ψ is specified, equal to 0, on a subset of A that is mapped by T to a set of full measure ν in $T(A)$.

Then, since ν is T-invariant, $\mathcal{L}_\nu(\mathbb{1}) = \mathbb{1}$, and for every ν-integrable function g we have $\int \mathcal{L}_\nu(g)\, d\nu = \int g\, d\nu$: compare (5.2.4).

Lemma 5.6.1. *Let $\psi : X \to \mathbb{R}$ be a continuous function such that $\mathcal{L}_\psi(\mathbb{1}) = \mathbb{1}$ (that is, for every x, $\sum_{y \in T^{-1}(x)} \exp(\psi(y)) = 1$), and let ν be an ergodic equilibrium state for ψ. Then J_ν is strong Jacobian and $J_\nu = \exp(-\psi)$ ν-almost everywhere.*

Proof. The proof is based on the following computation using the inequality $1 + \log(x) \leq x$, with the equality only for $x = 1$.

$$\begin{aligned} 1 &= \int \mathbb{1}\, d\nu \geq \int \mathcal{L}_\nu(J_\nu \exp \psi)\, d\nu = \int J_\nu \exp \psi\, d\nu \\ &\geq \int (1 + \log(J_\nu \exp \psi))\, d\nu = 1 + \int \psi d\nu + \int \log J_\nu\, d\nu \\ &= 1 + \int \psi d\nu + \mathrm{h}_\nu(T) \geq 1. \end{aligned}$$

To obtain the first inequality, write

$$\mathcal{L}_\nu(J_\nu \exp \psi)(x) = \sum_{y \in T^{-1}(x)} J_\nu(y)(\exp \psi(y))\Psi(y),$$

which is equal to 1 if $\Psi(y) > 0$ for all $y \in T^{-1}(x))$, or < 1 otherwise. This follows from (5.6.1) and from $\sum_{y \in T^{-1}(x)} \exp \psi(y) = 1$.

The last inequality follows from

$$\int \psi d\nu + \mathrm{h}_\nu(T) = \mathrm{P}(\psi) \geq \limsup_{n\to\infty} \frac{1}{n} \log \sum_{y\in T^{-n}(x)} \exp S_n\psi(y) = 0,$$

(see Theorem 3.3.2), since all points in $T^{-n}(x)$ are (n,η)-separated, for $\eta > 0$ defined in Chapter 4.

Therefore all the inequalities in this proof must become equalities. Thus the Jacobian $\Phi_x \neq 0$ for each branch T_x^{-1} and $J_\nu = \exp(-\psi)$, ν- a.e. ♣

Note that we have not assumed above that ψ is Hölder. Now, we shall assume it.

Theorem 5.6.2. *There exists exactly one equilibrium state for each Hölder continuous potential ϕ.*

Proof. Let ν be an equilibrium state for ϕ. As in Section 5, set $\psi = \phi - \mathrm{P}(T,\phi) + \log u_\phi \circ T - \log u_\phi$, and ν is also the equilibrium state for ψ. Then by Lemma 5.6.1 its Jacobian is strong Jacobian, equal to $\exp(-\psi)$. Hence

$$\begin{aligned}\nu\big(T_z^{-n}(B(T^n(z),\xi))\big) &= \int_{B(T^n(z),\xi)} \exp\,\big(S_n\psi(T_z^{-n}(x))\big)\,d\nu(x)\\ &= \int_{B(T^n(z),\xi)} \frac{u_\phi(x)}{u_\phi(T^n(x))}\exp(S_n\phi - n\mathrm{P}(T,\phi))(T_z^{-n}(x))\,d\nu(x).\end{aligned}$$

So, by the *Pre-Bounded Distortion Lemma* (Lemma 4.4.2),

$$\frac{\inf|u_\phi|}{\sup|u_\phi|}BC^{-1} \leq \frac{\nu\big(T_z^{-n}(B(T^n(z),\xi))\big)}{\exp(S_n\phi - nP(T,\phi))(z)} \leq \frac{\sup|u_\phi|}{\inf|u_\phi|}C,$$

where $B = \inf\{\nu(B(y,\xi)\}$. It is positive, by Proposition 5.2.10.

Therefore ν is an invariant Gibbs state for ϕ, unique by Corollary 5.2.14. ♣

Remark 5.6.3. The knowledge that $\exp(-\psi)$ is weak Jacobian automatically implies that it is a strong Jacobian. Indeed, by the invariance of ν we have

$$\sum_{y\in T^{-1}(x)} \Phi_y(y) = 1 = \sum_{y\in T^{-1}(x)} \exp\psi(y),$$

and each non-zero summand on the left is equal to a corresponding summand on the right. So there are no summands equal to 0.

Uniqueness. Proof II. We shall provide the second proof of Lemma 5.6.1. It is not so elementary as the previous one, but it exhibits a relation with the Finite Variational Principle in the Introduction.

For every $y \in X$ put $A(y) := T^{-1}\Big(T(\{y\})\Big)$. Let $\{\nu_A\}$ denote the canonical system of conditional measures for the partition of X into the sets $A = A(y)$: see Section 2.6.

Since there exists a finite one-sided generator (see Lemma 3.5.5), with the use of Theorem 2.9.7 we obtain

$$0 = \mathrm{P}(T, \psi) = \mathrm{h}_\nu(T) + \int \psi \, d\nu = \mathrm{H}_\nu\left(\varepsilon \mid T^{-1}(\varepsilon)\right) + \int \psi \, d\nu$$
$$= \int \Big(\sum_{z\in A(y)} \nu_{A(y)}(\{z\})\big(- \log\left(\nu_{A(y)}(\{z\})\right) + \psi(z)\big)\Big) \, d\nu(y).$$

The latter expression is always negative except for the case $\nu_{A(y)}(z) = \exp\psi(z)$ ν-a.e. by the Finite Variational Principle. So for a set $Y = T^{-1}(T(Y))$ of full measure ν, for every $y \in Y$ we have

$$\nu_{A(y)}(\{y\}) = \exp\psi(y), \quad \text{and in particular} \quad \nu_{A(y)}(\{y\}) \neq 0. \tag{5.6.2}$$

Hence, for every Borel set $Z \subset Y$ such that T is 1–to–1 on it, we can repeat the calculation in the proof of Theorem 2.9.6 and get

$$\nu(T(Z)) = \int_Z 1/\nu_{A(y)}(\{y\}) \, d\nu(y).$$

So our Jacobian for $T|_Y$ is equal to $1/\nu_{A(y)}$, and hence to $\exp(-\psi)$ by (5.6.2), and it is strong on Y. Observe finally that $\nu\big(T(X\setminus Y)\big) = 0$ because $X \setminus Y = T^{-1}\big(T(X\setminus Y)\big)$ and ν is T-invariant. So $\exp(-\psi)$ is a strong Jacobian on X.

Uniqueness. Proof III. Because of Corollary 3.6.7 it is sufficient to prove the differentiability of the pressure function $\phi \mapsto \mathrm{P}(T,\phi)$ at Hölder continuous ϕ, in a set of directions dense in the weak topology on $C(X)$.

Lemma 5.6.4. *Let $\phi : X \to \mathbb{R}$ be a Hölder continuous function with exponent α, and let μ_ϕ denote the invariant Gibbs measure for ϕ. Let $F : X \to \mathbb{R}$ be an arbitrary continuous function. Then, for every $x \in X$,*

$$\lim_{n\to\infty} \frac{\sum_{y\in T^{-n}(x)} \frac{1}{n} S_n F \exp(S_n\phi)(y)}{\sum_{y\in T^{-n}(x)} \exp(S_n\phi)(y)} = \int F \, d\mu_\phi. \tag{5.6.3}$$

In addition, the convergence is uniform for an equi-continuous family of F's and for ϕ's in a bounded subset of the Banach space of Hölder functions $\mathcal{H}_\alpha(X)$.

Proof. The left-hand side of (5.6.3) can be written in the form

$$\lim_{n\to\infty} \frac{\frac{1}{n}\sum_{j=0}^{n-1} \mathcal{L}_\phi^n(F\circ T^j)(x)}{\mathcal{L}_\phi^n(1\!\!1)(x)} = \lim_{n\to\infty} \frac{\frac{1}{n}\sum_{j=0}^{n-1} \mathcal{L}^{n-j}(F\cdot\mathcal{L}^j(1\!\!1))(x)}{\mathcal{L}^n(1\!\!1)(x)}, \tag{5.6.4}$$

where $\mathcal{L} = \mathcal{L}_0 = e^{-P(T,\phi)}\mathcal{L}_\phi$.

Since by Theorem 5.3.5 $F\cdot\mathcal{L}^j(1\!\!1)$ is an equi-continuous family of functions, we obtain the uniform convergence

$$\mathcal{L}^{n-j}(F\cdot\mathcal{L}^j(1\!\!1))(x) \to u_\phi(x)\int F\cdot\mathcal{L}^j(1\!\!1)\, dm_\phi$$

as $n - j \to \infty$: see Theorem 5.4.5.

Therefore we can continue (5.6.4) to get

$$\lim_{n\to\infty} \frac{\frac{1}{n}\sum_{j=0}^{n-1} u_\phi(x) \int F \cdot \mathcal{L}^j(\mathbb{1})\, dm_\phi}{u_\phi(x)} = \lim_{n\to\infty} \frac{1}{n}\sum_{j=0}^{n-1} \int F \cdot \mathcal{L}^j(\mathbb{1})\, dm_\phi = \int F\, d\mu_\phi,$$

since $\mathcal{L}^j(\mathbb{1})$ uniformly converges to u_ϕ and $\mu_\phi = u_\phi m_\phi$. ♣

Now we shall calculate the derivative $d\,\mathrm{P}(T, \phi + t\gamma)/dt$ for all Hölder continuous functions $\phi, \gamma : X \to \mathbb{R}$ at every $t \in \mathbb{R}$. In particular, this will give differentiability at $t = 0$. Thus our dense set of directions is spanned by the Hölder continuous functions γ.

Theorem 5.6.5. *We have*

$$\frac{d}{dt}\,\mathrm{P}(T, \phi + t\gamma) = \int \gamma\, d\mu_{\phi+t\gamma} \tag{5.6.5}$$

for all $t \in \mathbb{R}$.

Proof. Write

$$P_n(t) := \frac{1}{n} \log \sum_{y \in T^{-n}(x)} \exp(S_n(\phi + t\gamma))(y) \tag{5.6.6}$$

and

$$Q_n(t) := \frac{dP_n}{dt}(t) = \frac{\frac{1}{n}\sum_{y\in T^{-n}(x)} S_n\gamma(y) \exp(S_n(\phi + t\gamma))(y)}{\sum_{y\in T^{-n}(x)} \exp(S_n(\phi + t\gamma)(y)}. \tag{5.6.7}$$

By Lemma 5.6.4, $\lim_{n\to\infty} Q_n(t) = \int \gamma\, d\mu_{\phi+t\gamma}$ and the convergence is locally uniform with respect to t. Since, in addition, $\lim_{n\to\infty} P_n(t) = \mathrm{P}(T, \phi + t\gamma)$, we conclude that $\mathrm{P}(T, \phi + t\gamma) = \lim_{n\to\infty} P_n(t)$ is differentiable, and the derivative is equal to the limit of derivatives: $\lim_{n\to\infty} Q_n(t) = \int \gamma\, d\mu_{\phi+t\gamma}$. ♣

Note that the differential (Gateaux) operator $\gamma \mapsto \int \gamma\, d\mu_\phi$ is indeed the one from Proposition 3.6.6. Note also that *a posteriori*, by Corollary 3.6.7, we have proved that for ϕ Hölder continuous, $\mathrm{P}(T, \phi)$ is differentiable in the direction of every continuous function. This is obvious in general: two different supporting functionals are differently restricted to any dense subspace.

5.7 Probability laws and $\sigma^2(u, v)$

The exponential convergences in Section 5.4 allow us to prove the probability laws.

Theorem 5.7.1. *Let $T : X \to X$ be an open, distance-expanding topologically exact map and μ be the invariant Gibbs measure for a Hölder continuous function $\phi : X \to \mathbb{R}$. Then if $g : X \to \mathbb{R}$ satisfies*

$$\sum_{n=0}^{\infty} \|\hat{\mathcal{L}}^n(g - \mu(g))\|_2 < \infty, \tag{5.7.1}$$

and in particular if g is Hölder continuous, then the Central Limit Theorem (CLT) holds for g. If g is Hölder continuous then the Law of Iterated Logarithm (LIL) holds.

Proof. We first show how CLT can be deduced from Theorem 2.11.5. We can assume $\mu(g) = 0$. Let $(\tilde{X}, \tilde{\mathcal{F}}, \tilde{\mu})$ be the natural extension (see Section 2.7). Recall that $\tilde{X}$ can be viewed as the set of all T-trajectories $(x_n)_{n\in\mathbb{Z}}$ (or backward trajectories), $\tilde{T}((x_n)) = (x_{n+1})$ and $\pi_n((x_n)) = x_n$. It is sufficient now to check (2.11.14) for the automorphism $\tilde{T}$ the function $\tilde{g} = g \circ \pi_0$ and $\tilde{\mathcal{F}}_0 = \pi_0^{-1}(\mathcal{F})$ for the completed Borel σ-algebra $\mathcal{F}$. Since $\tilde{g}$ is measurable with respect to $\tilde{\mathcal{F}}_0$ it is also measurable with respect to all $\tilde{\mathcal{F}}_n = \tilde{T}^{-n}(\tilde{\mathcal{F}}_0)$ for $n \le 0$: hence $\tilde{g} = E(\tilde{g}|\tilde{\mathcal{F}}_n)$. So we need only to prove $\sum_{n\ge0}^{\infty} \|E(\tilde{g}|\tilde{\mathcal{F}}_n)\|_2 < \infty$.

Since for $n \ge 0$ we have $\pi_0 \circ \tilde{T}^n = \tilde{T}^n \circ \pi_0$, we have $E(\tilde{g}|\tilde{\mathcal{F}}_n) = E(g|\mathcal{F}_n) \circ \pi_0$. So we need to prove that $\sum_{n\ge0}^{\infty} \|E(g|\mathcal{F}_n)\|_2 < \infty$.

Let us start with a general fact concerning an arbitrary probability space $(X, \mathcal{F}, \mu)$ and an endomorphism T preserving μ.

Lemma 5.7.2. *Let U denote the unitary operator on $L^2(X, \mathcal{F}, \mu)$ associated to T, namely $U(f) = f \circ T$ (called the Koopman operator: see the beginning of Section 5.2 and Section 2.2). Let U^* be the operator conjugate to U, acting also on $L^2(X, \mathcal{F}, \mu)$. Then for every $k \ge 0$ the operator $U^k U^{*k}$ is the orthogonal projection of $H_0 = L^2(X, \mathcal{F}, \mu)$ onto $H_k = L^2(X, T^{-k}(\mathcal{F}), \mu)$.*

Proof. For each $k \ge 0$ the function $U^k(u) = u \circ T^k$ is measurable with respect to $T^{-k}(\mathcal{F})$, so the range of $U^k U^{*k}$ is indeed in $H_k = L^2(X, T^{-k}(\mathcal{F}), \mu)$.

For any $u, v \in H_0$ write $\int u \cdot v \, d\mu = \langle u, v\rangle$, the scalar product of u and v. For arbitrary $f, g \in H_0$ we calculate

$$\langle U^k U^{*k}(f), g \circ T^k\rangle = \langle U^k U^{*k}(f), U^k(g)\rangle$$

$$= \langle U^{*k}(f), g\rangle = \langle f, U^k(g)\rangle = \langle f, g \circ T^k\rangle.$$

It is clear that all functions in $H_k = L^2(X, T^{-k}(\mathcal{F}), \mu)$ are represented by $g \circ T^k$ with $g \in L^2(X, \mathcal{F}, \mu)$. Therefore, by the above equality for all $h \in H_k$, we obtain

$$\langle f - U^k U^{*k}(f), h\rangle = \langle f, h\rangle - \langle f, h\rangle = 0. \tag{5.7.2}$$

In particular, for $f \in H_k$ we conclude from (5.7.2) for $h = f - U^k U^{*k}(f)$ that $\langle f - U^k U^{*k}(f), f - U^k U^{*k}(f)\rangle = 0$: hence $U^k U^{*k}(f) = f$. Therefore $U^k U^{*k}$ is a projection onto H_k, which is orthogonal by (5.7.2). ♣

Since the conditional expectation value $f \mapsto E(f|T^{-k}(\mathcal{F}))$ is also the orthogonal projection onto H_k we conclude that $E(f|T^{-k}(\mathcal{F})) = U^k U^{*k}(f)$.

Now let us pass to our special situation of Theorem 5.7.1.

Lemma 5.7.3. *For every $f \in L^2(X, \mathcal{F}, \mu)$ we have $U^*(f) = \hat{\mathcal{L}}(f)$.*

Proof. $\langle U^* f, g\rangle = \langle f, Ug\rangle = \int f \cdot (g \circ T) \, d\mu = \int \hat{\mathcal{L}}(f \cdot (g \circ T)) \, d\mu = \int (\hat{\mathcal{L}}(f)) \cdot g \, d\mu = \langle \hat{\mathcal{L}}(f), g\rangle$: compare (5.2.7). ♣

Proof of Theorem 5.7.1. Conclusion. We can assume that $\mu(g)=0$. We have

$$\sum_{n\ge 0}^{\infty}\|E(g|\mathcal{F}_n)\|_2=\sum_{n\ge 0}^{\infty}\|U^nU^{*n}(g)\|_2=\sum_{n\ge 0}^{\infty}\|\hat{\mathcal{L}}^n(g)\|_2<\infty,$$

where the inequality been assumed in (5.7.1). Thus CLT has been proved by applying Theorem 2.11.5. If g is Hölder continuous it satisfies (5.7.1). Indeed $\hat{\mathcal{L}}^k(g)$ converges to 0 in the sup norm exponentially fast as $k\to\infty$ by Corollary 5.4.7 (see (5.4.8)). This implies the same convergence in L^2: hence the convergence of the above series. ♣

We have proved CLT and LIL using Theorem 2.11.5. Now let us show how to prove CLT and LIL with the use of Theorem 2.11.1, for Hölder continuous g. As in the proof of Corollary 5.4.11, let $\pi:\Sigma_A\to X$ be the coding map from a one-sided topological Markov chain with d symbols generated by a Markov partition: see Section 4.5. Since π is Hölder continuous, if g and ϕ are Hölder continuous, then the compositions $g\circ\pi,\phi\circ\pi$ are also Hölder continuous. π is an isomorphism between the measures $\mu_{\phi\circ\pi}$ on Σ_A and μ_ϕ on X: see Section 4.5 and Lemma 5.4.12. The function $g\circ\pi$ satisfies the assumptions of Theorem 2.11.1 with respect to the σ-algebra $\mathcal{F}$ associated to the partition of Σ_A into 0-th cylinders: see Theorem 5.4.10. ϕ-mixing follows from (5.4.10), and the estimate in (2.11.8) is exponential with an arbitrary δ, owing to the Hölder continuity of $g\circ\pi$. Hence, by Theorem 2.11.1, $g\circ\pi$, and therefore g satisfy CLT and LIL.

In Section 5.6 we calculated the first derivative of the pressure function. Here, using the same method, we calculate the second derivative, and we see that it is a dispersion (asymptotic variance) σ^2: see Section 2.11.

Theorem 5.7.4. *For all $\phi,u,v:X\to\mathbb{R}$ Hölder continuous functions there exists the second derivative, given by Ruelle's formula:*

$$\frac{\partial^2}{\partial s\partial t}\mathrm{P}(T,\phi+su+tv)|_{s=t=0}=\lim_{n\to\infty}\frac{1}{n}\int S_n(u-\mu_\phi u)S_n(v-\mu_\phi v)\,d\mu_\phi,\quad(5.7.3)$$

where μ_ϕ is the invariant Gibbs measure for ϕ. In particular,

$$\frac{\partial^2}{\partial t^2}\mathrm{P}(T,\phi+tv)|_{t=0}=\sigma^2_{\mu_\phi}(v)$$

(where $\sigma^2_{\mu_\phi}(v)$ is the asymptotic variance discussed in CLT, Section 2.11). In addition, the function $(s,t)\mapsto\mathrm{P}(T,\phi+su+tv)$ is C^2-smooth.

Proof. By Section 5.6 (see (5.6.3), (5.6.7)):

$$\frac{\partial^2}{\partial s\partial t}\mathrm{P}(T,\phi+su+tv)|_{t=0}$$

$$=\frac{\partial}{\partial s}\lim_{n\to\infty}\frac{\frac{1}{n}\sum_{y\in T^{-n}(x)}S_nv(y)\exp S_n(\phi+su)(y)}{\sum_{y\in T^{-n}(x)}\exp S_n(\phi+su)(y)}.\quad(5.7.4)$$

Now we change the order of $\partial/\partial s$ and lim. This will be justified if we prove the uniform convergence of the resulting derivative functions.

With $x \in X$ and n fixed we abbreviate further in the notation $\sum_{y \in T^{-n}(x)}$ to $\sum_y$ and compute

$$F_n(s) := \frac{\partial}{\partial s}\left(\frac{\sum_y S_n v(y) \exp S_n(\phi + su)(y)}{\sum_y \exp S_n(\phi + su)(y)}\right)$$
$$= \frac{\sum_y S_n u(y) S_n v(y) \exp S_n(\phi + su)(y)}{\sum_y \exp S_n(\phi + su)(y)}$$
$$- \frac{\left(\sum_y S_n u(y) \exp S_n(\phi + su)(y)\right)\left(\sum_y S_n v(y) \exp S_n(\phi + su)(y)\right)}{\left(\sum_y \exp S_n(\phi + su)(y)\right)^2}$$
$$= \frac{\mathcal{L}^n\left((S_n u)(S_n v)\right)(x)}{\mathcal{L}^n(\mathbb{1})(x)} - \frac{\mathcal{L}^n(S_n u)(x)}{\mathcal{L}^n(\mathbb{1})(x)} \frac{\mathcal{L}^n(S_n v)(x)}{\mathcal{L}^n(\mathbb{1})(x)}.$$

As in Section 5.6 we write here $\mathcal{L} = \mathcal{L}_0 = e^{-P(T, \phi + su)} \mathcal{L}_{\phi + su}$. It is useful to write the later expression for $F_n(s)$ in the form

$$F_n(s) = \int (S_n u)(S_n v)\, d\mu_{s,n} - \int (S_n u)\, d\mu_{s,n} \int (S_n v)\, d\mu_{s,n} \tag{5.7.5}$$

or

$$F_n(s) = \sum_{i,j=0}^{n-1} \int (u \circ T^i - \mu_{s,n}(u \circ T^i))(v \circ T^j - \mu_{s,n}(v \circ T^j))\, d\mu_{s,n}, \tag{5.7.6}$$

where $\mu_{s,n}$ is the probability measure distributed on $T^{-n}(x)$ according to the weights $\exp(S_n(\phi + su))(y)/\sum_y \exp S_n(\phi + su)(y)$.

Note that $\frac{1}{n} F_n(s)$, with $F_n(s)$ as in the formula (5.7.6), already resembles (5.7.3), because $\mu_{s,n} \to m_{\phi + su}$ in the weak*-topology: see (5.4.4). However, we need a little more work.

For each i, j denote the respective summand in (5.7.6) by $K_{i,j}$. To simplify the notation, denote $u \circ T^i$ by u_i and $v \circ T^j$ by v_j. We have

$$K_{i,j} = \frac{\mathcal{L}^n\left((u_i - \mu_{s,n} u_i)(v_j - \mu_{s,n} v_j)\right)(x)}{\mathcal{L}^n(\mathbb{1})(x)},$$

and for $0 \le i \le j < n$, using (5.2.7) twice,

$$K_{i,j} = \frac{\mathcal{L}^{n-j}\Big(\left(\mathcal{L}^{j-i}((u - \mu_{s,n} u_i)\mathcal{L}^i(\mathbb{1}))\right)(v - \mu_{s,n} v_j)\Big)(x)}{\mathcal{L}^n(\mathbb{1})(x)}. \tag{5.7.7}$$

By Corollary 5.4.7 for $\tau < 1$ and Hölder norm $\|\cdot\|_{\mathcal{H}_\alpha}$ for an exponent $\alpha > 0$, transforming the integral as in the proof of Theorem 5.4.9, we get

$$\|\mathcal{L}^{j-i}((u - \mu_{s,n} u_i)\mathcal{L}^i(\mathbb{1})) - u_{\phi + su}\Big(\int u_i\, dm_{\phi + su} - \mu_{s,n} u_i)\Big)\|_{\mathcal{H}_\alpha} \le C\tau^{j-i},$$

where C depends only on Hölder norms of u and $\phi + su$. The difference in the large parentheses (denote it by $D_{i,n}$) is bounded by $C\tau^{n-i}$ in the Hölder norm, again by Corollary 5.4.7.

We conclude that for all j the functions

$$L_j := \sum_{i\le j} \mathcal{L}^{j-i}((u - \mu_{s,n}u_i)\mathcal{L}^i(\mathbb{1}))$$

are uniformly bounded in the Hölder norm $\|\cdot\|_{\mathcal{H}_\alpha}$ by a constant C, depending again only on $||u||_{\mathcal{H}_\alpha}$ and $||\phi+su||_{\mathcal{H}_\alpha}$. Hence summing over $i\le j$ in (5.7.7) and applying $\mathcal{L}^{n-j}$ we obtain

$$\left\|\sum_{i=0}^{j} K_{i,j} - \sum_{i=0}^{j}\int (u_i - \mu_{s,n}u_i)(v_j - \mu_{s,n}v_j)\, dm_{\phi+su}\right\|_\infty \le C\tau^{n-j}.$$

Here C also depends on $||v||_{\mathcal{H}_\alpha}$. We can replace the first sum by the second sum without changing the limit in (5.7.4), since after summing over $j = 0, 1, \dots, n-1$, dividing by n and passing with n to ∞, they lead to the same result. Let us now show that $\mu_{s,n}$ can be replaced by $m_{\phi+su}$ in the above estimate without changing the limit in (5.7.4). Indeed, using the formula $ab - a'b' = (a-a')b' + a(b-b')$, we obtain

$$\begin{aligned}\Bigg|\int &(u_i - m_{\phi+su}u_i)(v_j - m_{\phi+su}v_j)\, dm_{\phi+su} \\ &- \int (u_i - \mu_{s,n}u_i)(v_j - \mu_{s,n}v_j)\, dm_{\phi+su}\Bigg| \\ &\le |(\mu_{s,n}u_i - m_{\phi+su}u_i) \times (m_{\phi+su}v_j - \mu_{s,n}v_j)| \\ &+ \left|\int (u_i - m_{\phi+su}u_i) \times (\mu_{s,n}v_j - m_{\phi+su}v_j)\, dm_{\phi+su}\right|.\end{aligned}$$

Since $D_{i,n} \le C\tau^{n-i}$ and $D_{j,n} \le C\tau^{n-j}$, the first summand is bounded above by $\tau^{n-i}\tau^{n-j}$. Note that the second summand is equal to 0. Thus our replacement is justified.

The last step is to replace $m = m_{\phi+su}$ by the invariant Gibbs measure $\mu = \mu_{\phi+su}$.

Similarly as above we can replace m by μ in mu_i, mv_i. Indeed,

$$\begin{aligned}|mu_i - \mu u_i| &= \left|\int u\cdot\mathcal{L}^i(\mathbb{1})\, dm - \int u u_{\phi+su}\, dm\right| \\ &= \left|\int u\cdot(\mathcal{L}^i(\mathbb{1}) - u_{\phi+su})\, dm\right| \le Cm(u)\tau^i. \end{aligned} \tag{5.7.8}$$

Thus the resulting difference is bounded by $Cm(u)m(v)\tau^i\tau^j$. Finally we justify the replacement of m by μ at the second integral in the previous formula. To simplify the notation write $F = u - \mu u, G = v - \mu v$. Since $j \ge i$, using (5.7.8), we can write

$$\left|\int (F\circ T^i)(G\circ T^j)\,dm - \int (F\circ T^i)(G\circ T^j)\,d\mu\right| =$$

$$= \left|\int (F\cdot (G\circ T^{j-i}))\circ T^i\,dm - \int (F\cdot (G\circ T^{j-i}))\circ T^i\,d\mu\right|$$

$$\le C\tau^i \int |F\cdot (G\circ T^{j-i})|\,dm \le Cm(F)m(G)\tau^i\tau^{j-i} = Cm(F)m(G)\tau^j$$

by Theorem 5.4.9 (exponential decay of correlations), the latter C depending again on the Hölder norms of $u, v, \phi + su$. Summing over all $0 \le i \le j < n$ gives the bound $Cm(F)m(G)\sum_{j=0}^{n-1} j\tau^j$, and our replacement is justified. For $i > j$ we do the same replacements, changing the roles of u and v. The C^2-smoothness follows from the uniformity of the convergence of the sequence of the functions $F_n(s)$, for $\phi + tv$ in place of ϕ, with respect to the variables (s,t), resulting from the proof. ♣

Exercises

5.1. Prove that (5.1.1) with an arbitrary $0 < \xi' \le \xi$ in place of ξ implies (5.1.1) for every $0 < \xi' \le \xi$ (with C depending on ξ').

5.2. Let $A = (a_{ij})$ be a non-zero $k \times k$ matrix with all entries non-negative. Assume that A is *irreducible*: that is, that for any pair (i,j) there is some $n > 0$ such that the (i,j)-th entry a_{ij}^n of A^n is positive. Prove that there is a unique positive eigenvalue λ with left (row) and right (column) eigenvectors $v = (v_1, \ldots, v_k)$ and $u = (u_1, \ldots, u_k)$ with all coordinates positive. The eigenvalue λ is simple. All other eigenvalues have absolute values smaller than λ.

Check that the matrix $P = (p_{ij})$ with $p_{ij} := v_i a_{ij}/\lambda v_j$ is stochastic: that is, $0 \le p_{ij} \le 1$ and $\sum_i p_{ij} = 1$ for all $j = 1, \ldots, k$. (One interprets p_{ij} as the probability of i under the condition j. Caution: often the roles of i and j are opposite.) Note that $q = (u_1v_1, \ldots, u_kv_k)$ satisfies $Pq = q$ (it is the stationary probability distribution).

Prove that for each pair (i,j) $\lim_{n\to\infty} \lambda^{-n} a_{ij}^n \to u_i v_j$.

Identify the vectors u and v for a piecewise affine (piecewise increasing) mapping of the interval $T : [0,1] \to [0,1]$. More precisely, let $0 = x_1 < x_2 < \ldots < x_k < x_{k+1} = 1$, and for each $i = 1, \ldots, k$ let $x_i = y_{i,1} < y_{i,2} < \ldots < y_{i,k+1} = x_{i+1}$. Consider T affine on each interval $[y_{i,j}, y_{i,j+1})$ mapping it onto $[x_j, x_{j+1})$, such that $T' = \text{Const}\, a_{ij}$. Consider the potential function $\phi = -\log F'$. What is the eigen-measure of the related transfer operator? What is the invariant Gibbs measure? Compare Exercise 5.7.

5.3. Prove that if a probability measure m is T-invariant then $\mathcal{L}_m(L^2(m)) \subset L^2(m)$, and the norm of $\mathcal{L}$ in $L^2(m)$ is equal to 1.

5.4. A linear operator $Q : L^1(\mu) \to L^1(\mu)$ for a measure space $(X, \mathcal{F}, \mu)$ (we allow μ to be infinite, say σ-finite) is called a *Markov operator* if for all $u \ge 0$, $u \in L^1(\mu)$ the following two conditions hold:

(a) $Q(u) \ge 0$ (compare the notion of *positive operator*).

(b) $||Q(u)||_1 = ||u||_1$ (compare (5.2.4).

(This notion generalizes the notion of a transfer operator $\mathcal{L}_m$: see Section 5.2.)

Prove that

(1) $||Q(u)||_1 \leq ||u||_1$ for all $u \in L^1(\mu)$.

(2) If $|Q^n(u)| \leq g$ for some positive $u, g \in L^1(\mu)$ and all $n = 1, 2, \dots$, then there exists non-zero $u_* \in L^1(\mu)$ such that $Q(u_*) = u_*$.

(3) Suppose additionally that there is a compact set $A \subset L^1$ such that for d the distance in L^1, for every $u \in L^1(\mu)$ it holds that $d(Q^n(u), A) \to 0$ as $n \to \infty$. (This property is called *strongly constrictive.*)

Then there exist two finite sequences of non-negative functions $g_i \in L^1(\mu), k_i \in L^\infty(\mu), i = 1, \dots, p$, and a linear bounded operator $S : L^1(\mu) \to L^1(\mu)$ such that for every $u \in L^1(\mu)$

$$Q(u) = \sum_{i=1}^{r} \lambda_i(u) g_i + S,$$

where $\lambda_i(u) = \int u k_i \, d\mu$, the functions g_i have disjoint supports and are permuted by Q, and $||Q^n S(u)||_1 \to 0$ as $n \to \infty$.

Hint: See the Spectral Decomposition Theorem in [Lasota & Mackey 1985] and [Lasota, Li & Yorke 1984].

(4) Suppose additionally that there is $A \subset X$ with $\mu(A) > 0$ such that for every $u \in L^1(\mu)$ we have $Q^n(u) > 0$ on A for all n large enough. Then $\lim ||Q^n(u) - (\int u \, d\mu) u_*||_1 = 0$. (This property is called *asymptotically stable.*)

Hint: Deduce this from (3): see the Asymptotic Stability Section in [Lasota & Mackey 1985] .

(5) Prove that if a Markov operator Q satisfies the following *lower bound function* property:

There exists a non-negative integrable function $h : X \to \mathbb{R}$ such that $||h||_1 > 0$ and such that for every non-negative $u \in L^1(\mu)$ with $||u||_1 = 1$,

$$\lim_{n \to \infty} ||\min(Q^n(u) - h, 0)||_1 = 0,$$

then Q is asymptotically stable.

Hint: Consider $u \in L^1(\mu)$ with $\int u \, d\mu = 0$ and decompose it into positive and negative parts $u = u^+ + u^-$. Using h, observe that by iterating Q on u^+ and u^- we achieve the cancelling $h - h$. For details see [Lasota & Mackey 1985].

(Note that the existence of the lower bound function u replaces the assumption on the existence of the set A in (4) and allows us to get rid of the constrictivness assumption.)

5.5. Consider a measure space $(X, \mathcal{F}, \mu)$ and a measurable function $K : X \times X \to \mathbb{R}$, non-negative and such that $\int_X K(x, y) \, d\mu(x) = 1$ for all $y \in X$ (such a function is called a *stochastic kernel*). Define the associated *integral operator* by

$$P(u)(x) = \int_X K(x, y) u(y) \, d\mu(y), \quad \text{for } u \in L^1(\mu).$$

Consider the convolutions

$$K_n(x,y) := \int K(x,z_{n-1})K(z_{n-1},z_{n-2})\dots K(z_1,y)\,d\mu(z_{n-1})\dots d\mu(z_1).$$

Prove that if there exists $n \ge 0$ such that $\int_X \inf_y K_n(x,y)\,d\mu(x) > 0$, then Q is asymptotically stable.

5.6. Let $T : X \to X$ be a measurable backward quasi-invariant endomorphism of a measure space $(X, \mathcal{F}, m)$. Suppose there exist disjoint sets $A, B \subset X$, both of positive measure m, with $T(A) = X$.

Prove that for $\mathcal{L}_m : L^1(m) \to L^1(m)$ being the transfer operator as in Section 5.2, all λ with $|\lambda| < 1$ belong to its spectrum. In particular, 1 is not an isolated eigenvalue (there is no spectral gap: compare Remark 5.4.8). Is there a spectral gap for $\mathcal{L}_\phi$ for $T : X \to X$ an open expanding topologically transitive map, acting on $C(X)$ for Hölder continuous ϕ? (In Corollary 5.4.7 we restrict the domain of $\mathcal{L}_\phi$ to Hölder continuous functions.)

Hint: Prove that all $\lambda, |\lambda| < 1$ belong to the spectrum of the conjugate Koopman operator on $L^\infty(m)$. Indeed, $\mathcal{L}^* - \lambda\mathrm{Id}$ is not onto, and hence not invertible. No function that is identically 0 on A and non-zero on B is in the image.

5.7. Let $I_1, I_2, \dots, I_N$ be a partition of the unit interval $[0,1]$ into closed subintervals (up to end points, that is, every two neighbours have a common end point). Let $T : [0,1] \to [0,1]$ be a piecewise $C^{1+\epsilon}$ expanding mapping. This means that the restriction of T to each I_j has an extension to a neighbourhood of the closure of I_j that is differentiable with the first derivative Hölder continuous of the modulus larger than 1. Suppose that each $T(I_j)$ is a union of some I_i's (Markov property), and for each I_j there is n such that $T^n(I_j) = [0,1]$.

Prove the following so-called *Folklore Theorem*: There is an exact (in the measure-theoretic sense) T-invariant probability measure μ on $[0,1]$ equivalent to the length measure l, with $\frac{d\mu}{dl}$ bounded and bounded away from 0, Hölder continuous on each I_j.

Hint: Consider the potential function $\phi = -\log|T'|$.

5.8. For T as in Exercise 5.7 assume that T is C^2 on I_j's, but do not assume Markov property. Prove that there exists a finite number of T-invariant probability measure absolutely continuous with respect to the length measure l.

Hint: This is the famous Lasota–Yorke Theorem [Lasota & Yorke 1973]. Instead of Hölder continuous functions consider the transfer operator $\mathcal{L}$ on the functions of a bounded variation of the derivative.

In fact, by the Ionescu Tulcea, Marinescu Theorem 5.5.5 item (8), there exists a spectral gap (see Remark 5.4.8), so we can prove probability laws as in Theorem 5.7.1.

5.9. Prove the existence of an invariant Gibbs measure, Theorem 5.3.2, for ϕ satisfying the following *Bowen's condition*: there exist $\delta > 0$ and $C > 0$ with the property that whenever $\rho(T^i(x), T^i(z)) \le \delta$ for $0 \le i \le n-1$, then

$$\left|\sum_{i=0}^{n-1} \phi(T^i(x)) - \phi(T^i(z))\right| \le C,$$

and for $T : X \to X$ an open topologically transitive map of a compact metric space that is *non-contracting*: that is, there exists $\eta > 0$ such that for all $x, y \in X$ $\rho(x, y) \leq \eta$ implies $\rho(T(x), T(y)) \geq \rho(x, y)$.

For m satisfying $\mathcal{L}^*_\phi(m) = cm$ (see Theorem 5.2.8) prove the convergences (5.4.2) and (5.4.3).

Hint: See [Walters 1978].

Bibliographical notes

In writing Sections 5.1–5.4 we relied mainly on the books [Bowen 1975] and [Ruelle 1978a]. See also [Zinsmeister 1996] and [Baladi 2000].

References to the facts in Section 5.5 concerning almost periodic operators can be found in [Lyubich & Lyubich 1986] and [Lyubich 1983]. For the proof of the Ionescu Tulcea & Marinescu Theorem see [Ionescu Tulcea & Marinescu 1950]. For Markov operators see [Lasota & Mackey 1985].

As mentioned in the Introduction, the first, simplest, proof of uniqueness of equilibrium follows [Keller 1998]. The second is similar to one in [Przytycki 1990]. The idea is taken from Ledrappier's papers: see for example [Ledrappier 1984].

For the Perron–Frobenius theory for finite matrices (Exercise 5.2) see for example [Walters 1982] and the references therein.

The Folklore Theorem in Exercise 5.7 can be found for example in [Boyarsky & Góra 1997].

The consequences of holomorphic dependence of the operator on parameters (and in particular the holomorphic dependence of an isolated eigenvalue of multiplicity one: see Remark 5.4.8) are comprehensively written in [Kato 1966]. See our Section 6.4.

We owe Exercise 5.6 to R. Rudnicki.

In the following chapters we shall discuss special open distance-expanding maps (X, T) with X embedded in a smooth manifold and T smooth (C^r). Then the transfer operator for C^r or real-analytic potential can be restricted to $C^k, k \leq r$. The bigger k, the more continuous spectrum is lost. In the C^∞ and C^ω (real-analytic) cases the transfer operator has only a pointwise spectrum. For this rich theory and references see for example [Baladi 2000].

6

Expanding repellers in manifolds and in the Riemann sphere: preliminaries

In this chapter we shall consider a compact metric space X with an open, distance-expanding map T on it, embedded isometrically into a smooth Riemannian manifold M. We shall assume that T extends to a neighbourhood U of X to a mapping f of class $C^{1+\varepsilon}$ for some $0 < \varepsilon \leq 1$ or smoother, including real-analytic. $C^{1+\varepsilon}$ and more general $C^{r+\varepsilon}$ for $r = 1, 2, \ldots$ means that the r-th derivative is Hölder continuous with the exponent ε for $\varepsilon < 1$ and Lipschitz continuous for $\varepsilon = 1$. We shall also assume that there exists a constant $\lambda > 1$ such that for every $x \in U$ and for every non-zero vector v tangent to M at x, $||Df(v)|| > \lambda||v||$ holds, where $|| \cdot ||$ is the norm induced by the Riemannian metric. The pair (X, f) will be called an *expanding repeller* and f an *expanding map*. If f is of some class A, e.g. C^α or analytic, we shall say that the expanding repeller is of that class, or that this is an A-expanding repeller. In particular, if f is conformal we call (X, f) a *conformal expanding repeller*, abbreviated to CER. Finally, if we skip the assumption that $T = f|_X$ is open on X, we shall call (X, f) an *expanding set*. Sometimes, to distinguish the domain of f, we shall write (X, f, U).

In Sections 6.2 and 6.3 we provide some introduction to conformal expanding repellers, studying the transfer operator, postponing the main study to Chapters 9 and 10, where we shall use tools of geometric measure theory.

In Section 6.4 we discuss analytic dependence of the transfer operator on parameters.

6.1 Basic properties

For any expanding set there exist constants playing an analogous role to constants for (open) distance-expanding maps:

Lemma 6.1.1. *For any expanding set (X, f) with λ, U as in the definition and f differentiable, for every $\eta > 0$ small enough, there exists $U' \subset U$ a neighbourhood of X such that $B(X, \eta) \subset U'$, $B(U', \eta) \subset U$ and for every $x \in U'$ the map f is injective on $B(x, \eta)$. Moreover, $f(B(x, \eta)) \supset B(f(x), \eta)$ and f increases distances on $B(x, \eta)$ by at least the factor λ.*

Proof. We leave the proof to the reader as an easy exercise: compare the proof of Lemma 6.1.2. ♣

In the sequel we shall consider expanding sets together with the constants η, λ from Lemma 6.1.1. Write also $\xi := \lambda\eta$. For the expanding repeller (X, f) these constants satisfy the properties of the constants η, ξ, λ for the distance-expanding map $T = f|_X$ on X, provided η is small enough: compare Lemmas 4.1.2 and 4.1.4. For every $x \in X$ we can consider the branch f_x^{-1} on $B(f(x), \xi)$, mapping $f(x)$ to x, extending the branch T_x^{-1} defined on $B(f(x), \xi) \cap X$. Similarly, we can consider such branches of f^{-1} for $x \in U'$.

Now let X be a compact subset of M forward invariant, that is, $f(X) \subset X$, for a continuous mapping f defined on a neighbourhood U of X. We say that X is a *repeller* if there exists a neighbourhood U' of X in U such that for every $y \in U' \setminus X$ there exists $n > 0$ such that $f^n(y) \notin U'$. In other words,

$$X = \bigcap_{n>0} f^{-n}(U'). \tag{6.1.1}$$

In the lemma below we shall see that the extrinsic property of being a repeller is equivalent to the intrinsic property of being open for f on X. It is a topological lemma; no differentiability is invoked.

Lemma 6.1.2. *Let X be a compact subset of M forward invariant for a continuous mapping f defined on its neighbourhood U. Suppose that f is an open map on U. Then if X is a repeller, $f|_X$ is an open map in X. Conversely, if f is distance-expanding on a neighbourhood of X and $f|_X$ is an open map, then X is a repeller: that is, it satisfies (6.1.1).*

Proof. If $f|_X$ were not open there would exist a sequence of points $x_n \in X$ converging to $x \in X$ a point $y \in X$ such that $f(y) = x$ and an open set V in M containing y so that no x_n is in $f(V \cap X)$. But as f is open there exists a sequence $y_n \in V$, $y_n \to y$ and $f(y_n) = x_n$ for all n large enough. Thus the forward trajectory of each y_n stays in every U', even in X except y_n itself, which is arbitrarily close to X with n respectively large. This contradicts the repelling property.

Conversely, suppose that X is not a repeller. Then for $U'' = B(X, r) \subset U'$ (U' from Lemma 6.1.1) with an arbitrary $r < \xi$ there exists $x \in U'' \setminus X$ such that its forward trajectory is also in U''. Then there exists $n > 0$ such that $\text{dist}(f^n(x), X) < \lambda \, \text{dist}(f^{n-1}(x), X)$. Let y be a point in X closest to $f^n(x)$. Then by Lemma 6.1.1 there exists $y' \in B(f^{n-1}(x), \eta)$ such that $f(y') = y$ and by the construction $y' \notin X$. Thus, letting $r \to 0$, we obtain a sequence of points x_n not in X but converging to $x_0 \in X$ with images in X. So $f(x_n) \notin f|_X(B(x_0, \eta) \cap X)$ because they are f-images of $x_n \in B(x_0, \eta) \setminus X$ for n large enough and f is injective on $B(x_0, \eta)$. But $f(x_n) \to f(x_0)$. So $f(x_0)$ does not belong to the interior of $f|_X(B(x_0, \eta) \cap X)$, so $f|_X$ is not open. ♣

To complete the description we provide one more fact.

Proposition 6.1.3. *If (X, f) is an expanding set in a manifold M, then it is a repeller iff there exists U'' a neighbourhood of X in M such that for every sequence of points $x_n \in U''$, $n = 0, -1, -2, \dots, -N$, where N is any positive number or ∞, such that $f(x_n) = x_{n+1}$ for $n < 0$, that is, for every backward trajectory in U'', there exists a backward trajectory $y_n \in X$ such that $x_n \in B(y_n, \eta)$.*

Additionally, if (X, f) is an expanding repeller and f maps X onto X, then for every $x_0 \in U''$ there exist x_n and y_n as above.

Proof. If U'' is small enough, then by the openess of $f|_X$, if $f(y) = z \in X$ and $y \in U''$ then $y \in X$: compare the Proof of Lemma 6.1.2. So, given x_n, defining $y_n = f_{x_n}^{-1}(y_{n+1})$, starting with $y_0 \in X$ such that $\rho(x_0, y_0) < \xi$, we prove that $y_n \in X$.

Conversely, if $f^n(x) \in U''$ for all $n = 0, 1, \dots$, then for each n we consider $f^n(x), f^{n-1}(x), \dots, x$ as a backward trajectory and find a backward trajectory $y(n)_0, y(n)_{-1}, \dots, y(n)_{-n}$ in X such that $f^{n-i}(x) \in B(y(n)_{-i}, \eta)$ for all $i = 0, \dots, n$. In particular, we deduce that $\rho(x, y(n)_{-n}) \le \eta\lambda^{-n}$. We conclude that the distance of x from X is arbitrarily small: that is, $x \in X$.

To prove the last assertion, given only x_0 we find $y_0 \in X$ close to x_0; next, take any backward trajectory $y_n \in X$ existing by the 'onto' assumption and find $x_n = f_{x_n}^{-1}(y_{n+1})$ by induction, analogously to finding y_n for x_n above. ♣

Remark 6.1.4. The condition after 'iff' in this proposition (for $N = \infty$) can be considered in the 'inverse limit', saying that every backward trajectory in U'' is in the 'unstable manifold' of a backward trajectory in X.

Now we shall prove a lemma corresponding to the Shadowing Lemma 4.2.4. For any two mappings F, G on the common domain A, to a metric space with a metric ρ, we write $\text{dist}(F, G) := \sup_{x \in A} \rho(f(x), g(x))$. Recall that we say that a sequence of points (y_i) β-shadows (x_i) if $\rho(x_i, y_i) \le \beta$.

Lemma 6.1.5. *Let (X, f, U) be an expanding set in a manifold M. Then for every $\beta : 0 < \beta < \eta$ there exist $\varepsilon, \alpha > 0$ (it is sufficient that $\alpha + \varepsilon < (\lambda - 1)\beta$) such that if a continuous mapping $g : U \to M$ is α-C^0-close to f, that is, $\text{dist}(f, g) \le \alpha$, then every $\varepsilon - f$-trajectory $x = x_0, x_1, \dots, x_n$ in U' can be β-shadowed by at least one g-trajectory. In particular, there exists X_g, a compact* *forward g-invariant set, that is, such that* $g(X_g) \subset X_g$*, and a continuous mapping* $h_{gf} : X_g \to M$ *such that* $\mathrm{dist}(h_{gf}, \mathrm{id}|_{X_g}) \le \beta$ *and* $h_{gf}(X_g) = X$.

If g is Lipschitz continuous, then h_{gf} *is Hölder continuous.*

Proof. This is similar to that of the Shadowing Lemma, but needs some more care. If we treat $x_i, i = 0, \dots, n$ as an $\alpha + \varepsilon$-trajectory for g we cannot refer to the proof of the Shadowing Lemma because we have not assumed that g is expanding.

However, let us make similar choices as there, for $\beta < \eta$ let $\alpha + \varepsilon = \beta(\lambda - 1)$. Then, by Lemma 6.1.1,

$$f(\mathrm{cl}\, B(x_i, \beta)) \supset \mathrm{cl}\, B(f(x_i), \lambda\beta), \quad i = 0, 1, \dots, n. \tag{6.1.2}$$

We left this in the proof of Lemma 6.1.1 as an exercise. One proof could use an integration.

We shall give, however, a standard topological argument, proving (6.1.2) with $\lambda' \le \lambda$ arbitrarily close to λ for β small enough, using the local approximation of f by Df, which will also be of use later on for g. This argument corresponds to Rouché's theorem in one complex variable (preservance of the index of a curve under small perturbation).

In the closed ball $B_1 = \mathrm{cl}\, B(x_i, \beta)$ for β small enough, f is $\beta(\lambda - \lambda')$-$C^0$-close to Df (locally it makes sense to compare f with Df using local charts on the manifold M). To get β independent of i we use f being C^1.

Hence f and Df are homotopic as maps of the sphere $S_1 = \partial B_1$ to $M \setminus z$ for any $z \in B(f(x_i), \lambda'\beta)$, just along the intervals joining the corresponding image points. If z were missing in $f(B_1)$ then we could project $f(B_1)$ to $S_2 = \partial B(f(x_i), \lambda'\beta)$ along the radii from z. Denote such a projection from any w by P_w.

$P_z \circ f|_{S_1} : S_1 \to S_2$ is not homotopic to a constant map, because it is homotopic to $P_z \circ Df|_{S_1}$, which is homotopic to $P_{f(x_i)} \circ DF|_{S_1}$ by using P_t, $t \in [z, f(x_i)]$, and finally $P_{f(x_i)} \circ Df|_{S_1}$ is not homotopic to a constant because Df is an isomorphism (otherwise, composing with Df^{-1} we would get the identity on S_1 homotopic to a constant map).

On the other hand, $P_z \circ f|_{S_1}$ is homotopic to a constant, since it extends to the continuous map $P_z \circ f|_{B_1}$.

Precisely the same topological argument shows that, setting $x_{n+1} = f(x_n)$,

$$g(\mathrm{cl}\, B(x_i, \beta)) \supset \mathrm{cl}\, B(f(x_i), \lambda'\beta - \alpha) \supset$$

$$\mathrm{cl}\, B(x_{i+1}, \lambda'\beta - \alpha - \varepsilon) \supset \mathrm{cl}\, B(x_{i+1}, \beta) \quad i = 0, 1, \dots, n. \tag{6.1.3}$$

So the intersection $A(x) := \bigcap_{j=0}^{n} g^{-j}(\mathrm{cl}\, B(x_j, \beta))$ is non-empty, and the forward g-trajectory of any point in $A(x)$ β-shadows $x_i, i = 0, \dots, n-1$.

The sequence $B(x_0, \beta) \to f(B(x_0, \beta)) \supset B(x_1, \beta) \to f(B(x_1, \beta)) \supset B(x_2, \beta) \to \dots$ is called a *telescope*. The essence of the proof was the existence and the stability of telescopes.

To prove the last assertion, let $\varepsilon = 0$, $x_i \in X$ and $n = \infty$. For the above sets $A(x) = A(x, g, \beta)$ set

$$X(g,\beta) = \bigcup_{x\in X} A(x,g,\beta) \qquad X_g = \operatorname{cl} X(g,\beta).$$

Suppose $\beta < \eta$. Then for $x, y \in X$ $x \neq y$ we have $A(x) \cap A(y) = \emptyset$, because the constant of expansiveness of f on X is 2η. This allows us to define $h_{gf}(y) = x$ for every $y \in A(x)$.

If $y \in \operatorname{cl} A(x)$ then, by the definition, for every $n \geq 0$ $f^n(y) \in \operatorname{cl} A(f^n(x))$. This proves the g-invariance of X_g, and $h_{gf} \circ g = f \circ h_{gf}$. The continuity of h_{gf} holds, because for an arbitrary n, if $y, y' \in X(g,\beta)$ and $\operatorname{dist}(y, y')$ is small enough, say less than $\varepsilon(n)$, then $\operatorname{dist}(g^j(y), g^j(y')) < \eta - \beta$ for every $j = 0, 1, \dots, n$. $\varepsilon(n)$ does not depend on y, y', since g is uniformly continuous on a compact neighbourhood of X in M. Then $\operatorname{dist}(f^j(x), f^j(x')) < 2\eta$, where $x = h_{gf}(y), x' = h_{gf}(y')$. Hence $\operatorname{dist}(x, x') < \lambda^{-n} 2\eta$. We obtain h_{gf} uniformly continuous on $X(g,\beta)$: hence it extends continuously to the closure X_g. ♣

If g is Lipschitz continuous with $\operatorname{dist}(g(y) - g(y')) \leq L \operatorname{dist}(y, y')$, then we set $\varepsilon(n) := (\eta - \beta)L^{-n}$. Then, for $\operatorname{dist}(y, y') \leq \varepsilon(n)$, we get $\operatorname{dist}(x, x') \leq \lambda^{-n} 2\eta = M\varepsilon(n)^{\log\lambda/\log L}$ for $M = (\frac{1}{\eta-\beta})^{\log\lambda/\log L} 2\eta$. In consequence, h_{gf} is Hölder continuous with exponent $\log\lambda/\log L$.

The existence of h_{gf} does not depend on the construction of X_g. That is, the following holds.

Proposition 6.1.6. *Let Y be a forward g-invariant subset of $U' \supset X$ defined in Lemma 6.1.1, for continuous $g : U \to M$ α-close to f. Then, for every $\beta : 0 < \beta < \eta$ and for every $\alpha : 0 < \alpha < \beta(\lambda - 1)$ there exists a unique transformation $h_{gf} : Y \to U$ such that $h_{gf} \circ g = f \circ h_{gf}$ and $\rho(h_{gf}, \mathrm{id}|_Y) < \beta$. (We call such a transformation h_{gf} a semiconjugacy to the image.) This transformation is continuous, and $\rho(h_{gf}, \mathrm{id}|_Y) \leq \frac{\alpha}{\lambda-1}$. If $g|_Y$ is positively expansive and 2β is less than the constant of expansiveness, then h_{gf} is injective (called a conjugacy to the image X_f). If X is a repeller, then $X_f \subset X$. If g is Lipschitz continuous, then h_{gf} is Hölder continuous.*

Proof. Each g-trajectory (y_n) in Y is an α-f-trajectory, and we can refer to Lemma 6.1.5 for α playing the role of ε and $g = f$. We find an f-trajectory (x_n) such that $\rho(x_n, y_n) \leq \alpha/(\lambda - 1)$, and define $h_{gf}(y_n) = x_n$. The uniqueness follows from the positive expansiveness of f with constant $2\eta > 2\beta$. The continuity can be proved as in Lemma 6.1.5. ♣

Proposition 6.1.7. *Let (X, f, U) be an expanding set. Then there exists $\mathcal{U}$ a neighbourhood of f in C^1 topology, that is, $\mathcal{U} = \{g : U \to M : g \in C^1, \rho(f, g) < \alpha, ||Df(x) - Dg(x)|| < \alpha \ \forall x \in U\}$, for a number $\alpha > 0$, such that for every $g \in \mathcal{U}$ there exists an expanding repeller X_g for g and a homeomorphism $h_{gf} : X_g \to X$ such that $h_{gf} \circ g = f \circ h_{gf}$ on X_g. Moreover, for each $x \in X$ the function $\mathcal{U} \to M$ defined by $g \mapsto x_g := h_{gf}^{-1}(x)$ is Lipschitz continuous, where $\mathcal{U}$ is considered with the metric $\rho(g_1, g_2)$ (in C^0 topology). All h_{gf} and their inverses are*

Hölder continuous, with the same exponent and common upper bound of their Hölder norms.

Proof. For $\mathcal{U}$ small enough, all $g \in \mathcal{U}$ are also expanding, with the constant λ possibly replaced by a smaller constant but also larger than 1 and η, with U' the same. Then X_g and h_{gf} exist, by Lemma 6.1.5. Since g is expanding, and hence expansive on X_g, h_{gf} is injective by Proposition 6.1.6.

To prove the Lipschitz continuity of x_g, consider $g_1, g_2 \in \mathcal{U}$. Then

$$h = h_{g_2 f}^{-1} \circ h_{g_1 f} : X_{g_1} \to X_{g_2}$$

is a homeomorphism, such that $\rho(g_1, g_2) < 2\alpha/(\lambda - 1) < \beta$ for appropriate α, where λ is taken to be common for all $g \in \mathcal{U}$.

On the other hand, by Proposition 6.1.6 applied to g_2 in place of f and g_1 in place of g, for the forward invariant set $Y = X_{g_1}$ there exists a homeomorphism $h_{g_1 g_2} : Y \to X_{g_2}$ conjugating g_1 to g_2. We have $\rho(h_{g_1 g_2}, \mathrm{id}|_{X_{g_1}}) < \rho(g_1, g_2)/(\lambda - 1)$. By the uniqueness in Lemma 6.1.5, we have $h = h_{g_1 g_2}$: hence $\rho(h, \mathrm{id}|_{X_{g_1}}) < \rho(g_1, g_2)/(\lambda - 1)$, which yields the desired Lipschitz continuity of x_g. ♣

The Hölder continuity of h_{gf} and h_{gf}^{-1} follows from Lemma 6.1.5. The uniform Hölder exponent results from the existence of a common Lipschitz constant and expanding exponent for g C^1-close to f. The uniformity of the Hölder norm follows from the formula, completing the proof of Lemma 6.1.5.

Examples

Example 6.1.8. Let $f : M \to M$ be an expanding mapping on a compact manifold M: that is, the repeller X is the whole manifold. Then f is C^1-structurally stable. This means that there exists $\mathcal{U}$, a neighbourhood of f in C^1 topology, such that for every $g : M \to M$ in $\mathcal{U}$ there exists a homeomorphism $h_{gf} : M \to M$ conjugating g and f.

This follows from Proposition 6.1.7. Note that $X_g = M$, since, being homeomorphic to M, it is a boundaryless manifold of the same dimension as M, and it is compact: hence, if $\mathcal{U}$ is small enough, it is equal to M (note that we have not assumed connectedness of M).

A standard example of an expanding mapping on a compact manifold is an expanding endomorphism of a torus $f : \mathbb{T}^d = \mathbb{R}^d/\mathbb{Z}^d \to \mathbb{T}^d$: that is, a linear mapping of $\mathbb{R}^d$ given by an integer matrix A, mod $\mathbb{Z}^d$.

Example 6.1.9. Let $f : \mathbb{C}^d \to \mathbb{C}^d$ be the Cartesian product of z^2's: that is, $f(z_1, \dots, z_d) = (z_1^2, \dots, z_d^2)$. Then the torus $\mathbb{T}^d = \{|z_i| = 1, i = 1, \dots, d\}$ is an expanding repeller. By Proposition 6.1.7 it is stable under small C^1 (particularly complex analytic) perturbations g. This means, in particular, that there exists a topological d-dimensional torus invariant under g close to $\mathbb{T}^d$.

Example 6.1.10. Let $f_c : \bar{\mathbb{C}} \to \bar{\mathbb{C}}$ be defined by $f_c(z) = z^2 + c, c \approx 0$: compare the Introduction and Chapter 1. As in Example 6.1.9, there exists a

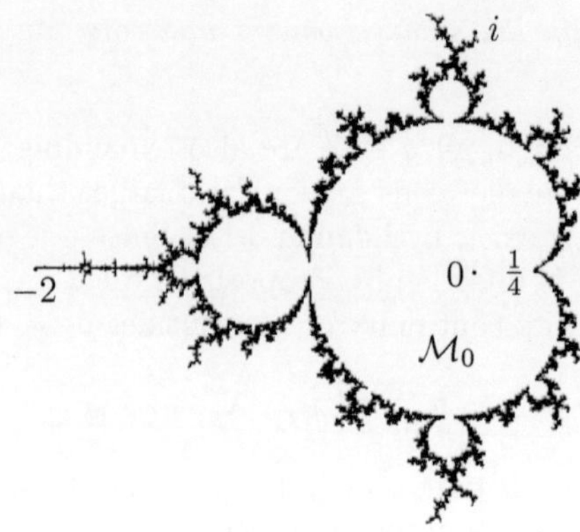

Figure 6.1 Mandelbrot set

Jordan curve X_{f_c} close to the unit circle that is an f_c-invariant repeller and a homeomorphic conjugacy $h_{f_c f_0}$. It is not hard to see that $X_{f_c} = J(f_c)$, the Julia set: see Example 1.5.

The equation of a fixed point for f_c is $z_c^2 + c = z_c$, and if we want a fixed point z_c to be attracting (note that there are two fixed points, except $c = 1/4$) we want $|f_c'(z_c)| = |2z_c| < 1$. This means that c is in the domain $\mathcal{M}_0$ bounded by the cardioid $c = -(\lambda/2)^2 + \lambda/2$ for λ in the unit circle (Figure 6.1).

It is not hard to prove that $\mathcal{M}_0$ is precisely the domain of c where the homeomorphisms $h_{f_c f_0}$, and in particular their domains X_{f_c}, exist. Each X_{f_c} is a Jordan curve, (X_{f_c}, f_c) is an expanding repeller, and the 'motion' $c \mapsto z_c := h_{f_c f_0}^{-1}(z)$ is holomorphic for each z in the unit circle X_{f_0}: see Section 6.2. In fact, X_{f_c} is equal to the Julia set $J(f_c)$ for f_c. At c in the cardioid, a self-pinching of X_{f_c} occurs at infinitely many points.

$\mathcal{M}_0$ is a part of the Mandelbrot set $\mathcal{M}$ where $J(f_c)$ is connected. When c leaves $\mathcal{M}$, the Julia set crumbles into a Cantor set. (However, in the Hausdorff distance between sets, it can explode: see [Douady, Sentenac & Zinsmeister, 1997].)

6.2 Complex dimension one; bounded distortion and other techniques

The basic property is the so-called *Bounded Distortion for Iteration.* We have have already seen that kind of lemma, Lemma 4.4.2 (Pre-Bounded Distortion Lemma for Iteration), used extensively in Chapter 5. Here it will finally get its geometric sense.

Definition 6.2.1. We say that V, an open subset of $\mathbb{C}$ or $\mathbb{R}$, has distortion with respect to $z \in V$ bounded by C if there exist $R > r > 0$ such that $R/r \le C$ and $B(z,r) \subset V \subset B(z,R)$.

Lemma 6.2.2 (Bounded Distortion Lemma for Iteration). *Let (X, f) be a $C^{1+\varepsilon}$ -expanding set in $\mathbb{R}$ or a conformal expanding set in $\mathbb{C}$. Then there exists a constant $C > 0$ such that:*

1. For every $x \in X$ and $n \geq 0$, for every $r \leq \xi$ the distortion of the set $f_x^{-n}(B(f^n(x), r))$ with respect to x is less than $\exp Cr$ *in the conformal case and less than* $\exp Cr^\varepsilon$ *in the real case.*

2. The same bound holds for the distortion of $f^n(B(x, r'))$ for any $r' > 0$ under the assumption $f^j(B(x, r')) \subset B(f^j(x), r)$ for every $j = 1, \dots, n$. Moreover,

3. If $y_1, y_2 \in B(x, r')$ then $\frac{\rho(f^n(y_1), f^n(x))}{\rho(f^n(y_2), f^n(x))} : \frac{\rho(y_1, x)}{\rho(y_2, x)} < \exp Cr$ *or* $\exp Cr^\varepsilon$.

Finally, in terms of derivatives,

4. $\exp Cr$ *and* $\exp Cr^\varepsilon$ *bound the fractions*

$$\frac{|(f^n)'(x)|}{r/\operatorname{diam} f_x^{-n}(B(f^n(x), r))} \quad \textit{and} \quad \frac{|(f^n)'(x)|}{\operatorname{diam} f^n(B(x, r'))/r'}$$

from above, and the inverses bound these fractions from below.

5. For $y_1, y_2 \in f_x^{-n}(B(f^n(x), r))$,

$$|\frac{(f^n)'(y_1)}{f^n)'(y_2)} - 1| < Cr \quad \textit{or} \quad Cr^\varepsilon.$$

Proof. Cr and Cr^ε bound the additive distortions of the functions $\log |(f^n)'|$ and $\operatorname{Arg}(f^n)'$ (in the complex holomorphic case) on the sets $f_x^{-n}(B(f^n(x), r))$ and $B(x, r')$. Indeed, these functions are of the form $S_n\psi$ for Hölder $\psi = \log |f'|$ or $\operatorname{Arg} f'$: see Chapter 4. We use the Pre-Bounded Distortion Lemma 4.4.2. To conclude the assertions involving diameters, integrate $|(f^n)'|$ or the inverse along curves. ♣

In the conformal situation, in $\overline{\mathbb{C}}$, instead of refering to Lemma 6.2.2, one can often refer to the Koebe Distortion Lemma, putting $g = f^n$ or inverse.

Lemma 6.2.3 (Koebe Distortion Lemma in the Riemann sphere). *Given $\varepsilon > 0$ there exists a constant $C = C(\varepsilon)$ such that for every $\lambda : 0 < \lambda < 1$ for every conformal (holomorphic univalent) map on the unit disc in $\mathbb{C}$ to the Riemann sphere $\overline{\mathbb{C}}$, $g : \mathbb{D} \to \overline{\mathbb{C}}$, such that* $\operatorname{diam}(\overline{\mathbb{C}} \setminus g(\mathbb{D})) \geq \varepsilon$, *for all $y_1, y_2 \in \lambda\mathbb{D}$,*

$$|g'(y_1)/g'(y_2)| \leq C(1 - \lambda)^{-4},$$

diameter and derivatives in the Riemann sphere metric.

One can replace $\mathbb{D}$ by any disc $B(x, r) \subset \overline{\mathbb{C}}$ with $\operatorname{diam} \overline{\mathbb{C}} \setminus B(x, r) \geq \varepsilon$ and $y_1, y_2 \in B(x, \lambda r)$.

This lemma follows easily (Exercise 6.2) from the classical lemma in the complex plane: see for example [Carleson & Gamelin, 1993, Section I.1].

Lemma 6.2.4 (Koebe Distortion Lemma). *For every holomorphic univalent function $g : \mathbb{D} \to \mathbb{C}$ for every $z \in \mathbb{D}$,*

$$\frac{1 - |z|}{(1 + |z|)^3} \leq \frac{|g'(z)}{g'(0)} \leq \frac{1 + |z|}{(1 - |z|)^3}.$$

Remark 6.2.5. In the situation of Lemma 6.2.3, with a fixed arbitrary $g : \mathbb{D} \to \overline{\mathbb{C}}$, there exists $C = C(g)$ such that for all $y, y_1, y_2 \in \lambda\mathbb{D}$

$$C^{-1}(1-\lambda) \leq |g'(0)/g'(y)| \leq C(1-\lambda)^{-1}$$

and and in particular

$$|g'(y_1)/g'(y_2)| \leq C(1-\lambda)^{-2}.$$

Of fundamental importance is the so-called holomorphic motion approach:

Definition 6.2.6. Let (X, ρ_X) and (Y, ρ_Y) be metric spaces. We call a mapping $f : X \to Y$ *quasi-symmetric* if there exists a constant $M > 0$ such that for all $x, y, z \in X$ if $\rho_X(x, y) = \rho_X(x, z)$, then $\rho_Y(f(x), f(y)) \leq M\rho_Y(f(x), f(z))$.

In the case where X is an open subset of a Euclidean space of dimension at least 2, the name *quasi-conformal* is usually used. In this case several equivalent definitions are used.

Definition 6.2.7. Let A be a subset of $\overline{\mathbb{C}}$. A mapping $i_\lambda(z)$, for $\lambda \in \mathbb{D}$ the unit disc and $z \in A$, is called a *holomorphic motion* of A if:

(i) for every $\lambda \in \mathbb{D}$ the mapping i_λ is an injection;
(ii) for every $z \in A$ the mapping $\lambda \mapsto i_\lambda(z)$ is holomorphic;
(iii) i_0 is the identity (i.e. inclusion of A in $\overline{\mathbb{C}}$).

Lemma 6.2.8 (Mañé, Sad, Sullivan's λ-lemma: see [Mañé, Sad & Sullivan 1983]). *Let $i_\lambda(z)$ be a holomorphic motion of $A \subset \overline{\mathbb{C}}$. Then every i_λ has a quasi-symmetric extension $i_\lambda : \overline{A} \to \overline{\mathbb{C}}$, which is an injection, for every $z \in \overline{A}$ the mapping $\lambda \to i_\lambda(z)$ is holomorphic, and the map $\mathbb{D} \times \overline{A} \ni (\lambda, z) \mapsto i_\lambda(z)$ is continuous.*

Note that the assumption that the domain of lambdas is complex is substantial. If, for example, the motion is only for $\lambda \in \mathbb{R}$, then the lemma is false. Consider, for example, the motion of $\mathbb{C}$ such that the lower half-plane moves in one direction, $i_\lambda(z) = z + \lambda$, and the upper (closed) half-plane moves in the opposite direction, $i_\lambda(z) = z - \lambda$. Then i_λ is not even continuous. However, this motion cannot be extended to complex lambdas, to injections.

Proof. The proof is based on the following. Any holomorphic map of $\mathbb{D}$ to the triply punctured sphere $\overline{\mathbb{C}} \setminus \{0, 1, \infty\}$ is distance non-increasing for the hyperbolic metrics on $\mathbb{D}$ and $\overline{\mathbb{C}} \setminus \{0, 1, \infty\}$ (Schwarz Lemma). Choose three points of A and renormalize i (i.e. for each i_λ compose it with a respective homography) so that the images by i_λ of these three points are constantly $0, 1$ and ∞. (We can assume A is infinite; otherwise the Lemma is trivial.)

For any three other points $x, y, z \in A$, consider the functions

$$x(\lambda) = i_\lambda(x),\ y(\lambda) = i_\lambda(y),\ z(\lambda) = i_\lambda(z), w(\lambda) = (y(\lambda) - x(\lambda))/y(\lambda).$$

These functions avoid 0, 1, ∞. Fix any $0 < m < M < \infty$. Let $y(0) \in A$ and $y(0)$ be in the ring $P(m, M) = \{m \leq |y| \leq M\}$. Then $|x(0) - y(0)|$ small implies

$w(0)$ small: hence for any $\lambda \leq R$ for an arbitrary constant $R : 0 < R < 1$ the hyperbolic distance between $w(0)$ and $w(\lambda)$ in $\overline{\mathbb{C}} \setminus \{0, 1, \infty\}$ is less than R, and hence $w(\lambda)$ is small. Therefore $x(\lambda) - y(\lambda)$ is small.

Thus each i_λ is uniformly continuous on $A \cap P(m, M)$. Moreover, the family i_λ is equi-continuous for $|\lambda| \leq R$

The annulus $P(m, M)$ for $m < 1 < M$ contains 1, so permuting the roles of 0, 1, ∞ we see that the annuli cover the sphere. So i_λ has a continuous extension to $\overline{A}$. The extensions for $|\lambda| \leq R$ are equi-continuous.

Similarly we prove that if $|x(0) - y(0)|$ is large, then $x(\lambda) - y(\lambda)$ is large. Therefore these extensions are injections.

To prove i_λ is quasi-symmetric consider $g(\lambda) = \frac{x(\lambda)-y(\lambda)}{x(\lambda)-z(\lambda)}$. This function also omits 0, 1, ∞. Assume $|g(0)| = 1$. Then for $|\lambda| \leq R < 1$ the hyperbolic distance of $g(\lambda)$ from the unit circle is not larger than R. Therefore $|g(\lambda)|$ is uniformly bounded for $|\lambda| \leq R$.

Note finally that for each $x \in \overline{A}$ the map $\lambda \mapsto i_\lambda(x)$ is holomorphic, as the limit of holomorphic functions $i_\lambda(z)$, $z \to x, z \in A$. In particular, it is continuous. So, owing to the equicontinuity of the family i_λ, i is continuous on $\mathbb{D} \times \overline{A}$. ♣

Remark 6.2.9. For $X \subset \mathbb{C}$ a topologically transitive expanding repeller for f holomorphic, the λ-lemma gives a new proof of stability under holomorphic perturbations of f to g: see Lemma 6.1.5 and Proposition 6.1.6. One can choose a periodic orbit $P \subset X$ and consider $A = \bigcup_{n=0}^{\infty}(f|_X)^{-n}(P)$. By Theorem 4.3.12(2), A is dense in X. By the Implicit Function Theorem P moves holomorphically under small holomorphic perturbations $g = g_\lambda$ of f. So A moves holomorphically, staying close to X (by the repelling property of (X, f)). So h_{fg}'s can be defined as $i_\lambda : X \to X_g$. Owing to the λ-lemma we conclude they are quasi-symmetric.

Remark 6.2.10. The maps i_λ of the holomorphic motion in Lemma 6.2.8 are Hölder continuous. Moreover, for $\lambda \in A$ any compact subset of D, they have a common Hölder exponent $\beta = \beta_A$ and a common norm in $\mathcal{H}_\beta$.

This follows from Slodkowski's theorem [Slodkowski 1991], which says that the motion $i_\lambda(z)$ extends to the whole Riemann sphere: see also [Astala, Iwaniec & Martin 2009]. Then we refer to the fact that each quasi-symmetric (K-quasi-conformal) homeomorphism is Hölder continuous with exponent $1/K$ and uniformly bounded Hölder norm: see [Ahlfors 1966].

6.3 Transfer operator for conformal expanding repeller with harmonic potential

We consider a conformal expanding repeller: that is, an expanding repeller (X, f) for $X \subset \overline{\mathbb{C}}$ and f conformal on a neighbourhood of X. This is a preparation for a study in Chapters 9 and 10.

We mainly consider potentials of the form $\phi = -t\log|f'|$ for all t real and related transfer operators $\mathcal{L}_\phi$ on (continuous) real functions on X: see Section 5.2. We proved in Chapter 5 that $\mathcal{L}$ has a unique positive eigenfunction u_ϕ, and there exist m_ϕ on X the eigenmeasure of $\mathcal{L}^*$, and μ_ϕ the invariant Gibbs measure equivalent to it, and u_ϕ is the Radon–Nikodym derivative $\frac{d\mu_\phi}{dm_\phi}$. Our aim is to prove that u_ϕ has a real-analytic extension to a neighbourhood of X.

We begin with the following.

Definition 6.3.1. A conformal expanding repeller $f : X \to X$ is said to be real-analytic if X is contained in a finite union of real-analytic curves with closures pairwise disjoint.

The union of these curves will be denoted by $\Gamma = \Gamma_f$. Frequently in such a context we will alternatively speak about the real analyticity of the set X.

Theorem 6.3.2. *If $f : X \to X$ is an orientation-preserving conformal expanding repeller, $X \subset C$, then the Radon–Nikodym derivative $u = u_\phi = d\mu_\phi/dm_\phi$ has a real-analytic real-valued extension on a neighbourhood of X in $\mathbb{C}$. If f is real-analytic, then u has a real-analytic extension on a neighbourhood of X in Γ and a complex-valued complex analytic extension on a neighbourhood in $\overline{\mathbb{C}}$.*

Proof. Since f is conformal and orientation preserving, f is holomorphic on a neighbourhood of X in $\mathbb{C}$. Take $r > 0$ so small that for every $x \in X$, every $n \geq 1$ and every $y \in f^{-n}(x)$ the holomorphic inverse branch $f_y^{-n} : B(x, 2r) \to \mathbb{C}$ sending x to y is well defined.

Suppose first that (X, f) is real-analytic. (We could deduce this case from the general case, but we separate it as it is simpler.)

Take an atlas of real-analytic maps (charts) $\phi_j : \Gamma_j \to \mathbb{R}$ for Γ_j the components of Γ; they extend complex-analytically to a neighbourhood of Γ in $\overline{\mathbb{C}}$. (If Γ_j is a closed curve we can use Arg.)

For $x \in \Gamma_j \cap J(f)$ we write $\Gamma_{j(x)}$ and $\phi_{j(x)}$. For all $k \geq 1$ and all $y \in f^{-k}(x)$ consider, for r small enough, the positive real-analytic function on $\phi_{j(x)}(B(x,r))$

$$z \mapsto |(f_y^{-k})'(\phi_{j(x)}^{-1}(z))|$$

for all $z \in \phi_{j(x)}(\Gamma_j(x) \cap B(x,r))$. Consider the following sequence of complex analytic functions on $z \in \phi_{j(x)}(B(x,r))$:

$$g_n(z) = \sum_{y \in f^{-n}(x)} \Big((f_y^{-n})'(\phi_{j(x)}^{-1}(z))\Big)^t \exp(-nP(t)),$$

where $P(t) = \mathrm{P}(f|_X, -t\log|f'|)$ denotes the pressure.

There is no problem here with raising to the t-th power, since $B(x,r)$, the domain of all $|(f_y^{-n})'|$, is simply connected. Since the latter functions are positive in $\mathbb{R}$, we can choose the branches of the t-th powers to also be positive in $\mathbb{R}$. By the Koebe Distortion Lemma (or Bounded Distortion for Iteration Lemma 6.2.2) for r small enough and every $w = B' := \phi_{j(x)}^{-1}(z) \in B(x,r)$, every $n \geq 1$ and every $y \in f^{-n}(x)$ we have $|(f_y^{-n})'(w)| \leq K|(f_y^{-n})'(x)|$. Hence $|g_n(z)| \leq Kg_n(x)$. Since, by (5.4.2) with $u \equiv 1$ and $c = P(t)$, the sequence $g_n(x)$ converges, we see that the functions $\{g_n|_{B'}\}_{n\geq 1}$ are uniformly bounded. So they form a normal family in the sense of Montel, and hence we can choose a convergent sub-sequence g_{n_j}. Since $g_n(z)$ converges to $u \circ \phi_{j(x)}^{-1}(z)$ for all $z \in X \cap B'$, it follows that $g_{n_j} \circ \phi_{j(x)}$ converges to a complex-analytic function on $B(x,r)$ extending u.

Let us pass now to the proof of the first part of this proposition. That is, we relax the X-th real analyticity assumption, and we want to construct a real-analytic real-valued extension of u to a neighbourhood of X in $\mathbb{C}$. Our strategy is to work in $\mathbb{C}^2$, to use an appropriate version of Montel's theorem and, in general, to proceed similarly as in the first part of the proof. So, fix $v \in X$. Now identify $\mathbb{C}$, where our f acts, to $\mathbb{R}^2$ with coordinates x, y, the real and complex parts of z. Embed this into $\mathbb{C}^2$ with x, y complex. Denote the above $\mathbb{C} = \mathbb{R}^2$ by $\mathbb{C}_0$. We may assume that $v = 0$ in $\mathbb{C}_0$. Given $k \geq 0$ and $v_k \in f^{-k}(v)$, define the function $\rho_{v_k} : B_{\mathbb{C}_0}(0,2r) \to \mathbb{C}$ (the ball in $\mathbb{C}_0$) by setting

$$\rho_{v_k}(z) = \frac{(f_{v_k}^{-k})'(z)}{(f_{v_k}^{-k})'(0)},$$

Since $B_{\mathbb{C}_0}(0,2r) \subset \mathbb{C}_0$ is simply connected and ρ_{v_n} nowhere vanishes, all the branches of logarithm $\log \rho_{v_n}$ are well defined on $B_{\mathbb{C}_0}(0,2r)$. Choose this branch that maps 0 to 0 and denote it also by $\log \rho_{v_n}$. By Koebe's Distortion Theorem, $|\rho_{v_k}|$ and $|\operatorname{Arg} \rho_{v_k}|$ are bounded on $B(0,r)$ by universal constants K_1, K_2 respectively. Hence $|\log \rho_{v_k}| \leq K = (\log K_1) + K_2$. We write

$$\log \rho_{v_k} = \sum_{m=0}^{\infty} a_m z^m,$$

and note that by Cauchy's inequalities

$$|a_m| \leq K/r^m. \tag{6.3.1}$$

We can write for $z = x + iy$ in $\mathbb{C}_0$

$$\operatorname{Re} \log \rho_{v_k} = \operatorname{Re} \sum_{m=0}^{\infty} a_m (x+iy)^m = \sum_{p,q=0}^{\infty} \operatorname{Re}\left(a_{p+q}\binom{p+q}{q} i^q\right) x^p y^q := \sum c_{p,q} x^p y^q.$$

In view of (6.3.1), we can estimate $|c_{p,q}| \leq |a_{p+q}| 2^{p+q} \leq K r^{-(p+q)} 2^{p+q}$. Hence $\operatorname{Re}\log \rho_{v_k}$ extends, by the same power series expansion $\sum c_{p,q} x^p y^q$, to the polydisc $D_{\mathbb{C}^2}(0, r/2)$, and its absolute value is bounded there from above by K. Now for every $k \geq 0$ consider a real-analytic function b_k on $B_{\mathbb{C}_0}(0,2r)$ by setting

$$b_k(z) = \sum_{v_k \in f^{-k}(0)} |(f_{v_k}^{-k})'(z)|^t \exp(-kP(t)).$$

By (5.4.2) the sequence $b_k(0)$ is bounded from above by a constant L. Each function b_k extends to the function

$$B_k(z) = \sum_{v_k \in f^{-k}(0)} |(f_{v_k}^{-k})'(0)|^t \, e^{t \operatorname{Re} \log \rho_{v_k}(z)} \exp(-kP(t)).$$

whose domain, similarly to the domains of the functions $\operatorname{Re}\log\rho_{v_k}$, contains the polydisc $D_{\mathbb{C}^2}(0,r/2)$. Finally, we get for all $k\geq 0$ and all $z\in D_{\mathbb{C}^2}(0,r/2)$

$$\begin{aligned}|B_k(z)| &= \sum_{v_k\in f^{-k}(0)} |(f_{v_k}^{-k})'(0)|^t\, \mathrm{e}^{\operatorname{Re}(t\operatorname{Re}\log\rho_{v_k}(z))}\exp(-kP(t)) \\ &\leq \sum_{v_k\in f^{-k}(0)} |(f_{v_k}^{-k})'(0)|^t\, \mathrm{e}^{t|\operatorname{Re}\log\rho_{v_k}(z)|}\exp(-kP(t)) \\ &\leq \mathrm{e}^{Kt}\sum_{v_k\in f^{-k}(0)} |(f_{v_k}^{-k})'(0)|^t\exp(-kP(t)) \leq \mathrm{e}^{Kt}\, L.\end{aligned}$$

Now by Cauchy's integral formula (in $D_{\mathbb{C}^2}(0,r/2)$) for the second derivatives we prove that the family B_n is equi-continuous on, say, $D_{\mathbb{C}^2}(0,r/3)$. Hence we can choose a uniformly convergent sub-sequence, and the limit function G is complex analytic and extends u on $X\cap B(0,r/3)$, by (5.4.2). Thus we have proved that u extends to a complex analytic function in a neighbourhood of every $v\in X$ in $\mathbb{C}^2$, that is, real-analytic in $\mathbb{C}_0$. These extensions coincide on the intersections of the neighbourhoods, otherwise X is real-analytic and we are in the case considered at the beginning of the proof. See Chapter 10, Lemma 10.1.4, for more details. ♣

In Theorem 6.3.2 we wanted to be concrete, and considered the potential function $-t\log|f'|$ (normalized). In fact, we proved the following more general theorem.

Theorem 6.3.3. *If $f: X\to X$ is an orientation-preserving conformal expanding repeller, $X\subset C$, and ϕ is a real-valued function on X that extends to a harmonic function on a neighbourhood of X in $\overline{\mathbb{C}}$, then $u_\phi = d\mu_\phi/dm_\phi$ has a real-analytic real-valued extension on a neighbourhood of X in $\mathbb{C}$.*

Proof. We can assume that pressure $\mathrm{P}(f,\phi)=0$. As in the previous proof, choose $0\in X$. Assume that r is small enough that all $v_k\in f^{-k}(0)$ and all $k=1,2,\dots$ all the branches f^{-k} on $B(0,r)$ and the compositions $\phi\circ f^{-k}$ exist, and are bounded by a constant $K>0$. They are harmonic as the compositions of holomorphic functions with harmonic ϕ. We have

$$b_k(z) = \sum_{v_k\in f^{-k}(0)} e^{S_k(\phi)(z)} \leq e^{2K}\sum_{v_k\in f^{-k}(0)} e^{S_k(\phi(0))} \leq e^{2K}L,$$

where $L=\sup_k {\mathcal{L}_\phi}^k(\mathbb{1})(0)$. We have used the estimate (6.3.1) for harmonic functions $u_{v_k} = S_k(\phi)(z) - S_k(\phi)(0) = \sum_{m=0}^{\infty} a_m z^m$, where for each v_k we define $S_k(\phi)(z) = \sum_{i=0}^{k-1}\phi(f^i(f_{v_k}^{-k}(z))$.

This version of (6.3.1) follows from the Poisson formula for harmonic functions u_{v_k}, which are uniformly bounded on $B(0,r)$ owing to the uniform exponential convergence to 0 of $|f^{-i}(0) - f^{-k}(z)|$ as $i\to\infty$. See for example Harnack's inequalities in [Hayman & Kennedy, 1976, Section 1.5.6, Example 2]. ♣

Remark 6.3.4. The proofs of Theorem 6.3.2 and Theorem 6.3.3 are the same. In Theorem 6.3.2 we explicitly write complex analytic power series extension in $\mathbb{C}^2$ of $\log|(f^{-k})'|$, whereas in Theorem 6.3.3 we observe that a general harmonic function is real-analytic, and discuss in particular its domain in complex extension. For a more precise description of domains of complex extensions of harmonic functions (in any dimension) see [Hayman & Kennedy, 1976, Section 1.5.3]; more references are provided there.

Remark 6.3.5. A version of Theorem 6.3.3 holds in the real case (say if X is in a finite union of pairwise disjoint circles and straight lines), with finite smoothness.

That is, if we assume the potential ϕ is $C^{r+\varepsilon}$ for $r \geq 1, 0 \leq \varepsilon \leq 1$ and $r+\varepsilon > 1$, then for m the density u_ϕ of the invariant Gibbs measure μ_ϕ with respect to the eigenmeasure m_ϕ of $\mathcal{L}$ (see the beginning of this section) is $C^{r+\varepsilon}$.

For a sketch of the proof see Chapter 7, Section 7.4, and Exercise 7.5. See also [Boyarsky & Góra 1997]. This is related to the $C^{r+\varepsilon}$-rigidity: see Exercise 6.1.

6.4 Analytic dependence of transfer operator on potential function

In this section we prove a fundamental theorem about the real-analytic dependence of transfer operators acting on the Banach space of Hölder continuous functions, with respect to the vector space of real-valued Hölder continuous potentials, and then we derive some consequences concerning the real-analytic dependence of pressure with respect to potential and a conformal expanding map (repeller) depending pointwise complex-analytically on a complex parameter. We shall apply Mañé, Sad and Sullivan's λ-lemma: see Section 6.2. In Chapter 9 we shall deduce the real-analytic dependence of the Hausdorff dimension of a conformal expanding repeller, on a parameter.

Let T be a continuous open topologically mixing distance-expanding map on a compact metric space (X, ρ): cf. Chapters 4 and 5. For every point $x \in X$ define $\mathcal{H}_{\beta;x}$ to be the Banach space of complex-valued Hölder continuous functions with exponent β, whose domain is the ball $B(x, \delta)$ with $\delta > 0$ so small that all the inverse branches of T are well defined on $B(x, \delta)$: for example $\delta \leq \xi$ in Section 4.1. The Hölder variation ϑ_β and the Hölder norm $||\cdot||_\beta = ||\cdot||_{\mathcal{H}_\beta}$ are defined in the standard way: see Chapter 4.

Let $L(F)$ and $L(F_1, F_2)$ denote the spaces of continuous linear operators from F to itself or from F_1 to F_2, respectively, for F, F_1, F_2 Banach spaces.

For every function $\Phi : G \to L(\mathcal{H}_\beta)$ for any set of parameters G and for every $x \in X$ define the function $F_x : G \to L(\mathcal{H}_\beta, \mathcal{H}_{\beta;x})$ by the formula

$$\Phi_x(\lambda)(\psi) = \Phi(\lambda)(\psi)|_{B(x,\delta)}.$$

Sometimes we write $\Phi(\lambda)_x$.

We start with the following.

Lemma 6.4.1. *Let G be an open subset of a complex plane $\mathbb{C}$, and fix a function $\Phi : G \to L(\mathcal{H}_\beta)$. If for every $x \in X$ the function $\Phi_x : G \to L(\mathcal{H}_\beta, \mathcal{H}_{\beta;x})$ is complex analytic and $\sup\{||\Phi_x(\lambda)||_\beta : x \in X : \lambda \in G\} < +\infty$, then the function $\Phi : G \to L(\mathcal{H}_\beta)$ is complex analytic.*

Proof. Fix $\lambda^0 \in G$ and take $r > 0$ so small that the disc centred at λ^0 of radius r, $D(\lambda^0, r)$, is contained in G. Then for each $x \in X$ and some $a_{x,n} \in L(\mathcal{H}_\beta, \mathcal{H}_{\beta;x})$,

$$\Phi_x(\lambda) = \sum_{n=0}^{\infty} a_{x,n}(\lambda - \lambda^0)^n \quad \lambda \in D(\lambda^0, r),$$

with convergence in the operator norm.

Put $M = \sup\{||\Phi_x(\lambda)||_\beta : x \in X : \lambda \in G\} < +\infty$. It follows from Cauchy's inequalities that

$$||a_{x,n}||_\beta \leq Mr^{-n}. \tag{6.4.1}$$

Now, for every $n \geq 0$, define the operator a_n by

$$a_n(\phi)(z) = a_{z,n}(\phi)(z), \quad \phi \in \mathcal{H}_\beta, \ z \in X.$$

Then

$$||a_n(\phi)||_\infty \leq ||a_{z,n}||_\infty ||\phi||_\infty \leq ||a_{z,n}||_\beta ||\phi||_\beta. \tag{6.4.2}$$

Now, if $|z - x| < \delta$, then for every $\phi \in \mathcal{H}_\beta$ and every $w \in D(x, \delta) \cap D(z, \delta)$,

$$\begin{aligned}\sum_{n=0}^{\infty} a_{x,n}(\phi)(w)(\lambda - \lambda^0)^n &= (\Phi_x(\lambda)(\phi))(w) = (\Phi(\lambda)(\phi))(w) = (\Phi_z(\lambda)(\phi))(w) \\ &= \sum_{n=0}^{\infty} a_{z,n}(\phi)(w)(\lambda - \lambda^0)^n\end{aligned}$$

for all $\lambda \in D(\lambda^0, r)$. The uniqueness of the coefficients of a Taylor series expansion implies that, for all $n \geq 0$,

$$a_{x,n}(\phi)(w) = a_{z,n}(\phi)(w).$$

Since $x, z \in D(x, \delta) \cap D(z, \delta)$, we thus get, using (6.4.1),

$$\begin{aligned}|a_n(\phi)(z) - a_n(\phi)(x)| &= |a_{z,n}(\phi)(z) - a_{x,n}(\phi)(x)| = |a_{x,n}(\phi)(z) - a_{x,n}(\phi)(x)| \\ &\leq ||a_{x,n}(\phi)||_\beta |x - z|^\beta \leq ||a_{x,n}||_\beta ||\phi||_\beta |x - z|^\beta \\ &\leq Mr^{-n} ||\phi||_\beta |x - z|^\beta.\end{aligned}$$

Consequently, $\vartheta_\beta(a_n(\phi)) \leq Mr^{-n}||\phi||_\beta$. Combining this with (6.4.2), we obtain $||a_n(\phi)||_\beta \leq 2Mr^{-n}||\phi||_\beta$. Thus $a_n \in L(\mathcal{H}_\beta)$ and $||a_n||_\beta \leq 2Mr^{-n}$. Thus the series

$$\sum_{n=0}^{\infty} a_n(\lambda - \lambda^0)^n$$

converges absolutely uniformly on $D(\lambda^0, r/2)$ and $||\sum_{n=0}^{\infty} a_n(\lambda-\lambda^0)^n||_\beta \le 2M$ for all $\lambda \in D(\lambda^0, r/2)$. Finally, for every $\phi \in \mathcal{H}_\beta$ and every $z \in X$,

$$\begin{aligned}\left(\sum_{n=0}^{\infty} a_n(\lambda-\lambda^0)^n\right)(\phi)(z) &= \sum_{n=0}^{\infty} a_n(\phi)(z)(\lambda-\lambda^0)^n = \sum_{n=0}^{\infty} a_{z,n}(\phi)(z)(\lambda-\lambda^0)^n \\ &= \left(\sum_{n=0}^{\infty} a_{z,n}(\lambda-\lambda^0)^n\right)(\phi)(z) = \Phi_z(\lambda)(\phi)(z) \\ &= (\Phi(\lambda)\phi)(z).\end{aligned}$$

So $\Phi(\lambda)(\phi) = \left(\sum_{n=0}^{\infty} a_n(\lambda-\lambda^0)^n\right)(\phi)$ for all $\phi \in \mathcal{H}_\beta$, and consequently, $\Phi(\lambda) = \sum_{n=0}^{\infty} a_n(\lambda-\lambda^0)^n$, $\lambda \in D(\lambda^0, r/2)$. The proof is complete. ♣

The main technical result of this section concerning the analytic dependence of the transfer operator $\mathcal{L}_{\phi_\lambda}$ on the parameter λ is the following.

Theorem 6.4.2. *Suppose that G is an open subset of the complex space $\mathbb{C}^d$ with some $d \ge 1$. If, for every $\lambda \in G$, $\phi_\lambda : X \to \mathbb{C}$ is a β-Hölder complex-valued potential, $H = \sup\{||\phi_\lambda||_\beta : \lambda \in G\} < \infty$, and for every $z \in X$ the function $\lambda \mapsto \phi_\lambda(z)$, $\lambda \in G$, is holomorphic, then the map $\lambda \mapsto \mathcal{L}_{\phi_\lambda} \in L(\mathcal{H}_\beta)$, $\lambda \in G$, is holomorphic.*

Proof. We have for all $\lambda \in G$ and all $v \in X$ that

$$||\exp(\phi_\lambda \circ T_v^{-1})||_\infty \le e^H, \tag{6.4.3}$$

where T_v^{-1} is the branch of T^{-1} on $B(T(v), \delta)$ mapping $T(v)$ to v: compare the notation in Section 4.1. By virtue of Hartog's Theorem, in order to prove our theorem we may assume without loss of generality that $d = 1$: that is, $G \subset \mathbb{C}$. Now fix $\lambda^0 \in G$ and take a radius $r > 0$ so small that $\overline{B}(\lambda^0, r) \subset G$. By our assumptions the function $\lambda \mapsto \exp(\phi_\lambda \circ T_v^{-1}(z))$ is holomorphic for every $z \in B(T(v), \delta)$. Consider its Taylor series expansion

$$\exp(\phi_\lambda \circ T_v^{-1}(z)) = \sum_{n=0}^{\infty} a_{v,n}(z)(\lambda-\lambda^0)^n, \quad \lambda \in B(\lambda^0, r).$$

In view of Cauchy's inequalities and (6.4.3) we get

$$|a_{v,n}(z)| \le e^H r^{-n}, \tag{6.4.4}$$

and, for $w, z \in B(T(v), \delta)$, also using Cauchy's inequalities,

$$\begin{aligned}|a_{v,n}(w) - a_{v,n}(z)| &\le r^{-n} \sup_{\lambda \in G} |\exp(\phi_\lambda \circ T_v^{-1}(w)) - \exp(\phi_\lambda \circ T_v^{-1}(z))| \\ &\le \hat{c} r^{-n} |w-z|^\beta, \end{aligned} \tag{6.4.5}$$

where $\hat{c}$ is a constant that depends only on T and H. Take an arbitrary $\phi \in \mathcal{H}_\beta$ and consider the product $a_{v,n}(z) \cdot \phi(T_v^{-1}(z))$. In view of (6.4.4) and (6.4.5) we obtain

$$\begin{aligned}|a_{v,n}(w)\phi(T_v^{-1}(w)) - a_{v,n}(z)\phi(T_v^{-1}(z))| \\ &\leq |a_{v,n}(w) - a_{v,n}(z)| \cdot ||\phi||_\infty + |a_{v,n}(z)| \cdot ||\phi||_\beta L^\beta |w-z|^\beta \\ &\leq r^{-n}(\hat{c} + e^H L^\beta)||\phi||_\beta |w-z|^\beta \\ &= \hat{c}_1 r^{-n} ||\phi||_\beta |w-z|^\beta,\end{aligned}$$

where $\hat{c}_1 = \hat{c}(1+L^\beta)$ and L is a common Lipschitz constant for all branches T_v^{-1} coming from the expanding property. Combining this and (6.4.4), we conclude that the formula $N_{v,n}\phi(z) = a_{v,n}(z)\phi(T_v^{-1}(z))$ defines a bounded linear operator $N_{v,n} : \mathcal{H}_\beta \to \mathcal{H}_{\beta;x}$, where $x = T(v)$, and

$$||N_{v,n}||_\beta \leq (e^H + \hat{c}_1) r^{-n}.$$

Consequently the function $\lambda \mapsto N_{v,n}(\lambda - \lambda^0)^n$, $\lambda \in B(\lambda^0, r/2)$, is analytic, and $||N_{v,n}(\lambda - \lambda^0)^n||_\beta \leq 2^{-n}(e^H + \hat{c}_1)$. Thus the series

$$A_{\lambda,v} = \sum_{n=0}^{\infty} N_{v,n}(\lambda - \lambda^0)^n, \ \lambda \in D(\lambda^0, r/2),$$

converges absolutely uniformly in the Banach space $L(\mathcal{H}_\beta, \mathcal{H}_{\beta,x})$, $||A_{\lambda,v}||_\beta \leq 2(e^H + \hat{c}_1)$, and the function $\lambda \mapsto A_{\lambda,v} \in L(\mathcal{H}_\beta, \mathcal{H}_{\beta;x})$, $\lambda \in B(\lambda^0, r/2)$, is analytic. Hence $\mathcal{L}_{\lambda,x} = \sum_{v \in f^{-1}(x)} A_{\lambda,v} \in L(\mathcal{H}_\beta, \mathcal{H}_{\beta;x})$,

$$||\mathcal{L}_{\lambda,x}||_\beta \leq 2N(T)(e^H + \hat{c}_1),$$

where $N(T)$ is the number of pre-images of a point in X, and the function $\lambda \mapsto \mathcal{L}_{\lambda,x}$, $\lambda \in D(\lambda^0, r/2)$, is analytic. Since $\mathcal{L}_{\lambda,x} = (\mathcal{L}_{\phi_\lambda})_x$, invoking Lemma 6.4.1 concludes the proof. ♣

Note that a function from a complex vector space to a complex Banach space is called holomorphic if its restriction to any complex finite dimensional affine subspace is holomorphic: see [Dunford & Schwartz, 1958, Definition VI.10.5]. So Theorem 6.4.2 yields the analyticity of

$$\mathcal{H}_\beta \ni \phi \mapsto \mathcal{L}_\phi \in L(\mathcal{H}_\beta)$$

mentioned in the introduction to this section: complex analyticity, and also real analyticity after restricting the function to the real space $\mathcal{H}_b$.

Remark 6.4.3. In the proof one can omit reference to Lemma 6.4.1 by considering the operators $\mathcal{L}_{\phi_\lambda}$ directly, rather than considering individual branches T_v^{-n} and the operators $A_{\lambda,v}$ first.

Now consider an expanding conformal repeller (X, f, U), in $\overline{\mathbb{C}}$, $f : U \to \overline{\mathbb{C}}$ conformal, preserving X, $T = f|_X$, and holomorphic perturbations $f_\lambda : U \to \overline{\mathbb{C}}$, $\lambda \in \Lambda$, where Λ is an open subset of $\mathbb{C}^d$, $f_{\lambda_0} = f$ for some $\lambda_0 \in \Lambda$. Let $i_\lambda : X \to X_\lambda$ be the corresponding holomorphic motion coming from Lemma 6.2.8 and Remark 6.2.9 ($d > 1$ does not cause problems).

Our goal is to prove that the pressure function

$$(\lambda, t) \mapsto \mathrm{P}(\lambda, t) = \mathrm{P}(f_\lambda, -t \log|f'_\lambda|) \in \mathbb{R} \quad \text{for } t \in \mathbb{R}$$

is real-analytic. The idea is to consider potentials $\phi_{\lambda,t} = -t \log|f'_\lambda| \circ i_\lambda : X \to \mathbb{R}$, $(\lambda, t) \in \Lambda \times \mathbb{R}$, to embed them into a holomorphic family to satisfy the assumptions of Theorem 6.4.2, and then to use Kato's Theorem for perturbations of linear operators. Indeed, by Lemma 6.2.8, for every $z \in X$ the function $\lambda \mapsto \Psi_z(\lambda) = \log|f'_\lambda(i_\lambda(z))| - \log|f'(z)|$ is harmonic on Λ, and $\Psi_z(\lambda_0) = 0$. Fix $r > 0$ so small that $B(\lambda_0, 2r) \subset \Lambda$. Then

$$M = \sup\{|\Psi_z| : (z, \lambda) \in X \times B(\lambda_0, r)\} < +\infty.$$

So each function Ψ_z extends holomorphically to $\lambda \in B_{\mathbb{C}^{2d}}(\lambda_0, r/2)$. We shall use the same symbol Ψ_z for this extension, and

$$M_1 = \sup\{|\Psi_z(\lambda)| : (z, \lambda) \in X \times B_{\mathbb{C}^{2d}}(\lambda_0, r/2)\} < +\infty.$$

Since all the functions i_λ, $\lambda \in B(\lambda_0, r)$, are Hölder continuous with a common Hölder exponent, say β, and a common Hölder norm for the exponent β (see Proposition 6.1.7 or Remark 6.2.10), an easy application of Cauchy inequalities gives that for all $\lambda \in B_{\mathbb{C}^{2d}}(\lambda_0, r/2)$ the function $z \mapsto \Psi_z(\lambda)$ is Hölder continuous with exponent β, and the corresponding Hölder norms are uniformly bounded, say by M_2. Thus the potentials

$$\phi_{\lambda,t}(z) = -t\Psi_z(\lambda) + t \log|f'(z)|, \ (\lambda, t) \in B_{\mathbb{C}^{2d}}(\lambda_0, r/2) \times U,$$

for any bounded $U \subset \mathbb{C}$ satisfy the assumptions of Theorem 6.4.2, and for all $(\lambda, t) \in B(\lambda_0, r/2) \times \mathbb{R}$, we have $\phi_{\lambda,t} = -t \log|f'_\lambda \circ i_\lambda|$. As an immediate application of this theorem, we get the following.

Lemma 6.4.4. *The function*

$$(\lambda, t) \mapsto \mathcal{L}_{\phi_{\lambda,t}} \in L(\mathcal{H}_\beta), \ (\lambda, t) \in B_{\mathbb{C}^{2d}}(\lambda_0, r/2) \times \mathbb{C},$$

is holomorphic.

Since for all $(\lambda, t) \in B(\lambda_0, r/2) \times \mathbb{R}$, $\exp(\mathrm{P}(\lambda, t)$ is a simple isolated eigenvalue of $\mathcal{L}_{\phi_{\lambda,t}} \in L(\mathcal{H}_\beta)$ depending continuously on (λ, t), it follows from Lemma 6.4.4 and Kato's perturbation theorem for linear operators that there exists a holomorphic function $\gamma : B_{\mathbb{C}^{2d}}(\lambda_0, R) \times \mathbb{C} \to \mathbb{C}$ ($R \in (0, r/2]$ sufficiently small) such that $\gamma(\lambda, t)$ is an eigenvalue of the operator $\mathcal{L}_{\phi_{\lambda,t}}$ for all $(\lambda, t) \in B_{\mathbb{C}^{2d}}(\lambda_0, R) \times \mathbb{C}$ and $\gamma(\lambda, t) = \exp(\mathrm{P}(\lambda, t))$ for all $(\lambda, t) \in B(\lambda_0, R) \times \mathbb{R}$. Consequently, the function $(\lambda, t) \mapsto \mathrm{P}(\lambda, t)$, $(\lambda, t) \in B(\lambda_0, R) \times \mathbb{R}$, is real-analytic, and as real analyticity is a local property, we finally get the following.

Theorem 6.4.5. *The pressure function* $(\lambda, t) \mapsto \mathrm{P}(f_\lambda, -t \log|f'_\lambda|)$, $t \in \mathbb{R}$, *is real-analytic.*

Exercises

6.1. Let $f, g : S^1 \to S^1$ be two $C^{1+\varepsilon}$-expanding maps of the circle $0 < \varepsilon \le 1$. Prove that if there is a conjugacy h, that is, a homeomorphism $h : S^1 \to S^1$, such that $g \circ h = h \circ f$ and h has at least one point x of differentiability and $h'(x) \neq 0, \pm\infty$, then $h \in C^{1+\varepsilon}$. If $r = 2, 3, ..., \infty, \omega$ (the latter means real-analytic) and $0 \le \varepsilon \le 1$, and if $f, g \in C^{r+\varepsilon}$, then $h \in C^{r+\varepsilon}$.

Hint: The proof can follow the lines of the Cantor repellers case (see Chapter 7), or the proof of Theorem 9.5.5 (the second method) in the analytic case.

6.2. Conclude Lemma 6.2.3 from Lemma 6.2.4.

Hint (due to K. Barański): Given y_1, y_2, change the coordinates on $\overline{\mathbb{C}}$ by a spherical isometry such that $\infty \notin g(\mathbb{D})$ and dist $(g(y_i), \infty) \ge \varepsilon/4$, for $i = 1, 2$.

Bibliographical notes

Lemma 6.1.2 and Proposition 6.1.3, establishing the equivalence of various properties of being a repeller for an expanding set, correspond to the equivalence for hyperbolic subsets of properties 'local product structure' being 'isolated' and 'unstable set' being the union of unstable manifolds of 'individual trajectories': see [Katok & Hasselblatt, 1995, Section 18.4]. For the theory of hyperbolic endomorphisms, particularly in the inverse limit (backward trajectories) language, as in Remark 6.1.4, see [Przytycki 1976] and [Przytycki 1977]. In [Przytycki 1977] some examples of Axiom A endomorphisms, whose basic sets are expanding repellers, are discussed. Example 6.1.9 was studied by M. Denker and S.-M. Heinemann in [Denker & Heinemann 1998]. Theorem 6.3.2 was stated, and applied as in our Chapter 10, in [Sullivan 1986]. Compare [Krzyżewski 1982]. For Section 6.4 compare [Urbański & Zinsmeister 2001] or [Mauldin & Urbański, 2003, Section 2.6.]. Theorem 6.4.5 holds in a setting more general than expanding: see Section 12.5, [Stratmann & Urbański 2003] and [Przytycki & Rivera-Letelier 2008]. For Exercise 6.1 and related considerations see in particular [Shub & Sullivan 1985], [Jiang 1996] and [Cui 1996]. See also the recent [Jordan et al. 2010] for the multifractal analysis of the conjugacy h in the case where it is not differentiable, with f, g piecewise expanding: compare Section 9.2.

7

Cantor repellers in the line; Sullivan's scaling function; application in Feigenbaum universality

After the very general previous chapters we want for a while to concentrate on the real one-dimensional situation, that is, fractals in the line. In Exercise 6.1 we discussed expanding maps of the circle. The aim of this chapter is to study thoroughly Cantor sets in the line with expanding maps on them (generalizing, but in some features more difficult than the whole circle case: see Remark 7.1.11). Starting from Chapter 9 we shall work mainly with the one-dimensional complex case (conformal fractals), the main aim of this book. Some consideration from this section will be continued, including the complex case, in Chapter 10.

In Section 7.1 we supply a one-sided shift space Σ^d (see Chapter 1) with ambient real one-dimensional differentiable structures, basically $C^{1+\varepsilon}$ (Hölder continuous differentials). In Section 7.2 we ask when does the shift map extend $C^{1+\varepsilon}$ to a neighbourhood of the Cantor set being an embedding of Σ^d into a real line. In the case where it does, we have a $C^{1+\varepsilon}$ expanding repeller: see the definition at the beginning of Chapter 6. There a scaling function appears, which is a complete geometric invariant for $C^{1+\varepsilon}$-equivalence (conjugacy). It happens that scaling functions also classify $C^{r+\varepsilon}$ equivalence classes for $C^{r+\varepsilon}$ Cantor expanding repellers, for all $r = 1, 2, \ldots, \infty$, $0 \le \varepsilon \le 1, r + \varepsilon > 1$, and for the real-analytic case. Section 7.3 is devoted to this (for $\varepsilon > 0$, for $\varepsilon = 0$ see Section 7.4). However, scaling functions 'see' the smoothness of the Cantor repeller – that is, the smoother the differentiable structure, the less scaling functions can occur: see examples at the end of Section 7.2 and Section 7.4.

In Section 7.5 we define so-called generating families of expanding maps. This is a bridge towards Section 7.6, where Feigenbaum's universality, concerning the geometry of the Cantor set being the closure of the forward trajectory of the critical point of the quadratic-like map of the interval, will be discussed.

Whereas the proofs in Sections 7.1–7.5 are very detailed, Section 7.6 has a sketchy character. We do not involve much in the theory of iterations of maps of the interval. We refer the reader to [Collet & Eckmann 1980] and [de Melo & van Strien 1993].

For the universality see [Sullivan 1991] and [McMullen 1996], where the key theorem towards this, the exponential convergence of renormalizations, has been proved with the use of complex methods. (For more references see the notes at the end of this chapter.)

In Section 7.6 we just show how the exponential convergence yields the $C^{1+\varepsilon}$ equivalence of Cantor sets being closures of post-critical sets.

Most of this chapter is written on the basis of Dennis Sullivan's paper [Sullivan 1988], completed in [Przytycki & Tangerman 1996].

7.1 $C^{1+\varepsilon}$-equivalence

For simplicity we shall consider here only the class $\mathcal{H}$ of homeomorphic embeddings of Σ^d into the unit interval $[0,1] \subset \mathbb{R}$ such that the order is preserved: that is, for $h : \Sigma^d \to \mathbb{R}$, if $\alpha = (\alpha_0, \alpha_1, \dots), \beta = (\beta_0, \beta_1, \dots) \in \Sigma^d$, $\alpha_j = \beta_j$ for all $j < n$ and $\alpha_n < \beta_n$, then $h(\alpha) < h(\beta)$. In Section 7.5 we need to consider more general situations, but the basic facts stay precisely the same.

Consider an arbitrary $h \in \mathcal{H}$. For every $j_0, j_1, \dots, j_n \in \{1, \dots, d\}, n > 0$, denote by $I_{j_0,\dots,j_n}$ the closed interval with ends $h((j_0, j_1, \dots, j_n, 1, 1, 1, \dots))$ and $h((j_0, j_1, \dots, j_n, d, d, d, \dots))$. The interval $[h((1, 1, \dots)), h((d, d, \dots))]$ will be denoted by I. For $j_n < d$ denote by $G_{j_0,\dots,j_n}$ the open interval with ends $h((j_0, j_1, \dots, j_n, d, d, d, \dots))$ and $h((j_0, j_1, \dots, j_n + 1, 1, 1, 1, \dots))$: the letters G stand for gaps here because of the disjointness with $h(\Sigma^d)$. Denote $E_n = \bigcup_{(j_0,\dots,j_n)} I_{j_0,\dots,j_n}$. We see that $h(\Sigma^d)$ is a Cantor set $\bigcap_{n=0}^{\infty} E_n$.

Definition 7.1.1. Given $h \in \mathcal{H}$ and $w = (j_0, j_1, \dots, j_n)$, where each $j_t \in \{1, \dots, d\}$, we call the sequence of numbers $A_j(h, w) := \frac{|I_{w,\frac{j+1}{2}}|}{|I_w|}$ for j odd, $A_j(h, w) := \frac{|G_{w,\frac{j}{2}}|}{|I_w|}$ for j even, $j = 1, \dots, 2d-1$, the *ratio geometry* of w ($|\cdot|$ denote lengths here). The ratio geometry is the function $w \mapsto (A_j(h, w), j = 1, \dots 2d-1)$.

Definition 7.1.2. We say $h \in \mathcal{H}$ has *bounded geometry* if the ratios $A_j(h, w)$ are uniformly, for all w, j, bounded away from zero. We denote the space of h's from $\mathcal{H}$ with the bounded geometry by $\mathcal{H}$b. We say $h \in \mathcal{H}$ has *exponential geometry* if $|I_{j_0,\dots,j_n}|$ converge to 0 uniformly exponentially fast in n, and not faster. We denote the space of h's from $\mathcal{H}$ with exponential geometry by $\mathcal{H}$e. Observe that $\mathcal{H}\text{e} \supset \mathcal{H}\text{b}$.

Definition 7.1.3. Given $h_1, h_2 \in \mathcal{H}$, we say they have *equivalent geometries* if $\frac{A_j(h_1,w)}{A_j(h_2,w)}$ converge to 1 uniformly in length of w. We say h_1, h_2 have *exponentially equivalent geometries* if the convergence is exponentially fast with the length of w.

One can easily check that exponential geometry is the property of the geometric equivalence classes, and that bounded geometry is the property of the exponentially equivalent geometry classes.

Definition 7.1.4. We say that $h_1, h_2 \in \mathcal{H}$ are $C^{1+\varepsilon}$-equivalent if there exists an increasing $C^{1+\varepsilon}$-diffeomorphism ϕ of a neighbourhood of $h_1(\Sigma^d)$ to a neighbourhood of $h_2(\Sigma^d)$ such that $\phi|_{h_1(\Sigma^d)} \circ h_1 = h_2$. We call $\phi|_{h_1(\Sigma^d)}$ *canonical conjugacy,* as it is uniquely determined by h_1 and h_2. So $C^{r+\varepsilon}$ means that the canonical conjugacy extends $C^{r+\varepsilon}$.

Each class of equivalence will be called a $C^{1+\varepsilon}$-structure for Σ^d. These definitions are valid also for $C^{1+\varepsilon}$ replaced by $C^{r+\varepsilon}$ for every $r = 0, 1, \ldots, \infty, \omega, 0 \leq \varepsilon \leq 1$. For $\varepsilon = 0$ this means the continuity of the r-th derivative: for $0 < \varepsilon < 1$ it is Hölder continuity, and for $\varepsilon = 1$ Lipschitz continuity. ω means real-analytic. (Compare this notation with Exercise 6.1.)

Proposition 7.1.5. *Let $h_1, h_2 \in \mathcal{H}$. Then if they are C^1-equivalent, they have equivalent geometries.*

We leave a simple proof to the reader. Also, the following holds.

Theorem 7.1.6. *Let $h_1, h_2 \in \mathcal{H}e$. Then h_1, h_2 are $C^{1+\varepsilon}$-equivalent for some $\varepsilon > 0$ if and only if h_1 and h_2 have exponentially equivalent geometries.*

Proof. We shall use the fact that a real function ϕ on a bounded interval is $C^{1+\varepsilon}$-smooth if and only if there exists a constant $C > 0$ such that for every $x < y < z$

$$\left| \frac{\phi(y) - \phi(x)}{y - x} - \frac{\phi(z) - \phi(y)}{z - y} \right| < C(z - x)^{\varepsilon} \tag{7.1.1}$$

(this is an easy calculus exercise).

Suppose there exists a diffeomorphism ϕ, as in the definition of equivalence. As ϕ is a diffeomorphism, we can write (7.1.1) for it in a multiplicative form, and obtain for each $w = (j_1, j_2, \ldots, j_n)$ and $j = 1, \ldots, d$ and intervals for h_1

$$\left| \frac{|\phi(I_{wj})|}{|I_{wj}|} / \frac{|\phi(I_w)|}{|I_w|} - 1 \right| < \text{Const}\, |I_w|^{\varepsilon} \tag{7.1.2}$$

and the analogous inequalities for the gaps. Changing order in this bifraction we obtain $\frac{A_j(h_1,w)}{A_j(h_2,w)}$ converging to 1 exponentially fast with n, the length of w. We have used here the assumption $h_1 \in \mathcal{H}e$ to get $|I_w| \leq \exp -\delta n$ for some $\delta > 0$. Thus we have proved the 'only if' part of the theorem. Using Sullivan's words, we have proved that the ratio geometry is determined exponentially fast in length of w by the $C^{1+\varepsilon}$-structure.

Now we shall prove the 'if' part of the theorem. Let us first fix some notation. For every $m \geq 0$ denote by $\mathcal{G}^m$ the set of all intervals $I_{j_0,j_1,\ldots,j_m}$ and $G_{j_0,j_1,\ldots,j_m}$.

We must extend the mapping $h_2 \circ h_1^{-1} : h_1(\Sigma^d) \to h_2(\Sigma^d)$ to a mapping ϕ on all the gaps G_w for h_1. (We could use the Whitney Extension Theorem – see Remark 7.1.7 – but we shall give a direct proof.) The extension will be denoted by ϕ. For each two points $u < v$ on which ϕ is already defined we denote $\frac{\phi(v)-\phi(u)}{v-u}$ by $R(u,v)$. We shall also use the notation $R(J)$ if u, v are ends of an interval J.

Given $G_{j_0,j_1,\ldots,j_n}$ with the ends $a < b$, we want to have the derivatives

$$\phi'(a) = \lim_{m\to\infty} R(J_m(a)), \qquad \phi'(b) = \lim_{m\to\infty} R(J_m(b)), \tag{7.1.3}$$

where $J_m(a), J_m(b) \in \mathcal{G}^m, m \geq n$, all $J_m(a)$ have the right end a and all $J_m(b)$ have the left end b.

It is easy to see that the limits exist, and are uniformly bounded and uniformly bounded away from 0 for all G's. This follows from the following distortion estimate (compare Section 6.2):

For every $j_0, \ldots, j_m$, if $J \subset I_{j_0,\ldots,j_m} = I$ and $J \in \mathcal{G}^k, k > m$, then

$$\left| \frac{R(J)}{R(I)} - 1 \right| \leq \text{Const} \exp -m\delta. \tag{7.1.4}$$

Here δ is the exponent of the assumed convergence in the notion of the exponential equivalence of geometries. This property can be called the *bounded distortion property*: compare Section 6.2.

To prove (7.1.4) observe that there is a sequence $I_{j_0,\ldots,j_m} = J_m \supset J_{m+1} \supset \cdots \supset J_k = J$ of intervals such that $J_j \in \mathcal{G}^j$, and by the assumptions of the theorem,

$$1 - \text{Const} \exp -(j-1)\delta \leq \left| \frac{R(I_j)}{R(I_{j-1})} \right| \leq 1 + \text{Const} \exp -(j-1)\delta.$$

We obtain (7.1.4) by multiplying these inequalities over $j = m+1, \ldots, k$.

If $x \in I_{j_0,\ldots,j_m} = I$ is the end point of any gap, then

$$\left| \frac{\phi'(x)}{R(I)} - 1 \right| \leq \text{Const} \exp -m\delta. \tag{7.1.5}$$

(In fact x can be any point of $h_1(\Sigma^d)$ in I, but there is no need here to define ϕ' at these points except the ends of gaps. Compare Remark 6.3.4.)

To get (7.1.5) one should consider an infinite sequence of intervals containing x, and consider the infinite product over $j = m+1, \ldots$

Later on we shall also use a constant $\varepsilon > 0$ such that, for every $s \geq 0$,

$$\exp -s\delta \leq \text{Const} \inf_{J \in \mathcal{G}^s} |J|^{\varepsilon}. \tag{7.1.6}$$

Such an ε exists because, by the exponential geometry assumption, $\inf_{J\in\mathcal{G}^s} |J|$ cannot converge to 0 faster than exponentially.

We can go back to our interval (a,b). We extend ϕ' linearly to the interval $[a, \frac{a+b}{2}]$ and linearly to $[\frac{a+b}{2}, b]$, continuously to $[a,b]$. Moreover, we are careful to choose $\phi'(\frac{a+b}{2}) = t$ such that $\int_{[a,b]} \phi'(x)dx = \phi(b) - \phi(a)$. But our gap $G_{j_0,\dots,j_n}$ is in the interval $I_{j_0,\dots,j_{n-1}}$, so, by (7.1.5),

$$\left| \frac{R(a,b)}{\phi'(a)} - 1 \right| < \text{Const} \exp -(n-1)\delta,$$

and the same for $\phi'(b)$. This, and the computation

$$\begin{aligned} R(a,b) &= \frac{\int_{[a,b]} \phi'(x)dx}{b-a} = \frac{1}{2}(b-a)\left(\frac{\phi'(a)+t}{2} + \frac{\phi'(b)+t}{2}\right)/(b-a) \\ &= \frac{1}{4}(\phi'(a) + \phi'(b) + 2t), \end{aligned}$$

show that

$$\left| \frac{t}{\phi'(a)} - 1 \right|, \left| \frac{t}{\phi'(b)} - 1 \right| < \text{Const} \exp -n\delta. \tag{7.1.7}$$

In particular, $t > 0$: hence ϕ is increasing.

Now we need to prove the property (7.1.1). It is sufficient to consider points x, y, z in gaps, because $h_1(\Sigma^d)$ is nowhere dense.

We shall construct a finite family $\mathcal{A}(x,y)$ of intervals in $\bigcup_{m=0}^{\infty} \mathcal{G}^m$ 'joining' the gaps in which x and y lie. Suppose $x < y$, and let n be the largest integer such that x, y belong to the same element of $\mathcal{G}^n$. If x, y belong to different elements of $\mathcal{G}^0$, we take $n = -1$.

If x, y belong to a gap $G_{j_0,\dots,j_n}$, then $\mathcal{A}(x,y)$ is empty. If they belong to $I_{j_0,\dots,j_n}$, then they belong to different intervals $J(x, n+1), J(y, n+1)$ of $\mathcal{G}^{n+1}$. We account to $\mathcal{A}(x,y)$ all the intervals in $\mathcal{G}^{n+1}$ lying between $J(x, n+1)$ and $J(y, n+1)$, excluding $J(x, n+1)$ and $J(y, n+1)$ themselves. We shall continue with $J(x, n+1)$; the procedure for $J(y, n+1)$ is analogous.

If $J(x, n+1)$ is a gap, we end the process: nothing new will be accounted to $\mathcal{A}(x,y)$ from this side. In the opposite case we account to $\mathcal{A}(x,y)$ all the intervals of $\mathcal{G}^{n+2}$ in $J(x, n+1)$ to the right of x not containing x, and denote the one that contains x by $J(x, n+2)$. We continue this procedure by induction until $J(x, m)$ is, for the first time, a gap.

Thus the 'joining' set $\mathcal{A}(x,y)$ has been constructed.

Consider first the case $\mathcal{A}(x,y) = \emptyset$. It is easy to see that both x and y belong to $G_{j_0,\dots,j_n}$. Suppose $x, y \in (a, \frac{a+b}{2}]$, where a, b are ends of the gap and t will be the value of ϕ' in the middle, as in the previous notation. For $u \in [x,y]$, by the linearity of ϕ' and using (7.1.7), we obtain

$$|\phi'(u) - \phi'(x)| \le \frac{2(u-x)}{b-a} |t - \phi'(a)| \le \text{Const} \frac{(u-x)}{b-a} \phi'(a) \exp -n\delta.$$

Next, using the fact that $\phi'(a)$ is uniformly bounded, and by (7.1.6), we get

$$\begin{aligned} R(x,y) = \int_{[x,y]} \phi'(u)du/(y-x) &\le \phi'(x)\left(1 + \operatorname{Const} \frac{y-x}{b-a}(b-a)^{\varepsilon}\right) \\ &\le \phi'(x)\left(1 + \operatorname{Const}(y-x)^{\varepsilon}\right) \qquad (7.1.8) \end{aligned}$$

and the analogous bound from below. Cases where x, y are to the right of $\frac{a+b}{2}$ can be dealt with similarly. We can also write $\phi'(y)$ instead of $\phi'(x)$ in (7.1.8). Finally, if $x < \frac{a+b}{2} < y$, we obtain (7.1.8) by summing up the estimates for $(x, \frac{a+b}{2}]$ and $[\frac{a+b}{2}, y)$.

Consider the case $\mathcal{A}(x,y) \neq \emptyset$. Let $m \ge n$ be the smallest integer such that there exists $J_{j_0,\dots,j_m} \in \mathcal{A}(x,y) \cap \mathcal{G}^m$, ($J$ can be I or G, which means it can be a gap or a non-gap).

Denote the right end of the gap containing x by x', and the left end of the gap containing y by y'.

We obtain, with the use of (7.1.4),

$$R(x',y') = \frac{\sum_{J\in\mathcal{A}(x,y)} |\phi(J)|}{\sum_{J\in\mathcal{A}(x,y)} |J|} \le R(I_{j_0,\dots,j_{m-1}})(1 + \operatorname{Const}\exp -(m-1)\delta). \qquad (7.1.9)$$

We have used the fact that all $J \in \mathcal{A}(x,y)$ are in $I_{j_0,\dots,j_{m-1}}$. By (7.1.7) we obtain

$$\phi'(x') \le R(I_{j_0,\dots,j_{m-1}})(1 + \operatorname{Const}\exp -m\delta).$$

From these and the analogous inequalities to the other side we finally obtain

$$\left|\frac{R(x',y')}{\phi'(x)} - 1\right| \le \operatorname{Const}\exp -m\delta \le \operatorname{Const}(y'-x')^{\varepsilon}. \qquad (7.1.10)$$

A similar inequality holds for $\phi'(y)$.

We shall now conclude. By (7.1.8) and (7.1.10), each two consecutive terms in the sequence

$$\phi'(x), R(x,x'), \phi'(x'), R(x',y'), \phi'(y'), R(y',y), \phi'(y)$$

have the ratio within the distance from 1 bounded by $\operatorname{Const}(y-x)^{\varepsilon}$. So

$$\left|\frac{R(x,y)}{\phi'(y)} - 1\right| < \operatorname{Const}(y-x)^{\varepsilon}. \qquad (7.1.11)$$

Recall now that to prove (7.1.1) we picked also a third point: $z > y$. If y, z play the role of the previous x, y, we obtain

$$\left|\frac{R(y,z)}{\phi'(y)} - 1\right| < \operatorname{Const}(z-y)^{\varepsilon}.$$

So

$$\left|\frac{R(x,y)}{R(y,z)} - 1\right| < \operatorname{Const}(z-x)^{\varepsilon}.$$

Using the uniform boundedness of R's we obtain this in the additive form, that is, (7.1.1). The theorem is proved. ♣

Remark 7.1.7. We can shorten the above proof by referring to the Whitney Extension Theorem (see for example [Stein 1970]).

Indeed, we can define $\phi'(x)$ for every $x \in h_1(\Sigma^d)$, $x = \bigcap_{m=0}^{\infty} I_{j_0,\dots,j_m}$, by the formula (as (7.1.3)) $\phi'(x) = \lim R(I_{j_0,\dots,j_m})$.

Then the estimate (7.1.8) for all $x, y \in h_1(\Sigma^d)$, rewritten as

$$\phi(y) = \phi(x) + \phi'(x)(y-x) + O(|y-x|^{1+\varepsilon}),$$

which, together with Hölder continuity of ϕ' with exponent ε (see (7.1.5) and (7.1.6)) are precisely the assumptions for the Whitney Theorem, which asserts that ϕ has a $C^{1+\varepsilon}$ extension.

Remark 7.1.8. It is essential to assume in Theorem 7.1.6 that the convergence $\frac{A_j(h_1,w)}{A_j(h_2,w)} \to 1$ is exponential: that is, that the geometries are exponentially equivalent. Otherwise $\phi'(a)$ in (7.1.3) may not exist.

To prove the existence of ϕ' on $h_1(\Sigma^d)$, the uniform convergence of the finite products (in the case where they end with expressions involving gaps) or infinite products $\prod_n \frac{A_{j_{n+1}}(h_1,(j_0,\dots,j_n))}{A_{j_{n+1}}(h_2,(j_0,\dots,j_n))}$ is sufficient.

Remark 7.1.9. For each $h_1, h_2 \in \mathcal{H}$ the order-preserving mapping $\phi : h_1(\Sigma^d) \to h_2(\Sigma^d)$ is quasi-symmetric (see Definition 6.2.6). The equivalence of the geometries is equivalent to the 1-quasi-symmetric equivalence: cf. Exercise 7.2.

Example 7.1.10. It can occur that above $\phi : h_1(\Sigma^d) \to h_2(\Sigma^d)$ is Lipschitz continuous but all extensions are non-differentiable at every point in $h_1(\Sigma^d)$.

Let $h_i : \Sigma^3 \to \mathbb{R}$ be defined by $h_1((j_0,\dots)) = a =: .a_1a_2\dots$ in the development of a in base 6, where

$a_s = 0$ if $j_s = 1$, $a_s = 2$ if $j_s = 2$ and $a_s = 5$ if $j_s = 3$ for h_1

and

$a_s = 0$ if $j_s = 1$, $a_s = 3$ if $j_s = 2$ and $a_s = 5$ if $j_s = 3$ for h_2.

See Figure 7.1.

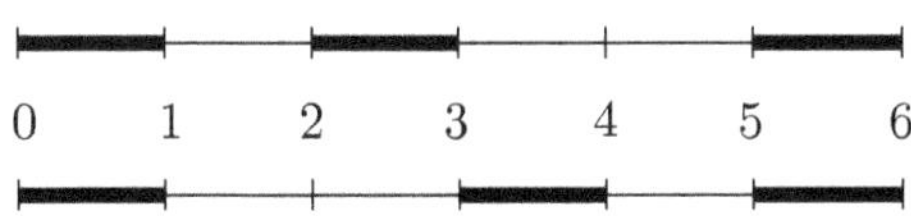

Figure 7.1 'Generators' of two differentiably different Cantor sets.

Remark 7.1.11. In the case where ϕ conjugates expanding maps belonging to $C^{1+\varepsilon}$ on the circle, this cannot happen. For example, Lipschitz conjugacy has points of differentiability: hence by the expanding property of, say, analytic maps involved, it is analytic (see Chapter 9). For Cantor sets, as above, if they are non-linear (see Chapter 10 for definition), then ϕ Lipschitz implies ϕ analytic. However, for linear sets, as in this example, an additional invariant is needed to describe classes of $C^{1+\varepsilon}$-equivalence, the so-called *scaling function*: see Section 7.2.

7.2 Scaling function: $C^{1+\varepsilon}$-extension of the shift map

So far we have not discussed dynamics. Recall, however, that we have on Σ^d the left-side shift map $s(j_0, j_1, \dots) = (j_1, \dots)$. We seek a condition about the ratio geometry for $h \in \mathcal{H}$ under which s, or more precisely $h \circ s \circ h^{-1}$, extends $C^{1+\varepsilon}$ to a neighbourhood of $h(\Sigma^d)$.

Definition 7.2.1. For the ratio geometry of $h \in \mathcal{H}$ we consider the sequence of functions to $\mathbb{R}^{2d-1}$:

$$S_n(j_{-n}, \dots, j_{-1}) = (S_n(j_{-n}, \dots, j_{-1})_j, j = 1, \dots, 2d-1) := \\ (A_j(h, (j_{-n}, \dots, j_{-1})), j = 1, \dots, 2d-1).$$

We call this a *scaling sequence* of functions. The limit

$$S(\dots, j_{-2}, j_{-1}) = \lim_{n\to\infty} S_n(j_{-n}, \dots, j_{-1}),$$

if it exists, is called a *scaling function*. By the definition,

$$\sum_{j=1}^{2d-1} S_n(\cdot)_j \equiv \sum_{j=1}^{2d-1} S(\cdot)_j \equiv 1.$$

Let us now discuss the domain of S_n, S. These functions are defined on one-sided sequences of symbols from $\{1, \dots, d\}$, and so formally on Σ^d. We want to be more precise, however.

Consider the natural extension of Σ^d: that is, a two-sided shift space $\tilde{\Sigma}^d = \{(\dots, j_{-1}, j_0, j_1, \dots)\}$. Then S can be considered as a function on $\tilde{\Sigma}^d$, but for each $(\dots, j_{-1}, j_0, j_1, \dots)$ depending only on the past $(\dots, j_{-2}, j_{-1})$. The functions S_n depend only on the finite past.

Definition 7.2.2. The domain of S and S_n is the factor of $\tilde{\Sigma}^d$, where we forget about the present and future: that is, we forget about the coordinates $j_0, j_1, \dots$. We call this factor a *dual Cantor set*, and denote it by Σ^{d*}. The range of S and S_n is the $2d-2$-dimensional simplex Simp_{2d-2} being the convex hull of the $2d-1$ points $(0, \dots, 1, \dots, 0)$, with 1 at the position $j = 1, 2, \dots, 2d-1$.

Thus S is not a function on $h(\Sigma^d)$, but if we consider $h(\Sigma^d)$ with the shift map $h \circ s \circ h^{-1}$ then we can see the dual Cantor set, that is, the domain of S and S_n, as the set of all infinite choices of consecutive branches of $(h \circ s \circ h^{-1})^{-1}$ on $h(\Sigma^d)$.

Note that if instead of $(h(\Sigma^d), h \circ s \circ h^{-1})$ we considered an arbitrary, say a distance-expanding, repeller, we could define backward branches only locally: that is, there would be no natural identification of fibres of the past over two different distant points of the repeller.

Proposition 7.1.5 yields the following.

Proposition 7.2.3. *If $h_1, h_2 \in \mathcal{H}$ are C^1-equivalent, and there exists a scaling function S for h_1, then h_2 has also a scaling function, equal to the same S.*

This says, in particular, that C^1-equivalence preserves the scaling function. Note that this is not the case for Lipschitz equivalence: see Example 7.1.10.

From Theorem 7.1.6 we easily deduce the following.

Theorem 7.2.4. *If $h \in \mathcal{H}e$ and $h \circ s \circ h^{-1}$ extends to a $C^{1+\varepsilon}$-mapping s_h on a neighbourhood of $h(\Sigma^d)$, then*

$$\frac{S_n}{S_{n+1}} \to 1; \quad \text{the convergence is uniformly exponentially fast.} \qquad (7.2.1)$$

Conversely, if $h \in \mathcal{H}$ and (7.2.1) is satisfied, then $h \in \mathcal{H}e$, and $h \circ s \circ h^{-1}$ extends to a $C^{1+\varepsilon}$-mapping.

Proof. Consider the sets $\Sigma_i^d = \{\alpha \in \Sigma^d : \alpha_0 = i\}$ for $i = 1, \dots, d$. Each Σ_i^d can be identified with Σ^d by $L_i((\alpha_0, \alpha_1, \dots)) = (i, \alpha_0, \alpha_1, \dots)$. Of course, $h_i := h \circ L_i \in \mathcal{H}$. Denote $h \circ s \circ h^{-1} : h(\Sigma_i^d) \to h(\Sigma^d)$ by s_i. We have $s_i \circ h_i = h$. So, by Theorem 7.1.6, all s_i extend $C^{1+\varepsilon}$ if and only if $\frac{A_j(h,w)}{A_j(h_i,w)}$ converge to 1 uniformly exponentially fast in length of w. These ratios are equal to $\frac{A_j(h,w)}{A_j(h,iw)}$ – that is, $\frac{S_n}{S_{n+1}}$, n being the length of w. So we obtain precisely the assertion of our theorem.

To apply Theorem 7.1.6 we used the observation that (7.2.1) easily implies $h \in \mathcal{H}b$ (by a sort of bounded distortion for iterates of the $h \circ s \circ h^{-1}$ property). In particular, $h \in \mathcal{H}e$: see Proposition 7.2.9. ♣

Example 7.2.5. Note that s_h of class $C^{1+\varepsilon}$ (even C^ω) does not imply $h \in \mathcal{H}e$. Indeed, consider h such that s_h has a parabolic point, for example $s_h(x) = x + 6x^2$ for $0 \le x \le 1/3$ and $s_h(x) = 1 - 3(1 - x)$ for $2/3 \le x \le 1$.

Remark 7.2.6. The assertions of Theorems 7.1.6 and 7.2.4 stay true if each Cantor set is constructed with the help of the intervals $I_{j_0,\dots,j_n}$ as before, but we do not assume that the left end of $I_{j_0,\dots,j_n,1}$ coincides with the left end of $I_{j_0,\dots,j_n}$, or that the right end of $I_{j_0,\dots,j_n,d}$ coincides with the right end of $I_{j_0,\dots,j_n}$.

So there might be some 'false' gaps in $I_{j_0,\dots,j_n}$ to the left of $I_{j_0,\dots,j_n,1}$ and to the right of $I_{j_0,\dots,j_n,d}$. In the definitions of bounded and exponential geometry we do not assume anything about these gaps; they may shrink faster than exponentially as $n \to \infty$. But wherever ratios are involved – in $A_j(h_1, w), A_j(h_2, w)$ in Theorem 7.1.6 or S, S_n in Theorem 7.2.4 – we take these gaps into account, so $j = 0, 1, \dots, 2d$.

The condition sufficient in Theorem 7.1.6 to $C^{1+\varepsilon}$-equivalence is that $A_j(h_1, w) - A_j(h_2, w) \to 0$ exponentially fast.

The condition sufficient in Theorem 7.2.4 to the $C^{1+\varepsilon}$-extentiability of $h \circ s \circ h^{-1}$ is that $S_n \to S$ exponentially fast.

To prove these assertions, observe that if we extend gaps of the $n+1$-th generation (between $I_{j_0,\dots,j_n,j}$ and $I_{j_0,\dots,j_n,j+1}$, $j = 1, \dots, d-1$) by false gaps of higher generations to get real gaps of the resulting Cantor set, then they and the remaining intervals satisfy the assumptions of Theorems 7.1.6 and 7.2.4.

This remark will be used in Section 7.4.

Definition 7.2.7. We say that $h \in \mathcal{H}$ satisfying (7.2.1) has an *exponentially determined geometry*. The set of such h's will be denoted by $\mathcal{H}$ed.

Definition 7.2.8. Let $j = (j_n)_{n=\dots,-2,-1}, j' = (j'_n)_{n=\dots -2,-1} \in \Sigma^{d*}$. Denote by $j \cap j'$ the sequence $(j_{-N}, \dots, j_{-1})$ with $N = N(j, j')$ the largest integer (or ∞), such that $j_{-n} = j'_{-n}$ for all $n \le N$. For an arbitrary $\delta > 0$ define the metric ρ_δ on Σ^{d*} by

$$\rho_\delta(j, j') = \exp -\delta N(j, j').$$

Let us make the following simple observation.

Proposition 7.2.9. (a) $\mathcal{H}e \supset \mathcal{H}b \supset \mathcal{H}ed$.

(b) *If $h \in \mathcal{H}ed$, then the scaling function S exists as Hölder continuous with respect to any metric ρ_δ (see Definition 7.2.8), and $S(\cdot)_i$ are bounded away from 0 and 1.*

(Observe, however, that the converse is false. One can take each S_n constant, and hence S constant, but $\frac{S_n}{S}$ converging to 1 slower than exponentially, so $h \notin \mathcal{H}ed$.)

(c) *If $h \in \mathcal{H}ed$, then $(h(\Sigma^d), s_h)$ is a $C^{1+\varepsilon}$ expanding repeller. (We shall also use the words $C^{1+\varepsilon}$-Cantor repeller in the line.)*

Proof. We leave (a) and (b) to the reader (the second inclusion in (a) has already been noted in the proof of Theorem 7.2.4) and prove (c). Similar to the way we proved property (7.1.5) in Theorem 7.1.6, we obtain the existence of a constant $C > 0$ such that, for $x = h((j_0, j_1, \dots)) \in \Sigma^d$ and $n \ge 0$,

$$C^{-1} < |(s_h^n)'(x)| / \frac{|I|}{|I_{j_0,\dots,j_n}|} < C.$$

As $h \in \mathcal{H}$e, and in particular $|I_{j_0,\dots,j_n}| \to 0$ uniformly, we obtain $|(s_h^n)'(x)| > 1$ for all n large enough and all x.

It follows from Theorem 7.1.6 that classes of $C^{1+\varepsilon}$-equivalence in $\mathcal{H}$ed are parametrized by Hölder continuous functions on Σ^{d*} (as scaling functions). To have one-to-one correspondence we need only to prove the existence theorem.

Theorem 7.2.10. *For every Hölder continuous function $S : \Sigma^{d*} \to \mathbb{R}_+^{2d-1}$ such that*

$$\sum_{j=1}^{2d-1} S(\cdot)_j \equiv 1 \tag{7.2.2}$$

there exists $h \in \mathcal{H}ed$ such that S is the scaling function of h.

First let us state the existence lemma:

Lemma 7.2.11. *Given numbers $A_{w,j} > 0$ for every $w = (j_0, \dots, j_n), n = 0, 1, \dots$, $j = 1, \dots 2d - 1$, such that $\sum_{j=1}^{2d-1} A_{w,j} = 1$, there exists $h \in \mathcal{H}$ such that $A_{w,j} = A_j(h, w)$: that is, h has the prescribed ratio geometry.*

Proof of Lemma 7.2.11. One builds a Cantor set by removing gaps of consecutive generations, from each I_w gaps of lengths $A_{w,j}|I_w|$, j even, so that the intervals not removed have lengths $A_{w,j}$, j odd, $j = 1, \ldots, 2d-1$. ♣

Proof of Theorem 7.2.10. Let $A_{w,j} := S((\ldots,1,1,w))_j$. By (7.2.2), $\sum_{j=1}^{2d-1} A_{w,j} = 1$, so we can apply Lemma 7.2.11. Property (6.4.11) (exponential convergence) follows immediately from the Hölder continuity of S and the fact that S is bounded away from 0 as positive continuous on the compact space Σ^{d*}. ♣

Summary: $C^{1+\varepsilon}$-structures in $\mathcal{H}$ed are in a one-to-one correspondence with the Hölder continuous scaling functions on the dual Cantor set.

Until now we have not been interested in ε in $C^{1+\varepsilon}$. It occurs, however, that scaling functions 'see' ε. First we introduce a metric ρ_S on Σ^{d*} depending only on a scaling function S, so that for a constant $K > 0$, for every j, j',

$$\frac{1}{K} \leq \frac{|I_{j \cap j'}|}{\rho_S(j, j')} \leq K. \tag{7.2.3}$$

Definition 7.2.12.

$$\rho_S(j, j') = \sup_w \prod_{t=1}^{n=N(j \cap j')} S(wj_{-n}j_{-n+1} \ldots j_{-t-1})_{j_t}$$

supremum over all w left infinite sequences of symbols in $\{1, \ldots, d\}$.

The estimate (7.2.3) follows easily from the exponential determination of geometry: we leave the details to the reader.

Theorem 7.2.13. *Fix $0 < \epsilon \leq 1$. The following are equivalent:*

1. There exists $h \in \mathcal{H}ed$, a $C^{1+\epsilon}$ embedding: that is, $h \circ s \circ h^{-1}$ extends to s_h being $C^{1+\epsilon}$, with scaling function S.

2. The scaling S is C^ϵ on (Σ^{d*}, ρ_S). (Here C^1 means Lipschitz.)

Proof. Substituting $\phi = s_h$, we can write (7.1.2), for all $n > N$ and all $i = 1, 2, \ldots, 2d-1$, in the form

$$|S_n(j_{-n}, \ldots, j_{-1})_i - S_{n-1}(j_{-n}, \ldots, j_{-1})_i| \leq \text{Const}\, |I_{j_{-n}, \ldots, j_{-1}}|^\varepsilon.$$

Summing up this geometric series for an arbitrary $j \in \Sigma^{d*}$ over $n = N, N+1, \ldots$ for $N = N(j, j')$, doing the same for another $j' \in \Sigma^{d*}$, and noting that $|S_N(j_{-N}, \ldots, j_{-1}) = |S_N(j'_{-N}, \ldots, j_{-1})$, yields

$$|S(j)_i - S(j')_i| \leq \text{Const}\, |I_{j \cap j'}|^\varepsilon.$$

Applying (7.2.3) to the right-hand side, we see that S is Hölder continuous with respect to ρ_S.

For the proof to the other side see the proof of Theorem 7.2.10. The construction gives the property (6.4.1a) for $\phi = s_h$, the extension as in the proof of Theorem 7.1.6. ♣

Example 7.2.14. For every $0 < \varepsilon_1 < \varepsilon_2 \le 1$ there exists S admitting a $C^{1+\varepsilon_1}$ embedding $h \in \mathcal{H}$ed, but not $C^{1+\varepsilon_2}$. We find S as follows. For an arbitrary (small) $\nu : 0 < \nu < (\varepsilon_1 - \varepsilon_2)/2$ we can easily find a function $S : \Sigma^{d*} \to \mathrm{Simp}_{2d-2}$ that is $C^{\varepsilon_1+\nu}$ but is not $C^{\varepsilon_2-\nu}$, in the metric ρ_δ, $\delta > \log d$ (Definition 7.2.8).

We can in fact find S so that for every $j \in \Sigma^{d*}$ and $i = 1, 3, \dots, 2d-1$ we have $|-\log S(j)_i/\delta - 1| < \nu/3$. (If $d \ge 3$ we can even have $S(j)_i = \log \delta$ constant for $i = 1, 3, \dots, 2d-1$, changing only gaps, i even.) Then, for all j, j' and $N = N(j, j')$, and a constant $K > 0$,

$$K^{-1}(\exp -N\delta)^{1-\nu/2} \le |I_{j\cap j'}| \le K(\exp -N\delta)^{1+\nu/3}.$$

Since $\rho_\delta(j, j') = \exp -N\delta$ we conclude that S is at best $(1+\varepsilon_2-\nu)/(1-\nu/3) < \varepsilon_2$-Hölder with respect to ρ_S. Hence s_h cannot be $C^{1+\varepsilon_2}$, by Theorem 7.2.13. Meanwhile, a construction as in the proof of Theorem 7.2.10 gives S being ε_1-Hölder: hence s_h is $C^{1+\varepsilon_1}$.

7.3 Higher smoothness

Definition 7.3.1. For every $r = 1, 2, \dots, \infty, \omega$ and $0 \le \varepsilon \le 1$ we can consider in $\mathcal{H}$e the subset $C^{r+\varepsilon}\mathcal{H}$ of such h's that $h \circ s \circ h^{-1}$ extends to a neighbourhood of $h(\Sigma^d)$ to a function of class $C^{r+\varepsilon}$.

By Theorem 7.2.4, for $r + \varepsilon > 1$,

$$C^{r+\varepsilon}\mathcal{H} \subset \mathcal{H}\mathrm{ed}.$$

Theorem 7.3.2 (On $C^{r+\varepsilon}$-rigidity). *If $h_1, h_2 \in C^{r+\varepsilon}\mathcal{H}$, $0 \le \varepsilon \le 1, r+\varepsilon > 1$, have equivalent geometries, then h_1, h_2 are $C^{r+\varepsilon}$-equivalent. That is, there exists a $C^{r+\varepsilon}$-diffeomorphism ϕ of a neighbourhood of $h_1(\Sigma^d)$ to a neighbourhood of $h_2(\Sigma^d)$ such that*

$$\phi|_{h_1(\Sigma^d)} \circ h_1 = h_2. \tag{7.3.1}$$

In other words, the canonical conjugacy extends $C^{r+\varepsilon}$.

We shall prove this theorem here for $\varepsilon > 0$. A different proof in Section 7.4 will also contain the case of $\varepsilon = 0$.

Remark 7.3.3. For h_1, h_2 in the class in $\mathcal{H}$ of functions having a scaling function, the condition that h_1, h_2 have equivalent geometries means that the scaling functions are the same. In the more narrow class $\mathcal{H}$ed it means that the canonical conjugacy ϕ extends $C^{1+\delta}$ for some $\delta > 0$: see Theorem 7.1.6. The virtue of Theorem 7.3.2 is that the more narrow the class, the better ϕ is forced to be. This is again a Livshic-type theorem.

Before proving Theorem 7.3.2, let us make a general calculation.

For any sequence of C^r real maps $F_j, j = 1, \dots, m$, consider the r-th derivative of the composition $(F_m \circ \dots \circ F_1)^{(r)}$, supposing that the maps can be composed – that is, that the range of each F_j is in the domain of F_{j+1}. We start with

$$(F_m \circ \cdots \circ F_1)'(z) = \prod_{j=1}^{m} F_j'(z_{j-1}),$$

where $z_0 = z$ and $z_j = F_j(z_{j-1})$.

Differentiating again, we see that

$$(F_m \circ \cdots \circ F_1)''(z) = \sum_{j=1}^{m} \Big(\prod_{i=1}^{j-1} (F_i'(z_{i-1}))^2 \Big(\prod_{i=j+1}^{m} F_i'(z_{i-1})\Big) \big(F_j''(z_{j-1})\big).$$

By induction we obtain

$$(F_m \circ \cdots \circ F_1)^{(r)}(z) = \sum_{1 \le j_1, \dots, j_{r-1} \le m} W^r_{j_1,\dots,j_{r-1}}(z),$$

where

$$W^r_{j_1,\dots,j_{r-1}}(z) = \Phi_{j_1,\dots,j_{r-1}}(z) P_{j_1,\dots,j_{r-1}}(z),$$

where for $j_1', \dots, j_{r-1}'$ denoting a permutation of $j_1, \dots, j_{r-1}$ so that $j_1' \le \cdots \le j_{r-1}'$ we denote

$$\Phi_{j_1,\dots,j_{r-1}}(z) := \Big(\prod_{i=1}^{j_1'-1} (F_i'(z_{i-1}))\Big)^r \Big(\prod_{i=j_1'+1}^{j_2'-1} (F_i'(z_{i-1}))\Big)^{r-1} \cdots \Big(\prod_{i=j_{r-1}'+1}^{m} (F_i'(z_{i-1}))\Big) \tag{7.3.2}$$

and

$$P_{j_1,\dots,j_{r-1}}(z) = \prod_{i=1}^{r'-1} P_{j_i}(z),$$

where each P_{j_i} is the sum of at most $(r-1)!$ terms of the form $\prod_{\sum t_s = r, \max t_s \ge 2} F_{j_i}^{(t_s)}(z_{j_i - 1})$. Above we replaced r by $r' \le r$, since if some j_s repeats, we consider it in the above product only once.

This can be seen by considering, for each $j_1, \dots, j_{r-1}$, tree graphs with vertices at m levels, $0, \dots, m-1$ – that is, derivatives at $z_0, \dots, z_{m-1}$ – each vertex (except for level 0) joined to the previous level vertices by the number of edges equal to the order of the derivative. Φ gathers levels with only first derivatives, P the remaining ones.

By induction, when we consider the first derivative of the product related to the tree $\mathcal{T}$ corresponding to the r-th derivative, we obtain a sum of expressions corresponding to trees, each received from the $\mathcal{T}$ by adding a branch from a vertex in $\mathcal{T}$ of a level $j_r - 1$, composed of new vertices v_i at levels $0 \le i < j_r - 1$ and edges e_i joining v_i to v_{i+1}. Since the number of vertices in T at each level, and in particular level $j_r - 1$, is at most r, we have at most r graphs that arise from T by differentiating at the level $j_r - 1$.

Proof of Theorem 7.3.2. The method, passing to small and then to large scale, is similar to the method of the second proof of Theorem 9.5.5.

Choose an arbitrary sequence of branches of $s_{h_1}^{-n}$ on a neighbourhood of $h_1(\Sigma^d)$, and denote them by g_n, $n = 1, 2, \dots$.

We have ψ a diffeomorphism, assuring $C^{1+\delta}$-equivalence: see Remark 7.3.3 above. (In fact we shall use only C^1.) We define on a neighbourhod of $h_1(\Sigma^d)$

$$\phi_n = s_{h_2}^n \circ \psi \circ g_n.$$

Of course, $\phi_n = \psi$ on $h_1(\Sigma^d)$. However s_{h_1}, s_{h_2} are defined only on neighbourhoods $U_\nu = B(h_\nu(\Sigma^d), \varepsilon)$ of $h_\nu(\Sigma^d)$, for some $\varepsilon > 0$, $\nu = 1, 2$. As s_{h_ν} are expanding, we can assume $s_{h_\nu}^{-1}(U_\nu) \subset U_\nu$, so all the maps g_n are well defined. We shall now explain why all the ϕ_n above are well defined.

Observe first that owing to the assumption that ψ is a C^1 diffeomorphism (7.3.1), and that $h_1(\Sigma^d)$ has no isolated points, there exists a constant $C > 0$ such that for every $x \in h_1(\Sigma^d)$, $j \geq 0$,

$$C^{-1} < |(s_{h_1}^j)'(x)/|(s_{h_2}^j)'(\psi(x))| < C. \tag{7.3.3}$$

So by the bounded distortion property for iterates of s_{h_ν} (following from the expanding property and the $C^{1+\varepsilon}$-smoothness: see Lemma 6.2.2), for every $j = 0, 1, \dots, n$, if we already know that $s_{h_2}^j \circ \psi \circ g_n$ is defined on $B := B(h_1(\Sigma^d), \eta/(2C^2 \sup \psi'))$, we obtain

$$s_{h_2}^j \psi g_n(B) \subset B(h_2(\Sigma^d), \eta). \tag{7.3.4}$$

So $s_{h_2}^{j+1} \circ \psi \circ g_n$ is defined on B, and so on, up to $j = n$. (The 2 in the denominator of the radius of B is a bound, taking care about the distortions, sufficient for η small enough. Note the possibility that U_i is not connected, but this has no influence on the proof.)

We shall find a conjugacy ϕ from the assertion of the theorem being the limit of a uniformly convergent sub-sequence of ϕ_n, so it will also be ψ on $h_1(\Sigma^d)$: hence (7.2.2) will hold.

Choose a sequence $x_n \in g_n(h_1(\Sigma^d))$. Instead of ϕ_n, consider

$$\tilde{\phi}_n = s_{h_2}^n \circ L_n \circ g_n,$$

where $L_n(w) = \psi(x_n) + \psi'(x_n)(w - x_n)$.

Observe first that

$$\mathrm{dist}_{C^0}(\phi_n, \tilde{\phi}_n) \to 0 \quad \text{for } n \to \infty. \tag{7.3.5}$$

Indeed,

$$\phi_n(z) - \tilde{\phi}_n(z) = s_{h_2}^n(\psi(g_n(z))) - s_{h_2}^n(L_n(g_n(z))).$$

As $|g_n(z) - x_n| \to 0$ for $n \to \infty$, we have, by the C^1-smoothness of ψ,

$$\frac{\psi(g_n(z)) - L_n(g_n(z))}{g_n(z) - x_n} \to 0.$$

So because of the bounded distortion property for the iterates of s_{h_2}, and using also the property $\psi'(x_0) \neq 0$, we get

$$\frac{s_{h_2}^n(\psi(g_n(z))) - s_{h_2}^n(L_n(g_n(z)))}{s_{h_2}^n(\psi(g_n(z))) - s_{h_2}^n(\psi(x_n))} \to 0,$$

and hence (7.3.5). We have also proved that all $\tilde{\phi}_n$ are well defined on a neighbourhood of $h_1(\Sigma^d)$, similarly to the way we obtained (7.3.4).

Thus we can consider $\tilde{\phi}_n$'s, all of which are $C^{r+\varepsilon}$. We need to prove that their r-th derivatives are uniformly bounded in C^ε. Then, by the Arzela–Ascolli theorem, we can choose a sub-sequence $\tilde{\phi}_{n_k}^{(r)}$ uniformly convergent to a C^ε function. (Here we use $\varepsilon > 0$.) By the calculus theorem that the limit of derivatives is the derivative of the limit, we shall obtain the assertion that a uniformly convergent sub-sequence of $\tilde{\phi}_n$ has the limit $C^{r+\varepsilon}$smooth.

We shall use our calculations of $(F_m \circ \cdots \circ F_1)^{(r)}$ preceding the proof of Theorem 7.3.2. We can assume that $r \geq 2$, as for $r = 1$ the theorem has already been proved (see Theorem 7.1.6). For $m = 2n + 1$ we set as $F_1, \ldots, F_n$ the branches of $s_{h_1}^{-1}$, which composition gives g_n. We set $F_{n+1} = L_n$. Finally, for $j = n + 2, \ldots, 2n + 1$ we set $F_j = s_{h_2}$.

For every sequence $j_1, \ldots, j_{r-1}$ we assign the number

$$T(j_1, \ldots, j_{r-1}) = \sum\{j_i : j_i \leq n\} + \sum\{m - j_i : j_i \geq n + 1\}.$$

For any x, z in a neighbourhood of $h_1(\Sigma^d)$ sufficiently close to each other, and $\alpha = (j_1, \ldots, j_{r-1})$ we have

$$|W_\alpha^r(x) - W_\alpha^r(z)| = \left| \left(\frac{\Phi_\alpha(x)}{\Phi_\alpha(z)} - 1 \right) P_\alpha(x) + (P_\alpha(x) - P_\alpha(z))\Phi_\alpha(z) \right|. \quad (7.3.6)$$

By (7.3.2), organizing the products there in $\prod_{i=1}^{j_1'-1} \prod_{i=1}^{j_2'-1} \cdots \prod_{i=1}^{m}$ (after multiplying by the missing terms $F'_{j'_s}$), using bounded distortion of iterates of $s_{h_n u}, \nu = 1, 2$, we obtain

$$\left| \left(\frac{\Phi_\alpha(x)}{\Phi_\alpha(z)} - 1 \right) \right| \leq \text{Const}\, r |x - z|^\varepsilon.$$

Observe also that, using $|x_j - z_j| \leq \text{Const}\, |x - z|$,

$$|P_\alpha(x) - P_\alpha(z)| \leq \text{Const}\, |x - z|^\varepsilon,$$

and $P_\alpha(x)$ is bounded by a constant independent of n (depending only on r). Finally we have

$$|\Phi_\alpha(z)| \leq \text{Const}\, \lambda^{T(\alpha)}, \quad (7.3.7)$$

where λ is an arbitrary constant such that $1 < \lambda^{-1} < \inf |s'_{h_1}|, \inf |s'_{h_2}|$.

We have here used (7.3.2). The crucial observation leading from (7.3.2) to (7.3.7) was the existence of a constant $C > 0$ such that, for every $0 < i \leq j \leq n$,

$$C^{-1} < (F_i \circ \cdots \circ F_j)'(z_{j-1}) \cdot (F_{m-i} \circ \cdots \circ F_{m-j})'(z_{m-j-1}) < C,$$

following from (7.3.3). We also need to refer again to the bounded distortion property for the iterates of s_{h_ν}, as, z's do not need to belong to $h_\nu(\Sigma^d)$, the unlike the x's in (7.3.3).

Thus, by (7.3.6) and the estimates following it, we obtain

$$\begin{aligned} &|(s^n_{h_2} \circ L_n \circ g_n(x) - s^n_{h_2} \circ L_n \circ g_n(z))^{(r)}| \\ &\leq \sum_{j_1,\dots,j_{r-1}} \text{Const}\,|x-z|^\varepsilon \lambda^{T(j_1,\dots,j_{r-1})} \\ &\leq \text{Const}\,|x-z|^\varepsilon \sum_{T=0}^{\infty} 2T^r\lambda^T \leq \text{Const}\,|x-z|^\varepsilon \end{aligned}$$

because $\text{Card}\{(j_1,\dots,j_{r-1}) : T(j_1,\dots,j_{r-1}) \leq T\} \leq 2T^r$.

The proof of Theorem 7.3.2 in the $C^{r+\varepsilon}$ case for every $r = 1, 2, \dots, \infty$ is now complete. We need to consider the C^ω case separately. The maps s_{h_ν} extend holomorphically to neighbourhoods of $h_\nu(\Sigma^d)$ in $\mathbb{C}$, the complex plane in which the interval I is embedded. As in the $C^{r+\varepsilon}$, $r = 1, \dots, \infty$, case, we see that there are neighbourhoods U_ν of $h_\nu(\Sigma^d)$ in $\mathbb{C}$ such that $\tilde{\phi}_n$ are well defined on U_1 and $\tilde{\phi}_n(U_1) \subset U_2$. By the definition, they are holomorphic. Now we can use Montel's Theorem. So there exists a sub-sequence $\tilde{\phi}_{n_j}$, $n_j \to \infty$ as $j \to \infty$, uniformly convergent on compact subsets of U_1 to a holomorphic map. The proof is complete; it was simpler for $r = \omega$ than for $r \neq \omega$. For similar considerations see also Section 9.5. ♣

Summary. We have the following situation. In $\mathcal{H}$ above the equivalence of geometries and even the exponential equivalence of geometries do not induce any reasonable smoothness. In $\mathcal{H}$e the exponential equivalence of geometries does work: it implies $C^{1+\varepsilon}$-equivalence. In $\mathcal{H}$b even the equivalence of geometries starts to work: it implies that the canonical conjugacy is 1-quasisymmetric. This we have not discussed: see Exercise 7.2. In $\mathcal{H}$ed the equivalence of geometries, which now means the same as exponential equivalence, yields $C^{1+\varepsilon}$-equivalence. The higher smoothness of $\mathcal{H}$ then forces the same smoothness of the conjugacy.

We shall show in Chapter 10 that in C^ω in a subclass of non-linear Cantor sets even a weaker equivalence of geometries, not taking gaps into account, forces C^ω-equivalence (we have mentioned this already at the end of Section 7.1).

7.4 Scaling function and smoothness; Cantor set valued scaling function

The question arises as to which scaling functions appear in which classes $C^{r+\varepsilon}$ (compare Example 7.2.14). We shall give some answers below.

For simplification we assume $I = [0, 1]$.

Definition 7.4.1. Scaling with values in Cantor sets. Given a scaling function S on Σ^{d*}, we define a scaling function $\hat{S}$ with values in $\mathcal{H}$ rather than Simp_{2d-2}. For each $j = (\dots, j_{-2}, j_{-1}) \in \Sigma^{d*}$ we define $\hat{S}(j) \in \mathcal{H}$ by induction, as follows.

Suppose for every $j \in \Sigma^{d*}$ and $i_0, \dots, i_n$ the interval $I(j)_{i_0,\dots,i_n}$ is already defined. (For an empty string we set $[0,1]$.) Then for every $i_{n+1} = 1, 2, \dots, d$ we define $I(j)_{i_0,\dots,i_n,i_{n+1}}$ as the $2i_{n+1} - 1$'th interval of the partition of $I(j)_{i_0,\dots,i_n}$ determined by the proportions $S(j, i_0, \dots, i_n)_i, i = 1, 2, \dots, 2d-1$. We conclude with $\hat{S}(j)(i_0, i_1, \dots) = \bigcap_{n=0}^{\infty} I(j)_{i_0,\dots,i_n}$.

Denote the Cantor set $\hat{S}(j)(\Sigma^d)$ by $\mathrm{Can}(j)$.

Theorem 7.4.2. *For a scaling function S and $r = 1, 2, \dots, \infty, \varepsilon : 0 \leq \varepsilon \leq 1$ with $r + \varepsilon > 1$, or $r = \omega$, the following conditions are equivalent:*

(1) There exists a $C^{r+\varepsilon}$, or C^{ω} (real-analytic) embedding $h \in \mathcal{H}ed$ with scaling function S (we assume here that in the definition of $C^{r+\varepsilon}$, see Definition 7.3.1, s_h maps each component of its domain diffeomorphically onto $[0,1]$).

(2) For every $j, j' \in \Sigma^{d*}$ there exists a $C^{r+\varepsilon}$, or C^{ω} respectively, diffeomorphism $F_{j'|j} : [0,1] \to [0,1]$ mapping $\mathrm{Can}(j)$ to $\mathrm{Can}(j')$.

Proof. Let us prove (1) $\Rightarrow$ (2) For any $j \in \Sigma^{d*}$ and $n \geq 1$ denote $j(n) = (j_{-n}, \dots, j_{-1})$. Write

$$F_{j(n)} := ((s_h)^n|_{I_{j_{-n},\dots,j_{-1}}}) \circ A^{-1}_{j(n)},$$

where $A_{j(n)}$ is the affine rescaling of $I_{j_{-n},\dots,j_{-1}}$ to $[0,1]$. Given $j, j' \in \Sigma^{d*}$ and $n, n' \geq 1$, define

$$F_{j'(n')|j(n)} := F^{-1}_{j'(n')} \circ F_{j(n)}.$$

Finally, define

$$F_{j'|j} := \lim_{n,n' \to \infty} F_{j'(n')|j(n)}.$$

The convergence, even exponential, easily follows from $s_h \in C^{1+\varepsilon}$. The fact that $F_{j'|j}$ maps $\mathrm{Can}(j)$ to $\mathrm{Can}(j')$ follows from the definitions.

In the case of C^{ω} there is a neighbourhood U of $[0,1]$ in the complex plane so that all $(s_h)^n|^{-1}_{I_{j_{-n},\dots,j_{-1}}}$ extend holomorphically, injectively, to U. This is so since $(h(\Sigma^d), \hat{s}_h)$, where $\hat{s}_h$ is a holomorpic extension of s_h, is a conformal expanding repeller. With the use of the Koebe Distortion Lemma (Chapter 6), one concludes that all $F_{j'(n')|j(n)}$ have a common domain in $\mathbb{C}$, containing $[0,1]$, on which they are uniformly bounded. So, for a given j, j', a sub-sequence is convergent to a holomorphic function: hence $F_{j'|j}$ is analytic.

Consider now the $C^{r+\varepsilon}$ case.

Let us prove first the following claim.

Claim 7.4.3. Let $F_1, F_2, \dots$ be C^{r+1} maps of the unit interval $[0,1]$ for $r \geq 1, 0 \leq \varepsilon \leq 1, r + \varepsilon > 1$. Assume all F_m are uniform contractions: that is, there exist $0 < \lambda_1 \leq \lambda_2 < 1$ such that for every m and every $x \in [0,1]$ it holds that $\lambda_1 \leq |F'_m(x)| \leq \lambda_2$. Then there exists $C > 0$ such that, for all m,

$$||F_m \circ \dots \circ F_1||_{C^{r+\varepsilon}} \leq C||F_m \circ \dots \circ F_1||_{C^1}.$$

(We set the convention that we omit the supremum of the modulus of the functions in the norms in $C^{r+\varepsilon}$; we consider only derivatives.)

Proof of the claim. Consider first $\varepsilon > 0$. We use (7.3.6) and the estimates that follow it. (7.3.7) is replaced by

$$|\Phi_\alpha(z)| \le \text{Const}\,|(F_m \circ \cdots \circ F_1)'(z)|\lambda_2^{\hat{T}(\alpha)},$$

where for $\alpha = (j_1, \ldots, j_{r-1})$ we define $\hat{T}(\alpha) = j_1 + .. + j_{r-1}$. We conclude for $r \ge 2$ with

$$\begin{aligned}|(F_m \circ \cdots \circ F_1)^{(r)}(x) &- (F_m \circ \cdots \circ F_1)^{(r)}(z)| \\ &\le \text{Const}\,|x-z|^\varepsilon |(F_m \circ \cdots \circ F_1)'(z)| \sum_\alpha \lambda_2^{\hat{T}(\alpha)} \\ &\le \text{Const}\,|x-z|^\varepsilon |(F_m \circ \cdots \circ F_1)'(z)| \sum_{T=0}^{\infty} T^r \lambda_2^T \\ &\le \text{Const}\,|(F_m \circ \cdots \circ F_1)'(z)||x-z|^\varepsilon.\end{aligned}$$

For $r = 1$ there is no summation over α, and the assertion is immediate.

For $\varepsilon = 0$ we get

$$\begin{aligned}|(F_m \circ \cdots \circ F_1)^{(r)}(z)| &\le \sum_\alpha |\Phi_\alpha(z) P_\alpha(z)| \le \text{Const} \sum_\alpha |(F_m \circ \cdots \circ F_1)'(z)|\lambda_2^{\hat{T}(\alpha)} \\ &\le \text{Const}\,|(F_m \circ \cdots \circ F_1)'(z)|.\end{aligned}$$

The claim is proved. ♣

We apply the claim to $F_1, F_2, \ldots$ being inverse branches of s_h on $[0,1]$. Let λ be the supremum of the contraction rate $|s_h'|^{-1}$. Given $j \in \Sigma^{d*}$ and integers $n, m \ge 0$, we get for $z \in I_{j_{-n},\ldots j_{-1}}$

$$||(s_h^m|_{I_{j_{-(n+m)},\ldots j_{-1}}})^{-1}||_{C^{r+\varepsilon}} \le C|((s_h^m|_{I_{j_{-(n+m)},\ldots j_{-1}}})^{-1})'(z)|.$$

If we rescale the domain and range to $[0,1]$ we obtain, using bounded distortion of s_h^m,

$$\begin{aligned}||F_{j(n+m)|j(n)}||_{C^{r+\varepsilon}} &\le C \frac{|I_{j_{-n},\ldots,j_{-1}}|^{r+\varepsilon}}{|I_{j_{-(n+m)},\ldots,j_{-1}}|} |((s_h^m|_{I_{j_{-(n+m)},\ldots j_{-1}}})^{-1})'(z)| \\ &\le \text{Const}\,|I_{j_{-n},\ldots j_{-1}}|^{r+\varepsilon-1}. \qquad (7.4.1)\end{aligned}$$

The right-hand expression in this estimate does not depend on m and tends (exponentially quickly) to 0 as $n \to \infty$, for $r > 1$.

Note that

$$F_{j(n+m)}^{-1} = F_{j(n+m)|j(n)} \circ F_{j(n)}^{-1}. \qquad (7.4.2)$$

Therefore for the sequence $F_{j(n)}^{-1}$ we have verified a condition that is reminiscent of Cauchy's condition. However, to conclude convergence in $C^{r+\varepsilon}$, we still need to do some work.

For $r = 1$ we have uniform exponential convergence of $|(F_{j(n)}^{-1})'(z)|$, since $|(F_{j(n+m)|j(n)})'| \to 1$ uniformly exponentially quickly as $n \to \infty$. This holds since $F_{j(n)}([0,1]) = [0,1]$, by integration of the second derivative, or, in the case of merely $C^{1+\varepsilon}$, since distortion of $F_{j(n+m)|j(n)}$ tends exponentially to 1 as $n \to \infty$.

For $r > 1, \varepsilon = 0$ the derivatives of $F_{j(n)}^{-1}$ of orders $2, \dots, r$ tend uniformly to 0, since $F_{j(n+m)|j(n)}$ tend uniformly to identity in C^r as $n \to \infty$. One can see this using our formula for the composition of two maps, as in (7.4.2), or, more simply, by substituting a Taylor expansion series up to order r of one map in the other.

For $\varepsilon > 0$ the sequence $F_{j(n)}^{-1}$ has been proved in (7.4.1) to be uniformly bounded in $C^{r+\varepsilon}$, and every convergent sub-sequence has the same limit, being the limit in C^r. Therefore this is a limit in $C^{r+\varepsilon}$.

If we denote the limit by G_j, we conclude that

$$F_{j'|j} = G_{j'} \circ G_j^{-1}, \tag{7.4.3}$$

defined above as $C^{r+\varepsilon}$.

The proof of (2) $\Rightarrow$ (1). The embedding h in the proof of Theorem 7.2.10 is the correct one. Indeed, $\hat{S}(\dots,1,1)$ coincides with h by construction, and $s_h = s_{\hat{S}} = F_{(\dots,1,1)|(\dots,1,i)} \circ A_i$, where A_i is rescaling to $[0,1]$ of $I_i, i = 1,\dots,d$ in the ratio geometry of $\hat{S}(\dots,1,1)$. ♣

Remark 7.4.4. Theorem 7.4.2 (or, more precisely, smoothness of G_j in (7.4.3)) yields a new proof of Theorem 7.3.2, in full generality, that includes the case $\varepsilon = 0$. Indeed, one can define $\phi = G_j(h_2)^{-1} \circ G_j(h_1)$ for an arbitrary $j \in \Sigma^{d*}$, where $G_j(h_i), i = 1,2$ means G_j for h_i.

In the case where the ranges of s_{h_1}, s_{h_2} are not the whole $[0,1]$, we define G_j as the limit of $F_{j(n+n_0)|j(n_0)}$, so ϕ is defined only on some $I_{j_{-n_0},\dots,j_{-1}}$, for n_0 large enough that this F makes sense. Then we define ϕ on a neighbourhood of $h_1(\Sigma^d)$ as $s_{h_2}^{n_0} \circ \phi \circ (s_{h_1}|I_{j_{-n_0},\dots,j_{-1}})^{-n_0}$.

Theorem 7.4.5. *For every $r = 1,2,\dots$ and $\varepsilon : 0 \le \varepsilon < 1$ with $r + \varepsilon > 1$ there is a scaling function S such that there is $h \in \mathcal{H}ed$, a $C^{r+\varepsilon}$ embedding with the scaling function S, but there is no $C^{r+\varepsilon'}$ embedding with $\varepsilon' > \varepsilon$. There is also S admitting a C^∞ embedding, but not real-analytic.*

This theorem addresses Example 7.2.14 in particular, giving a different approach.

Proof. Consider $d > 1$ disjoint closed intervals I_j in $[0,1]$, with I_1 having 0 as an end point, and f mapping each I_j onto $[0,1]$, so that $f|_{I_j}$ is affine for each $j = 2,\dots,d$ and $C^{r+\varepsilon}$ on I_1 but not $C^{r+\varepsilon'}$, say at 0 (or C^∞ but not analytic at 0). This produces $h \in C^{r+\varepsilon}\mathcal{H}$. Choose any sequence $j \in \Sigma^{d*}$ not containing 1's, say $j = (\dots,2,2,2)$. Then, for the arising scaling function S, we have

$$\hat{S}(j) = h \ \text{ and } \ \hat{S}(j1) = A \circ (f|_{I_1})^{-1} \circ h,$$

where A is the affine rescaling of I_1 to $[0,1]$.

So $f|_{I_1} \circ A^{-1} : [0,1] \to [0,1]$ maps the Cantor set Can($j1$) to Can(j). Its restriction to Can($j1$) cannot extend $C^{r+\varepsilon'}$, since its derivatives up to order r are already computable on C and $f^{(r)}$ is not ε'-Hölder, by construction. So S cannot admit $C^{r+\varepsilon'}$ embedding by Theorem 7.4.2. The case of C^∞ but not analytic is dealt with similarly. ♣

7.5 Cantor set generating families

We shall discuss here a general construction of a $C^{1+\varepsilon}$ Cantor repeller in $\mathbb{R}$, which will be used in the next section.

Definition 7.5.1. We call a family of maps $F = \{f_{n,j} : n = 0,1,\ldots,\ j = 1,...,d\}$ of a closed interval $I \subset \mathbb{R}$ into itself a *Cantor set generating family* if the following conditions are satisfied.

All $f_{n,j}$ are $C^{1+\varepsilon}$-smooth and uniformly bounded in the $C^{1+\varepsilon}$-norm; they preserve an orientation in $\mathbb{R}$. There exist numbers $0 < \lambda_1 < \lambda_2 < 1$ such that for every n, j $\lambda_1 < |(f_{n,j})'|$ and $\frac{|f_{n,j}(I)|}{|I|} < \lambda_2$ (a natural stronger assumption would be $|f_{n,j})'| < \lambda_2$, but we need the weaker one for later use).

For every n all the intervals $f_{n,j}(I)$ are pairwise disjoint and ordered according to j's, and the gaps between them are bounded away from 0.

Given a Cantor set generating family $F = \{f_{n,j} : n = 0,1,\ldots,\ j = 1,...,d\}$, we write

$$I_{j_0,\ldots,j_n}(F) := (f_{0,j_0} \circ \cdots \circ f_{n,j_n})(I)$$

Then we obtain the announced Cantor set as

$$C(F) := \bigcap_{n=0}^{\infty} E_n(F), \quad \text{where } E_n(F) = \bigcup_{(j_0,\ldots,j_n)} I_{j_0,\ldots,j_n}(F)$$

and the corresponding coding $h(F)$ defined by

$$h(F)((j_0, j_1, \ldots)) = \bigcap_{n\to\infty} I_{j_0,\ldots,j_n}(F).$$

It is easy to see that $h(F)$ has bounded geometry (we leave it as an exercise for the reader).

Theorem 7.5.2. *Let $F_\nu = \{f_{\nu,n,j} : n = 0,1,\ldots,\ j = 1,...,d\}$ be two Cantor set generating families, for $\nu = 1$ and $\nu = 2$. Suppose that for each $j = 1,\ldots,d$*

$$\lim_{n\to\infty} \operatorname{dist}_{C^0}(f_{1,n,j}, f_{2,n,j}) = 0$$

and the convergence is exponential.

Then $h(F_1)$ and $h(F_2)$ are $C^{1+\varepsilon}$-equivalent.

Proof. Observe that the notation is consistent with that at the beginning of Section 7.1, except that the situation is more general; it is like that in Remark 7.2.6.

For every $s \leq t$, $\nu = 1, 2$, we denote $(f_{\nu,(s,j_s)} \circ \cdots \circ f_{\nu,(t,j_t)})(I)$ by $I_{\nu,(s,j_s),\ldots,(t,j_t)}$. For every such I_w we denote the left end by lI_w and the right end by rI_w. Observe that, although we have not assumed $|f'_{\nu,n,j}| < \lambda_2$, we can deduce from our weaker assumptions (using the bounded distortion property for the iterates) that there exists $k \geq 1$ so that for every $l \geq 0$ and $j_i, i = 0, \ldots, k-1$ we have $|(f_{\nu,l+k-1,j_{k-1}} \circ \cdots \circ f_{\nu,l+i,j_i} \circ \cdots \circ f_{\nu,l,j_0})'| < \lambda_2 < 1$. In future, to simplify our notation we assume, however, that $k = 1$. The general case can be dealt with, for example, by considering the new family of k compositions of the maps of the original family.

For every $w = ((s, j_s), (s+1, j_{s+1}), \ldots, (t, j_t))$, $w' = ((s+1, j_{s+1}), \ldots, (t, j_t))$ we have

$$\begin{aligned} |lI_{1,w} - lI_{2,w}| &\leq |f_{1,s,j_s}(lI_{1,w'}) - f_{1,s,j_s}(lI_{2,w'})| \\ &\quad + |f_{1,s,j_s}(lI_{2,w'}) - f_{2,s,j_s}(lI_{2,w'})| \\ &\leq |lI_{1,w'} - lI_{2,w'}| + \text{Const} \exp -\delta s \end{aligned} \tag{7.5.1}$$

for some $\delta > 0$ lower bound of the exponential convergence in the assumptions of the theorem.

Thus for every $w = ((m, j_m), \ldots, (n, j_n))$ we obtain for $t = n$, by induction for $s = n-1, n-2, \ldots, m$,

$$|lI_{1,w)} - lI_{2,w}| \leq \text{Const} \exp -\delta m. \tag{7.5.2}$$

For every $j = 1, \ldots, d$ we obtain a similar estimate with w replaced by $w, (n+1), j)$. We also obtain similar inequalities for the right ends.

As a result of all this we obtain

$$\left| \frac{|I_{1,w,(n+1,j)}|}{|I_{1,w}|} - \frac{|I_{2,w,(n+1,j)}|}{|I_{1,w}|} \right| \leq \text{Const}\, \lambda_1^{-(n-m)} \exp -\delta m.$$

Now, iterating by $f_{\nu,m-1}, f_{\nu,m-2}, \ldots, f_{\nu,0}$ for $\nu = 1, 2$ hardly changes the proportions, as we are already in a small scale; more precisely we get

$$\left| \frac{I_{j_0,\ldots,j_n,j}(F_1)}{I_{j_0,\ldots,j_n}(F_1)} - \frac{I_{j_0,\ldots,j_n,j}(F_2)}{I_{j_0,\ldots,j_n}(F_2)} \right| \leq \text{Const}\big((\lambda_1^{-(n-m)} \exp -\delta m) + \lambda_2^{(n-m)\varepsilon}\big). \tag{7.5.3}$$

The same holds for gaps in numerators, including 'false' gaps: that is, for $j = 0, \ldots, d$.

Now we pick $m = (1-\kappa)n$, where κ is a constant such that $0 < \kappa < 1$ and

$$\kappa \log \lambda_1^{-1} - (1-\kappa)\delta := \vartheta < 0.$$

Then the bound in (7.5.3) is replaced by $(\exp \vartheta n) + \lambda_2^{(\varepsilon\kappa)n}$, which converges to 0 exponentially fast for $n \to \infty$.

So our theorem follows from Theorem 7.1.6 or, more precisely, from its variant described in Remark 7.2.6. ♣

We have also the following theorem.

Theorem 7.5.3. *Let $F = \{f_{n,j} : n = 0,1,\dots,\ j = 1,...,d\}$ be a Cantor set generating family such that, for every j,*

$$f_{n,j} \to f_{\infty,j} \qquad \text{uniformly as } n \to \infty.$$

Then the shift map on the Cantor set $C(F)$ extends $C^{1+\varepsilon}$.

Proof. For any Cantor set generating family Φ, every $w = (j_0,\dots,j_n), j \in \{0,\dots,2d\}$, we use the notation $A_j(\Phi,w)$ as in Definition 7.1.1, that is, for j odd $A_j(\Phi,w) = \frac{I_{wj'}(\Phi)}{I_w(\Phi)}$, where $j' = \frac{j+1}{2}$. The similar definition is for j's even with gaps in the numerators. We are again in the situation of Remark 7.2.6, including $j = 0, d$.

Consider, together with F, the family $F' = \{f'_{n,j} : n = 0,1,\dots,\ j = 1,..,d\}$, where $f'_{n,j} = f_{n+1,j}$. For every $w = (j_0,\dots,j_n), j \in \{0,\dots,2d\}$, say j odd and $i \in \{1,\dots,d\}$, we rewrite the definitions for clarity:

$$A_(F,iw) = \frac{|I_{iwj'}(F)|}{|I_{iw}(F)|} = \frac{|f_{0,i} \circ f_{1,j_0} \circ \cdots \circ f_{n+1,j_n} \circ f_{n+2,j}(I)|}{|f_{0,i} \circ f_{1,j_0} \circ \cdots \circ f_{n+1,j_n}(I)|}$$

$$A_j(F',w) = \frac{|f_{1,j_0} \circ \cdots \circ f_{n+1,j_n} \circ f_{n+2,j}(I)|}{|f_{1,j_0} \circ \cdots \circ f_{n+1,j_n}(I)|}$$

We have

$$\left| \frac{A_j(F,iw)}{A_j(F',w)} - 1 \right| \le \text{Const} \exp -\delta n$$

for some constant $\delta > 0$ related to the distortion of $f_{0,i}$ on the interval $f_{1,j_0} \circ \cdots \circ f_{n+1,j_n}(I)$.

So

$$|A_j(F,iw) - A_j(F',w)| \le \text{Const} \exp -\delta n.$$

But

$$|A_j(F',w) - A_j(F,w)| \le \text{Const} \exp -\delta n$$

for some $\delta' > 0$, because the pair of the families F, F' satisfies the assumptions of Theorem 7.5.2.

Thus $|A_j(F,iw) - A_j(F,w)|$ converge to 0 uniformly exponentially fast in length of w. So we can apply Theorem 7.2.4, or more precisely the variant from Remark 7.2.6. The proof of Theorem 7.5.3 is complete. ♣

7.6 Quadratic-like maps of the interval; an application to Feigenbaum's universality

We show here how to apply the material of Section 7.5 to study 'attracting' Cantor sets, which are closures of forward orbits of critical points, appearing for Feigenbaum-like and more general so-called 'infinitely renormalizable' unimodal maps of the interval. The original map on such a Cantor set is not expanding at all, but one can view these sets almost as expanding repellers by constructing for them so-called generating families of expanding maps.

We finish this chapter with a beautiful application: *Feigenbaum's First Universality*. It was numerically discovered by M. J. Feigenbaum and independently by P. Coullet and Ch. Tresser.

Rigorously, this universality has been explained 'locally' by O. Lanford, who proved the existence of the renormalization operator fixed point, and later on for large classes of maps by D. Sullivan, who applied quasi-conformal maps techniques, and then, more completely, by several other mathematicians, particularly the fundamental contribution by C. McMullen [McMullen 1996]. We refer the reader to Sullivan's breakthrough paper [Sullivan 1991]: 'Bounds, quadratic differentials and renormalization conjectures'. Fortunately a small piece of this can be easily explained with the use of the elementary Theorems 7.5.2 and 7.5.3; we shall explain it below.

Let us start with a standard example called a *logistic family* (Figure 7.2): the one-parameter family of maps of the interval $I = [0,1]$ into itself $f_\lambda(x) = \lambda x(1-x)$. For $1 < \lambda < 3$ there are two fixed points in $[0,1]$, a source at 0 i.e. $|f_\lambda'(0)| > 1$ and a sink x_λ, $|f_\lambda'(x_\lambda)| < 1$, attracting all the points except 0,1 under iterations of f_λ. For $\lambda = 3$ this sink changes to a neutral fixed point, namely $|f_\lambda'(x_\lambda)| = 1$, or more precisely $f_\lambda'(x_\lambda) = -1$. For λ growing beyond 3 this point changes to a source, and nearby an attracting periodic orbit of period 2 arises. f^2 maps the interval $I_0 = [x_\lambda', x_\lambda]$ into itself (x_λ' denotes the point symmetric to x_λ with respect to the critical point 1/2).

If λ continues to grow, the left point of this period 2 orbit crosses 1/2, and the derivative of f_λ^2 at this point changes from positive to negative until it reaches the value -1. The periodic orbit starts to repel, and an attracting periodic orbit of period 4 arises. For f^2 on I_0 this means the same bifurcation as before: a periodic orbit of period 2 arises. The respective interval containing 1/2 invariant for f^4 will be denoted by I_2, and soon. Denote the values of λ where the

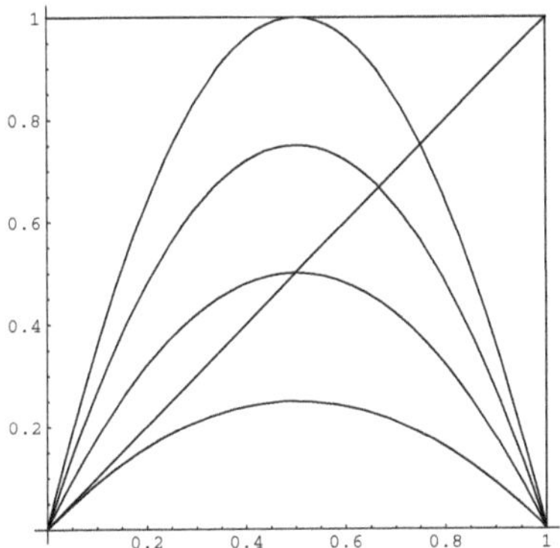

Figure 7.2 Logistic family.

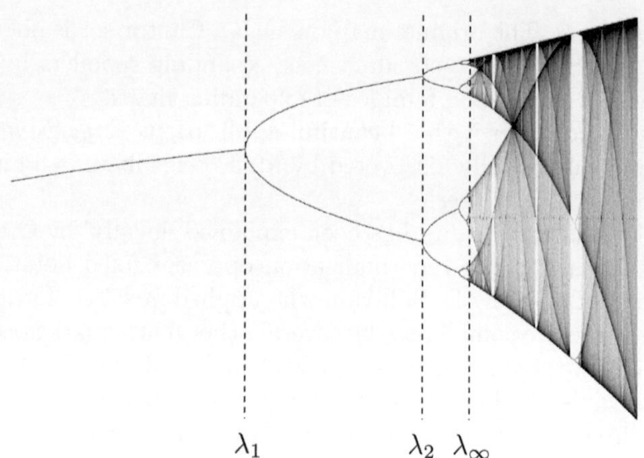

Figure 7.3 Bifurcation diagram.

consecutive orbits of periods 2^n arise, by λ_n. For the limit parameter $\lambda_\infty = \lim_{n\to\infty} \lambda_n$ there are periodic orbits of all periods 2^n, all of them sources: see Figure 7.3.

In effect, for λ_∞ we obtain a Cantor set $C(f_{\lambda_\infty}) = \bigcap_{n=1}^{\infty} \bigcup_{k=0}^{2^n-1} f_{\lambda_\infty}^k(I_n)$. This Cantor set attracts all points except the above-mentioned sources. It contains the critical point 1/2, and is precisely the closure of its forward orbit.

Instead of the quadratic polynomials one can consider a quite arbitrary one-parameter family g_λ of C^2 maps of the unit interval with one critical point where the second derivative does not vanish, and such that $g_\lambda(0) = g_\lambda(1) = 0$ so that, roughly, the parameter raises the graph. Again, one obtains period doubling bifurcations, and for the limit parameter $\lambda_\infty(g)$ one obtains the same topological picture as above. We say the map is *Feigenbaum-like.* Feigenbaum's and Coullet, Tresser's numerical discovery was that the deeper ratios in the Cantor set the weaker dependence of the ratios on the family, and that the ratios at the critical point stabilize with the growing magnifications. Moreover, the limit quantities do not depend on g.

Another numerical discovery, which will not be discussed here (see for example [Avila, Lyubich & de Melo 2003] for a rigorous explanation) was that λ_n/λ_{n+1} has a limit as $n \to \infty$. Moreover, this limit does not depend on g. We call it *Feigenbaum's Second Universality.*

Let us pass to the description of a general situation:

Definition 7.6.1. For any closed interval $[a, b]$ we call a mapping $f : [a, b] \to [a, b]$ *smooth quadratic-like* if $f(a) = f(b) = a$ and f can be decomposed into $f = Q \circ h$, where Q is a quadratic polynomial and h is a smooth diffeomorphism of I. The word 'smooth' will be applied below for C^2. Here we allow $a > b$: in such a case the interval $[b, a]$ is under consideration, of course; its right end rather than the left is a fixed point, and the map has a minimum at the critical point. If $a = 0, b = 1$ we say that f is *normalized.*

We call f *infinitely renormalizable* if there exists a decreasing sequence of intervals $I_n, n = 0, 1, 2, \ldots$ all containing the critical point c_f and a sequence of integers $d_n \geq 2$ such that for every n all $f^j(I_n)$ have pairwise disjoint interiors for $j = 0, 1, \ldots, D_n - 1$, where $D_n := \prod_{i=0,\ldots,n} d_i$ and $f^{D_n}(I_n)$ maps I_n into itself.

We call the numbers d_n and the order in which the intervals $f^j(I_n)$ are placed in I a *combinatorics* of f. Finally, we say that an infinitely renormalizable f has *bounded combinatorics* if all d_n are uniformly bounded. We write $C(f) = \bigcap_{n=0}^{\infty} \bigcup_{k=0}^{D_n - 1} f^k(I_n)$.

It may happen that the maps f^{D_n} on I_n are not quadratic-like because the assumption $f(a) = f(b) = a$ is not satisfied.

Consider, however, an arbitrary $f : [a, b] \to [a, b]$ that is smooth quadratic-like and *renormalizable*, which means that there exists $I_0 \subset [a, b]$ containing c_f and an integer $d > 1$ such that all $f^j(I_0)$ have pairwise disjoint interiors for $j = 0, 1, \ldots, d - 1$ and $f^d(I_0) \subset I_0$. Then I_0 can be extended to an interval I_0' for which all $f^j(I_0')$ still have pairwise disjoint interiors, f^d maps I_0' into itself, and f^d on I_0' is quadratic-like. The proof is not hard: the reader can do it as an exercise, or look into [Collet & Eckmann 1980]. The periodic end of I_0' is called a *restrictive central point.*

We define the *rescaling map* R_f as an affine map that transforms I_0' onto I, and

$$f_1 := R_f \circ f^d \circ R_f^{-1} \tag{7.6.1}$$

is normalized. We call the operator $f \mapsto f_1$ the *renormalization operator*, and denote it by $\mathcal{R}$. (Caution: d, I_0' and so $\mathcal{R}$ have not been uniquely defined, but this will not affect the correctness of the considerations that follow; in particular, in the infinitely renormalizable case, $C(f)$ does not depend on these objects, as the closure of the forward orbit of the critical point: see the remark ending the proof of Theorem 7.6.3.)

Now, for an arbitrary smooth quadratic-like map f of $I = [0, 1]$, infinitely renormalizable with a bounded combinatorics, we consider a sequence of maps f_n defined by induction: $f_0 = f$, $f_n = \mathcal{R}(f_{n-1})$. The domain I_0' for the renormalization of f_n is denoted by I^n, and we have the affine rescaling map $R_n := R_{f_n}$ from I^n onto I and $f_{n+1} = \mathcal{R}(f_n) = R_n \circ f^{d_n} \circ R_n^{-1}$.

Now we can formulate the fundamental Sullivan–McMullen's Theorem.

Theorem 7.6.2. *Suppose f and g are two C^2-quadratic-like maps of $I = [0, 1]$, both infinitely renormalizable, with the same bounded combinatorics. Then $\operatorname{dist}(R_{f_n}, R_{g_n}) \to 0$ as $n \to \infty$. Moreover, both sequences f_n and g_n stay uniformly bounded as $C^{1+\varepsilon}$-quadratic-like maps (that is, h's and h^{-1}'s in the $Q \circ h$ decomposition stay uniformly bounded in $C^{1+\varepsilon}$), and*

$$\operatorname{dist}_{C^0}(f_n, g_n) \to 0.$$

In the case where f, g are real-analytic, the convergence is exponentially fast, even in the C^0-topology in complex functions on a neighbourhood of I in $\mathbb{C}$.

The intuitive meaning of the above is that the larger the magnification of a neighbourhood of 0 is, the more similar the respective iterates of f and g look. The same geometry of the depths of the Cantor sets would mean that the similar looks close to zero propagate to the Cantor sets.

Now we can fulfil our promise and, relying on the results of this section, prove this propagation property. That is, relying on Theorem 7.6.2, we can prove rigorously Feigenbaum's First Universality.

Theorem 7.6.3. *Suppose f and g are two C^2-quadratic-like maps of $I = [0,1]$, both infinitely renormalizable with the same bounded combinatorics. Suppose also that the convergences in the assertion of Theorem 7.6.2 are exponential. Then $C(f)$ and $C(g)$ are $C^{1+\varepsilon}$-equivalent Cantor sets.*

Proof. Related to f, we define a generating family (Definition 7.5.1) $F = \{f_{n,j}, n = 0, 1, \ldots, j = 1, \ldots, d_n\}$. That is, we define

$$f_{n,j} = f_n^{-(d_n-j+1)} \circ R_n^{-1}, \tag{7.6.2}$$

where each $f_n^{-(d_n-j+1)}$ means the branch leading to an interval containing $f^{j-1}(I^n)$.

The $C^{1+\varepsilon}$ uniform boundedness of $f_{n,j}$'s follows immediately from the boundedness asserted in Theorem 7.6.2 if we know (see the next paragraph) that all I^n's have lengths bounded away from 0. Indeed, if we denote $f_n = Q \circ h_n$, we have

$$f_{n,j} = h_n^{-1} \circ Q^{-1} \circ \cdots \circ h_n^{-1} \circ Q^{-1} \circ R_n^{-1}$$

with all h_n^{-1} uniformly bounded in $C^{1+\varepsilon}$ and Q^{-1} as well, because their domains are far from the critical value $f(c_{f_n})$. Also, $|(R_n)'|$'s are uniformly bounded.

Now $f_n^{d_n}(I^n) \subset I^n$ with I^n arbitrarily small and d_n's uniformly bounded, together with the uniform boundedness of f_n's asserted in Theorem 7.6.2, would result in the existence of a periodic sink attracting c_f. Indeed, $|(f_n)'|$ would be small on I^n, as $|(f^{d_n-1})'|$ is bounded on $f_n(I^n)$ by a constant not depending on n. So $|(f^{d_n})'|$ on I^n would be small: hence its graph has a unique intersection with the diagonal, which is a sink attracting I^n.

This is almost the end of the proof, because we construct the analogous generating family G for g and refer to Theorem 7.1.6. The convergence assumed there can be proved similarly to the way we proved the uniform $C^{1+\varepsilon}$-boundedness above. This also concerns the assumptions involving λ_1 and λ_2 in the definition of the generating families. Still, however, some points should be explained:

1. For each n the intervals $f_{n,j}(I)$ in Definition 7.5.1 were ordered in $\mathbb{R}$ according to the order $<$ in the integers j. Here this is not so. Moreover, $f_{n,j}$ here do not all preserve the orientation in $\mathbb{R}$. Finally, the d_n's are not all equal to the same integer d. Fortunately, all done we have before is also correct in this situation.

2. The intervals $f_{n,j}(I)$ may have common ends here: in particular, the assumption about gaps in the definition of the generating family may not be satisfied. In this case we replace I by a slightly smaller interval and restrict all $f_{n,j}$'s to it. We can do this because for each $J = f_{n,j}(I)$ we have $\operatorname{dist}(C(f), \text{ends of } J) >$ Const > 0. This is so because for every normalized renormalizable f, if $f(c_f)$ is close to 1, then a very large d is needed in order to have $f^d(c_f) \in I_0$ unless $\sup|f'|$ is very large. But all d_n are uniformly bounded in our infinitely renormalizable case, and the derivatives $|(f_n)'|$ are uniformly bounded. So, for every n, $f_n(c_{f_n})$ is not very close to 1, $f_n^2(c_{f_n})$ is not very close to 0, and $C(f_n) \subset [f_n(c_{f_n}), f_n^2(c_{f_n})]$.

We managed to present $C(f)$ and $C(g)$ as subsets of Cantor sets $C(F), C(G)$ for generating families. But, by the construction, every interval $I_{j_0,\dots,j_n}(F)$ in the definition of $C(F)$ contains an interval of the form $f^j(I_n)$, $0 \le j < D_n$: hence every component of $C(F)$ contains a component of $C(f)$. So $C(f) = C(F)$ and similarly $C(g) = C(G)$. Hence $C(f)$ and $C(g)$ are Cantor sets indeed, and everything we have proved concerning $C(F), C(G)$ concerns them as well. Observe that, by the definition, every $f^j(I_n)$ contains $f^j(c_f)$: hence $C(f)$ can be defined in the intrinsic way, independently of the choice of I_n's, as $\operatorname{cl} \bigcup_{j=0}^{\infty} f^j(c_f)$.

The following specification of Theorems 7.6.2 and 7.6.3 holds.

Theorem 7.6.4. *Let f be a C^2 quadratic-like map of $[0,1]$, infinitely renormalizable with a bounded combinatorics. Suppose it is periodic: that is, that for some $n_2 > n_1 \ge 0$ f_{n_1} and f_{n_2} have the same combinatorics. Then there exists g, a real-analytic quadratic-like map of $[0,1]$, such that for $t := n_2 - n_1$, $\mathcal{R}^t(g) = g$ and $\operatorname{dist}_{C^0}(f_n, g_{n-n_1}) \to 0$ as $n \to \infty$.*

If the convergence is exponential, then the shift map on $C(f^t_{n_1})$ extends $C^{1+\varepsilon}$.

On the proof. The existence of g is another fundamental result in this theory, which we shall not prove in this book. (The first, computer-assisted, proof was provided by O. Lanford [Lanford 1982] for $d = 2$, i.e. for the Feigenbaum-like class.) Then the convergence follows from Theorem 6.4.29. Indeed, from $\mathcal{R}^t(g) = g$ we obtain the convergence of $(f^t_{n_1})_{n,j}$ to g. If the convergences are exponential (which is the case if f is real-analytic) then the shift map extends $C^{1+\varepsilon}$ because of Theorem 7.5.3. Note that instead of $C(f)$ we consider $C(f^t_{n_1})$. This is so because

$$\prod_{j=d_{n_1}}^{d_{n_1}+t-1} d_j = \prod_{j=d_{n_1}+t}^{d_{n_1}+2t-1} d_j = \cdots := d,$$

and it makes sense to speak about the shift map on Σ^d. For f itself, if we denote the Cartesian product $\prod_{j=0}^{\infty}\{1,\dots,d_j\}$ by $\Sigma(d_0, d_1, \dots)$, we can speak only about the left-side shift map from $\Sigma(d_0, d_1, \dots)$ to $\Sigma(d_1, \dots)$.

Observe again that the embedding of Σ^d into I does not need to preserve the order, but this does not affect the validity of Theorem 7.5.3.

The set $C(f)$ is presented as the union of $D = \prod_{j=0}^{n_1-1} d_j$ Cantor sets, which are embeddings of Σ^d, of the form $f^j(C(f^t_{n_1})), j = 0, \ldots, n_1 - 1$, each of which has an exponentially determined geometry and Hölder continuous scaling function.

Remarks 7.6.5. (1) Observe for f being any smooth quadratic-like infinitely renormalizable map of I and the corresponding generating family F, that as some $f_{n,j}$ may change the orientation, the corresponding intervals $I_{j_0,\ldots,j_n,j}(F)$ have their order in $I_{j_0,\ldots,j_n}(F)$ the same as or opposite to that of the $I_j(F)$'s in I, according to whether there is an even or odd number of $j_i, i = 0, \ldots, n$ such that f_{i,j_i} changes the orientation.

(2) Recall that $C(f)$ has a 1-to-1 coding $h : \Sigma(d_0, d_1, \ldots) \to C(f)$, defined by $h(j_0, j_1, \ldots) = \bigcap_{n\to\infty} I_{j_0,\ldots,j_n}(F)$. Let us write here $j = 0, \ldots, d_n - 1$ rather than $j = 1, \ldots, d_n$. Then f yields on $\Sigma(d_0, d_1, \ldots)$ the map $\Phi(f)(j_0, j_1, \ldots) = (0, 0, \ldots, j_i + 1, j_{i+1}, \ldots)$, where i is the first integer such that $j_i \neq d_i - 1$, or $\Phi(f)(j_0, j_1, \ldots) = (0, 0, 0 \ldots)$ if for all i we have $j_i = d_i - 1$. Φ is sometimes called the *adding machine*: compare Example 1.4. If all $d_n = p$ the map Φ is just the adding of the unity in the group of p-adic numbers. For d_n different we have the group structure on $\Sigma(d_0, d_1, \ldots)$ of the inverse limit of the system $\cdots \to \mathbb{Z}_{d_2 d_1 d_0} \to \mathbb{Z}_{d_1 d_0} \to \mathbb{Z}_{d_0}$, and $\Phi(f)$ is also the adding of the unity.

If we denote the shift map from $\Sigma(d_0, d_1, \ldots)$ to $\Sigma(d_1, \ldots)$ by s, we obtain the equality

$$\Phi(f_1) \circ s = s \circ \Phi(f)^{d_0}$$

(the indexing in (7.6.2) has been adjusted to ensure this). On I^0 this corresponds to (7.6.1).

(3) The combinatorics of an infinitely renormalizable f is determined by the so-called *kneading sequence* $K(f)$, defined as a sequence of letters L and R, where $n = 1, 2, \ldots$ such that at the n'th place we have L or R depending on whether $f^n(c_f)$ is left or right of c_f in $\mathbb{R}$ (we leave this as an exercise for the reader). So in Theorems 7.6.2–7.6.4 we can write 'the same kneading sequences', instead of 'the same combinatorics'.

Also, the property renormalizable (and hence infinitely renormalizable) can be inferred from the look of the kneading sequence. A renormalization with I_0 and $f^d(I_0) \subset I_0$ implies, of course, that the kneading sequence is of the form $AB_1AB_2AB_3\ldots$, where each B_i is L or R, and A is a block built from L's and R's of length $d-1$. The converse is also true; the proof is related to the proof of the existence of the restrictive central point. One can do this as an exercise, or look into [Collet & Eckmann 1980].

(4) Let us now return to the example f_{λ_∞}, or more generally $g_{\lambda_\infty(g)}$, mentioned at the beginning of this section. We have $d_n = 2$ for all n (R_n change orientation). So we can apply Theorems 7.6.3 and 7.6.4, which explain Feigenbaum's and Coullet-Tresser's discoveries.

Observe that $g_{\lambda_\infty(g)}$ is exceptional among smooth quadratic-like infinitely renormalizable maps: except for a sequence of periodic sources, every point is attracted to $C(f)$. The topological entropy is equal to 0. For every infinitely renormalizable map with a different kneading sequence there is an invariant repelling Cantor set (in fact, some of its points can be blown up to intervals). Topological entropy is positive on it. One says that such a map is already *chaotic*, whereas $g_{\lambda_\infty(g)}$ is on the *boundary of chaos.*

Exercises

7.1. For maps as in Definition 7.6.1, prove the existence of the restrictive central point.

Hint: Consider the so-called Guckenheimer set,

$$\begin{aligned} G_d =& \{x : \mathrm{dist}_f(f^d(x), c_f) < \mathrm{dist}_f(x, c_f) \quad \text{and} \\ & \mathrm{dist}_f(f^j(x), c_f) > \mathrm{dist}_f(x, c_f) \quad \text{for} \quad j = 1, \ldots, d-1\}, \end{aligned}$$

where $\mathrm{dist}_f(x, y) = |h(x) - h(y)|$ in the decomposition $f = Q \circ h$.

7.2. Suppose f and g are smooth quadratic-like maps of $I = [0, 1]$, both infinitely renormalizable, with the same bounded combinatorics as in Theorem 7.6.2. Using the fact asserted there that $\mathrm{dist}_{C^0}(f_n, g_n) \to 0$, but not assuming that the convergence is exponential, prove that the standard conjugacy ϕ between $C(f)$ and $C(g)$ is 1-quasi-symmetric, or more precisely that for every $x, y, z \in C(f), x > y > z$, $|x-y|/|y-z| <$ Const we have $\frac{|\phi(x)-\phi(y)|/|\phi(y)-\phi(z)|}{|x-y|/|y-z|} \to 1$ as $x - z \to 0$. In particular, if the scaling function $S(f)$ exists for f, then it exists for g and $S(f) = S(g)$, for the scaling functions for the related generating families.

Hint: One can modify the proof of Theorem 7.5.2. Instead of $\exp -\delta s$ in (7.5.1) one has some a_n converging to 0 as $n \to \infty$. Then in (7.5.2) we estimate by $\sum_{s=m}^{n} a_s$ and then consider $m = m(n)$ so that $n - m \to \infty$ but $\sum_{s=m}^{n} a_s \to 0$ as $n \to \infty$.

7.3. Let f and g be unimodal maps of the interval $[0, 1]$ (f unimodal means continuous, having a unique critical point c, being strictly increasing to the left of it and strictly decreasing to the right of it, $f(c) = 1$), having no interval J on which all iterates are monotone. Prove that f and g are topologically conjugate if and only if they have the same kneading sequences (see Remark 7.6.5, item 3.)

7.4. Prove that for $f \in C^3$ a unimodal map of the interval with no attracting (from both or one side) periodic orbit, if the Schwarzian derivative Sf is negative, then there are no homtervals, that is, intervals on which all iterates of f are monotone.

Hint: First prove that there is no homterval whose forward orbit is disjoint with a neighbourhood of the critical point c_f (A. Schwartz's Lemma. One does not use $Sf < 0$ here; C^{1+1} is sufficient.)

Next use the property implied by $Sf < 0$, that for all n and every interval J on which $(f^n)'$ is non-zero, $(f^n)''$ is monotone on J.

For details see for example [Collet & Eckmann 1980].

7.5. Prove the $C^{r+\varepsilon}$ version of the so-called Folklore Theorem, saying that if $0 = a_0 < a_1 < \cdots < a_{n-1} < a_n = 1$ and for each $i = 0, \ldots, n-1$, f_i :

$[a_i, a_{i+1}] \to [0,1]$ onto, each f_i is $C^{r+\varepsilon}$ for $r \geq 2, 0 \leq \varepsilon \leq 1, r+\varepsilon > 2$ and $|f_i'| \geq \text{Const} > 1$, then for f defined as f_i on each (a_i, a_{i+1}), there exists an f invariant probability μ equivalent to the Lebesgue measure, with the density bounded away from 0, of class $C^{r-1+\varepsilon}$.

Formulate and prove an analogous version for Cantor sets $h(\Sigma^d)$ with 'shifts' $h \circ s \circ h^{-1}$, as in Section 7.2.

Hint: The existence of μ follows from the Hölder property of the potential function $\phi = -\log|f'|$: see Chapter 5. μ is the invariant Gibbs measure. Its density is $\lim_{n\to\infty} \mathcal{L}_\phi^n(1\!\!1)(x) = \sum_{y \in f^{-n}(x)} |(f^n)'(y)|^{-1}$. Each summand considered along an infinite backward branch, after rescaling, converges in $C^{r-1+\varepsilon}$: see Theorem 7.4.2, smoothness of G_j.

A slightly different proof can be found for example in [Boyarsky & Góra 1997].

Bibliographical notes

As we have already mentioned, this chapter is based mainly on [Sullivan 1988], [Sullivan 1991] and [Przytycki & Tangerman 1996]. A weak version of Theorem 7.5.2 was independently proved by W. Paluba in [Paluba 1989].

A version of most of the theory presented in this chapter was also published by A. Pinto and D. Rand in [Pinto & Rand 1988], [Pinto & Rand 1992]. They proved, moreover, that the canonical conjugacy between two real-analytic quadratic-like maps as in Theorem 7.6.2 is $C^{2+0.11}$, using numerical results for the speed of contraction (leading eigenvalues) of the renormalization operator at the Feigenbaum fixed point.

In the proof of Theorem 7.6.3 we proved for the families F and G that the assumptions of Theorem 6.4.26 were satisfied: that is, that F and G were Cantor set generating families. This implied that the Cantor sets $C(F), C(G)$, or more precisely $h(F), h(G)$, have bounded geometries. In fact this can be proved directly (though it is by no mean easy) without referring to the difficult Theorem 7.6.2: see for example [de Melo & van Strien 1993].

Moreover, this bounded geometry phenomenon is in fact the first step (the real part) in the proof of the fundamental Theorem 7.6.2 (which also includes a complex part): see [Sullivan 1991]. Instead of C^2 one can assume the weaker $C^{1+\mathrm{z}}$, z for Zygmund: see [Sullivan 1991] or [de Melo & van Strien 1993].

The exponential convergence in Theorem 7.6.2 for analytic f, g was proved by C. McMullen in [McMullen 1996]. Recently, the exponential convergence for $f, g \in C^2$ has been proved by W. de Melo and A. Pinto: see [de Melo & Pinto 1999].

The first proof of the existence of g as in Theorem 7.6.4 and the exponential convergence of $\mathcal{R}^n(f)$ to g was provided by O. Lanford [Lanford 1982] for f very close to g, in the case of Feigenbaum-like maps. His proof did not use the bounded geometry of the Cantor set.

For each $k \geq 3$ the convergence in C^k in Theorem 7.6.2 for all $f, g \in C^k$ (symmetric) was proved in [Avila, Martens & de Melo 2001].

For deep studies concerning the hyperbolicity of the renormalization operator at periodic points, and laminations explaining the second Feigenbaum universality, see the recent papers [Lyubich 1999], [Avila, Lyubich & de Melo 2003], [Smania 2005] and [de Faria *et al.* 2006].

8

Fractal dimensions

In the first section of this chapter we provide a more complete treatment of outer measure, begun in Chapter 2. The rest of the chapter is devoted to presentation of basic definitions and facts related to Hausdorff and packing measures, Hausdorff and packing dimensions of sets and measures, and ball (or box) -counting dimensions.

8.1 Outer measures

In Section 2.1 we introduced the abstract notion of measure. At the beginning of this section we want to show how to construct measures starting with functions of sets called *outer measures*, which are required to satisfy much weaker conditions. Our exposition of this material is brief, and the reader should find its complete treatment in all handbooks of geometric measure theory (see for example [Rogers 1970], [Falconer 1985], [Falconer 1997], [Mattila 1995] or [Pesin 1997]). This approach has already been applied in Chapter 2: see Theorem 2.7.2.

Definition 8.1.1. An *outer measure* on a set X is a function μ defined on all subsets of X taking values in $[0, \infty]$ such that

$$\mu(\emptyset) = 0, \tag{8.1.1}$$

$$\mu(A) \le \mu(B) \quad \text{if} \quad A \subset B \tag{8.1.2}$$

and

$$\mu\Big(\bigcup_{n=1}^{\infty} A_n\Big) \le \sum_{n=1}^{\infty} \mu(A_n) \tag{8.1.3}$$

for any countable family $\{A_n : n = 1, 2, \ldots\}$ of subsets of X.

A subset A of X is called *μ-measurable* or simply *measurable with respect to the outer measure* μ if and only if

$$\mu(B) \ge \mu(B \cap A) + \mu(B \setminus A) \tag{8.1.4}$$

for all sets $B \subset X$. Check that the opposite inequality follows immediately from (8.1.3). Check also that if $\mu(A) = 0$ then A is μ-measurable.

Theorem 8.1.2. *If μ is an outer measure on X, then the family $\mathcal{F}$ of all μ-measurable sets is a σ-algebra, and the restriction of μ to $\mathcal{F}$ is a measure.*

Proof. Obviously $X \in \mathcal{F}$. By symmetry of (8.1.4), $A \in \mathcal{F}$ if and only if $A^c \in \mathcal{F}$. So the conditions (2.1.1) and (2.1.2) of the definition of σ-algebra are satisfied. To check the condition (2.1.3) that $\mathcal{F}$ is closed under countable union, suppose that $A_1, A_2, \ldots \in \mathcal{F}$ and let $B \subset X$ be any set. Applying (8.1.4) in turn to $A_1, A_2, \ldots$ we get for all $k \geq 1$

$$\begin{aligned}
\mu(B) &\geq \mu(B \cap A_1) + \mu(B \setminus A_1) \\
&\geq \mu(B \cap A_1) + \mu((B \setminus A_1) \cap A_2) + \mu(B \setminus A_1 \setminus A_2) \\
&\geq \ldots \\
&\geq \sum_{j=1}^{k} \mu\Big(\Big(B \setminus \bigcup_{i=1}^{j-1} A_i\Big) \cap A_j\Big) + \mu(B \setminus \bigcup_{j=1}^{k} A_j) \\
&\geq \sum_{j=1}^{k} \mu\Big(\Big(B \setminus \bigcup_{i=1}^{j-1} A_i\Big) \cap A_j\Big) + \mu(B \setminus \bigcup_{j=1}^{\infty} A_j)
\end{aligned}$$

and therefore

$$\mu(B) \geq \sum_{j=1}^{\infty} \mu\Big(\Big(B \setminus \bigcup_{i=1}^{j-1} A_i\Big) \cap A_j\Big) + \mu(B \setminus \bigcup_{j=1}^{\infty} A_j). \tag{8.1.5}$$

Since

$$B \cap \bigcup_{j=1}^{\infty} A_j = \bigcup_{j=1}^{\infty} \Big(B \setminus \bigcup_{i=1}^{j-1} A_i\Big) \cap A_j,$$

using (8.1.3) we thus get

$$\mu(B) \geq \mu\Big(\bigcup_{j=1}^{\infty} \Big(B \setminus \bigcup_{i=1}^{j-1} A_i\Big) \cap A_j\Big) + \mu(B \setminus \bigcup_{j=1}^{\infty} A_j).$$

Hence condition (2.1.3) is also satisfied, and $\mathcal{F}$ is a σ-algebra. To see that μ is a measure on $\mathcal{F}$, that is, that condition (2.1.4) is satisfied, consider mutually disjoint sets $A_1, A_2, \ldots \in \mathcal{F}$ and apply (8.1.5) to $B = \bigcup_{j=1}^{\infty} A_j$. We get

$$\mu\Big(\bigcup_{j=1}^{\infty} A_j\Big) \geq \sum_{j=1}^{\infty} \mu(A_j).$$

Combining this with (8.1.3) we conclude that μ is a measure on $\mathcal{F}$. ♣

Now, let (X, ρ) be a metric space. An outer measure μ on X is said to be a *metric outer measure* if

$$\mu(A \cup B) = \mu(A) + \mu(B) \tag{8.1.6}$$

for all *positively separated* sets $A, B \subset X$: that is, satisfying the following condition

$$\rho(A, B) = \inf\{\rho(x, y) : x \in A, y \in B\} > 0.$$

We assume the convention that $\rho(A, \emptyset) = \rho(\emptyset, A) = \infty$.

Recall that the Borel σ-algebra on X is the σ-algebra generated by open, or equivalently closed, sets. We want to show that if μ is a metric outer measure then the family of all μ-measurable sets contains this σ-algebra. The proof is based on the following version of Carathéodory's Lemma.

Lemma 8.1.3. *Let μ be a metric outer measure on (X, ρ). Let $\{A_n : n = 1, 2, \ldots\}$ be an increasing sequence of subsets of X, and denote $A = \bigcup_{n=1}^{\infty} A_n$. If $\rho(A_n, A \setminus A_{n+1}) > 0$ for all $n \geq 1$, then $\mu(A) = \lim_{n\to\infty} \mu(A_n)$.*

Proof. By (8.1.2) it is sufficient to show that

$$\mu(A) \leq \lim_{n\to\infty} \mu(A_n). \tag{8.1.7}$$

If $\lim_{n\to\infty} \mu(A_n) = \infty$, there is nothing to prove. So, suppose that

$$\lim_{n\to\infty} \mu(A_n) = \sup_n \mu(A_n) < \infty. \tag{8.1.8}$$

Let $B_1 = A_1$ and $B_n = A_n \setminus A_{n-1}$ for $n \geq 2$. If $n \geq m + 2$, then $B_m \subset A_m$ and $B_n \subset A \setminus A_{n-1} \subset A \setminus A_{m+1}$. Thus B_m and B_n are positively separated, and applying (8.1.6) we get for every $j \geq 1$

$$\mu\Big(\bigcup_{i=1}^{j} B_{2i-1}\Big) = \sum_{i=1}^{j} \mu(B_{2i-1}) \quad \text{and} \quad \mu\Big(\bigcup_{i=1}^{j} B_{2i}\Big) = \sum_{i=1}^{j} \mu(B_{2i}). \tag{8.1.9}$$

We also have for every $n \geq 1$

$$\begin{aligned} \mu(A) &= \mu\Big(\bigcup_{k=n}^{\infty} A_k\Big) = \mu\Big(A_n \cup \bigcup_{k=n+1}^{\infty} B_k\Big) \\ &\leq \mu(A_n) + \sum_{k=n+1}^{\infty} \mu(B_k) \leq \lim_{l\to\infty} \mu(A_l) + \sum_{k=n+1}^{\infty} \mu(B_k). \end{aligned} \tag{8.1.10}$$

Since the sets $\bigcup_{i=1}^{j} B_{2i-1}$ and $\bigcup_{i=1}^{j} B_{2i}$ appearing in (8.1.9) are both contained in A_{2j}, it follows from (8.1.8) and (8.1.9) that the series $\sum_{k=1}^{\infty} \mu(B_k)$ converges. Therefore (8.1.7) follows immediately from (8.1.10). The proof is complete. ♣

Theorem 8.1.4. *If μ is a metric outer measure on (X, ρ), then all Borel subsets of X are μ-measurable.*

Proof. Since the Borel sets form the least σ-algebra containing all closed subsets of X, it follows from Theorem 8.1.2 that it is enough to check (8.1.4) for every non-empty closed set $A \subset X$ and every $B \subset X$. For all $n \geq 1$, let $B_n = \{x \in B \setminus A : \rho(x, A) \geq 1/n\}$. Then $\rho(B \cap A, B_n) \geq 1/n$, and by (8.1.6)

$$\mu(B \cap A) + \mu(B_n) = \mu((B \cap A) \cup B_n) \leq \mu(B). \tag{8.1.11}$$

The sequence $\{B_n\}_{n=1}^{\infty}$ is increasing and, since A is closed, $B \setminus A = \bigcup_{n=1}^{\infty} B_n$. In order to apply Lemma 8.1.3 we shall show that

$$\rho(B_n, (B \setminus A) \setminus B_{n+1}) > 0$$

for all $n \geq 1$. And indeed, if $x \in (B \setminus A) \setminus B_{n+1}$, then there exists $z \in A$ with $\rho(x, z) < 1/(n+1)$. Thus, if $y \in B_n$, then

$$\rho(x, y) \geq \rho(y, z) - \rho(x, z) > 1/n - 1/(n+1) = \frac{1}{n(n+1)},$$

and consequently $\rho(B_n, (B \setminus A) \setminus B_{n+1}) > 1/n(n+1) > 0$. Applying now Lemma 8.1.3 with $A_n = B_n$ shows that $\mu(B \setminus A) = \lim_{n \to \infty} \mu(B_n)$. Thus (8.1.4) follows from (8.1.11). The proof is complete. ♣

8.2 Hausdorff measures

Let $\phi : [0, \infty) \to [0, \infty)$ be a non-decreasing function continuous at 0, positive on $(0, \infty)$, and such that $\phi(0) = 0$. Let (X, ρ) be a metric space. For every $\delta > 0$ define

$$\Lambda_\phi^\delta(A) = \inf\Big\{\sum_{i=1}^{\infty} \phi(\operatorname{diam}(U_i))\Big\}, \tag{8.2.1}$$

where the infimum is taken over all countable covers $\{U_i : i = 1, 2, \ldots\}$ of A of diameter not exceeding δ. Conditions (8.1.1) and (8.1.2) are obviously satisfied with $\mu = \Lambda_\phi^\delta$. To check (8.1.3) let $\{A_n : n = 1, 2, \ldots\}$ be a countable family of subsets of X. Given $\varepsilon > 0$ for every $n \geq 1$ we can find a countable cover $\{U_i^n : i = 1, 2, \ldots\}$ of A_n of diameter not exceeding δ such that $\sum_{i=1}^{\infty} \phi(\operatorname{diam}(U_i^n)) \leq \Lambda_\phi^\delta(A_n) + \varepsilon/2^n$. Then the family $\{U_i^n : n \geq 1, i \geq 1\}$ covers $\bigcup_{n=1}^{\infty} A_n$, and

$$\Lambda_\phi^\delta\Big(\bigcup_{n=1}^{\infty} A_n\Big) \leq \sum_{n=1}^{\infty}\sum_{i=1}^{\infty} \phi(\operatorname{diam}(U_i^n)) \leq \sum_{n=1}^{\infty} \Lambda_\phi^\delta(A_n) + \varepsilon.$$

Thus, letting $\varepsilon \to 0$, (8.1.3) follows, proving that Λ_ϕ^δ is an outer measure. Define

$$\Lambda_\phi(A) = \lim_{\delta \to 0} \Lambda_\phi^\delta(A) = \sup_{\delta > 0} \Lambda_\phi^\delta(A). \tag{8.2.2}$$

The limit exists, but may be infinite, since $\Lambda_\phi^\delta(A)$ increases as δ decreases. Since all Λ_ϕ^δ are outer measures, the same argument also shows that Λ_ϕ is an outer measure. Moreover, Λ_ϕ turns out to be a metric outer measure, since if A and B are two positively separated sets in X, then no set of diameter less than $\rho(A,B)$ can intersect both A and B. Consequently

$$\Lambda_\phi^\delta(A \cup B) = \Lambda_\phi^\delta(A) + \Lambda_\phi^\delta(B)$$

for all $\delta < \rho(A,B)$, and letting $\delta \to 0$ we get the same formula for Λ_ϕ, which is just (8.1.6) with $\mu = \Lambda_\phi$. The metric outer measure Λ_ϕ is called the *Hausdorff outer measure* associated to the function ϕ. Its restriction to the σ-algebra of Λ_ϕ-measurable sets, which by Theorem 8.1.4 includes all the Borel sets, is called the *Hausdorff measure* associated to the function ϕ.

As an immediate consequence of the definition of the Hausdorff measure and the properties of the function ϕ, we get the following.

Proposition 8.2.1. *The Hausdorff measure Λ_ϕ is non-atomic.*

Remark 8.2.2. A particular role is played by functions ϕ of the form $t \mapsto t^\alpha$, $t, \alpha > 0$, and in this case the corresponding outer measures are denoted by Λ_α^δ and Λ_α.

Remark 8.2.3. Note that if ϕ_1 is another function, but such that ϕ_1 and ϕ restricted to an interval $[0, \varepsilon)$, $\varepsilon > 0$, are equal, then the outer measures Λ_{ϕ_1} and Λ_ϕ are also equal. So, in fact, it is sufficient to define the function ϕ only on an arbitrarily small interval $[0, \varepsilon)$.

Remark 8.2.4. Note that we get the same values for $\Lambda_\phi^\delta(A)$, and consequently also for $\Lambda_\phi(A)$, if the infimum in (8.2.1) is taken only over covers consisting of sets contained in A. This means that the Hausdorff outer measure $\Lambda_\phi(A)$ of A is its intrinsic property: that is, it does not depend on which space contains the set A. If we treated A as the metric space $(A, \rho|_A)$ with the metric $\rho|_A$ induced from ρ, we would get the same value for the Hausdorff outer measure.

If, however, we took the infimum in (8.2.1) only over covers consisting of balls, we could get a different 'Hausdorff measure' which (depending on ϕ) need not even be equivalent to the Hausdorff measure just defined. To ensure this last property ϕ is from now on assumed to satisfy the following condition.

There exists a function $C : (0, \infty) \to [1, \infty)$ such that, for every $a \in (0, \infty)$ and every $t > 0$ sufficiently small (depending on a),

$$C(a)^{-1}\phi(t) \le \phi(at) \le C(a)\phi(t). \tag{8.2.3}$$

Since $(ar)^t = a^t r^t$, all functions ϕ of the form $r \mapsto r^t$, considered in Remark 7.2.2, satisfy (8.2.3) with $C(a) = a^t$. Check that all functions $r \mapsto r^t \exp(c\sqrt{\log 1/r \log\log\log 1/r}$, $c \ge 0$ also satisfy (8.2.3) with a suitable function C.

Definition 8.2.5. A countable collection $\{(x_i, r_i) : i = 1, 2, \ldots\}$ of pairs $(x_i, r_i) \in X \times (0, \infty)$ is said to *cover* a subset A of X if $A \subset \bigcup_{i=1}^\infty B(x_i, r_i)$, and is said to be *centred* at the set A if $x_i \in A$ for all $i = 1, 2, \ldots$. The *radius* of this collection is defined as $\sup_i r_i$, and its diameter as the diameter of the family $\{B(x_i, r_i) : i = 1, 2, \ldots\}$.

For every $A \subset X$ and every $r > 0$ let

$$\Lambda_\phi^{Br}(A) = \inf\Big\{\sum_{i=1}^{\infty} \phi(r_i)\Big\}, \tag{8.2.4}$$

where the infimum is taken over all collections $\{(x_i, r_i) : i = 1, 2, \ldots\}$ centred at the set A, covering A and of radii not exceeding r. Let

$$\Lambda_\phi^{B}(A) = \lim_{r\to 0} \Lambda_\phi^{Br}(A) = \sup_{r>0} \Lambda_\phi^{Br}(A). \tag{8.2.5}$$

The limit exists by the same argument as used for the limit in (8.2.2). We shall prove the following.

Lemma 8.2.6. *For every set $A \subset X$,*

$$1 \le \frac{\Lambda_\phi(A)}{\Lambda_\phi^B(A)} \le C(2)$$

Proof. Since the diameter of any ball does not exceed its double radius, since the diameter of any collection $\{(x_i, r_i) : i = 1, 2, \ldots\}$ also does not exceed its double radius, and since the function ϕ is non-decreasing and satisfies (8.2.3), we see that, for every $r > 0$ small enough,

$$\sum_{i=1}^{\infty} \phi(\mathrm{diam}(B(x_i, r_i))) \le \sum_{i=1}^{\infty} \phi(2r_i) \le C(2) \sum_{i=1}^{\infty} \phi(r_i),$$

and therefore $\Lambda_\phi^{2r}(A) \le C(2)\Lambda_\phi^{Br}(A)$. Thus, letting $r \to 0$,

$$\Lambda_\phi(A) \le C(2)\Lambda_\phi^B(A). \tag{8.2.6}$$

On the other hand, let $\{U_i : i = 1, 2, \ldots\}$ be a countable cover of A consisting of subsets of A. For every $i \ge 1$ choose $x_i \in U_i$, and put $r_i = \mathrm{diam}(U_i)$. Then the collection $\{(x_i, r_i) : i = 1, 2, \ldots\}$ covers A, is centred at A, and

$$\sum_{i=1}^{\infty} \phi(r_i) = \sum_{i=1}^{\infty} \phi(\mathrm{diam}(U_i)),$$

which implies that $\Lambda_\phi^{B\delta}(A) \le \Lambda_\phi^{\delta}(A)$ for every $\delta > 0$. Thus $\Lambda_\phi^B(A) \le \Lambda_\phi(A)$, which combined with (8.2.6) completes the proof. ♣

Remark 8.2.7. The function of sets Λ_ϕ^B need not be an outer measure, since condition (8.1.2) need not be satisfied. Since we shall never be interested in exact computation of a Hausdorff measure, only in establishing its positiveness or finiteness, or in comparing the ratio of its value with some other quantities up to bounded constants, we shall be dealing mostly with $\Lambda_\phi^{B\delta}$ and Λ_ϕ^B, always using the symbols $\Lambda_\phi^\delta(A)$ and $\Lambda_\phi(A)$.

8.3 Packing measures

As in the previous section, let $\phi : [0,\infty) \to [0,\infty)$ be a non-decreasing function such that $\phi(0) = 0$, and let (X,ρ) be a metric space. A collection $\{(x_i, r_i) : i = 1, 2, \ldots\}$ centred at a set $A \subset X$ is said to be a *packing* of A if and only if, for any pair $i \neq j$,

$$\rho(x_i, x_j) \geq r_i + r_j.$$

This property is not generally equivalent to the requirement that all the balls $B(x_i, r_i)$ be mutually disjoint. It is obviously so if X is a Euclidean space. For every $A \subset X$ and every $r > 0$ let

$$\Pi_\phi^{*r}(A) = \sup\Big\{\sum_{i=1}^{\infty} \phi(r_i)\Big\}, \tag{8.3.1}$$

where the supremum is taken over all packings $\{(x_i, r_i) : i = 1, 2, \ldots\}$ of A of radius not exceeding r. Let

$$\Pi_\phi^*(A) = \lim_{r\to 0} \Pi_\phi^{*r}(A) = \inf_{r>0} \Pi_\phi^{*r}(A). \tag{8.3.2}$$

The limit exists since $\Pi_\phi^{*r}(A)$ decreases as r decreases. In contrast to Λ_ϕ^B, the function Π_ϕ^* satisfies condition (8.1.2), but it also need not be an outer measure, since this time condition (8.1.3) need not be satisfied. To obtain an outer measure we put

$$\Pi_\phi(A) = \inf\Big\{\sum \Pi_\phi^*(A_i)\Big\}, \tag{8.3.3}$$

where the supremum is taken over all covers $\{A_i\}$ of A. The reader can check easily, with arguments similar to the case of Hausdorff measures, that Π_ϕ is already an outer measure and, what is more, a metric outer measure on X. It will be called the *outer packing measure* associated to the function ϕ. Its restriction to the σ-algebra of Π_ϕ-measurable sets, which by Theorem 8.1.4 includes all the Borel sets, will be called the *packing measure* associated to the function ϕ.

Proposition 8.3.1. *For every set $A \subset X$ it holds that $\Lambda_\phi(A) \leq C(2)\Pi_\phi(A)$.*

Proof. First we shall show that, for every set $A \subset X$ and every $r > 0$,

$$\Lambda_\phi^{2r}(A) \leq C(2)\Pi_\phi^{*r}(A) \tag{8.3.4}$$

Indeed, if there is no finite maximal (in the sense of inclusion) packing of the set A of the form $\{(x_i, r)\}$, then for every $k \geq 1$ there exists a packing $\{(x_i, r) : i = 1, \ldots, k\}$ of A, and therefore $\Pi_\phi^{*r}(A) \geq \sum_{i=1}^k \phi(r) = k\phi(r)$. Since $\phi(r) > 0$, this implies that $\Pi_\phi^{*r}(A) = \infty$, and (8.3.4) holds. Otherwise, let $\{(x_i, r) : i = 1, \ldots, l\}$ be a maximal packing of A. Then the collection $\{(x_i, 2r) : i = 1, \ldots, l\}$ covers A, and therefore

$$\Lambda_\phi^{2r}(A) \leq \sum_{i=1}^{l} \phi(2r) \leq C(2)l\phi(r) \leq C(2)\Pi_\phi^{*r}(A).$$

That is, (8.3.4) is satisfied. Thus, letting $r \to 0$, we get

$$\Lambda_\phi(A) \le C(2)\Pi_\phi^*(A). \tag{8.3.5}$$

So, if $\{A_n\}_{n\ge1}$ is a countable cover of A, then

$$\Lambda_\phi(A) \le \sum_{n=1}^\infty \Lambda_\phi(A_i) \le C(2)\sum_{n=1}^\infty \Pi_\phi^*(A_i).$$

Hence, applying (8.3.3), the lemma follows. ♣

8.4 Dimensions

As in the two previous sections, let (X,ρ) be a metric space. Recall (compare Remark 8.2.2) that Λ_t, $t>0$, is the Hausdorff outer measure on X associated to the function $r \mapsto r^t$, and all Λ_t^δ are of corresponding meaning. Fix $A \subset X$. Since for every $0<\delta\le 1$ the function $t \mapsto \Lambda_t^\delta(A)$ is non-increasing, so is the function $t \to \Lambda_t(A)$. Furthermore, if $s<t$, then for every $0<\delta$

$$\Lambda_s^\delta(A) \ge \delta^{s-t}\Lambda_t^\delta(A),$$

which implies that, if $\Lambda_t(A)$ is positive, then $\Lambda_s(A)$ is infinite. Thus there is a unique value, $\mathrm{HD}(A)$, called the *Hausdorff dimension* of A, such that

$$\Lambda_t(A) = \begin{cases} \infty & \text{if } 0 \le t < \mathrm{HD}(A) \\ 0 & \text{if } \mathrm{HD}(A) < t < \infty \end{cases} \tag{8.4.1}$$

Note that, similar to Hausdorff measures (compare Remark 8.2.4), the Hausdorff dimension is consequently also an intrinsic property of sets, and does not depend on their complements. The following is an immediate consequence of the definitions of the Hausdorff dimension and the outer Hausdorff measures.

Theorem 8.4.1. *The Hausdorff dimension is a monotonic function of sets: that is, if $A \subset B$ then $\mathrm{HD}(A) \le \mathrm{HD}(B)$.*

We shall prove the following.

Theorem 8.4.2. *If $\{A_n\}_{n\ge1}$ is a countable family of subsets of X, then*

$$\mathrm{HD}(\cup_n A_n) = \sup_n\{\mathrm{HD}(A_n)\}.$$

Proof. Inequality $\mathrm{HD}(\cup_n A_n) \ge \sup_n\{\mathrm{HD}(A_n)\}$ is an immediate consequence of Theorem 8.4.1. Thus, if $\sup_n\{\mathrm{HD}(A_n)\} = \infty$, there is nothing to prove. So, suppose that $s = \sup_n\{\mathrm{HD}(A_n)\}$ is finite, and consider an arbitrary $t>s$. In view of (8.4.1), $\Lambda_t(A_n) = 0$ for every $n \ge 1$, and therefore, since Λ_t is an outer measure, $\Lambda_t(\cup_n A_n) = 0$. Hence, by (8.4.1) again, $\mathrm{HD}(\cup_n A_n) \le t$. The proof is complete. ♣

As an immediate consequence of this theorem, Proposition 8.2.1 and formula (8.4.1) we get the following.

Proposition 8.4.3. *The Hausdorff dimension of any countable set is equal to* 0.

In exactly the same way as the Hausdorff dimension HD, one can define the *packing* dimension* PD* and *packing dimension* PD using respectively $\Pi_t^*(A)$ and $\Pi_t(A)$ instead of $\Lambda_t(A)$. The reader can check easily that results analogous to Theorem 8.4.1, Theorem 8.4.2 and Proposition 8.4.3 are also true in these cases. As an immediate consequence of these definitions and Proposition 8.3.1 we get the following.

Lemma 8.4.4. $\mathrm{HD}(A) \le \mathrm{PD}(A) \le \mathrm{PD}^*(A)$ *for every set* $A \subset X$.

Now we shall define the third basic dimension – the *ball-counting dimension*, frequently also called the box-counting dimension, Minkowski dimension or (limit) capacity. Let A be an arbitrary subset of the metric space (X, ρ). We first need the following.

Definition 8.4.5. For every $r > 0$ consider the family of all collections $\{(x_i, r_i)\}$ (see Definition 8.2.5) of radius not exceeding r, which cover A and are centred at A. Put $N(A, r) = \infty$ if this family is empty. Otherwise define $N(A, r)$ to be the minimum of all cardinalities of elements of this family. Note that one gets the same number if one considers the subfamily of collections of radius exactly r and even only its subfamily of collections of the form $\{(x_i, r)\}$.

Now the *lower ball-counting dimension* and *upper ball-counting dimension* of A are defined respectively by

$$\underline{\mathrm{BD}}(A) = \liminf_{r\to 0} \frac{\log N(A,r)}{-\log r} \quad \text{and} \quad \overline{\mathrm{BD}}(A) = \limsup_{r\to 0} \frac{\log N(A,r)}{-\log r}. \tag{8.4.2}$$

If $\underline{\mathrm{BD}}(A) = \overline{\mathrm{BD}}(A)$, the common value is called simply the ball-counting dimension, and is denoted by $\mathrm{BD}(A)$. The reader can easily prove the next theorem, which explains the reason for the name 'box-counting dimension'. The other names will not be discussed here.

Proposition 8.4.6. *Fix* $n \ge 1$. *For every* $r > 0$ *let* $\mathcal{L}(r)$ *be any partition (up to the boundaries) of* $\mathbb{R}^n$ *into closed cubes of sides of length* r. *For any set* $A \subset \mathbb{R}^n$ *let* $L(A, r)$ *denote the number of cubes in* $\mathcal{L}(r)$ *that intersect* A. *Then*

$$\underline{\mathrm{BD}}(A) = \liminf_{r\to 0} \frac{\log L(A,r)}{-\log r} \quad \text{and} \quad \overline{\mathrm{BD}}(A) = \limsup_{r\to 0} \frac{\log L(A,r)}{-\log r}.$$

Remark 8.4.7. The ball-counting dimension has properties that distinguish it qualitatively from the Hausdorff and packing dimensions. For instance, $\underline{\mathrm{BD}}(\overline{A}) = \underline{\mathrm{BD}}(A)$ and $\overline{\mathrm{BD}}(\overline{A}) = \overline{\mathrm{BD}}(A)$. So, in particular, there exist countable sets of positive ball-counting dimensions, for example the set of rational numbers in the interval $[0, 1]$. Furthermore, there exist compact countable sets with this property, such as the set $\{1, 1/2, 1/3, \ldots, 0\} \subset \mathbb{R}$. On the other hand, in many cases (see Theorem 8.6.7) all these dimensions coincide.

Now we shall provide other characterizations of ball-counting dimension, which in particular will be used to prove Lemma 8.4.9 and consequently Theorem 8.4.10, which establishes most general relations between the dimensions considered in this section.

Let $A \subset X$. For every $r > 0$ define $P(A, r)$ to be the supremum of cardinalities of all packings of the set A of the form $\{(x_i, r)\}$. First we shall prove the following.

Lemma 8.4.8. *For every set $A \subset \mathbb{R}^n$ and every $r > 0$*

$$N(A, 2r) \le P(A, r) \le N(A, r).$$

Proof. Let us start with the proof of the first inequality. If $P(A, r) = \infty$, there is nothing to prove. Otherwise, let $\{(x_i, r) : i = 1, \dots, k\}$ be a packing of A with $k = P(A, r)$. Then this packing is maximal in the sense of inclusion, and therefore the collection $\{(x_i, 2r) : i = 1, \dots, l\}$ covers A. Thus $N(A, 2r) \le l = P(A, r)$. The first part of Lemma 8.4.8 is proved.

If $N(A, r) = \infty$, the second part is obvious. Otherwise consider a finite packing $\{(x_i, r) : i = 1, \dots, k\}$ of A and a finite cover $\{(y_j, r) : j = 1, \dots, l\}$ of A centred at A. Then for every $1 \le i \le k$ there exists $1 \le j = j(i) \le l$ such that $x_i \in B(y_j(i), r)$, and every ball $B(y_j, r)$ can contain at most one element of the set $\{x_i : i = 1, \dots, k\}$. So the function $i \mapsto j(i)$ is injective, and therefore $k \le l$. The proof is complete. ♣

As an immediate consequence of Lemma 8.4.8 we get the following.

$$\underline{\mathrm{BD}}(A) = \liminf_{r \to 0} \frac{\log P(A, r)}{-\log r} \text{ and } \overline{\mathrm{BD}}(A) = \limsup_{r \to 0} \frac{\log P(A, r)}{-\log r}. \tag{8.4.3}$$

Now we are in a position to prove the following.

Lemma 8.4.9. *For every set $A \subset X$ we have* $\mathrm{PD}^*(A) = \overline{\mathrm{BD}}(A)$.

Proof. Take $t < \overline{\mathrm{BD}}(A)$. In view of (8.4.3) there exists a sequence $\{r_n : n = 1, 2, \dots\}$ of positive reals converging to zero and such that $P(A, r_n) \ge r_n^{-t}$ for every $n \ge 1$. Then $\Pi_t^{*r_n}(A) \ge r^t P(A, r_n) \ge 1$, and consequently $\Pi_t^*(A) \ge 1$. Hence $t \le \mathrm{PD}^*(A)$, and therefore $\overline{\mathrm{BD}}(A) \le \mathrm{PD}^*(A)$.

In order to prove the converse inequality consider $s < t < \mathrm{PD}^*(A)$. Then $\Pi_t^*(A) = \infty$, and therefore for every $n \ge 1$ there exists a finite packing $\{(x_{n,i}, r_{n,i}) : i = 1, \dots, k(n)\}$ of A of radius not exceeding 2^{-n} and such that

$$\sum_{i=1}^{k(n)} r_{n,i}^t > 1. \tag{8.4.4}$$

Now for every $m \ge n$ let

$$l_{n,m} = \#\{i \in \{1, \dots, k(n)\} : 2^{-(m+1)} < r_{n,i} \le 2^{-m}\}.$$

Then by (8.4.4)

$$\sum_{m=n}^{\infty} l_{n,m} 2^{-mt} > 1. \tag{8.4.5}$$

Suppose that $l_{n,m} < 2^{ms}(1 - 2^{(s-t)})$ for every $m \geq n$. Then

$$\sum_{m=n}^{\infty} l_{n,m} 2^{-mt} < (1 - 2^{(s-t)}) \sum_{m=0}^{\infty} 2^{m(s-t)} = 1,$$

which contradicts (8.4.5). Thus for every $n \geq 1$ there exists $m = m(n) \geq n$ such that

$$l_{n,m} \geq 2^{ms}(1 - 2^{(s-t)}).$$

Hence $P(A, 2^{-(m+1)}) \geq 2^{ms}(1 - 2^{(s-t)})$, and so

$$\frac{\log P(A, 2^{-(m+1)})}{(m+1)\log 2} \geq \frac{sm \log 2 + \log(1 - 2^{s-t})}{(m+1)\log 2}.$$

Thus, letting $n \to \infty$ (then also $m = m(n) \to \infty$), we obtain $\overline{\mathrm{BD}}(A) \geq s$. ♣

Combining Lemma 8.4.4 and Lemma 8.4.9 and checking easily that $\mathrm{HD}(A) \leq \underline{\mathrm{BD}}(A)$, we obtain the following main general relation connecting all the dimensions under consideration.

Theorem 8.4.10. *For every set $A \subset X$*

$$\mathrm{HD}(A) \leq \min\{\mathrm{PD}(A), \underline{\mathrm{BD}}(A)\} \leq \max\{\mathrm{PD}(A), \underline{\mathrm{BD}}(A)\} \leq \overline{\mathrm{BD}}(A) = \mathrm{PD}^*(A).$$

We finish this section with the following definition.

Definition 8.4.11. Let μ be a Borel measure on (X, ρ). We write

$$\mathrm{HD}_\star(\mu) = \inf\{\mathrm{HD}(Y) : \mu(Y) > 0\} \quad \text{and} \quad \mathrm{HD}^\star(\mu) = \inf\{\mathrm{HD}(Y) : \mu(X \setminus Y) = 0\}.$$

In the case where $\mathrm{HD}_\star(\mu) = \mathrm{HD}^\star(\mu)$, we call it the *Hausdorff dimension* of the measure μ and write it $\mathrm{HD}(\mu)$.

An analogous definition can be formulated for the packing dimension, with notation $\mathrm{PD}_\star(\mu), \mathrm{PD}^\star(\mu), \mathrm{PD}(\mu)$, and the name *packing dimension* of the measure μ.

8.5 Besicovitch Covering Theorem; Vitali Theorem and density points

In this section the main result is the Besicovitch Covering Theorem. Although this theorem often seems to be omitted in classical geometric measure theory, we consider it one of most powerful geometric tools when dealing with some aspects of fractal sets. We refer the reader to Section 8.6 to verify our opinion. We deduce also easily two other fundamental classical theorems: the *Vitali-type Covering Theorem*, and the *Density Points Theorem*.

Theorem 8.5.1 (Besicovitch Covering Theorem). *Let $n \geq 1$ be an integer. Then there exists a constant $b(n) > 0$ such that the following claim is true.*

If A is a bounded subset of $\mathbb{R}^n$ then for any function $r : A \to (0, \infty)$ there exists $\{x_k : k = 1, 2, \ldots\}$ a countable subset of A such that the collection $\mathcal{B}(A, r) = \{B(x_k, r(x_k)) : k \geq 1\}$ covers A and can be decomposed into $b(n)$ packings of A.

In particular, it follows from Theorem 8.5.1 that $\#\{B \in \mathcal{B} : x \in B\} \leq b(n)$. Exactly the same proof (word by word) goes through if open balls in Theorem 8.5.1 are replaced by closed ones.

For any $x \in \mathbb{R}^n$, any $0 < r \leq \infty$ and any $0 < \alpha < \pi$ by $\mathrm{Con}(x, \alpha, r)$ we shall denote any solid central cone with vertex x, radius r and angle α, that is for an arbitrary straight half-line l starting at x $\mathrm{Con}(x, \alpha, r) = \mathrm{Con}(l, x, \alpha, r) := \{y \in \mathbb{R}^n : 0 < |y - x| < r, \angle(y - x, l) \leq \alpha\} \cup \{x\}$.

The proof of Theorem 8.5.1 is based on the following obvious geometric observation.

Observation 8.5.2. Let $n \geq 1$ be an integer. Then there exists $\alpha(n) > 0$ so small that the following holds. If $x \in \mathbb{R}^n$, $0 < r < \infty$, if $z \in B(x, r) \setminus B(x, r/3)$ and $x \in \mathrm{Con}(z, \alpha(n), \infty)$, then the set $\mathrm{Con}(z, \alpha(n), \infty) \setminus B(x, r/3)$ consists of two connected components, one of z and one of '∞', and that one containing z is contained in $B(x, r)$.

Proof of Theorem 8.5.1. In the sequel we consider balls in $\mathbb{R}^n$. We shall construct the sequence $\{x_k : k = 1, 2, \ldots\}$ inductively. Let

$$a_0 = \sup\{r(x) : x \in A\}.$$

If $a_0 = \infty$, then one can find $x \in A$ with $r(x)$ so large that $B(x, r(x)) \supset A$, and the proof is complete.

If $a_0 < \infty$, choose $x_1 \in A$ so that $r(x_1) > a_0/2$. Fix $k \geq 1$ and assume that the points $x_1, x_2, \ldots, x_k$ have been already chosen. If $A \subset B(x_1, r(x_1)) \cup \ldots \cup B(x_k, r(x_k))$, then the selection process is finished. Otherwise put

$$a_k = \sup\{r(x) : x \in A \setminus \big(B(x_1, r(x_1)) \cup \ldots \cup B(x_k, r(x_k))\big)\}$$

and take

$$x_{k+1} \in A \setminus \big(B(x_1, r(x_1)) \cup \ldots \cup B(x_k, r(x_k))\big) \tag{8.5.1}$$

such that

$$r(x_{k+1}) > a_k/2. \tag{8.5.2}$$

In order to shorten the notation from now on throughout this proof we shall write r_k for $r(x_k)$. By (8.5.1) we have $x_l \notin B(x_k, r_k)$ for all pairs k, l with $k < l$. Hence

$$\|x_k - x_l\| \geq r(x_k). \tag{8.5.3}$$

It follows from the construction of the sequence (x_k) that

$$r_k > a_{k-1}/2 \geq r_l/2 \tag{8.5.4}$$

and therefore $r_k/3+r_l/3 < r_k/3+2r_k/3 = r_k$. Joining this and (8.5.3) we obtain

$$B(x_k, r_k/3) \cap B(x_l, r_l/3) = \emptyset \tag{8.5.5}$$

for all pairs k, l with $k \neq l$, since then either $k < l$ or $l < k$.

Now we shall show that the balls $\{B(x_k, r_k) : k \geq 1\}$ cover A. Indeed, if the selection process stops after finitely many steps, this claim is obvious. Otherwise it follows from (8.5.5) that $\lim_{k\to\infty} r_k = 0$, and if $x \notin \bigcup_{k=1}^{\infty} B(x_k, r_k)$ for some $x \in A$ then by construction $r_k > a_{k-1}/2 \geq r(x)/2$ for every $k \geq 1$. The contradiction obtained proves that $\bigcup_{k=1}^{\infty} B(x_k, r_k) \supset A$.

The main step of the proof is given by the following.

Claim. For every $z \in \mathbb{R}^n$ and any cone $\mathrm{Con}(z, \alpha(n), \infty)$ ($\alpha(n)$ given by Observation 8.5.2),

$$\#\{k \geq 1 : z \in B(x_k, r_k) \setminus B(x_k, r_k/3) \text{ and } x_k \in \mathrm{Con}(z, \alpha(n), \infty)\} \leq 12^n.$$

Denote by Q the set of integers whose cardinality is to be estimated. If $Q = \emptyset$, there is nothing to prove. Otherwise let $i = \min Q$. If $k \in Q$ and $k \neq i$, then $k > i$ and therefore $x_k \notin B(x_i, r_i)$. In view of this, Observation 8.5.2 applied with $x = x_i$, $r = r_i$, and the definition of Q, we get $\|z - x_k\| \geq 2r_i/3$, whence

$$r_k \geq \|z - x_k\| \geq 2r_i/3. \tag{8.5.6}$$

On the other hand, by (8.5.4) we have $r_k < 2r_i$, and therefore $B(x_k, r_k/3) \subset B(z, 4r_k/3) \subset B(z, 8r_i/3)$. Thus, using (8.5.5), (8.5.6) and the fact that the n-dimensional volume of balls in $\mathbb{R}^n$ is proportional to the nth power of radii, we obtain $\#Q \leq (8r_i/3)^n/(2r_i/9)^n = 12^n$. The proof of the claim is complete.

Clearly there exists an integer $c(n) \geq 1$ such that for every $z \in \mathbb{R}^n$ the space $\mathbb{R}^n$ can be covered by at most $c(n)$ cones of the form $\mathrm{Con}(z, \alpha(n), \infty)$. Therefore it follows from the claim that, for every $z \in \mathbb{R}^n$,

$$\#\{k \geq 1 : z \in B(x_k, r_k) \setminus B(x_k, r_k/3)\} \leq c(n)12^n.$$

Thus, applying (8.5.5),

$$\#\{k \geq 1 : z \in B(x_k, r_k) \leq 1 + c(n)12^n. \tag{8.5.7}$$

Since the ball $\overline{B}(0, 3/2)$ is compact, it contains a finite subset P such that $\bigcup_{x\in P} B(x, 1/2) \supset \overline{B}(0, 3/2)$. Now for every $k \geq 1$ consider the composition of the map $\mathbb{R}^n \ni x \mapsto r_k x \in \mathbb{R}^n$ and the translation determined by the vector from 0 to x_k. Denote by P_k the image of P under this affine map. Then $\#P_k = \#P$, $P_k \subset \overline{B}(x_k, 3r_k/2)$, and

$$\bigcup_{x\in P_k} B(x, r_k/2) \supset \overline{B}(0, 3r_k/2). \tag{8.5.8}$$

Consider now two integers $1 \leq k < l$ such that

$$B(x_k, r_k) \cap B(x_l, r_l) \neq \emptyset. \tag{8.5.9}$$

Let $y \in \mathbb{R}^n$ be the only point lying on the interval joining x_l and x_k at the distance $r_k - r_l/2$ from x_k. As $x_l \notin B(x_k, r_k)$, by (8.5.9) we have $\|y - x_l\| \leq r_l + r_l/2 = 3r_l/2$, and therefore by (8.5.8) there exists $z \in P_l$ such that $\|z-y\| < r_l/2$. Consequently $z \in B(x_k, r_l/2+r_k-r_l/2) = B(x_k, r_k)$. Thus, applying (8.5.7) with z being the elements of P_l, we obtain the following:

$$\#\{1 \leq k \leq l-1 : B(x_k, r_k) \cap B(x_l, r_l) \neq \emptyset\} \leq \#P(1 + c(n)12^n) \qquad (8.5.10)$$

for every $l \geq 1$.

Putting $b(n) = \#P(1 + c(n)12^n) + 1$, this property allows us to decompose the set $\mathbb{N}$ of positive integers into $b(n)$ subsets $\mathbb{N}_1, \mathbb{N}_2, \ldots, \mathbb{N}_{b(n)}$ in the following inductive way. For every $k = 1, 2, \ldots, b(n)$ set $\mathbb{N}_k(b(n)) = \{k\}$, and suppose that for every $k = 1, 2, \ldots, b(n)$ and some $j \geq b(n)$ mutually disjoint families $\mathbb{N}_k(j)$ have been already defined, so that

$$\mathbb{N}_1(j) \cup \ldots \cup \mathbb{N}_{b(n)}(j) = \{1, 2, \ldots, j\}.$$

Then by (8.5.10) there exists at least one $1 \leq k \leq b(n)$ such that $B(x_{j+1}, r_{j+1}) \cap B(x_i, r_i) = \emptyset$ for every $i \in \mathbb{N}_k(j)$. We set $\mathbb{N}_k(j+1) = \mathbb{N}_k(j) \cup \{j+1\}$ and $\mathbb{N}_l(j+1) = \mathbb{N}_l(j)$ for all $l \in \{1, 2, \ldots, b(n)\} \setminus \{k\}$. Putting now for every $k = 1, 2, \ldots, b(n)$

$$\mathbb{N}_k = \mathbb{N}_k(b(n)) \cup \mathbb{N}_k(b(n)+1) \cup \ldots,$$

we see from the inductive construction that these sets are mutually disjoint, that they cover $\mathbb{N}$, and that for every $k = 1, 2, \ldots, b(n)$ the families of balls $\{B(x_l, r_l) : l \in \mathbb{N}_k\}$ are also mutually disjoint. The proof of the Besicovitch Covering Theorem is complete. ♣

We should like to emphasize here once more that the same statement remains true if open balls are replaced by closed ones. It also remains true if instead of balls one considers n-dimensional cubes, but then the proof, based on the same idea, is technically considerably easier.

We can easily deduce from the Besicovitch Covering Theorem some other fundamental facts.

Theorem 8.5.3 (Vitali-type Covering Theorem). *Let μ be a probability Borel measure on $\mathbb{R}^n$, let $A \subset \mathbb{R}^n$ be a Borel set, and let $\mathcal{B}$ be a family of closed balls such that each point of A is the centre of arbitrarily small balls of $\mathcal{B}$: that is,*

$$\inf\{r : B(x,r) \in \mathcal{B}\} = 0 \text{ for all } x \in A.$$

Then there is a finite or countable infinite collection $\mathcal{B}(A)$ of disjoint balls $B_i \in \mathcal{B}$ such that

$$\mu(A \setminus \bigcup_i B_i) = 0.$$

Proof. (See [Mattila 1995].) We assume A is bounded, leaving the unbounded case to the reader. We may assume $\mu(A) > 0$. The measure μ restricted to a compact ball $B(0, R)$ such that $A \subset B(0, R/2)$ is Borel and hence regular: see the comments preceding Theorem 3.1.2. Hence there exists an open set $U \subset \mathbb{R}^n$ containing A and such that

$$\mu(U) \leq (1 + (4b(n))^{-1})\mu(A),$$

where $b(n)$ is as in the Besicovitch Covering Theorem 8.5.1. By that theorem, applied for closed balls, we can decompose $\mathcal{B}$ into packings $\mathcal{B}_1, ..., \mathcal{B}_{b(n)}$ of A contained in U: that is, each $\mathcal{B}_i$ consists of disjoint balls and

$$A \subset \bigcup_{i=1}^{b(n)} \bigcup \mathcal{B}_i \subset U.$$

Then $\mu(A) \leq \sum_{i=1}^{b(n)} \mu(\bigcup \mathcal{B}_i)$, and consequently there exists an i such that

$$\mu(A) \leq b(n)\mu\big(\bigcup \mathcal{B}_i\big).$$

Further, for some finite subfamily $\mathcal{B}_i'$ of $\mathcal{B}_i$,

$$\mu(A) \leq 2b(n)\mu\big(\bigcup \mathcal{B}_i'\big).$$

Letting $A_1 = A \setminus (\bigcup \mathcal{B}_i')$, we get

$$\begin{aligned}\mu(A_1) &\leq \mu(U \setminus \bigcup \mathcal{B}_i') = \mu(U) - \mu(\bigcup \mathcal{B}_i') \\ &\leq \big(1 + \frac{1}{4}(b(n))^{-1} - \frac{1}{2}(b(n))^{-1}\big)\mu(A) = u\mu(A)\end{aligned}$$

with $u := 1 - \frac{1}{4}(b(n))^{-1} < 1$.

Next, consider A_1 in the role of A before. Since $A_1 \subset \mathbb{R}^n \setminus \big(\bigcup \mathcal{B}_i'\big)$, which is open, we find a packing, playing the role of $\mathcal{B}_i'$ contained in it, so disjoint with $\mathcal{B}_i'$. We get the measure of a non-covered remnant bounded above by $u^2\mu(A)$. We can continue, exhausting the whole A except at most a set of measure 0. ♣

Theorem 8.5.4 (On points of density). *Let μ be a probability Borel measure on $\mathbb{R}^n$ and $A \subset \mathbb{R}^n$ be a Borel set. Then the limit*

$$\lim_{r \to 0} \frac{\mu(A \cap B(x,r))}{\mu(B(x,r))}$$

exists and is equal to 1 for μ-almost every $x \in \mathbb{R}^n$.

Proof. Suppose the set of points in A where the limit above is not 1 (or does not exist) has positive measure. Then there exists $a < 1$ and Borel $A' \subset A$ of positive measure μ such that for every $x \in A'$ there is a sequence $r_i \searrow 0$ such that $\mu(A' \cap B(x, r_i))/\mu(B(x, r_i)) < a$. Let $\mathcal{B}(A')$ be the collection of balls whose existence is asserted in Theorem 8.5.3, contained in an arbitrary open set U containing A'. Then

$$\mu(A') = \sum_{B \in \mathcal{B}(A')} \mu(A' \cap B) < a \sum_{B \in \mathcal{B}(A')} \mu(B) \leq a\mu(U).$$

This gives a contradiction for U sufficiently small. ♣

These theorems are an introduction to 'differentiation' theory: compare Exercise 2.6.

8.6 Frostman-type lemmas

In this section we shall explain how some knowledge about a measure of small balls versus diameter yields information about dimensions of support of the measure.

Let a function $\phi : [0,\infty) \to [0,\infty)$ satisfy the same conditions as in Section 8.2, including (8.2.3), and moreover let ϕ be continuous. We start with the following.

Theorem 8.6.1. *Let $n \geq 1$ be an integer, and let $b(n)$ be the constant claimed in Theorem 8.5.1 (Besicovitch Covering Theorem). Assume that μ is a Borel probability measure on $\mathbb{R}^n$, and A is a bounded Borel subset of $\mathbb{R}^n$. If there exists $C \in (0,\infty]$, $(1/\infty = 0)$, such that*

(a) for all (but countably many maybe) $x \in A$

$$\limsup_{r\to 0} \frac{\mu(B(x,r))}{\phi(r)} \geq C$$

then $\Lambda_\phi(E) \leq \frac{b(n)}{C}\mu(E)$ for every Borel set $E \subset A$. In particular, $\Lambda_\phi(A) < \infty$.

or

(b) for all $x \in A$

$$\limsup_{r\to 0} \frac{\mu(B(x,r))}{\phi(r)} \leq C < \infty$$

then $\mu(E) \leq C\Lambda_\phi(E)$ for every Borel set $E \subset A$.

Proof. (a) In view of Proposition 8.2.1 we can assume that E does not intersect the exceptional countable set. Fix $\varepsilon > 0$ and $r > 0$. Since μ is a regular measure, there exists an open set $G \supset E$ such that $\mu(G) \leq \mu(E) + \varepsilon$. By openness of G, and by assumption (a), for every $x \in E$ there exists $0 < r(x) < r$ such that $B(x, r(x)) \subset G$ and $(1/C + \varepsilon)\mu(B(x,r)) \geq \phi(r)$. Let $\{(x_k, r(x_k)) : k \geq 1\}$ be the cover of E obtained by applying Theorem 8.5.1 (Besicovitch Covering Theorem) to the set E. Then

$$\begin{aligned}\Lambda_\phi^r(E) &\leq \sum_{k=1}^{\infty} \phi(r(x_k)) \leq \sum_{k=1}^{\infty}(C^{-1}+\varepsilon)\mu(B(x_k, r(x_k))) \\ &\leq b(n)(C^{-1}+\varepsilon)\mu\Big(\bigcup_{k=1}^{\infty} B(x_k, r(x_k))\Big) \leq b(n)(C^{-1}+\varepsilon)(\mu(E)+\varepsilon)\end{aligned}$$

Letting $r \to 0$ we thus obtain $\Lambda_\phi(E) \leq b(n)(1/C + \varepsilon)(\mu(E) + \varepsilon)$, and therefore letting $\varepsilon \to 0$ part (a) follows (note that the proof is correct with $C = \infty$!).

(b) Fix an arbitrary $s > C$. Since for every $r > 0$ the function $x \mapsto \mu(B(x,r))/\phi(r)$ is measurable, and since the supremum of a countable sequence of measurable functions is also a measurable function, we conclude that for every $k \geq 1$ the function $\psi_k : A \to \mathbb{R}$ is measurable, where

$$\psi_k(x) = \sup\left\{\frac{\mu(B(x,r))}{\phi(r)} : r \in Q \cap (0, 1/k]\right\},$$

and Q denotes the set of rational numbers. For every $k \geq 1$ let $A_k = \psi_k^{-1}((0,s])$. In view of the measurability of the functions ψ_k, all the sets A_k are measurable. Take an arbitrary $r \in (0, 1/k]$. Then there exists a sequence $\{r_j : j = 1, 2, \ldots\}$ of rational numbers converging to r from above. Since the function ϕ is continuous and the function $t \mapsto \mu(B(x,t))$ is non-decreasing, we have, for every $x \in A_k$,

$$\frac{\mu(B(x,r))}{\phi(r)} \leq \lim_{j\to\infty} \frac{\mu(B(x,r_j))}{\phi(r_j)} \leq s.$$

So, if $F \subset A_k$ is a Borel set and if $\{(x_i, r_i) : i = 1, 2, \ldots\}$ is a collection centred at the set F, covering F and of radius not exceeding $0 < r \leq 1/k$, then

$$\sum_{i=1}^{\infty} \phi(r_i) \geq s^{-1} \sum_{i=1}^{\infty} \mu(B(x_i, r_i)) \geq s^{-1}\mu(F).$$

Hence $\Lambda_\phi^r(F) \geq s^{-1}\mu(F)$, and letting $r \to 0$ we get

$$\Lambda_\phi(E) \geq \Lambda_\phi(F) \geq s^{-1}\mu(F).$$

By the assumption of (b), $\cup_k A_k = A$, and therefore, putting $B_k = A_k \setminus (A_1 \cup A_2 \cup \ldots \cup A_{k-1})$, $k \geq 1$, we see that the family $\{B_k : k \geq 1\}$ is a countable partition of A into Borel sets. Therefore if $E \subset A$ then

$$\Lambda_\phi(E) = \sum_{k=1}^{\infty} \Lambda_\phi(E \cap A_k) \geq s^{-1} \sum_{k=1}^{\infty} \mu(E \cap A_k) = s^{-1}\mu(E).$$

So letting $s \searrow C$ completes the proof. ♣

In an analogous way, using the Besicivitch Covering Theorem, decomposition into packings, one can prove the following.

Theorem 8.6.2. *Let $n \geq 1$ be an integer, and let $b(n)$ be the constant claimed in Theorem 8.5.1 (Besicovitch Covering Theorem). Assume that μ is a Borel probability measure on $\mathbb{R}^n$ and A is a bounded subset of $\mathbb{R}^n$. If there exists $C \in (0, \infty]$, $(1/\infty = 0)$, such that*

(a) for all $x \in A$

$$\liminf_{r\to 0} \frac{\mu(B(x,r))}{\phi(r)} \leq C$$

then $\mu(E) \leq b(n)C\Pi_\phi(E)$ for every Borel set $E \subset A$.

or

(b) for all $x \in A$

$$\liminf_{r\to 0} \frac{\mu(B(x,r))}{\phi(r)} \geq C < \infty$$

then $\Pi_\phi(E) \leq C^{-1}\mu(E)$ for every Borel set $E \subset A$. In particular, $\Pi_\phi(A) < \infty$.

Note that each Borel measure μ defined on a Borel subset B of R^n can, in a canonical way, be considered as a measure on $\mathbb{R}^n$ by putting $\mu(A) = \mu(A \cap B)$ for every Borel set $A \subset \mathbb{R}^n$.

As a simple consequence of Theorem 8.6.1 we shall prove the following.

Theorem 8.6.3 (Frostman's Lemma). *Suppose that μ is a Borel probability measure on $\mathbb{R}^n$, $n \geq 1$, and A is a bounded Borel subset of $\mathbb{R}^n$.*

(a) If $\mu(A) > 0$ and there exists θ_1 such that for every $x \in A$

$$\liminf_{r \to 0} \frac{\log \mu(B(x,r))}{\log r} \geq \theta_1$$

then $\mathrm{HD}(A) \geq \theta_1$.

(b) If there exists θ_2 such that for every $x \in A$

$$\liminf_{r \to 0} \frac{\log \mu(B(x,r))}{\log r} \leq \theta_2$$

then $\mathrm{HD}(A) \leq \theta_2$.

Proof. (a) Take any $0 < \theta < \theta_1$. Then, by the assumption, $\limsup_{r \to 0} \mu(B(x,r))/r^\theta \leq 1$. Therefore, applying Theorem 8.6.1(b) with $\phi(t) = t^\theta$, we obtain $\Lambda_\theta(A) \geq \mu(A) > 0$. Hence $\mathrm{HD}(A) \geq \theta$ by definition (8.4.1), and consequently $\mathrm{HD}(A) \geq \theta_1$.

(b) Take now an arbitrary $\theta > \theta_2$. Then by the assumption $\limsup_{r \to 0} \mu(B(x,r))/r^\theta \geq 1$. Therefore, applying Theorem 8.6.1(a) with $\phi(t) = t^\theta$ we obtain $\Lambda_\theta(A) < \infty$, whence $\mathrm{HD}(A) \leq \theta$ and consequently $\mathrm{HD}(A) \leq \theta_2$. The proof is complete. ♣

Similarly we can prove a consequence of Theorem 8.6.2.

Theorem 8.6.4. *Suppose that μ is a Borel probability measure on $\mathbb{R}^n$, $n \geq 1$, and A is a bounded Borel subset of $\mathbb{R}^n$.*

(a) If $\mu(A) > 0$ and there exists θ_1 such that for every $x \in A$

$$\limsup_{r \to 0} \frac{\log \mu(B(x,r))}{\log r} \geq \theta_1$$

then $\mathrm{PD}(A) \geq \theta_1$.

(b) If there exists θ_2 such that for every $x \in A$

$$\limsup_{r \to 0} \frac{\log \mu(B(x,r))}{\log r} \leq \theta_2$$

then $\mathrm{PD}(A) \leq \theta_2$.

Let μ be a Borel probability measure on a metric space X. For every $x \in X$ we define the *lower and upper pointwise dimension* of μ at x by putting respectively

$$\underline{d}_\mu(x) = \liminf_{r \to 0} \frac{\log \mu(B(x,r))}{\log r} \quad \text{and} \quad \overline{d}_\mu(x) = \limsup_{r \to 0} \frac{\log \mu(B(x,r))}{\log r}.$$

Suppose now that $X \subset \mathbb{R}^d$ with Euclidean metric. Then the following theorem on the Hausdorff and packing dimensions of μ, defined in Definition 7.4.10, follows easily from Theorems 8.6.3 and 8.6.4.

Theorem 8.6.5.

$$\mathrm{HD}_\star(\mu) = \operatorname{ess\,inf} \underline{d}_\mu, \quad \mathrm{PD}_\star(\mu) = \operatorname{ess\,inf} \overline{d}_\mu \quad \textit{and}$$
$$\mathrm{HD}^\star(\mu) = \operatorname{ess\,sup} \underline{d}_\mu(x), \quad \mathrm{PD}^\star(\mu) = \operatorname{ess\,sup} \overline{d}_\mu(x).$$

Proof. Recall that the μ-essential infimum (ess inf) of a measurable function ϕ and the μ-essential supremum (ess sup) are defined by

$$\operatorname{ess\,inf}(\phi) = \sup_{\mu(N)=0} \inf_{x \in X \setminus N} \phi(x) \quad \text{and} \quad \operatorname{ess\,sup}(\phi) = \inf_{\mu(N)=0} \sup_{x \in X \setminus N} \phi(x).$$

So, to begin with, for $\theta_1 := \operatorname{ess\,inf} \phi$ we have

$$\mu\{x : \phi(x) < \theta_1\} = 0 \quad \text{and} \quad (\forall \theta > \theta_1) \mu\{x : \phi(x) < \theta\} > 0.$$

Indeed, if $\mu\{x : \phi(x) < \theta_1\} > 0$ then there exists $\theta < \theta_1$ with $\mu\{x : \phi(x) \le \theta\} > 0$: hence for every N with $\mu(N) = 0$ we have $\inf_{X \setminus N} \phi \le \theta$, and hence $\operatorname{ess\,inf} \phi \le \theta$, which is a contradiction. If there exists $\theta > \theta_1$ with $\mu\{x : \phi(x) < \theta\} = 0$ then for $N = \{x : \phi(x) < \theta\}$ we have $\inf_{X \setminus N} \phi \ge \theta$, and hence hence $\operatorname{ess\,inf} \phi \ge \theta$, a contradiction.

This, applied to $\phi = \underline{d}_\mu$, yields for every A with $\mu(A) > 0$ the existence of $A' \subset A$ with $\mu(A') = \mu(A) > 0$ such that for every $x \in A'$ $\underline{d}_\mu(x) \ge \theta_1$: hence $\mathrm{HD}(A) \ge \mathrm{HD}(A') \ge \theta_1$ by Theorem 8.6.3(a), and hence $\mathrm{HD}_\star(\mu) \ge \theta_1$.

On the other hand, for every $\theta > \theta_1$ $\mu\{x : \underline{d}_\mu(x) < \theta\} > 0$, and by Theorem 8.6.3(b) $\mathrm{HD}(\{x : \underline{d}_\mu(x) < \theta\}) \le \theta$: therefore $\mathrm{HD}_\star(\mu) \le \theta$. Letting $\theta \to \theta_1$ we get $\mathrm{HD}_\star(\mu) \le \theta_1$. We conclude that $\mathrm{HD}_\star(\mu) = \theta_1$.

Similarly one proceeds to prove $\mathrm{HD}^\star(\mu) = \operatorname{ess\,sup} \underline{d}_\mu(x)$ and to deal with the packing dimension, refering to Theorem 8.6.4. ♣

Then by the definition of ess inf there exists $Y \subset X \subset \mathbb{R}^n$, a Borel set such that $\mu(Y) = 1$ and for every $x \in Y$ $\underline{d}_\mu(x) \ge \theta_1$. Hence for every $A \subset X$ with $\mu(A) > 0$ we have $\mu(A \cap Y) > 0$, and for every $x \in A \cap Y$, $\underline{d}_\mu(x) \ge \theta_1$. So, using Theorem 8.6.3(a), we get $\mathrm{HD}(A) \ge \mathrm{HD}(A \cap Y) \ge \theta_1$. Hence by the definition of $\mathrm{HD}_\star$ we get $\mathrm{HD}_\star \ge \theta_1$. Other parts of Theorem 8.6.5 follow from the definitions, and Theorems 8.6.4 and 8.6.3(a) similarly. ♣

Definition 8.6.6. Let X be a Borel bounded subset of $\mathbb{R}^n$, $n \ge 1$. A Borel probability measure on X is said to be a *geometric measure* with an exponent $t \ge 0$ if and only if there exists a constant $C \ge 1$ such that

$$C^{-1} \le \frac{\mu(B(x,r))}{r^t} \le C$$

for every $x \in X$ and every $0 < r \le 1$.

We shall prove the following.

Theorem 8.6.7. *If X is a Borel bounded subset of $\mathbb{R}^n$, $n \geq 1$, and μ is a geometric measure on X with an exponent t, then* $\mathrm{BD}(X)$ *exists and*

$$\mathrm{HD}(X) = \mathrm{PD}(X) = \mathrm{BD}(X) = t$$

Moreover, the three measures μ, Λ_t and Π_t on X are equivalent, with bounded Radon–Nikodym derivatives.

Proof. The last part of the theorem follows immediately from Theorem 8.6.1 and Theorem 8.6.2 applied for $A = X$. Consequently also $t = \mathrm{HD}(X) = \mathrm{PD}(X)$ and therefore, in view of Theorem 8.4.10, we need only show that $\overline{\mathrm{BD}}(X) \leq t$. And, indeed, let $\{(x_i, r) : i = 1, \dots, k\}$ be a packing of X. Then

$$kr^t \leq C \sum_{i=1}^{k} \mu(B(x_i, r)) \leq C,$$

and therefore $k \leq Cr^{-t}$. Thus $P(X, r) \leq Cr^{-t}$, whence $\log P(X, r) \leq \log C - t \log r$. Applying formula (8.4.3) completes the proof. ♣

In particular, it follows from this theorem that every geometric measure admits exactly one exponent. Numerons examples of geometric measures will be provided in subsequent chapters.

Bibliographical notes

The history of the notions and development of the geometric measure theory is very long, rich and complicated, and its outline exceeds the scope of this book. We refer the interested reader to the books [Falconer 1997] and [Mattila 1995] and other books mentioned in the introduction to this chapter.

9

Conformal expanding repellers

Conformal expanding repellers (abbreviation CERs) have already been defined in Chapter 6, and some basic properties of expanding sets and repellers in dimension one were discussed in Section 6.2. A more advanced geometric theory in the real one-dimensional case was covered in Chapter 7.

Now we have a new tool: the Frostman Lemma and related facts from Chapter 8. Equipped with the theory of Gibbs measures, and with the pressure function, we are able to develop a geometric theory of CERs, with Hausdorff measures and dimension playing the crucial role. We shall present this theory for $C^{1+\varepsilon}$ conformal expanding repellers in $\mathbb{R}^d$. The main case of our interest will be $d = 2$. Recall (Section 6.2) that the assumed conformality forces for $d = 2$ that f is holomorphic or anti-holomorphic, and for $d \geq 3$ that f is locally a Möbius map. Conformality for $d = 1$ is meaningless, so we assume $C^{1+\varepsilon}$ in order to be able to rely on the Bounded Distortion for Iteration lemma.

We shall outline a theory of Gibbs measures from the point of view of multifractal spectra of dimensions (Section 9.2) and pointwise fluctuations due to the Law of Iterated Logarithm (Section 9.3).

For $d = 2$ we shall apply this theory to study the boundary $\operatorname{Fr}\Omega$ of a simply connected domain Ω, and in particular a simply connected immediate basin of attraction to a sink for a rational mapping of the Riemann sphere.

To simplify our considerations we shall usually restrict them to cases where the mapping on the boundary is expanding, and sometimes we assume that the boundary is a Jordan curve, for example for the mapping $z \mapsto z^2 + c$ for $|c|$ small.

In Section 9.5 we study the harmonic measure on $\operatorname{Fr}\Omega$. We adapt the results of Section 9.3 to study its pointwise fluctuations, and we prove that, except for special cases, these fluctuations occur. We shall derive from this information about the fluctuations of the radial growth of the derivative of the Riemann mapping R from the unit disc $\mathbb{D}$ to Ω. In Section 9.6 we discuss integral means

$\int_{\partial\mathbb{D}} |R'(rz)|^t \, |dz|$ as $r \nearrow 1$. In Section 9.7 we provide other examples of Ω, the von Koch snowflake and Carleson's (generalized) snowflakes.

9.1 Pressure function and dimension

Let $f : X \to X$ be a topologically mixing (equivalently: topologically exact) conformal expanding repeller in $\mathbb{R}^d$. As before, we abbreviate the notation for the pressure $\mathrm{P}(f, \phi)$, to $\mathrm{P}(\phi)$. We start with the following technical lemma.

Lemma 9.1.1. *Let m be a Gibbs state (not necessarily invariant) on X, and let $\phi : X \to \mathbb{R}$ be a Hölder continuous function. Assume $\mathrm{P}(\phi) = 0$. Then there is a constant $E \geq 1$ such that for all r small enough and all $x \in X$ there exists $n = n(x, r)$ such that*

$$\frac{\log E + S_n\phi(x)}{-\log E - \log |(f^n)'(x)|} \leq \frac{\log m(B(x,r))}{\log r} \leq \frac{-\log E + S_n\phi(x)}{\log E - \log |(f^n)'(x)|}. \tag{9.1.1}$$

Proof. Take an arbitrary $x \in X$. Fix $r \in (0, C^{-1}\xi)$, and let $n = n(x, r) \geq 0$ be the largest integer so that

$$|(f^n)'(x)| rC \leq \xi, \tag{9.1.2}$$

where $C = C_{\mathrm{MD}}$ is the multiplicative distortion constant (corresponding to the Hölder continuous function $\log |f'|$), as in the Distortion Lemma for Iteration (Lemma 6.2.2): see Definition 6.2.1. Then

$$f_x^{-n}(B(f^n(x), \xi)) \supset B(x, \xi |(f^n)'(x)|^{-1} C^{-1}) \supset B(x, r). \tag{9.1.3}$$

Now take n_0 such that $\lambda^{n_0 - 1} \geq C^2$. We then obtain

$$|(f^{n+n_0})'| rC^{-1} \geq \xi. \tag{9.1.4}$$

Hence, again by the Distortion Lemma for Iteration,

$$f_x^{-n-n_0}(B(f^{n+n_0}(x), \xi)) \subset B\big(x, \xi |(f^{n+n_0})'(x)|^{-1} C\big) \subset B(x, r). \tag{9.1.5}$$

By the Gibbs property of the measure m (see (5.1.1)), for a constant $E \geq 1$ (the constant C in (5.1.1)) we can write

$$E^{-1} \leq \frac{\exp S_n\phi(x)}{m(f_x^{-n}(B(f^n(x), \xi)))} \quad \text{and} \quad \frac{\exp S_{n+n_0}\phi(x)}{m(f_x^{-(n+n_0)}(B(f^{n+n_0}(x), \xi)))} \leq E.$$

Using this, (9.1.3), (9.1.5), the inequality $S_{n+n_0}\phi(x) \geq S_n\phi(x) + n_0 \inf \phi$, and finally increasing E so that the new $\log E$ is larger than the old $\log E - n_0 \inf \phi$, we obtain

$$\log E + S_n\phi(x) \geq \log m(B(x, r)) \geq -\log E + S_n\phi(x). \tag{9.1.6}$$

Using (9.1.2) and (9.1.4), denoting $L = \sup |f'|$, and applying logarithms, we now obtain

$$\frac{\log E + S_n\phi(x)}{\log |(f^n)'(x)|^{-1} - n_0 \log L + \log \xi} \leq \frac{\log m(B(x, r)}{\log r} \leq \frac{-\log E + S_n\phi(x)}{\log |(f^n)'(x)|^{-1}\xi}.$$

Increasing E further so that $\log E \geq n_0 \log L - \log \xi$, we can rewrite it in the 'symmetric' form of (9.1.1). ♣

When we studied the pressure function $\phi \mapsto \mathrm{P}(\phi)$ in Chapters 3 and 5 the linear functional $\psi \mapsto \int \psi d\mu_\phi$ appeared. This was the Gateaux differential of P at ϕ (Theorem 3.6.5, Proposition 3.6.6 and (5.6.5)). Here the presence of an ambient smooth structure (one-dimensional or conformal) distingushes ψ's of the form $-t \log|f'|$. We obtain a link between the ergodic theory and the geometry of the embedding of X into $\mathbb{R}^d$.

Definition 9.1.2. Let μ be an ergodic f-invariant probability measure on X. Then by Birkhoff's Ergodic Theorem, for μ-almost every $x \in X$, the limit $\lim_{n\to\infty} \frac{1}{n} \log|(f^n)'(x)|$ exists and is equal to $\int \log|f'| d\mu$. We call this number the *Lyapunov characteristic exponent* of the map f with respect to the measure μ, and we denote it by $\chi_\mu(f)$. In our case of expanding maps considered in this chapter we obviously have $\chi_\mu(f) > 0$.

This definition does not demand the expanding property. It makes sense for an arbitrary invariant subset X of $\mathbb{R}^d$ or the Riemann sphere $\bar{\mathbb{C}}$, for f conformal (or differentiable in the real case) defined on a neighbourhood of X. There is no problem with the integrability because $\log|f'|$ is upper bounded on X. We do not exclude the possibility that $\chi_\mu = -\infty$. The notion of a Lyapunov characteristic exponent will play a crucial role in subsequent chapters, where non-expanding invariant sets will be studied.

Theorem 9.1.3 (Volume Lemma, expanding map and Gibbs measure case). *Let m be a Gibbs state for a topologically mixing conformal expanding repeller $X \in \mathbb{R}^d$ and a Hölder continuous potential $\phi : X \to \mathbb{R}$. Then for m-almost every point $x \in X$ there exists the limit*

$$\lim_{r\to 0} \frac{\log m(B(x,r))}{\log r}.$$

Moreover, this limit is almost everywhere constant and is equal to $\mathrm{h}_\mu(f)/\chi_\mu(f)$, where μ denotes the only f-invariant probability measure equivalent to m.

Proof. We can assume that $\mathrm{P}(\phi) = 0$. We can achieve this by subtracting $\mathrm{P}(\phi)$ from ϕ; the Gibbs measure class will stay the same (see Proposition 5.1.4). In view of the Birkhoff Ergodic Theorem, for μ-a.e $x \in X$ we have

$$\lim_{n\to\infty} \frac{1}{n} S_n\phi(x) = \int \phi\, d\mu \quad \text{and} \quad \lim_{n\to\infty} \frac{1}{n} \log|(f^n)'(x)| = \chi_\mu(f).$$

Combining these equalities with (9.2.1), along with the observation that $n = n(x,r) \to \infty$ as $r \to 0$, and using also the equality $\mathrm{h}_\mu(f) + \int \phi d\mu = \mathrm{P}(\phi) = 0$, we conclude that

$$\lim_{r\to 0} \frac{\log \mu(B(x,r))}{\log r} = \frac{\mathrm{h}_\mu(f)}{\chi_\mu(f)}.$$

The proof is complete. ♣

As an immediate consequence of this lemma and Theorem 8.6.5 we get the following.

Theorem 9.1.4. *If μ is an ergodic Gibbs state for a conformal expanding repeller $X \in \mathbb{R}^d$ and a Hölder continuous potential ϕ on X, then there exist Hausdorff and packing dimensions of μ and*

$$\mathrm{HD}(\mu) = \mathrm{PD}(\mu) = \mathrm{h}_\mu(f)/\chi_\mu(f). \tag{9.1.7}$$

Using the above technique, we can find a formula for the Hausdorff dimension and other dimensions of the whole set X. This is the solution of the non-linear problem, corresponding to the formula for the Hausdorff dimension of the linear Cantor sets discussed in the introduction.

Definition 9.1.5 (Geometric pressure). Let $f : X \to X$ be a topologically mixing conformal expanding repeller in $\mathbb{R}^d$. We call the pressure function

$$\mathrm{P}(t) := \mathrm{P}(-t \log |f'|)$$

a *geometric pressure* function.

As f is Lipschitz continuous (or as f is forward expanding), the function $\mathrm{P}(t)$ is finite (see comments at the beginning of Section 3.6). As $|f'| \geq \lambda > 1$, it follows directly from the definition that $\mathrm{P}(t)$ is strictly decreasing from $+\infty$ to $-\infty$. In particular, there exists exactly one parameter t_0 such that $\mathrm{P}(t_0) = 0$ (Figure 9.1).

We first prove the following.

Theorem 9.1.6 (Existence of geometric measures). *Let t_0 be defined by $\mathrm{P}(t_0) = 0$. Write ϕ for $-t_0 \log |f'|$ restricted to X. Then each Gibbs state m corresponding to the function ϕ is a geometric measure with the exponent t_0. In particular, $\lim_{r \to 0} \frac{\log m(B(x,r))}{\log r} = t_0$ for every $x \in X$.*

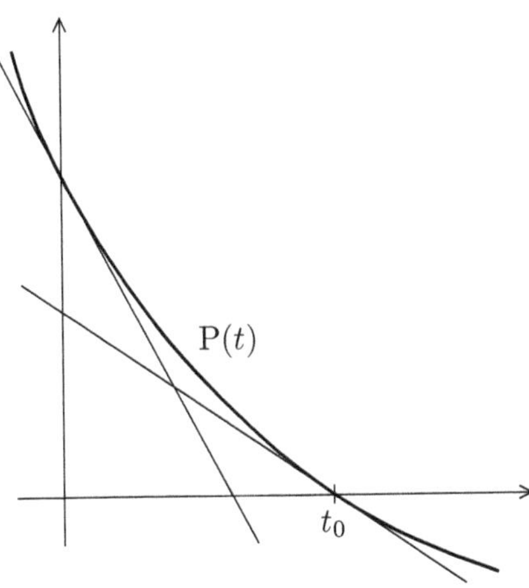

Figure 9.1 Geometric pressure function.

Proof. We put in (9.1.1) $\phi = -t_0 \log|f'|$. Then using (9.1.2), (9.1.4) and $\sup|f'| \le L$ to replace $|(f^n)'(x)|^{-1}$ by r, we obtain

$$\frac{\log E + t_0 \log r}{-\log E + \log r} \le \frac{\log m(B(x,r))}{\log r} \le \frac{-\log E + t_0 \log r}{\log E + \log r}$$

with a corrected constant E. Hence

$$\frac{\log E + t_0 \log r}{\log r} \le \frac{\log m(B(x,r))}{\log r} \le \frac{-\log E + t_0 \log r}{\log r} \tag{9.1.8}$$

for further corrected E. In consequence

$$t_0 \le \frac{\log\bigl(m(B(x,r))/E\bigr)}{\log r} \quad \text{and} \quad \frac{\log\bigl(Em(B(x,r))\bigr)}{\log r} \le t_0,$$

and hence

$$m(B(x,r))/E \le r^{t_0} \text{ and } Em(B(x,r)) \ge r^{t_0}.$$

(In the denominators we passed in the proof of Lemma 9.1.1 from r to $|(f^n)'(x)|^{-1}$ and here we passed back, so at this point the proof could be shortened. That is, we could deduce (9.1.8) directly from (9.1.6). However, we also needed to pass from $|(f^n)'(x)|^{-1}$ to r in numerators, and this point could not be simplified.) ♣

As an immediate consequence of this theorem and Theorem 8.6.7 we get the following.

Corollary 9.1.7. *The Hausdorff dimension of X is equal to t_0. Moreover, it is equal to the packing and Minkowski dimensions. All Gibbs states corresponding to the potential $\phi = -t_0 \log|f'|$, as well as t_0-dimensional Hausdorff and packing measures, are mutually equivalent with bounded Radon–Nikodym derivatives.*

Remarks and summary

The straight line tangent to the graph of $\mathrm{P}(t)$ at each $t \in \mathbb{R}$ is the graph of the affine function

$$L_t(s) := \mathrm{h}_{\mu_t}(f) + s\chi_{\mu_t}(f),$$

where μ_t is the invariant Gibbs measure for the potential $-t\log|f'|$. Indeed, by the Variational Principle (Theorem 3.4.1) $L_t(t) = \mathrm{P}(t)$ and $L_t(s) \le \mathrm{P}(t)$ for all $s \in \mathbb{R}$: compare Section 3.6. The points where the graph of L_t intersects the domain and range axes are respectively $\mathrm{HD}(\mu_t)$, by (9.1.7) and Theorem 5.6.5, and $\mathrm{h}_{\mu_t}(f)$. The derivative is equal to $-\chi_{\mu_t}(f)$. Corollary 9.1.7 says in particular that $\mathrm{HD}(\mu_{t_0}) = \mathrm{HD}(X)$.

For example, in Figure 9.1 the tangent through the point of intersection of the graph of $\mathrm{P}(t)$ with the range axis intersects the domain axis at the Hausdorff dimension of the measure with maximal entropy.

As in Theorem 9.1.6, we can prove that for every $x \in X$ and $t \in \mathbb{R}$ we have for all r small

$$\mu_t(B(x,r)) \sim r^t \varepsilon^{-\mathrm{P}(t)},$$

where $\sim$ means that the mutual ratios are bounded. Compare (5.1.1). This justifies the name 'geometric pressure'. This topic will be developed further in the next section on multifractal spectra. In Section 12.5 we shall introduce geometric pressure in the case where a Julia set contains critical points.

More on the Volume Lemma

We end this section with a version of the Volume Lemma for a Borel probability invariant measure on the expanding repeller (X, f). In Chapter 11 we shall prove this without the expanding assumption, assuming only positivity of the Lyapunov exponent (although also assuming ergodicity), and the proof will be difficult. So we first prove a simpler version, which will be needed in the next section. We start with a simple fact following from the Lebesgue Differentiation Theorem: see for example [Lojasiewicz 1988, Theorem 7.1.4], [Mattila 1995], and compare also Exercise 2.6(b). We provide a proof, since it is very much in the spirit of Chapter 8.

Lemma 9.1.8. *Every non-decreasing function $k : I \to \mathbb{R}$ defined on a bounded closed interval $I \subset \mathbb{R}$ is Lipschitz continuous at Lebesgue almost every point in I. In other words, for every $\varepsilon > 0$ there exist $L > 0$ and a set $A \subset I$ such that $|I \setminus A| < \varepsilon$, where $|\cdot|$ is the Lebesgue measure in $\mathbb{R}$, and at each $r \in A$ the function k is Lipschitz continuous with the Lipschitz constant L.*

Proof. Suppose, on the contrary, that

$$B = \{x \in I : \sup\{y \in I : x \neq y, \ \frac{|k(x) - k(y)|}{|x - y|}\} = \infty\}$$

has positive Lebesgue measure. Write $I = [a, b]$. We can assume, by taking a subset, that B is compact and contains neither a nor b. For every $x \in B$ choose $x' \in I, x' \neq x$ such that

$$\frac{|k(x) - k(x')|}{|x - x'|} > 2\frac{k(b) - k(a)}{|B|}. \tag{9.1.9}$$

Replace each pair x, x' by y, y' with $(y, y') \supset [x, x']$, and y, y' so close to x, x' that (9.1.9) still holds for y, y' instead of x, x'. (In the case where x or x' equals a or b we do not make the replacement.) We shall use for y, y' the old notation x, x', assuming $x < x'$.

Now from the family of intervals (x, x') choose a finite family $\mathcal{I}$ covering our compact set B. From this family it is possible to choose a subfamily of intervals whose union still covers B and which consists of two subfamilies $\mathcal{I}^1$ and $\mathcal{I}^2$ of pairwise disjoint intervals. Start with $I_1 = (x_1, x_1') \in \mathcal{I}$ with minimal possible $x = x_1$ and maximal in $\mathcal{I}$ in the sense of inclusion. Having found $I_1 = (x_1, x_1'), \ldots, I_n = (x_n, x_n')$, we choose I_{n+1} as follows. Consider $\mathcal{I}_{n+1} := \{(x, x') \in \mathcal{I} : x \in \bigcup_{i=1,\ldots,n} I_i, x' > \sup_{i=1,\ldots,n} x_i'\}$. If $\mathcal{I}_{n+1}$ is non-empty, we set (x_{n+1}, x_{n+1}') so that $x_{n+1}' = \max\{x' : (x, x') \in \mathcal{I}_{n+1}\}$. If $\mathcal{I}_{n+1} = \emptyset$, we set (x_{n+1}, x_{n+1}') so that x_{n+1} is minimal possible to the right of $\max\{x_i' : i = 1, \ldots, n\}$ or equal to it, and maximal in $\mathcal{I}$.

In this construction the intervals (x_n, x'_n) with even n are pairwise disjoint, since each (x_{n+2}, x'_{n+2}) has not been a member of $\mathcal{I}_{n+1}$. The same is true for odd n's. We define $\mathcal{I}^i$ for $i = 1, 2$ as the family of (x_n, x'_n) for even, respectively odd, n.

In view of the pairwise disjointness of the intervals of the families $\mathcal{I}^1$ and $\mathcal{I}^2$, monotonicity of k and (9.1.9), we get that

$$k(b) - k(a) \geq \sum_{n \in \mathcal{I}^1} k(x'_n) - k(x_n) > 2\frac{k(b) - k(a)}{|B|} \sum_{n \in \mathcal{I}^1} (x'_n - x_n)$$

and the similar inequality for $n \in \mathcal{I}^2$. Hence, taking into account that $\mathcal{I}^1 \cup \mathcal{I}^2$ covers B, we get

$$2(k(b)-k(a)) > 2\frac{k(b) - k(a)}{|B|} \sum_{n \in \mathcal{I}^1 \cup \mathcal{I}^2} (x'_n - x_n) \geq 2\frac{k(b) - k(a)}{|B|} |B| = 2(k(b)-k(a)),$$

which is a contradiction. ♣

Corollary 9.1.9. *For every Borel probability measure ν on a compact metric space (X, ρ) and for every $r > 0$ there exists a finite partition $\mathcal{P} = \{P_t, t = 1, \ldots, M\}$ of X into Borel sets of positive measure ν and with* $\operatorname{diam}(P_t) < r$ *for all t, and there exists $C > 0$ such that for every $a > 0$*

$$\nu(\partial_{\mathcal{P},a}) \leq Ca, \tag{9.1.10}$$

where $\partial_{\mathcal{P},a} := \bigcap_t \Big(\bigcup_{s \neq t} B(P_s, a)\Big)$.

Proof. Let $\{x_1, \ldots, x_N\}$ be a finite $r/4$-net in X. Fix $\varepsilon \in (0, r/4N)$. For each function $t \mapsto k_i(t) := \nu(B(x_i, t))$, $t \in I = [r/4, r/2]$, apply Lemma 9.1.8 and find appropriate L_i and A_i, for all $i = 1, \ldots, N$. Let $L = \max\{L_i, i = 1, \ldots, N\}$ and let $A = \bigcap_{i=1,\ldots,N} A_i$. The set A has positive Lebesgue measure by the choice of ε. So we can choose its point r_0 different from $r/4$ and $r/2$. Therefore, for all $a < a_0 := \min\{r_0 - r/4, r/2 - r_0\}$ and for all $i \in \{1, 2, \ldots, n\}$, we have $\nu(B(x_i, r_0 + a) \setminus B(x_i, r_0 - a)) \leq 2La$. Hence, putting

$$\Delta(a) = \bigcup_i (B(x_i, r_0 + a) \setminus B(x_i, r_0 - a)),$$

we get $\nu(\Delta(a)) \leq 2LNa$. Define $\mathcal{P} = \{\bigcap_{i=1}^N B^{\kappa(i)}(x_i, r_0)\}$ as a family over all functions $\kappa : \{1, \ldots, N\} \to \{+, -\}$, where $B^+(x_i, r_0) := B(x_i, r_0)$ and $B^-(x_i, r_0) := X \setminus B(x_i, r_0)$, except κ yielding sets of measure 0, in particular except empty intersections. After removing from X a set of measure 0, the partition $\mathcal{P}$ covers X. Since $r_0 \geq r/4$, the balls $B(x_i, r_0)$ cover X. Hence, for each non-empty $P_t \in \mathcal{P}$, at least one value of κ is equal to $+$. Hence $\operatorname{diam}(P_t) \leq 2r_0 < r$.

Note now that $\partial_{\mathcal{P},a} \subset \Delta(a)$. Let $x \in \partial_{\mathcal{P},a}$. Since $\mathcal{P}$ covers X there exists t_0 such that $x \in P_{t_0}$, so $x \notin P_t$ for all $t \neq t_0$. However, since $x \in \bigcup_{t \neq t_0} B(P_t, a)$, there exists $t_1 \neq t_0$ such that $\mathrm{dist}(x, P_{t_1}) < a$. Let $B = B(x_i, r_0)$ be such that $P_{t_0} \subset B^+$ and $P_{t_1} \subset B^-$, or *vice versa*. In the case where $x \in P_{t_0} \subset B^+$, by the triangle inequality $\rho(x, x_i) > r_0 - a$, and since $\rho(x, x_i) < r_0$ we get $x \in \Delta(a)$. In the case where $x \in P_{t_0} \subset B^-$ we have $x \in B(x_i, r_0 + a) \setminus B(x_i, r_0) \subset \Delta(a)$.

We conclude that $\nu(\partial_{\mathcal{P},a}) \leq \nu(\Delta(a)) \leq 2LNa$ for all $a < a_0$. For $a \geq a_0$ it suffices to take $C \geq 1/a_0$. So the corollary is proved, with $C = \max\{2LN, 1/a_0\}$. ♣

Remark. If X is embedded, for example, in a compact manifold Y, then we can view ν as a measure on Y; we find a partition $\mathcal{P}$ of Y, and then $\partial_{\mathcal{P},a} = B(\bigcup_{t=1,,,.M} \partial P_t, a)$, provided $M \geq 2$. This justifies the notation $\partial_{\mathcal{P},a}$.

Corollary 9.1.10. *Let ν be a Borel probability measure on a compact metric space (X, ρ), and let $f : X \to X$ be an endomorphism measurable with respect to the Borel σ-algebra on X and preserving measure ν. Then for every $r > 0$ there exists a finite partition $\mathcal{P} = \{P_t, t = 1, \ldots, M\}$ of X into Borel sets of positive measure ν and with* $\mathrm{diam}(P_t) < r$ *such that for every $\delta > 0$ and ν-a.e. $x \in X$ there exists $n_0 = n_0(x)$ such that, for every $n \geq n_0$,*

$$B(f^n(x), \exp(-n\delta)) \subset \mathcal{P}(f^n(x)). \tag{9.1.11}$$

Proof. Let $\mathcal{P}$ be the partition from Corollary 9.1.9. Fix an arbitrary $\delta > 0$. Then, by Corollary 9.1.9,

$$\sum_{n=0}^{\infty} \nu(\partial_{\mathcal{P},\exp(-n\delta)})) \leq \sum_{n=0}^{\infty} C \exp(-n\delta) < \infty.$$

Hence, by the f-invariance of ν, we obtain

$$\sum_{n=0}^{\infty} \nu(f^{-n}(\partial_{\mathcal{P},\exp(-n\delta)})) < \infty.$$

Applying now the Borel–Cantelli Lemma for the family $\{f^{-n}(\partial_{\mathcal{P},\exp(-n\delta)})\}_{n=1}^{\infty}$, we conclude that for ν-a.e $x \in X$ there exists $n_0 = n_0(x)$ such that for every $n \geq n_0$ we have $x \notin f^{-n}(\partial_{\mathcal{P},\exp(-n\delta)})$, so $f^n(x) \notin \partial_{\mathcal{P},\exp(-n\delta)}$. Hence, by the definition of $\partial_{\mathcal{P},\exp(-n\delta)}$, if $f^n(x) \in P_t$ for some $P_t \in \mathcal{P}$, then $f^n(x) \notin \bigcup_{s \neq t} B\big(P_s, \exp(-n\delta)\big)$. Thus

$$B(f^n(x), \exp -n\delta) \subset P.$$

The proof is complete. ♣

Theorem 9.1.11 (Volume Lemma, expanding map and an arbitrary measure case)**.** *Let ν be an f-invariant Borel probability measure on a topologically exact conformal expanding repeller (X, f), where $X \subset \mathbb{R}^d$. Then*

$$\mathrm{HD}_\star(\nu) \leq \frac{\mathrm{h}_\nu(f)}{\chi_\nu(f)} \leq \mathrm{HD}^\star(\nu). \tag{9.1.12}$$

If in addition ν is ergodic, then for ν-a.e. $x \in X$

$$\lim_{r\to 0} \frac{\log \nu(B(x,r))}{\log r} = \frac{\mathrm{h}_\nu(f)}{\chi_\nu(f)} = \mathrm{HD}(\nu). \tag{9.1.13}$$

Proof. Fix the partition $\mathcal{P}$ coming from Corollary 9.1.9 with $r = \min\{\xi, \eta\}$, where $\eta >$ was defined in (4.0.1). Then, as we saw in Chapter 5,

$$\mathcal{P}^{n+1}(x) \subset f_x^{-n}(B(f^n(x),\xi)) \tag{9.1.14}$$

for every $x \in X$ and all $n \geq$. We shall now work to get a sort of opposite inclusion. Consider an arbitrary $\delta > 0$ and x so that (9.1.11) from Corollary 9.1.10 is satisfied for all $n \geq n_0(x)$. For every $0 \leq i \leq n$ define $k(i) = [i\frac{\delta}{\log\lambda} + \frac{\log \xi}{\log \lambda}] + 1$, $\lambda > 1$ being the expanding constant for $f : X \to X$ (see (4.0.1)). Hence $\exp(-i\delta) \geq \xi\lambda^{-k}$, and therefore $f_{f^i(x)}^{-k}(B(f^{i+k}(x),\xi)) \subset B(f^i(x), \exp -i\delta)$. So, using (9.1.11) for i in place of n, we get

$$f_x^{-(i+k)}(B(f^{i+k}(x),\xi)) \subset f_x^{-i}(\mathcal{P}(f^i(x))$$

for all $i \geq n_0(x)$. From this estimate for all $n_0 \leq i \leq n$, we conclude that

$$f_x^{-(n+k(n))}(B(f^{n+k(n)}(x),\xi) \subset \mathcal{P}_{n_0}^{n+1}(x).$$

Notice that for ν-a.e. x there is $a > 0$ such that $B(x,a) \subset \mathcal{P}^{n_0}(x)$, by the definition of $\partial_{\mathcal{P}}$. Therefore, for all n large enough,

$$f_x^{-(n+k(n))}(B(f^{n+k(n)}(x),\xi)) \subset \mathcal{P}^n(x). \tag{9.1.15}$$

It follows from (9.1.15) and (9.1.14), with $n + k(n)$ in place of n, that

$$\begin{aligned}\lim_{n\to\infty} -\frac{1}{n}\log\nu(\mathcal{P}^n(x)) &\leq \liminf_{n\to\infty} \frac{-\log\nu(f_x^{-(n+k(n))}(B(f^{n+k(n)}(x),\xi)))}{n} \\ &\leq \limsup_{n\to\infty} \frac{-\log\nu(f_x^{-(n+k(n))}(B(f^{n+k(n)}(x),\xi)))}{n} \\ &\leq \lim_{n\to\infty} -\frac{1}{n}\log\nu(\mathcal{P}^n(x))(\mathcal{P}^{n+k(n)+1})(x).\end{aligned}$$

The limits on the far left and far right-hand sides of these inequalities exist for ν-a.e. x by the Shennon–McMillan–Breiman Theorem (Theorem 2.5.4: see also (2.5.2)), and their ratio is equal to $\lim_{n\to\infty} \frac{n}{n+k(n)} = 1 + \frac{\delta}{\log\lambda}$. Letting $\delta \to 0$, we obtain the existence of the limit and the equality

$$\mathrm{h}_\nu(f,\mathcal{P},x) := \lim_{n\to\infty} -\frac{1}{n}\log\nu(\mathcal{P}^n(x)) = \lim_{n\to\infty} \frac{-\log\nu(f_x^{-n}(B(f^n(x),\xi)))}{n}. \tag{9.1.16}$$

In view of Birkhoff's Ergodic Theorem, the limit

$$\chi_\nu(f,x) := \lim_{n\to\infty} \frac{1}{n}\log|(f^n)'(x)| \tag{9.1.17}$$

exists for ν-a.e. $x \in X$. Dividing the left and right sides of (9.1.16) by the corresponding sides of (9.1.17) and using (9.1.2)–(9.1.5), we get

$$\lim_{r\to 0} \frac{\log \nu(B(x,r))}{\log r} = \frac{\mathrm{h}_\nu(f,\mathcal{P},x)}{\chi_\nu(f,x)}.$$

Since, by the Shennon–McMillan–Breiman Theorem (formula (2.5.3)) and Birkhoff's Ergodic Theorem,

$$\frac{\int \mathrm{h}_\nu(f,\mathcal{P},x)\,d\nu(x)}{\int \chi_\nu(f,x)\,d\nu} = \frac{\mathrm{h}_\nu(f,\mathcal{P})}{\chi_\nu(f)} = \frac{\mathrm{h}_\nu(f)}{\chi_\nu(f)}.$$

The latter equality holds since f is expansive and $\mathrm{diam}(\mathcal{P})$ is less than the expansivness constant of f, which exceeds η.

There thus exists a positive measure set where $\frac{\mathrm{h}_\nu(f,\mathcal{P},x)}{\chi_\nu(f,x)} \le \frac{\mathrm{h}_\nu(f)}{\chi_\nu(f)}$, and a positive measure set where the opposite inequality holds. Therefore

$$\lim_{r\to 0} \frac{\log \nu(B(x,r))}{\log r} \le \frac{\mathrm{h}_\nu(f)}{\chi_\nu(f)},$$

and the opposite inequality also holds on a positive measure set. In view of the definitions of $\mathrm{HD}_\star$ and $\mathrm{HD}^\star$ (Definition 8.4.11), and by Theorem 8.6.5, this finishes the proof of the inequalities (9.1.12) in our theorem. In the ergodic case $\mathrm{h}_\nu(f,\mathcal{P},x) = \mathrm{h}_\nu(f)$ and $\chi_\nu(f,x) = \chi_\nu(f)$ for ν-a.e. $x \in X$. So (9.1.13) holds. ♣

9.2 Multifractal analysis of Gibbs state

In the previous section we linked to a (Gibbs) measure only one dimension number, $\mathrm{HD}(m)$. Here one of our aims is to introduce one-parameter families of dimensions, so-called spectra of dimensions. In these definitions we do not need the mapping f. Let ν be a Borel probability measure on a metric space X. Recall from Section 8.6 that, given $x \in X$, we defined the lower and upper pointwise dimension of ν at x by putting respectively

$$\underline{d}_\nu(x) = \liminf_{r\to 0} \frac{\log \nu(B(x,r))}{\log r} \text{ and } \overline{d}_\nu(x) = \limsup_{r\to 0} \frac{\log \nu(B(x,r))}{\log r}.$$

If $\underline{d}_\nu(x) = \overline{d}_\nu(x)$, we call the common value the pointwise dimension of ν at x, and we denote it by $d_\nu(x)$. The function d_ν is called the *pointwise dimension* of the measure ν: compare Chapter 8. For any $\alpha \le 0 \le \infty$ write

$$X_\nu(\alpha) = \{x \in X : d_\nu(x) = \alpha\}.$$

The domain of d_ν, that is, the set $\bigcup_\alpha X_\nu(\alpha)$, is called *a regular part* of X, and its complement $\hat{X}$ a *singular part*. The decomposition of the set X as

$$X = \bigcup_{0\le\alpha\le\infty} X_\nu(\alpha) \cup \hat{X}$$

is called the *multifractal decomposition with respect to the pointwise dimension.*

Define the $F_\nu(\alpha)$-*spectrum for pointwise dimensions* (another name: *dimension spectrum for pointwise dimensions*), a function related to the Hausdorff dimension, by

$$F_\nu(a) = \mathrm{HD}(X_\nu(\alpha)),$$

where we define the domain of F_ν as $\{\alpha : X_\nu(\alpha) \neq \emptyset\}$.

Note that by Theorem 9.1.6, if (X, f) is a topologically exact expanding conformal repeller and $\nu = \mu_{-\mathrm{HD}(X)\log|f'|}$, then all $X_\nu(\alpha)$ are empty except $X_\nu(\mathrm{HD}(X))$. In particular, the domain of F_ν is in this case just one point, $\mathrm{HD}(X)$.

Let for every real $q \neq 1$

$$R_q(\nu) := \frac{1}{q-1} \lim_{r\to 0} \frac{\log \sum_{i=1}^N \nu(B_i)^q}{\log r},$$

where $N = N(r)$ is the total number of boxes B_i of the form $B_i = \{(x_1, \dots, x_d) \in \mathbb{R}^d : rk_j \leq x_j \leq r(k_j+1),\ j = 1, \dots, d\}$ for integers $k_j = k_j(i)$ such that $\nu(B_i) > 0$. This function is called the *Rényi spectrum for dimensions*, provided the limit exists. It is easy to check (Exercise 9.1) that it is equal to the *Hentschel–Procaccia spectrum*

$$HP_q(\nu) := \frac{1}{q-1} \lim_{r\to 0} \frac{\log \inf_{\mathcal{G}_r} \sum_{B(x_i,r)\in \mathcal{G}_r} \nu(B(x_i, r))^q}{\log r},$$

where infimum is taken over all $\mathcal{G}_r$ being finite or countable coverings of the (topological) support of ν by balls of radius r centred at $x_i \in X$, or

$$HP_q(\nu) := \frac{1}{q-1} \lim_{r\to 0} \frac{\log \int_X \nu(B(x,r))^{q-1} d\nu(x)}{\log r},$$

provided the limits exist. For $q = 1$ we define the *information dimension* $I(\nu)$ as follows. Set

$$H_\nu(r) = \inf_{\mathcal{F}_r}\Big(-\sum_{B\in\mathcal{F}_r} \nu(B)\log\nu(B)\Big),$$

where infimum is taken over all partitions $\mathcal{F}_r$ of a set of full measure ν into Borel sets B of diameter at most r. We define

$$I(\nu) = \lim_{r\to 0} \frac{H_\nu(r)}{-\log r},$$

provided the limit exists. A complement to Theorem 8.6.5 is that

$$\mathrm{HD}_\star(\nu) \leq I(\nu) \leq \mathrm{PD}^\star(\nu). \tag{9.2.1}$$

For the proof see Exercise 9.5. Note that for Rényi and HP dimensions it does not make any difference whether we consider coverings of the topological support (the smallest closed set of full measure) of a measure or any set of full measure, since all balls have the same radius r, so we can always choose locally finite (number independent of r) subcovering. These are 'box type' dimension quantities.

A priori there is no reason for the function $F_\nu(\alpha)$ to behave nicely. If ν is an f-invariant ergodic measure for (X, f), a topologically exact conformal expanding repeller, then at least we know that for $\alpha_0 = \mathrm{HD}(\nu)$, we have $d_\nu(x) = \alpha_0$ for ν-a.e. x (by the Volume Lemma: Theorem 9.1.3 and Theorem 9.1.4 for a Gibbs measure ν of a Hölder continuous function and by Theorem 9.1.11 in the general case). So, in particular, we know at least that the domain of $F_\alpha(\nu)$ is non-empty. However, for $\alpha \neq \alpha_0$ we then have $\nu(X_\nu(\alpha)) = 0$, so $X_\nu(\alpha)$ are not visible for the measure ν. Whereas the function $HP_q(\nu)$ can be determined by the statistical properties of a ν-typical (a.e.) trajectory, the function $F_\nu(\alpha)$ seems intractable. However, if $\nu = \mu_\phi$ is an invariant Gibbs measure for a Hölder continuous function (potential) ϕ, then miraculously the above spectra of dimensions happen to be real-analytic functions and $-F_{\mu_\phi}(-p)$ and $HP_q(\mu_\phi)$ are mutual Legendre transforms. Compare this with the pair of Legendre–Fenchel transforms, pressure and entropy (Remark 3.6.3). Thus fix an invariant Gibbs measure μ_ϕ corresponding to a Hölder continuous potential ϕ. We can assume without losing generality that

$$\mathrm{P}(\phi) = 0.$$

Indeed, starting from an arbitrary ϕ, we can achieve this without changing μ_ϕ by subtracting from ϕ its topological pressure (as at the beginning of the proof of Theorem 9.1.3). Having fixed ϕ, in order to simplify the notation we denote $X_{\mu_\phi}(\alpha)$ by X_α and F_{μ_ϕ} by F. We define a two-parameter family of auxiliary functions $\phi_{q,t} : X \to \mathbb{R}$ for $q, t \in \mathbb{R}$, by setting

$$\phi_{q,t} = -t \log |f'| + q\phi.$$

Lemma 9.2.1. *For every $q \in \mathbb{R}$ there exists a unique $t = T(q)$ such that* $\mathrm{P}(\phi_{q,t}) = 0$.

Proof. This lemma follows from the fact that the function $t \mapsto \mathrm{P}(\phi_{q,t})$ is decreasing from ∞ to $-\infty$ for every q (see comments preceding Theorem 9.1.6 and at the beginning of Section 3.6) and the Darboux theorem. ♣

We deal with invariant Gibbs measures $\mu_{\phi_{q,T(q)}}$, which we denote for abbreviation by μ_q, and with the measure μ_ϕ, so we need to know a relation between them. This is explained in the following.

Lemma 9.2.2. *For every $q \in \mathbb{R}$ there exists $C > 0$ such that for all $x \in X$ and $r > 0$*

$$C^{-1} \le \frac{\mu_q(B(x,r))}{r^{T(q)}\mu_\phi(B(x,r))^q} \le C. \tag{9.2.2}$$

Proof. Let $n = n(x,r)$ be defined as in Lemma 9.1.1. Then, by (9.1.6), (9.1.2) and (9.1.4), the ratios

$$\frac{\mu_\phi(B(x,r))}{\exp S_n\phi(x)}, \quad \frac{\mu_q(B(x,r))}{|(f^n)'(x)|^{-T(q)} \exp qS_n\phi(x)}, \quad \frac{r}{|(f^n)'(x)|^{-1}}$$

are bounded from below and above by positive constants independent of x, r. This yields the estimates (9.2.2). ♣

Let us prove the following.

Lemma 9.2.3. *For any f-invariant ergodic probability measure τ on X and for τ-a.e. $x \in X$, we have*

$$d_{\mu_\phi}(x) = \frac{\int \phi d\tau}{-\int \log|f'| d\tau}.$$

Proof. Using formula (9.1.1) in Lemma 9.1.1 and Birkhoff 's Ergodic Theorem, we get

$$d_{\mu_\phi}(x) = \lim_{n\to\infty} \frac{S_n\phi(x)}{\log|(f^n)'(x)|^{-1}} = \frac{\lim_{n\to\infty} \frac{1}{n} S_n\phi(x)}{\lim_{n\to\infty} \frac{1}{n}\log|(f^n)'(x)|^{-1}} = \frac{\int \phi d\tau}{-\int \log|f'| d\tau}.$$

♣

One can conclude from this that the singular part $\hat{X}$ of X has zero measure for every f-invariant τ. Yet the set $\hat{X}$ is usually large: see Exercise 9.4.

On the Legendre transform. Let $k = k(q) : I \to \mathbb{R} \cup \{-\infty, \infty\}$ be a continuous convex function on $I = [\alpha_1(k), \alpha_2(k)]$, where $-\infty \le \alpha_1(k) \le \alpha_2(k) \le \infty$, except for the case where I is only $-\infty$ or only ∞. That is, I is either a point in $\mathbb{R}$, or a closed interval, or a closed semi-line jointly with $-\infty$ or with ∞, or else $\mathbb{R} \cup \{-\infty, \infty\}$. We also assume that k on (α_1, α_2) is finite.

The Legendre transform of k is the function g of a new variable p, defined by

$$g(p) = \sup_{q\in I}\{pq - k(q)\}.$$

Its domain is defined as the closure in $\mathbb{R} \cup \{-\infty, \infty\}$ of the set of points p in $\mathbb{R}$, where $g(p)$ is finite, and g is extended to the boundary by the continuity.

It can be easily proved (Exercise 9.2) that the domain of g is also either a point, or a closed interval, or a semi-line, or $\mathbb{R}$ (with $\pm\infty$). More precisely, the domain is $[\alpha_1(g), \alpha_2(g)]$, where $\alpha_1(g) = -\infty$ if $\alpha_1(k)$ is finite, or $\alpha_1(g) = \lim_{x\to -\infty} k'(x)$ if $\alpha_1(k) = -\infty$. The derivative means here a one-sided derivative, it does not matter whether left or right.

Similarly, one describes $\alpha_2(g)$ replacing $-\infty$ by ∞.

It is also easy to show that g is a continuous convex function (on its domain), and that the Legendre transform is involutive. We then say that the functions k and g form a *Legendre transform pair.*

Proposition 9.2.4. *If two convex functions k and g form a Legendre transform pair, then $g(k'(q)) = qk'(q) - k(q)$, where $k'(q)$ is any number between the left- and right-hand side derivative of k at q, defined as $-\infty, \infty$ at $q = \alpha_1(k)$, $\alpha_2(k)$ respectively, if $\alpha_1(k), \alpha_2(k)$ are finite. The formula also holds at $\alpha_i(g)$ if the arising $0 \cdot \infty$ and $\infty - \infty$ are defined appropriately.*

Note that if k is C^2 with $k'' > 0$, and therefore strictly convex, then also $g'' > 0$ at all points $k'(q)$ for $\alpha_1(k) < q < \alpha_2(k)$: therefore g is strictly convex on $[k'(\alpha_1(k)), k'(\alpha_2(k))]$. Outside this interval g is affine in its domain. If the domain of k is one point then g is affine on $\mathbb{R}$, and *vice versa*.

We are now in a position to formulate our main theorem in this section, gathering in particular some facts already proven.

Theorem 9.2.5.

(a) *The pointwise dimension $d_{\mu_\phi}(x)$ exists for μ_ϕ-almost every $x \in X$ and*

$$d_{\mu_\phi}(x) = \frac{\int \phi \, d\mu_\phi}{-\int \log|f'| d\mu_\phi} = \mathrm{HD}(\mu_\phi) = \mathrm{PD}(\mu_\phi).$$

(b) *The function $q \mapsto T(q)$ for $q \in \mathbb{R}$ is real-analytic, $T(0) = \mathrm{HD}(X)$, $T(1) = 0$,*

$$T'(q) = \frac{\int \phi \, d\mu_q}{\int \log|f'| d\mu_q} < 0$$

and $T''(q) \geq 0$.

(c) *For all $q \in \mathbb{R}$ we have $\mu_q(X_{-T'(q)}) = 1$, where μ_q is the invariant Gibbs measure for the potential $\phi_{q,T(q)}$, and $\mathrm{HD}(\mu_q) = \mathrm{HD}(X_{-T'(q)})$.*

(d) *For every $q \in \mathbb{R}$, $F(-T'(q)) = T(q) - qT'(q)$: that is, $p \mapsto -F(-p)$ is the Legendre transform of $T(q)$. In particular, F is continuous at $-T'(\pm\infty)$ the boundary points of its domain, as the Legendre transform is, and for $\alpha \notin [-T'(\infty), -T'(-\infty)]$ the sets $X_{\mu_\phi}(\alpha)$ are empty, so these α's do not lie in the domain of F (see the definition), as they do not belong to the domain of the Legendre transform.*

(This emptiness property is called completness of the F-spectrum.)

If the measures μ_ϕ and $\mu_{-\mathrm{HD}(X)\log|f'|}$ (the latter discussed in Theorem 9.1.6 and Corollary 9.1.7) do not coincide, then $T'' > 0$ and $F'' < 0$: that is, the functions T and F are respectively strictly convex on $\mathbb{R}$ and strictly concave on $[-T'(\infty), -T'(-\infty)]$, which is a bounded interval in $\mathbb{R}^+ = \{\alpha \in \mathbb{R} : \alpha > 0\}$. If $\mu_\phi = \mu_{-\mathrm{HD}(X)\log|f'|}$ then T is affine, and the domain of F is one point $-T'$.

(e) *For every $q \neq 1$ the HP and Rényi spectra exist (i.e. limits in the definitions exist) and $\frac{T(q)}{1-q} = HP_q(\mu_\phi) = R_q(\mu_\phi)$. For $q = 1$ the information dimension $I(\mu_\phi)$ exists, and*

$$\lim_{q \to 1, q \neq 1} \frac{T(q)}{1-q} = -T'(1) = \mathrm{HD}(\mu_\phi) = \mathrm{PD}(\mu_\phi) = I(\mu_\phi).$$

For outlines of the graphs of T and F see Figures 9.2 and 9.3. See also Exercise 9.3. Compare [Pesin 1997, Figures 17a, 17b].

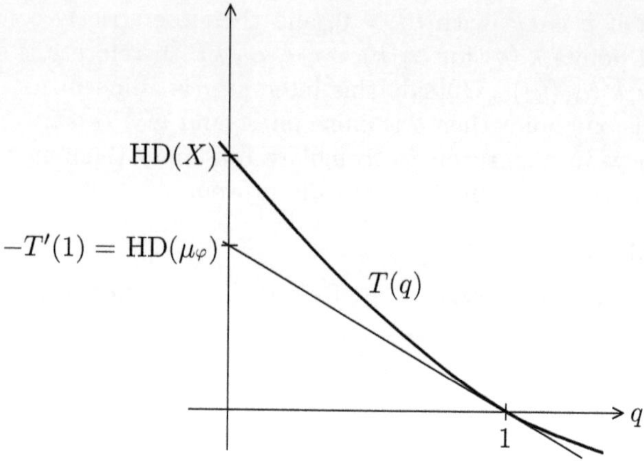

Figure 9.2 Graph of T.

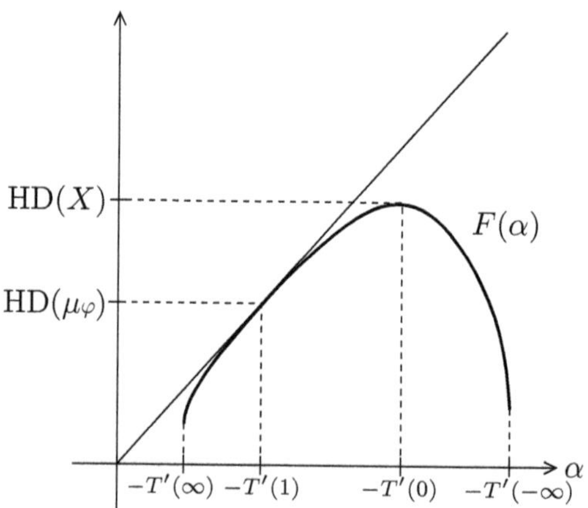

Figure 9.3 Graph of F.

Proof. **1.** Since $\mathrm{P}(\phi) = 0$, part (a) is an immediate consequence of Theorem 9.1.3, and its second and third equalities follow from Theorem 9.1.4. The first equality is also a special case of Lemma 9.2.3 with $\tau = \mu_\phi$.

2. We shall prove some statements of part (b). The function $\phi_{q,t} = -t \log |f'| + q\phi$, from $\mathbb{C}^2$ to $C^\theta(X)$, where θ is a Hölder exponent of the function ϕ, is affine. Since by [Ruelle, 1978a, Corollary 7.10], or our Section 6.4, the topological pressure function $\mathrm{P} : C^\theta \to \mathbb{R}$ is real-analytic, then the composition that we denote $\mathrm{P}(q,t)$ is real-analytic. Hence the real-analyticity of $T(q)$ follows immediately from the Implicit Function Theorem once we verify the non-degeneracy assumption. In fact, C^2-smoothness of $\mathrm{P}(q,t)$ is sufficient to proceed further (here only C^1), which has been proved in Theorem 5.7.4.

Indeed, owing to Theorem 5.6.5, for every $(q_0, t_0) \in \mathbb{R}^2$

$$\frac{\partial \mathrm{P}(q,t))}{\partial t}|_{(q_0,t_0)} = -\int_X \log|f'|d\mu_{q_0,t_0} < 0, \tag{9.2.3}$$

where μ_{q_0,t_0} is the invariant Gibbs state for the function ϕ_{q_0,t_0}. Differentiating the equality $\mathrm{P}(q,t) = 0$ with respect to q, we obtain

$$0 = \frac{\partial \mathrm{P}(q,t)}{\partial t}|_{(q,T(q))} \cdot T'(q) + \frac{\partial \mathrm{P}(q,t))}{\partial q}|_{(q,T(q))}. \tag{9.2.4}$$

Hence we obtain the standard formula

$$T'(q) = -\frac{\partial \mathrm{P}(q,t))}{\partial q}|_{(q,T(q))} \Big/ \frac{\partial \mathrm{P}(q,t)}{\partial t}|_{(q,T(q))}.$$

Again, using (5.6.5) and $\mathrm{P}(\phi_{q,T(q)}) = 0$, we obtain

$$T'(q) = \frac{\int \phi d\mu_q}{\int \log|f'|d\mu_q} \leq \frac{-\mathrm{h}_{\mu_q}(f)}{\int \log|f'|d\mu_q} < 0, \tag{9.2.5}$$

the latter true since the entropy of any invariant Gibbs measure for a Hölder function is positive: see for example Theorem 5.2.12.

The equality $T(0) = \mathrm{HD}(X)$ is just Corollary 9.1.7. $T(1) = 0$ follows from the equality $\mathrm{P}(\phi) = 0$.

3. The inequality $T''(q) \geq 0$ follows from the convexity of $\mathrm{P}(q,t)$: see Theorem 3.6.2. Indeed, the assumption that the part of $\mathbb{R}^3$ above the graph of $\mathrm{P}(q,t)$ is convex implies that its intersection with the plane (q,t) is also convex. Since $\frac{\partial \mathrm{P}(q,t))}{\partial t}|_{(q_0,t_0)} < 0$, this is the part of the plane above the graph of T. Hence T is a convex function.

In the above consideration we avoided an explicit computation of T''. However, to discuss strict convexity (part of (d)) it is necessary to compute it.

Differentiating (9.2.4) with respect to q we obtain the standard formula

$$T''(q) = \frac{T'(q)^2 \frac{\partial^2 \mathrm{P}(q,t)}{\partial t^2} + 2T'(q)\frac{\partial^2 \mathrm{P}(q,t)}{\partial q \partial t} + \frac{\partial^2 \mathrm{P}(q,t)}{\partial q^2}}{-\frac{\partial \mathrm{P}(q,t)}{\partial t}} \tag{9.2.6}$$

with the derivatives of P taken at $(q, T(q))$. The numerator is equal to

$$\left(T'(q)\frac{\partial}{\partial t} + \frac{\partial}{\partial q}\right)^2 \mathrm{P}(q,t) = \sigma^2_{\mu_q}(-T'(q)\log|f'| + \phi)$$

by Theorem 5.7.4, since this is the second-order derivative of $P : C(X) \to \mathbb{R}$ in the direction of the function $-T'(q)\log|f'| + \phi$.

The inequality $\sigma^2 \geq 0$, true by definition, implies $T'' \geq 0$, since the denominator in (9.2.6) is positive by (9.2.3).

By Theorem 2.11.3 $\sigma^2_{\mu_q}(-T'(q)\log|f'|+\phi)=0$ if and only if the function $-T'(q)\log|f'|+\phi$ is co-homologous to a constant, say to a. It follows then from the equality in (9.2.5) that $a=\int a\,d\mu_q=\int(-T'(q)\log|f'|+\phi)d\mu_q=0$. Therefore $T'(q)\log|f'|$ is co-homologous to ϕ and, as $\mathrm{P}(\phi)=0$, also $\mathrm{P}(T'(q)\log|f'|)=0$. Thus, by Theorem 9.1.6 and Corollary 9.1.7, $T'(q)=-\mathrm{HD}(X)$ and consequently ϕ is co-homologous to the function $-\mathrm{HD}(X)\log|f'|$. This implies that $\mu_\phi=\mu_{-\mathrm{HD}(X)\log|f'|}$, the latter being the equilibrium (invariant Gibbs) state of the potential $-\mathrm{HD}(X)\log|f'|$. Therefore, in view of our formula for T'', if $\mu_\phi\neq\mu_{-\mathrm{HD}(X)\log|f'|}$, then $T''(q)>0$ for all $q\in\mathbb{R}$.

4. We prove (c). By Lemma 9.2.3 applied to $\tau=\mu_q$, there exists a set $\tilde{X}_q\subset X$, of full measure μ_q, such that for every $x\in\tilde{X}_q$ there exists

$$d_{\mu_\phi}(x)=\lim_{r\to0}\frac{\log\mu_\phi(B(x,r))}{\log r}=\frac{\int\phi d\mu_q}{-\int\log|f'|d\mu_q}=-T'(q),$$

the latter proved in (b). Hence $\tilde{X}_q\subset X_{-T'(q)}$. Therefore $\mu_q(X_{-T'(q)})=1$.

By Lemma 9.2.2, for every $B=B(x,r)$

$$|\log\mu_q(B)-T(q)\log r-q\log\mu_\phi(B)|<C$$

for some constant $C\in\mathbb{R}$. Hence

$$\left|\frac{\log\mu_q(B)}{\log r}-T(q)-q\frac{\log\mu_\phi(B)}{\log r}\right|\to0 \tag{9.2.7}$$

as $r\to0$.

Using (9.2.7), observe that for every $x\in X_{-T'(q)}$, in particular for every $x\in\tilde{X}_q$,

$$\lim_{r\to0}\frac{\log\mu_q(B)}{\log r}=T(q)+q\lim_{r\to0}\frac{\log\mu_\phi(B)}{\log r}=T(q)-qT'(q).$$

Although $\tilde{X}_q$ can be much smaller than $X_{-T'(q)}$, amazingly their Hausdorff dimensions coincide. Indeed, the measure μ_q restricted to either $\tilde{X}_q$ or to $X_{-T'(q)}$ satisfies the assumptions of Theorem 8.6.3 with $\theta_1=\theta_2=T(q)-qT'(q)$. Therefore

$$\mathrm{HD}(\tilde{X}_q)=\mathrm{HD}(X_{-T'(q)})=T(q)-qT'(q) \tag{9.2.8}$$

and consequently

$$F(-T'(q))=T(q)-qT'(q).$$

Remarks. (a) If we consider sets larger than $X_\nu(\alpha)$, replacing the pointwise dimension d_ν in the definition by the lower pointwise dimension $\underline{d}_\nu$, we obtain the same Hausdorff dimension spectra, again by Theorem 8.6.3. This means that the $F_\nu(\alpha)$ spectrum is the same: in particular, it is given by the same Legendre transform formula in the case $\nu=\mu_\phi$. There is no singular part.

(b) Note that (9.2.8) means that $\mathrm{HD}(X_{-T'(q)})$ is the value where the straight line tangent to the graph of T at $(q,T(q))$ intersects the range axis.

(c) Note that we used f-invariance of μ_ϕ only in estimating $\mathrm{HD}(X_{-T'(q)})$ from below (we used Birkhoff's Ergodic Theorem). In the estimate from above we used only (5.1.1). In a more general setting it is sufficient that this measure is conformal. See Chapter 12 and [Gelfert, Przytycki & Rams 2009].

In the next steps of the proof the following will be useful.

Claim (Variational Principle for T). For any f-invariant ergodic probability measure τ on X, consider the following linear equation of variables q, t:

$$\int \phi_{q,t} d\tau + \mathrm{h}_\tau(f) = 0.$$

That is,

$$t = t_\tau(q) = \frac{\mathrm{h}_\tau(f)}{\int \log|f'| d\tau} + q \frac{\int \phi d\tau}{\int \log|f'| d\tau}. \tag{9.2.9}$$

Then for every $q \in \mathbb{R}$

$$T(q) = \sup_\tau \{t_\tau(q)\} = t_{\mu_q}(q),$$

where the supremum is taken over all f-invariant ergodic probability measures τ on X.

Proof of the claim. Since $\int \phi_{q,t} d\tau + \mathrm{h}_\tau(f) \leq \mathrm{P}(\phi_{q,t})$, and since $\frac{\partial \mathrm{P}(q,t)}{\partial t} < 0$ (compare the proof of convexity of T), we obtain

$$t_\tau(q) \leq T(q).$$

On the other hand, by (9.2.9), and using $\mathrm{P}(\phi_{q,T(q)} = 0$, we obtain

$$t_{\mu_q}(q) = \frac{\mathrm{h}_{\mu_q}(f) + q\int \phi d\mu_q}{\int \log|f'| d\mu_q} = \frac{T(q)\int \log|f'| d\mu_q}{\int \log|f'| d\mu_q} = T(q).$$

The claim is proved. ♣

5. We continue the proof of Theorem 9.2.5. We shall prove the missing parts of (d). We have already proved that

$$F(-T'(q)) = \mathrm{HD}(X_{-T'(q)}) = \mathrm{HD}(\mu_q) = T(q) - qT'(q).$$

Note that $[-T'(\infty), -T'(-\infty)] \subset \mathbb{R}^+ \cup \{0, \infty\}$ since $T'(q) < 0$ for all q. Note finally that

$$-T'(-\infty) = \lim_{q \to -\infty} \frac{-\int \phi d\mu_q}{\int \log|f'| d\mu_q} \leq \frac{\sup(-\phi)}{\inf \log|f'|} < \infty$$

and

$$-T'(\infty) = \lim_{q \to \infty} \frac{-\int \phi d\mu_q}{\int \log|f'| d\mu_q}.$$

The expressions under lim are positive (see (9.2.5)). It is enough now to prove that they are bounded away from 0 as $q \to \infty$. To this end, choose q_0 such that

$T(q_0) < 0$. By our claim (Variational Principle for T), $t_{\mu_q}(q_0) \le T(q_0)$. Since $t_{\mu_q}(0) \ge 0$, we get

$$-q_0 \frac{\int \phi d\mu_q}{\int \log|f'|d\mu_q} = t_{\mu_q}(0) - t_{\mu_q}(q_0) \ge |T(q_0)|.$$

Hence $\frac{-\int \phi d\mu_q}{\int \log|f'|d\mu_q} \ge |T(q_0)|/q_0 > 0$ for all q.

6. To complete the proof of (d) we need to prove the formula for F at $-T'(\pm\infty)$ (in the case where T is not affine) and prove that for $\alpha \notin [-T'(\infty), -T'(-\infty)]$ the sets $X_{\mu_\phi}(\alpha)$ are empty. First note the following.

6a. For any f-invariant ergodic probability measure τ on X, there exists $q \in \mathbb{R} \cup \{\pm\infty\}$ such that

$$\frac{\int \phi d\tau}{\int \log|f'|d\tau} = \frac{\int \phi d\mu_q}{\int \log|f'|d\mu_q} \tag{9.2.10}$$

($\lim_{q\to\pm\infty}$ in the $\pm\infty$ case).

Indeed, by the claim, the graphs of the functions $t_\tau(q)$ and $T(q)$ do not intersect transversally (they can be only tangent), and hence the first graph, which is a straight line, is parallel to a tangent to the graph of T at a point $(q_0, T(q_0))$, or one of its asymptotes, at $-\infty$ or $+\infty$. Now (9.2.10) follows from the same formula (9.2.9) for $\tau = \mu_{q_0}$, since the graph of $t_{\mu_{q_0}}$ is tangent to the graph of T just at $(q_0, T(q_0))$. (Note that the latter sentence proves the formula $T'(q) = \frac{\int \phi d\mu_q}{\int \log|f'|d\mu_q}$ in a different way than in 2., namely via the Variational Principle for T.).

6b. Proof that $X_\alpha = \emptyset$ for $\alpha \notin [-T'(\infty), -T'(-\infty)]$. Suppose there exists $x \in X$ with $\alpha := d_{\mu_\phi}(x) \notin [-T'(\infty), -T'(-\infty)]$. Consider any sequence of integers $n_k \to \infty$ and real numbers b_1, b_2 such that

$$\lim_{k\to\infty} \frac{1}{n_k} S_n\phi(x) = b_1, \quad \lim_{k\to\infty} \frac{1}{n_k}(-\log|(f^n)'(x)|) = b_2$$

and $b_1/b_2 = \alpha$. Let τ be any weak*-limit of the sequence of measures

$$\tau_{n_k} := \frac{1}{n_k} \sum_{j=0}^{n_k-1} \delta_{f^j(x)},$$

where $\delta_{f^j(x)}$ is the Dirac measure supported at $f^j(x)$: compare Remark 3.1.15. Then $\int \phi d\tau = b_1$ and $\int(-\log|f'|)d\tau = b_2$.

Because of the Choquet Theorem (Section 3.1) (or the Decomposition into Ergodic Components Theorem, Theorem 2.8.11) we can assume that τ is ergodic. Indeed, τ is an 'average' of ergodic measures. So among all ergodic measures ν involved in the average, there is ν_1 such that $\frac{\int \phi d\nu_1}{\int -\log|f'|d\nu_1} \le \alpha$ and ν_2 such that $\frac{\int \phi d\nu_2}{\int -\log|f'|d\nu_2} \ge \alpha$. If $\alpha < -T'(\infty)$ we consider ν_1 as an ergodic τ; if $\alpha > -T'(-\infty)$ we consider ν_2. For the ergodic τ found in this way the limit α can be different from that for the original τ, but it will not belong to $[-T'(\infty), -T'(-\infty)]$, and we shall use the same symbol α to denote it. By Birkhoff's Ergodic Theorem applied to the functions ϕ and $\log|f'|$, for τ-a.e. x we have $\lim_{n\to\infty} \frac{S_n(\phi)(x)}{-\log|(f^n)'(x)|} = \alpha$. Hence, applying Lemma 9.2.3, we get

$$\alpha = d_{\mu_\phi}(x) = \frac{\int \phi d\tau}{-\int \log|f'| d\tau}.$$

Finally, note that by (9.2.10) there exists $q \in \mathbb{R}$ such that $\alpha = \frac{\int \phi d\mu_q}{-\int \log|f'| d\mu_q}$, whence $\alpha \in [-T'(\infty), -T'(-\infty)]$. This contradiction completes the proof. ♣

Remark. We have in fact proved that, for all $x \in X$, any limit number of the quotients $\log \mu_\phi(B(x,r))/\log r$ as $r \to 0$ lies in $[-T'(\infty), -T'(-\infty)]$, the fact stronger than $d_{\mu_\phi}(x) \in [-T'(\infty), -T'(-\infty)]$ for all x in the regular part of X.

6c. On $F(-T'(\pm\infty))$. Consider any τ being a weak*-limit of a sub-sequence of μ_q as q tends to, say, ∞. We shall try to proceed with τ in the same way as we did with μ_q, although we shall meet some difficulties. We do not know whether τ is ergodic (and choosing an ergodic one from the ergodic decomposition we may loose the convergence $\mu_q \to \tau$). Nevertheless using the Birkhoff Ergodic Theorem and proceeding as in the proof of Lemma 9.2.3, we get

$$\begin{aligned}\frac{\int \lim_{n\to\infty} \frac{1}{n} S_n\phi(x)\, d\tau(x)}{-\int \lim_{n\to\infty} \frac{1}{n} \log|(f^n)'(x)|\, d\tau(x)} &= \frac{\int \phi\, d\tau}{-\int \log|f'|\, d\tau} = \lim_{q\to\infty} \frac{\int \phi d\mu_q}{-\int \log|f'| d\mu_q} \\ &= \lim_{q\to\infty} (-T'(q)) = -T'(\infty)\end{aligned}$$

with the convergence over a sub-sequence of q's. Since we know already that

$$d_{\mu_\phi}(x) = \frac{\lim_{n\to\infty} \frac{1}{n} S_n\phi(x)}{-\lim_{n\to\infty} \frac{1}{n} \log|(f^n)'(x)|} \geq -T'(\infty),$$

we obtain for every x in a set $\tilde{X}_\tau$ of full measure τ that the limit $d_\tau(x) = -T'(\infty)$. We conclude with $\tilde{X}_\tau \subset X_{-T'(\infty)}$.

Now we use the Volume Lemma for the measure τ. There is no reason for it to be Gibbs, nor ergodic, so we need to refer to the version of the Volume Lemma coming from Theorem 9.1.11. We obtain

$$\begin{aligned}\mathrm{HD}(X_{-T'(\infty)}) \geq \mathrm{HD}^\star(\tau) \geq \frac{\mathrm{h}_\tau(f)}{\int \log|f'|\, d\tau} &\geq \liminf_{q\to\infty} \frac{\mathrm{h}_{\mu_q}(f)}{\chi_{\mu_q}(f)} \\ &= \lim_{q\to\infty} T(q) - qT'(q) = F(-T'(\infty)).\end{aligned}$$

We have used here the upper semi-continuity of the entropy function $\nu \to \mathrm{h}_\nu(f)$ at τ owing to the expanding property (see Theorem 3.5.6).

It is only left to estimate $\mathrm{HD}(X_{-T'(\infty)})$ from above. As for μ_q, we have for every q and $x \in X_{-T'(\infty)}$ (see (9.2.7)) that

$$\lim_{r\to 0}\frac{\log \mu_q(B)}{\log r} = T(q) + q\lim_{r\to 0}\frac{\log \mu_\phi(B)}{\log r} = T(q) - qT'(\infty) \le T(q) - qT'(q).$$

Hence $\mathrm{HD}(X_{-T'(\infty)}) \le T(q) - qT'(q)$. Letting $q \to \infty$, we obtain $\mathrm{HD}(X_{-T'(\infty)}) \le F(-T'(\infty))$.

7. HP and Rényi spectra. Recall that topological supports of μ_ϕ and μ_q are equal to X, since these measures, as Gibbs states for Hölder functions, do not vanish on open subsets of X owing to Proposition 5.2.10. For every $\mathcal{G}_r$ a finite or countable covering X by balls of radius r of multiplicity at most C we have

$$1 \le \sum_{B\in\mathcal{G}_r} \mu_q(B) \le C.$$

Hence, by Lemma 9.2.2,

$$C^{-1} \le r^{T(q)} \sum_{B\in\mathcal{G}_r} \mu_\phi(B)^q \le C$$

with another appropriate constant C. Taking logarithms and, for $q \ne 1$, dividing by $(1-q)\log r$ yields (e) for $q \ne 1$.

8. Information dimension. For $q = 1$ we have $\lim_{q\to 1, q\ne 1} \frac{T(q)}{1-q} = -T'(1)$ by the definition of the derivative. It is equal to $\mathrm{HD}(\mu_\phi) = \mathrm{PD}(\mu_\phi)$ by (a) and (b) and equal to $I(\mu_\phi)$ by Exercise 9.5. ♣

9.3 Fluctuations for Gibbs measures

In Section 9.2, given an invariant Gibbs measure μ_ϕ, we studied a fine structure of X, a stratification into sets of zero measure (except the stratum of typical points), treatable with the help of Gibbs measures of other functions. Here we shall continue the study of typical (μ_ϕ-a.e.) points. We shall replace Birkhoff's Ergodic Theorem by a more refined one: the Law of Iterated Logarithm (Section 2.11), the Volume Lemma in a form more refined than Theorem 9.1.11, and the Frostman Lemma in the form of Theorem 8.6.1.

For any two measures μ, ν on a σ-algebra $(X, \mathcal{F})$, not necessarily finite, we use the notation $\mu \ll \nu$ for μ absolutely continuous with respect to ν, the same as in Section 2.1, and $\mu \perp \nu$ for μ singular with respect to ν: that is, if there exist measurable disjoint sets $X_1, X_2 \subset X$ of full measure, that is, $\mu(X \setminus X_1) = \nu(X \setminus X_2) = 0$, generalizing the notation for finite measures: see Section 2.2. We write $\mu \asymp \nu$ if the measures are equivalent: that is, if $\mu \ll \nu$ and $\nu \ll \mu$.

The symbol $\log_k$ means iteration of the log function k times. As in Chapter 8, Λ_α means the Hausdorff measure with the gauge function α, Λ_κ abbreviates Λ_{t^κ}, and HD means Hausdorff dimension.

Theorem 9.3.1. *Let $f : X \to X$ be a topologically exact conformal expanding repeller. Let $\phi : X \to \mathbb{R}$ be a Hölder continuous function and let μ_ϕ be its invariant Gibbs measure. Denote $\kappa = \mathrm{HD}(\mu_\phi)$.*

Then either

(a) *the following conditions, equivalent to each other, hold:*

 (a1) *ϕ is co-homologous to $-\kappa \log|f'|$ up to an additive constant, i.e. $\phi + \kappa \log|f'|$ is co-homologous to a constant in Hölder functions; then this constant must be equal to the pressure $\mathrm{P}(f,\phi)$ so we can say that $\psi := \phi + \kappa \log|f'| - \mathrm{P}(f,\phi)$ is a coboundary (see Definition 2.11.2 and Remark 4.4.6),*

 (a2) *$\mu_\phi \asymp \Lambda_\kappa$ on X,*

 (a3) *$\kappa = \mathrm{HD}(X)$*

or

(b) *$\psi = \phi + \kappa \log|f'|$ is not co-homologous to a constant, $\mu_\phi \perp \Lambda_\kappa$, and moreover, there exists $c_0 > 0$, ($c_0 = \sqrt{2\sigma^2_{\mu_\phi}(\psi)/\chi_{\mu_\phi}(f)}$), such that with the gauge function $\alpha_c(r) = r^\kappa \exp(c\sqrt{\log 1/r \log_3 1/r})$:*

 (b1) *$\mu_\phi \perp \Lambda_{\alpha_c}$ for all $0 < c < c_0$, and*

 (b2) *$\mu_\phi \ll \Lambda_{\alpha_c}$ for all $c > c_0$.*

Remark. Also, $\mu_\phi \perp \Lambda_{\alpha_{c_0}}$ holds: see Exercise 9.8.

Proof. Note first that by Theorem 9.1.4

$$\int \psi \, d\mu_\phi = \int \phi \, d\mu_\phi + \mathrm{HD}(\mu_\phi)\chi_{\mu_\phi} - \mathrm{P}(\phi) = \int \phi \, d\mu_\phi + \mathrm{h}_{\mu_\phi}(f) - \mathrm{P}(\phi) = 0, \quad (9.3.1)$$

since μ_ϕ is an equilibrium state. If $\phi + \kappa \log|f'|$ is co-homologous to a constant a, then for a Hölder function u we have $\int(\phi + \kappa \log|f'| - a)d\mu_\phi = \int(u \circ f - u)d\mu_\phi = 0$ hence $a = \mathrm{P}(\phi)$. Therefore ψ is indeed a coboundary.

By Proposition 5.1.4 the property (a1) is equivalent to $\mu_\phi = \mu_{-\kappa \log|f'|}$ (the potentials co-homologous up to an additive constant have the same invariant Gibbs measures, and vice versa). Finally, since two co-homologous continuous functions have the same pressure (by definition), $\mathrm{P}(-\kappa \log|f'|) = \mathrm{P}(\phi - \mathrm{P}(\phi)) = 0$, so, by Corollary 9.1.7, $\kappa = \mathrm{HD}(X)$ and $\mu_\phi = \mu_{-\kappa \log|f'|} \asymp \Lambda_\kappa$. We have proved that (a1) implies (a2) and (a3).

(a2) implies that Λ_κ is non-zero finite on X: hence $\mathrm{HD}(X) = \kappa$ i.e. (a3). Finally (a3), i.e. $\kappa = \mathrm{HD}(\mu_\phi) = \mathrm{HD}(X)$ implies $\mathrm{h}_{\mu_\phi}(f) - \kappa\chi_{\mu_\phi}(f) = 0$ by Theorem 9.1.4 and $\mathrm{P}(-\kappa \log|f'|) = 0$ by Corollary 9.1.7. Hence μ_ϕ is an invariant equilibrium state for $-\kappa \log|f'|$. By the uniqueness of equilibrium measure (Chapter 5), $\mu_\phi = \mu_{-\kappa \log|f'|}$: hence (a1). (This implication can be called *uniqueness of the measure-maximizing Hausdorff dimension.*)

Let us now discuss part (b). Suppose that ψ is not co-homologous to a constant. In this case $\sigma^2_{\mu_\phi}(\psi) > 0$: see Theorem 2.11.3. We can assume that $\mathrm{P}(\phi) = 0$, because subtracting the constant $\mathrm{P}(\phi)$ from the original ϕ does not change the Gibbs measure.

Let us invoke (9.1.6) and the conclusion from (9.1.2) and (9.1.4), namely

$$K^{-1}\exp S_n\phi(x) \le \mu(B(x,r)) \le K\exp S_n\phi(x) \tag{9.3.2}$$

and

$$K^{-1}|(f^n)'(x)|^{-1} \le r \le K|(f^n)'(x)|^{-1} \tag{9.3.3}$$

for a constant $K \ge 1$ not depending on x and $r > 0$ and for $n = n(x,r)$ defined by (9.1.2).

Note that for $F(t) := \sqrt{t\log_2 t}$, for every $s_0 \ge 0$ there exists t_0 such that for all $s : -s_0 \le s \le s_0, s \ne 0$ and $t \ge t_0$ we have $|(F(t+s)-F(t))/s| < 1$. This follows from the Lagrange Theorem and $dF/dt \to 0$ as $t \to \infty$, easy to calculate.

Substituting $t = \log|(f^n)'(x)|$ and denoting $\sqrt{\log(|(f^n)'(x)|)\log_3(|(f^n)'(x)|)}$ by $g_n(x)$, we get for $r > 0$ small enough

$$\begin{aligned}\frac{\mu_\phi(B(x,r))}{r^\kappa\exp(c\sqrt{\log 1/r\log_3 1/r})} &\le \frac{K\exp(S_n\phi(x)}{(K|(f^n)'(x)|)^{-\kappa}\exp(cF(\log|(f^n)'(x)|-\log K))}\\ &= \frac{Q\exp(S_n\phi(x)}{|(f^n)'(x)|^{-\kappa}\exp(cg_n(x))}\end{aligned}$$

for $Q := K^{\kappa+1+c}\exp c$, and similarly

$$\frac{\mu_\phi(B(x,r))}{r^\kappa\exp(c\sqrt{\log 1/r\log_3 1/r})} \ge \frac{Q^{-1}\exp(S_n\phi(x)}{|(f^n)'(x)|^{-\kappa}\exp(cg_n(x))}.$$

We rewrite these inequalities in the form

$$\begin{aligned}-\log Q + g_n(x)&\left(\frac{S_n\phi(x)+\kappa\log|(f^n)'(x)|}{g_n(x)} - c\right)\\ &\le \log\left(\frac{\mu_\phi(B(x,r))}{r^\kappa\exp(c\sqrt{\log 1/r\log_3 1/r})}\right)\\ &\le \log Q + g_n(x)\left(\frac{S_n\phi(x)+\kappa\log|(f^n)'(x)|}{g_n(x)} - c\right).\end{aligned} \tag{9.3.4}$$

We have $S_n\phi+\kappa\log|(f^n)'| = S_n\psi$, so we need to evaluate the following upper limit:

$$\limsup_{n\to\infty}\frac{S_n\psi(x)}{\sqrt{\log|(f^n)'(x)|\log_3|(f^n)'(x)|}}.$$

By the Law of Iterated Logarithm (see (2.11.3)) and Theorems 5.7.1 and 2.11.1, for μ_ϕ-a.e. $x \in X$, and writing $\sigma^2 = \sigma^2_{\mu_\phi}$, we have

$$\limsup_{n\to\infty}\frac{S_n\psi(x)}{\sqrt{n\log_2(n)}} = \sqrt{2\sigma^2}. \tag{9.3.5}$$

Applying the Birkhoff Ergodic Theorem to the function $\log|f'|$, for μ_ϕ-a.e. $x \in X$, and writing $c_n = \log|(f^n)'(x)| = S_n \log|f'|(x)$, we obtain

$$\lim_{n\to\infty} \frac{\sqrt{c_n \log_2 c_n}}{\sqrt{n \log_2 n}} = \sqrt{\chi_\mu} \lim_{n\to\infty} \frac{\sqrt{\log_2 c_n}}{\sqrt{\log_2 n}} = \sqrt{\chi_{\mu_\phi}}. \tag{9.3.6}$$

Here $\lim_{n\to\infty} \frac{\sqrt{\log_2 c_n}}{\sqrt{\log_2 n}} = 1$, since

$$\frac{\log_2 c_n}{\log_2 n} - 1 = \frac{\log(\log(c_n)/\log(n))}{\log_2 n} \to 0.$$

This is true, since the numerator is bounded; in fact it tends to 0. Indeed, the assumption that $c_n/n \to \chi_\mu$, and in particular that c_n/n is bounded and bounded away from 0, implies $\log(c_n)/\log(n) \to 1$: hence its logarithm tends to 0.

Combining (9.3.5) with (9.3.6) we obtain for μ_ϕ-a.e. x the following formula:

$$\limsup_{n\to\infty} \frac{S_n\psi(x)}{\sqrt{\log|(f^n)'(x)| \log_3 |(f^n)'(x)|}} = \sqrt{\frac{2\sigma^2}{\chi_{\mu_\phi}}} = c_0. \tag{9.3.7}$$

Therefore, because $g_n \to \infty$ as $n \to \infty$, for $c < c_0$, both the left- and the right-hand-side expressions in (9.3.4) tend to ∞. Hence the middle expression in (9.3.4) also tends to ∞. Analogously for $c > c_0$ these expressions tend to $-\infty$. Applying exp, we get rid of the log and obtain

$$\limsup_{r\to 0} \frac{\mu_\phi(B(x,r))}{r^\kappa \exp(c\sqrt{\log 1/r \log_3 1/r})} = \begin{cases} \infty & \text{if } c < \sqrt{\frac{2\sigma^2}{\chi_{\mu_\phi}}} \\ 0 & \text{if } c > \sqrt{\frac{2\sigma^2}{\chi_{\mu_\phi}}} \end{cases} \tag{9.3.8}$$

Therefore, by Theorem 8.6.1, $\mu_\phi \perp \Lambda_{\alpha_c}$ for all $c < \sqrt{\frac{2\sigma^2}{\chi_{\mu_\phi}}}$ and $\mu_\phi \ll \Lambda_{g_c}$ for all $c > \sqrt{\frac{2\sigma^2}{\chi_{\mu_\phi}}}$. The proof is complete. ♣

Note that this proof is done without the use of Markov partitions, unlike the proof in [Przytycki, Urbański & Zdunik 1989], though it is virtually the same.

The last display, (9.3.8), is known as an LIL Refined Volume Lemma, here in the expanding map, Gibbs measure case: compare Theorem 9.1.3.

Above, (9.3.8) has been obtained from (9.3.7) via (9.3.4). Instead, using (9.3.2) and (9.3.3), one can obtain from (9.3.7) the following, equivalent to (9.3.8).

Lemma 9.3.2 (LIL Refined Volume Lemma). *For μ_ϕ-a.e. x*

$$\limsup_{r\to 0} \frac{\log(\mu_\phi(B(x,r))/r^\kappa)}{\sqrt{\log 1/r \log_3 1/r}} = \sqrt{\frac{2\sigma^2}{\chi_{\mu_\phi}}} = c_0.$$

9.4 Boundary behaviour of the Riemann map

In this and the next section we shall apply the results of Section 9.3 to the conformal expanding repeller (X, f) for X at least a two-points set, being the boundary $\operatorname{Fr}\Omega$ of a connected, simply connected open domain Ω in the Riemann sphere $\overline{\mathbb{C}}$. A model example is Ω, the immediate basin of attraction to an attracting fixed point for a rational mapping, and in particular a basin of attraction to ∞ for a polynomial with $X = J(f)$, the Julia set. We shall assume the expanding property only for technical reasons (and the nature of Chapter 9): for a more general case see [Przytycki 1986a], [Przytycki, Urbański & Zdunik 1989] and [Przytycki, Urbański & Zdunik 1991]. In this section we shall consider a large class of invariant measures. In the next section we shall apply the results to harmonic measure. We shall interpret the results in terms of the radial growth of $|R'(t\zeta)|$ for $R : \mathbb{D} \to \Omega$ a Riemann map, that is, a holomorphic bijection from the unit disc $\mathbb{D}$ to Ω , for a.e. $\zeta \in \partial\mathbb{D}$ and $t \nearrow 1$.

We start with some general useful facts. Let Ω be an arbitrary open connected simply connected domain in $\overline{\mathbb{C}}$. Denote by $R : \mathbb{D} \to \Omega$ a Riemann mapping, as above.

Lemma 9.4.1. *For any sequence $x_n \in \mathbb{D}$, $x_n \to \partial\mathbb{D}$ if and only if $R(x_n) \to \operatorname{Fr}\Omega$.*

Proof. If a subsequence of $R(x_n)$ does not converge to $\operatorname{Fr}\Omega$, then we find its convergent subsequence $R(x_{n_i}) \to y \in \mathbb{D}$. So $x_{n_i} \to R^{-1}(y) \in \mathbb{D}$, which contradicts $x_n \to \partial\mathbb{D}$. The converse implication can be proved similarly. ♣

Now let U be a neighbourhood of $\operatorname{Fr}\Omega$ in $\overline{\mathbb{C}}$, and $f : \Omega \cap U \to \Omega$ be a continuous map, which extends continuously on $\operatorname{cl}(\Omega \cap U)$, mapping $\operatorname{Fr}\Omega$ in $\operatorname{Fr}\Omega$. Define $g : R^{-1}(\Omega \cap U) \to \mathbb{D}$ by $g = R^{-1} \circ f \circ R$.

Lemma 9.4.2. *For any sequence $x_n \in R^{-1}(\Omega \cap U)$, $x_n \to \partial\mathbb{D}$ iff $g(x_n) \to \partial\mathbb{D}$.*

Proof. The implication to the right follows from Lemma 9.4.1 and the continuity of f at $\operatorname{Fr}\Omega$. Conversely, if $g(x_n) \to \partial\mathbb{D}$, then by Lemma 9.4.1 $R(g(x_n)) = f(R(x_n)) \to \operatorname{Fr}\Omega$. Hence $R(x_n) \to \operatorname{Fr}\Omega$; otherwise a sub-sequence of $R(x_n)$ converges to $x \in \Omega$ and $f(x) \in \operatorname{Fr}\Omega$, which contradicts $f(\Omega \cap U) \subset \Omega$. Hence, again by Lemma 9.4.1, $x_n \to \partial\mathbb{D}$. ♣

Proposition 9.4.3 (on desingularization). *Suppose f as above extends holomorphically to U, a neighbourhood of $\operatorname{Fr}\Omega$. Then $g = R^{-1} \circ f \circ R$ on $R^{-1}(\Omega \cap U)$ extends holomorphically to g_1 on a neighbourhood of $\partial\mathbb{D}$, satisfying $I \circ g_1(z) = g_1 \circ I$ for the inversion $I(z) = \bar{z}^{-1}$. This g_1 has no critical points in $\partial\mathbb{D}$.*

Proof. Let $W_1 = \{z : r_1 < |z| < 1\}$ for $r_1 < 1$ large enough that $\operatorname{cl} W_1 \subset R^{-1}(\Omega \cap U)$ and W_1 contains no critical points for g. It is possible, since f has only a finite number of critical points in every compact subset of U, and hence in a neighbourhood of $\operatorname{Fr}\Omega$: hence g has a finite number of critical points in a neighbourhood of $\partial\mathbb{D}$ in $\mathbb{D}$. Let $W_2 = \{z : r_2 < |z| < 1\}$ for $r_2 < 1$ large enough that if $z \in W_2 \cap g(x)$, then $x \in W_1$. Consider V, a component of $g^{-1}(W_2)$. By the above definitions g is a covering map on V. V contains a neighbourhood of $\partial\mathbb{D}$, since by Lemma 9.4.2 V contains points arbitrarily close to $\partial\mathbb{D}$, and if $x_1 \in V$ and $x_2 \in g^{-1}(W_2)$ with x_1, x_2 close enough to $\partial\mathbb{D}$ then an arc δ joining x_1 to x_2 (an interval in polar coordinates) is also close enough to $\partial\mathbb{D}$ that $g(\delta) \subset W_2$. Hence $x_2 \in V$. Let d be the degree of g on V. Then there exists a lift $\tilde{g} : V \to W_3 := \{z : r_2^{1/d} < |z| < 1\}$, that is, a univalent holomorpic mapping such that $(\tilde{g}(x))^d = g$.

The mapping $\tilde{g}$ extends continuously from V to $\tilde{g}_1$ on $V \cup \partial\mathbb{D}$ by Carathéodory's theorem (see for example [Collingwood & Lohwater, 1966, Chapter 3.2]). Formally, this theorem says that a holomorphic bijection between two Jordan domains extends to a homeomorphism between the closures. However, the proof for annuli (W_3 and V) is the same. We use the fact that $\partial\mathbb{D}$ are the corresponding components of the boundaries. (One can also intersect V with small discs B with origins in $\partial\mathbb{D}$: consider $g|_{B\cap V}$ on the topological discs $B \cap V$, and get the continuity of the extensions to $\partial\mathbb{D}$ directly from Carathéodory's theorem.) Finally, define the extension of g to $V \cup \partial\mathbb{D}$ by $g_1(x) = (\tilde{g}_1)^d$. It extends holomorphically to a neighbourhood of $\partial\mathbb{D}$ by the Schwarz reflection principle. $g_1'(z) \neq 0$ for $z \in \partial\mathbb{D}$ since $g_1(V) \subset \mathbb{D}$ and $g_1(V^*) \subset \mathbb{C} \setminus \mathrm{cl}\,\mathbb{D}$, where V^* is the image of V by the inversion I. ♣

Remark 9.4.4. We can consider g_1 on a neighbourhood of $\partial\mathbb{D}$ as stretching the possibly wild set $\mathrm{Fr}\,\Omega$ lifting (a part of) f and extending to g_1 not having critical points. The lemma on desingularization applies to all periodic simply connected Fatou domains for rational mappings, and in particular to Siegel discs and basins of attraction to periodic orbits. In the latter case the following applies.

Proposition 9.4.5. *Let $f : U \to \overline{\mathbb{C}}$ be a holomorphic mapping preserving $\mathrm{Fr}\,\Omega$ and mapping $U \cap \Omega$ in Ω as before. Assume also that*

$$\bigcap_{n\geq 0} f^{-n}(U \cap \mathrm{cl}\,\Omega) = \mathrm{Fr}\,\Omega. \tag{9.4.1}$$

Then the extension g_1 of $g = R^{-1} \circ f \circ R$ provided by Proposition 9.4.3 is expanding on $\partial\mathbb{D}$; moreover, $(\partial\mathbb{D}, g_1)$ is a conformal expanding repeller.

Proof. By (9.4.1), for every $x \in U \cap \Omega$ there exists $n > 0$ such that $f^n(x) \notin U$. Hence, owing to Lemma 9.4.1, there exists $r_1 < 1$ such that $W_1 = \{z : r_1 \leq |z| < 1\} \subset R^{-1}(U \cap \Omega)$, and for every $z \in W_1$ there exists $n > 0$, for which $g^n(z) \notin W_1$.

Next observe that by Lemma 9.4.2 there exists $r_2 : r_1 < r_2 < 1$ such that if $|g(z)| > r_2$ then $|z| > r_1$. Moreover, there exists $r_3 < 1$ such that if $r_3 < |z| < 1$ then for all $n \geq 0$ $|g^{-n}(z)| < r_2$. By $g^{-n}(z)$ we understand here any point in this set. Indeed, suppose there exist sequences $r_n \nearrow 1$ $r_n < |z_n| < 1$ and $m_n > 0$ such that $r_1 \leq |g^{-m_n}(z_n)| \leq r_2$ and $r_2 < |g^{-m}(z_n)|$ for all $0 \leq m < m_n$. Then for z_0 a limit point of the sequence $g^{-m_n}(z_n)$ we have $|g^m(z)| \geq r_1$ for all $m \geq 0$, which contradicts the first paragraph of the proof.

Moreover, for every $0 < r < 1$ there exists $n(r) > 0$ such that if $r_3 < |z| < 1$ and $n \geq n(r)$ then $|g^{-n}(z)| \geq r$. Otherwise a limit point $z_0 = \lim_{m_n \to \infty} g^{-m_n}(z_n)$ for $r_3 < |z_n|$ would satisfy $g^m(z_0) \leq r_2$ for all $m \geq 0$: a contradiction. By the symmetry given by I the same holds for $1 < |z| < r_3^{-1}$. Hence $\bigcap_{n \geq 0} g^{-n}(\{z : r_3 < |z| < r_3^{-1}\}) = \partial\mathbb{D}$, that is, $\partial\mathbb{D}$ is a repeller for g: see Chapter 4 S.1.

Let z_n be a g_1-trajectory in $\partial\mathbb{D}$, $g(z_n) = z_{n-1}, n = 0, -1, \ldots$. Then for all $n \geq 0$ there exist univalent branches g_1^{-n} on $B(z_0, r_3)$ mapping z_0 to z_{-n} and such that $g_1^{-n}(B(z_0, r_3)) \subset \{z : r_2 < |z| < r_2^{-1}\}$. Moreover, $g_1^{-n}(B(z_0, r_3)) \to \partial\mathbb{D}$. With z_0 fixed consider all branches $G_{z_0,\nu,n}$ of g_1^{-n} on $B(z_0, r_3)$ indexed by ν and n. This family is normal, and the limit functions have values in $\partial\mathbb{D}$. Since $\partial\mathbb{D}$ has an empty interior, all the limit functions are constant. Hence there exists $n(z_0)$ such that for all $n \geq n(z_0)$ and $G_{\nu,n}$ for all $z \in B(z_0, r_3/2)$ we have $|G'_{z_0\nu,n}(z)| < 1$. If we take a finite family of points z_0 such that the discs $B(z_0, r_3/2)$ cover $\partial\mathbb{D}$, then for all $G_{z_0,\nu,n}$ with $n \geq \max\{n(z_0)\}$ and $z \in B(z_0, r_3/2)$ $|G'_{z_0\nu,n}(z)| < 1$. Hence for all $z \in \partial\mathbb{D}$ and $n \geq \max\{n(z_0)\}$ $|(g^n)'(z)| > 1$, which is the expanding property. ♣

Now we pass to the main topic of this section, the boundary behaviour of R. We shall denote g_1, the extension of g, simply by g.

Definition 9.4.6. We say that for $z \in \partial\mathbb{D}$, $x \to z$ *non-tangentially* if $x \in \mathbb{D}$, x converges to z and there exists $0 < \alpha < \pi/2$ such that for x close enough to z, x belongs to the so-called Stoltz angle:

$$S_\alpha(z) = z \cdot (1 + \{x \in \mathbb{C} \setminus \{0\} : \pi - \alpha \leq \mathrm{Arg}(x) \leq \pi + \alpha\}).$$

We say that $x \to z$ *radially* if $x = tz$ for $t \nearrow 1$. For any ϕ a real or complex-valued function on $\mathbb{D}$ it is said that ϕ has a non-tangential or radial limit at $z \in \partial\mathbb{D}$ if $\phi(x)$ has a limit for $x \to z$ non-tangentially or radially respectively.

Theorem 9.4.7. *Assume that (X, f) is a conformal expanding repeller for $X = \mathrm{Fr}\,\Omega \subset \bar{\mathbb{C}}$ for a domain $\Omega \subset \mathbb{C}$. Let $R : \mathbb{D} \to \Omega$ be a Riemann mapping. Then*

$$\limsup_{|x| \to 1} \frac{\log |R'(x)|}{-\log(1 - |x|)} < 1.$$

(This is better than the generally true non-sharp inequality, following from Remark 6.2.5.) In particular, R extends to a Hölder continuous function on $\mathrm{cl}\,\mathbb{D}$. Denote the extension by the same symbol R. Let g be as before and let its extension (in Proposition 9.4.3) also be denoted by g. Then the equality $f \circ R = R \circ g$ extends to $\mathrm{cl}\,\mathbb{D}$.

If μ is a g-invariant ergodic probability measure on $\partial\mathbb{D}$, then the non-tangential limit

$$\lim_{x \to z} \frac{\log |R'(x)|}{-\log(1 - |x|)}$$

exists for μ-almost every point $z \in \partial\mathbb{D}$, and is constant almost everywhere. Denote this constant by $\chi_\mu(R)$. Then

$$\chi_\mu(R) = 1 - \frac{\chi_{\mu\circ R^{-1}}(f)}{\chi_\mu(g)}, \tag{9.4.2}$$

where the measure $\mu \circ R^{-1} = R_*(\mu)$ is well defined (and Borel) owing to the continuity of R on $\partial\mathbb{D}$.

Proof. Fix $\delta > 0$ such that for every $z \in \partial\mathbb{D}$ there exists a backward branch g_z^{-n} of g^{-n} on $B(g^n(z), \delta)$ mapping $g^n(z)$ to z. Such δ exists since g is expanding, by Proposition 9.4.5: compare the Proof of Proposition 9.4.3.

By the expanding property of g or by the Koebe Distortion Lemma we can assume that the distortion for all g_z^{-n} is bounded on $B(g^n(z), \delta)$ by a constant K.

For $x \in S_\alpha(z)$ and $|x - z| < \delta/2$ denote by $n = n(x, z, \delta)$ the least non-negative integer such that $|g^{n+1}(x) - g^{n+1}(z)| \geq \delta/2$. Such n exists if δ is small enough, again since g is expanding. We get for α, as in Definition 9.4.6,

$$\frac{1 - |g^n(x)|}{|g^n(z) - g^n(x)|} \geq \frac{1}{\pi}(\frac{\pi}{2} - \alpha)K^{-1}. \tag{9.4.3}$$

Otherwise there exists $w \in \partial\mathbb{D}$ such that $|w - g^n(x)| < |g^n(z) - g^n(x)|\alpha K^{-1} < \delta/2$. Then $w \in B(g^n(z), \delta)$ so w is in the domain of g_z^{-n} and we obtain $1 - |x| \leq |g_z^{-n}(w) - x| < |z - x|\alpha$, a contradiction. We used here the fact that $g_z^{-n}(g^n(x)) = x$, true since $|g^j(z) - g^j(x)| < \delta/2$ for all $j = 0, 1, \ldots, n$ and g is expanding (two different pre-images of a point are far from one another).

From the above bound of distortion it also follows that

$$K^{-1} \leq \frac{|(g^n)'(x)|}{|(g^n)'(z)|} \leq K, \tag{9.4.4}$$

and, writing $||g'|| = \sup_{1-\delta/2\leq|x|\leq 1} |g'(x)|$,

$$(K||g'||)^{-1}\delta/2 \leq |z - x| \cdot |(g^n)'(x)| \leq K\delta/2. \tag{9.4.5}$$

By $f^n \circ R = R \circ g^n$ we have $R'(x) = ((f^n)'(R(x)))^{-1}R'(g^n(x))(g^n)'(x)$. Because of (9.4.3), $1 - |g^n(x)| \geq \delta/2\,||g'||^{-1}\alpha K^{-1}$: hence there exists a constant $C > 0$ such that, for all $z \in \partial\mathbb{D}$ and x, n as above,

$$C^{-1} \leq |R'(g^n(x))| \leq C.$$

We conclude with

$$|R'(x)| \leq \lambda_f^{-n} C||g'||^n,$$

where λ_f is the expanding constant for f. Hence, with the use of (9.4.5) to the denominator, we obtain

$$\limsup_{|x|\to 1} \frac{\log|R'(x)|}{-\log 1 - |x|} \leq \limsup_{x\to z} \frac{-n\log\lambda_f + \log|(g^n)'(z)|}{\log|(g^n)'(z)|} \leq 1 - \frac{\log\lambda_f}{||g'||} < 1.$$

If we consider $x_1, x_2 \in \mathbb{D}$ close to each other and also close to $\partial\mathbb{D}$, we find $y \in \mathbb{D}$ and $z_1, z_2 \in \partial\mathbb{D}$ such that $|y| \le \min\{|x_1|, |x_2|\}$, $|x_i - y| \le 2|x_1 - x_2|$ for $i = 1, 2$ and the intervals joining x_i to y are in the Stoltz angles $S_{\pi/4}(z_i)$. By integration of $|R'|$ along these intervals one obtains Hölder continuity of R on $\mathbb{D}$ with an arbitrary exponent smaller than $a := 1 - \frac{\log \lambda_f}{\|g'\|}$ (a more careful consideration yields the exponent a) and a definite Hölder norm, thus Hölder extending to $\operatorname{cl}\mathbb{D}$.

Now we pass to $\chi_\mu(R)$. Since the Riemann map extends to $R : \operatorname{cl} D^1 \to \operatorname{Fr}\Omega$ uniformly continuous, $R(g^n(x))$ lies close to $R(g^n(z))$. Let $f^{-n}_{R(z)}$ be a holomorphic inverse branch of f^n defined on some small neighbourhood of $R(g^n(z))$, containing $R(g^n(x))$ and sending $R(g^n(z)) = f^n(R(z))$ to $R(z)$. Then $f^{-n}_{R(z)}(R(g^n(x)) = x$, and applying Koebe's Distortion Theorem or bounded distortion for iterates we obtain

$$\hat{K}^{-1} \le \frac{|(f^n)'(R(x))|}{|(f^n)'(R(z))|} \le \hat{K} \tag{9.4.6}$$

for some constant $\hat{K}$ independent of z, x and n.

By Birkhoff's Ergodic Theorem there exists a Borel set $Y \in \partial\mathbb{D}$ such that $\mu(Y) = 1$ and

$$\lim_{k\to\infty} \frac{1}{k} \log |(g^k)'(z)| = \chi_\mu(g) \quad \text{and} \quad \lim_{n\to\infty} \frac{1}{k} \log |(f^k)'(R(z))| = \chi_{\mu\circ R^{-1}}(f)$$

for all $z \in Y$.

We conclude that, for all $z \in Y$ and $x \in S_\alpha(z)$,

$$\lim_{x\to z} \frac{-\log |R'(x)|}{\log(1 - |x|)} = \lim_{x\to z} \frac{\log |(f^n)'(R(z))|^{-1} + \log |(g^n)'(x)|}{\log(1 - |x|)} = 1 - \frac{\chi_{\mu\circ R^{-1}}(f)}{\chi_\mu(g)}.$$

♣

9.5 Harmonic measure; 'fractal vs. analytic' dichotomy

We continue to study $\operatorname{Fr}\Omega \subset \mathbb{C}$, the boundary of a simply connected domain $\Omega \subset \bar{\mathbb{C}}$, and the boundary behaviour of a Riemann map $R : \mathbb{D} \to \Omega$, in the presence of a map f as in the previous section, with the use of *harmonic measure*, although most of the theory holds under the weak assumption that f is boundary repelling to the side of Ω, as in Proposition 9.4.5. We call such a domain an RB-domain. We assume in most of this section, for simplicity, a stronger property that f is expanding on $\operatorname{Fr}\Omega$, and sometimes that Ω is a Jordan domain: that is, $\operatorname{Fr}\Omega$ is a Jordan curve.

Harmonic measure $\omega(x, A) = \omega_\Omega(x, A)$, for $x \in \Omega$ and $A \subset \operatorname{Fr}\Omega$ Borel sets, is a harmonic function with respect to x and a Borel probability measure with respect to A, such that for every continuous $\phi : \partial\Omega \to \mathbb{R}$ the function $\tilde{\phi}(x) := \int \phi(z)\, d\omega(x, z)$ is a harmonic extension of ϕ to Ω, continuous on $\operatorname{cl}\Omega$. Its existence is called the solution of the Dirichlet problem. For simply connected Ω with a non-one-point boundary it always exists. If $R(0) = x_0$ then $\omega(x_0, \cdot) = R_*(l)$, where l is the normalized length measure on $\partial\mathbb{D}$. Of course, $R_*(l)$ makes sense if R is continuous on $\operatorname{cl}\mathbb{D}$. However, it also makes sense in general, if we consider the extension of R by the radial limit, which exists l-a.e. by the Fatou Theorem [Pommerenke 1992], [Collingwood & Lohwater 1966].

Since all the Riemann maps differ by compositions with homographies (Möbius maps) preserving the unit circle, all the harmonic measures $\omega(x, \cdot)$ for $x \in \Omega$ are equivalent, and the corresponding Radon–Nikodym derivatives are bounded away from zero and infinity. If we are interested only in this equivalence class we write ω without specifying the point x, and call it a harmonic measure, or a harmonic measure equivalence class on $\operatorname{Fr}\Omega$ viewed from Ω.

Harmonic measure $\omega(x, \cdot)$ can be defined as the probability distribution of the first hit of $\operatorname{Fr}\Omega$ by the Brownian motion starting from x. This is a very intuitive and inspiring point of view.

For more information about harmonic measures in $\mathbb{C}$ we refer the reader for example to [Pommerenke 1992] or [Tsuji 1959].

In the presence of f boundary repelling to the side of Ω, the lift g defined in Proposition 9.4.3, extended to $\partial\mathbb{D}$, is expanded by Proposition 9.4.5, and hence by Chapter 5 there exists a g-invariant measure μ equivalent to l, which is a Gibbs measure for the potential $-\log|g'|$ (with real-analytic density, see Chapter 6.2). So the equivalence class ω contains an f-invariant measure – that is, $R_*(\mu)$ – allowing us to apply ergodic theory.

If Ω is a simply connected basin of attraction to ∞ for a polynomial f of degree $d \geq 2$, then $\omega = \omega(\infty, \cdot)$ is a measure of maximal entropy, $\log d$: see [Brolin 1965]. This measure is often called *balanced measure.*

A major theorem is Makarov's Theorem [Makarov 1985] that $\operatorname{HD}(\omega) = 1$. This is a general result, true for any simply connected domain Ω as above, with no dynamics involved. We shall provide here a simple proof in the dynamical context, in the presence of expanding f for Jordan $\operatorname{Fr}\Omega$.

We start with a simple general observation.

Lemma 9.5.1. *If for l-a.e. $z \in \partial\mathbb{D}$ there exists a radial limit $\chi(R)(z) := \lim_{x\to z} \frac{-\log|R'(x)|}{\log(1-|x|)}$, then $\int \chi(R)(z)\, dl = 0$.*

(In fact the assumption of the existence of the limit for l-a.e. z, equal to 0, is always true by Makarov's Theorem.)

Proof. We have

$$\int \chi(R)\, dl = \int \lim_{r\to 1} \frac{\log|R'(rz)|}{-\log(1-r)}\, dl(z)$$
$$= \lim_{r\to 1} \frac{1}{-\log(1-r)} \int \log|R'(rz)|\, dl(z) = 0.$$

We could change the order of integral and limit above, owing to the bounds $-2 \leq \frac{\log|R'(rz)|}{-\log(1-r)} \leq 2$ for all r sufficiently close to 1, following from the Koebe

Distortion Lemma: see Section 6.2. The latter expression is equal to 0, since $\log|R'(rz)|$ is a harmonic function, so the integral is equal to $\log|R'(0)|$, which does not depend on r. ♣

Corollary 9.5.2. *Suppose that f is a holomorphic mapping preserving* $\operatorname{Fr}\Omega$ *repelling to the side of Ω. Then, for μ the g-invariant measure equivalent to the length measure l,*

$$\chi_{R_*(\mu)}(f) = \chi_\mu(g) > 0 \tag{9.5.1}$$

$$\mathrm{h}_{R_*(\mu)}(f) = \mathrm{h}_\mu(g) \tag{9.5.2}$$

and $\mathrm{HD}(\omega) = 1$, *for ω, the harmonic measure on* $\operatorname{Fr}\Omega$ *viewed from Ω.*

Proof. We prove this corollary only in the case where $(\operatorname{Fr}\Omega, f)$ is an expanding conformal repeller and Ω is a Jordan domain. Then, as we have already mentioned in the introduction to this section, $R_*(\mu)$ is a probability f-invariant measure in the class of harmonic measure ω.

In view of Theorem 9.4.7, $\chi(R)$ exists and is constant l-a.e. equal to $\chi_\mu(R)$: hence by Lemma 9.5.1 it is equal to 0. Hence by (9.4.2) we get (9.5.1). The property (8.5.2) is immediate in the Jordan case, since R is a homeomorphism from $\partial\mathbb{D}$ to $\operatorname{Fr}\Omega$ by Carathéodory's theorem, conjugating g with f.

Hence

$$\frac{\mathrm{h}_{R_*(\mu)}(f)}{\chi_{R_*(\mu)}} = \frac{\mathrm{h}_\mu(g)}{\chi_\mu(g)}.$$

Since $\mathrm{HD}(\mu) = 1$, an immediate application of Theorem 9.1.11 for f and g (the Volume Lemma) completes the proof. ♣

From now on we assume that Ω is Jordan and f expanding. Then the f-invariant measure $R_*(\mu)$, the R_*-image of the Gibbs g-invariant measure μ, is itself a Gibbs measure (see below), and we can apply the results of Section 9.3, that is, Theorem 9.3.1.

Theorem 9.5.3. *The harmonic measure class ω on* $\operatorname{Fr}\Omega$ *contains an f-invariant Gibbs measure for the map $f : \operatorname{Fr}\Omega \to \operatorname{Fr}\Omega$ and the Hölder continuous potential $-\log|g'| \circ R^{-1}$. The pressure satisfies $P(f, -\log|g'| \circ R^{-1}) = 0$.*

Proof. Recall that the Jacobian $J_l(g)$ of $g : \partial\mathbb{D} \to \partial\mathbb{D}$ with respect to the length measure l is equal to $|g'|$: hence l is a Gibbs measure for $\phi = -\log|g'|$ containing in its class a g-invariant Gibbs measure μ. The pressure satisfies $P = P(g, \phi) = 0$ by direct checking of the condition (5.1.1), or since $J_l(g) = e^{\phi - P} = |g'|e^{-P}$.

Since R is a topological conjugacy between $g : \partial\mathbb{D} \to \partial\mathbb{D}$ and $f : \operatorname{Fr}\Omega \to \operatorname{Fr}\Omega$, we automatically get the Gibbs property (5.1.1) for the measure $R_*(l)$ in the class of harmonic measure ω, for f and $\phi \circ R^{-1}$. We also get $P(f, \phi \circ R^{-1}) = P(g, \phi) = 0$. We obtain the Gibbs f-invariant measure $R_*(\mu)$ in the class of ω for the potential function $\phi \circ R^{-1}$, which is Hölder since R^{-1} is Hölder.

(Note that in Theorem 9.4.7 we proved that R is Hölder, not knowing a priori that R extends continuously to $\partial\mathbb{D}$. Here we assume that $\operatorname{Fr}\Omega$ is Jordan, so R extends to a homeomorphism by Carathéodory's theorem: hence R^{-1} makes sense. Therefore the proof that R^{-1} is Hölder is straightforward: go from small scale to large scale by f^n, then back on the R^{-1} image by g^{-n}, the appropriate branch, and use bounded distortion for the iterates, Chapter 6.2.) ♣

Theorem 9.5.4. *Let $f : \operatorname{Fr}\Omega \to \operatorname{Fr}\Omega$ be a conformal expanding repeller, where Ω is a Jordan domain. Then either*

(a) $\omega \asymp \Lambda_1$ *on $\partial\Omega$, which is equivalent to the property that the functions* $\log|g'|$ *and* $\log|f' \circ R|$ *are co-homologous, and else equivalent to* $\mathrm{HD}(\partial\Omega) = 1$, *or*

(b) $\omega \perp \Lambda_1$, *which implies the existence of $c_0 > 0$ such that with the gauge function $\alpha_c(t) = t\exp(c\sqrt{\log(1/t)\log_3(1/t)})$,*

$$\omega \perp \Lambda_{\alpha_c} \quad \textit{for all} \quad 0 \le c < c_0$$

and

$$\omega \ll \Lambda_{\alpha_c} \quad \textit{for all } c > c_0.$$

Proof. The property that $\log|g'|$ and $\log|f' \circ R|$ are co-homologous implies that the functions $-\log|g' \circ R^{-1}|$ and $-\log|f'|$ are co-homologous (with respect to the map $f : \partial\Omega \to \partial\Omega$). By Theorem 9.5.3 ω contains in its equivalence class an invariant Gibbs state of the potential $-\log|g' \circ R^{-1}|$. By Corollary 9.5.2 $\kappa = \mathrm{HD}(\omega) = 1$. Since $\mathrm{P}(f, -\log|f'|) = \mathrm{P}(f, -\log|g' \circ R^{-1}|) = \mathrm{P}(g, -\log|g'|) = 0$, it follows from Corollary 9.1.7 that the cohomology is equivalent to $\mathrm{HD}(\operatorname{Fr}\Omega) = 1$, and to the property that ω is equivalent to the one-dimensional Hausdorff measure on $\operatorname{Fr}\Omega$. So part (a) of Theorem 9.5.4 is proved.

Suppose now that $\log|g'|$ and $\log|f' \circ R|$ are not co-homologous. Then $-\log|g' \circ R^{-1}|$ and $-\log|f'|$ are not co-homologous. Let μ be the invariant Gibbs state of $-\log|g' \circ R^{-1}|$ in the class of ω. By $\kappa = 1$ we get part (b) immediately from Theorem 9.3.1(b) for $X = \operatorname{Fr}\Omega$. ♣

Now we shall take a closer look at case (a), of a rectifiable Jordan curve $\operatorname{Fr}\Omega$. In particular we shall conclude that this curve must be real-analytic.

Theorem 9.5.5. *If $f : \operatorname{Fr}\Omega \to \operatorname{Fr}\Omega$ is a conformal expanding repeller, Ω is a Jordan domain and* $\mathrm{HD}(\operatorname{Fr}\Omega) = 1$ *(or any other condition in case (a) in Theorem 9.5.4), then* $\operatorname{Fr}\Omega$ *is a real-analytic curve.*

If additionally we assume that f extends holomorphically onto $\overline{\mathbb{C}}$, that is, f is a rational function, and Ω is completely invariant, namely $f^{-1}(\Omega) = \Omega$, then R is a homography, $\operatorname{Fr}\Omega$ *is a geometric circle, and f is a finite Blaschke product in appropriate holomorphic coordinates on $\overline{\mathbb{C}}$. That is,*

$$f(z) = \theta \prod_{i=1}^{d} \frac{z - a_i}{1 - \bar{a}_i z},$$

with d the degree of f, $|\theta| = 1$ and $|a_i| < 1$.

Finally, if f is a polynomial and Ω is completely invariant, then in appropriate coordinates $f(z) = z^d$.

For a stronger version, where Ω is assumed only to be forward invariant rather than completely invariant, and for a counterexample, see Exercise 9.12.

Proof. Condition (a2) in Theorem 9.5.4 means that $R : \partial\mathbb{D} \to \mathrm{Fr}\,\Omega$ transports the length measure on $\partial\mathbb{D}$ to the measure equivalent to the Hausdorff measure Λ_1 on $\mathrm{Fr}\,\Omega$.

The idea now is to look at $\mathrm{Fr}\,\Omega$ from outside. We denote $D_1 = \mathbb{D}$, $R_1 =: \mathbb{D} \to \Omega$ and denote $S^1 = \partial\mathbb{D}$. Consider a Riemann map $R_2 : D_2 := \{z : |z| > 1\} \to \overline{\mathbb{C}} \setminus \mathrm{cl}\,\Omega = \Omega^*$. By Carathéodory's Theorem, R_2 (analogously to R_1) extends to a homeomorphism from $\mathrm{cl}\,D_2$ to $\mathrm{cl}\,\Omega^*$. Denote the extension to $\mathrm{cl}\,\mathbb{D}_i$ by the same symbols R_i.

The map g_1 extending $R_1^{-1} \circ f \circ R_1$ (see Proposition 9.4.3), as being expanding, is a local homeomorphism on a neighbourhood of S^1 in $\mathrm{cl}\,D_1$. Since Ω is a Jordan domain, R_1 is a homeomorphism between closures, $\mathrm{cl}\,D_1$ to $\mathrm{cl}\,\Omega$. So f is a local homeomorphism, an open neighborhood U of $\mathrm{Fr}\,\Omega$ in $\mathrm{cl}\,\mathrm{Fr}\,\Omega$. In conclusion, since f has no critical points in $\mathrm{Fr}\,\Omega$, there exists an open neighborhood U of $\mathrm{Fr}\,\Omega$ such that f is defined on $U \cap \Omega*$ and maps it into Ω^*.

Indeed, if $\Omega^* \ni z_n \to z \in \mathrm{Fr}\,\Omega$ and $\mathrm{cl}\,\Omega \ni f(z_n) \to f(z)$, then, since f is a local homeomorphism on a neighbourhood of $\mathrm{Fr}\,\Omega$ in $\mathrm{cl}\,\Omega$ (see the paragraph above), there exists $\mathrm{cl}\,\Omega \ni w_n \to z$ such that $f(w_n) \to f(z)$. This contradicts the assumption that f has no critical points in $\mathrm{Fr}\,\Omega$: that is, f is a local homeomorphism in a neighbourhood of $\mathrm{Fr}\,\Omega$ in $\mathbb{C}$.

Therefore, analogously to g_1, we can define $g_2 = R_2^{-1} \circ f \circ R_2$, the lift of f via the Riemann map R_2 on the set D_2 intersected with a sufficiently thin open annulus surrounding S^1, and consider the extensions of R_2 and g_2 to the closure $\mathrm{cl}\,D_2$: see Figure 9.4.

Set

$$h = R_2^{-1} \circ R_1|_{S^1} : S^1 \to S^1.$$

Composing, if necessary, R_2 with a rotation we may assume that $h(1) = 1$. Our first objective is to demonstrate that h is real-analytic.

Let $\mu_i = u_i l$ be g_i-invariant Gibbs measures for potentials $-\log|g_i'|$, that is, g_i-invariant measures equivalent to length measure l, for $i = 1, 2$ respectively. In view of Section 6.2, the densities u_1 and u_2 are both real-analytic.

Now we refer to the F. and M. Riesz Theorem (or Riesz–Privalov: see for example [Pommerenke, 1992, Chapter 6.3]), which says that $\mathrm{Fr}\,\Omega$ a rectifiable Jordan curve implies that the map $R_2 : \partial\mathbb{D} \to \mathrm{Fr}\,\Omega$ transports the length measure

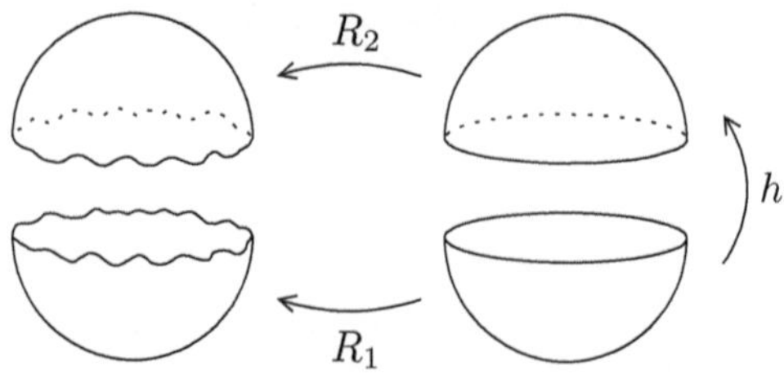

Figure 9.4 Broken egg argument.

on S^1 to the measure equivalent to Λ_1 on $\operatorname{Fr}\Omega$. (Recall that we stated a similar fact on R_1 at the beginning of the proof, which followed directly from the assumptions, without referring to the Riesz theorem.) We conclude that $h_*(\mu_1)$ is equivalent to μ_2. Since h establishes conjugacy between g_1 and g_2, the measure $h_*(\mu_1)$ is g_2-invariant.

Now comes the main point. The measures μ_2 and $h(\mu_1)$ are ergodic, and hence equal, by Theorem 2.2.6: that is,

$$h(\mu_1) = \mu_2.$$

Therefore, writing $a(t) = \frac{1}{2\pi} \log h(e^{2\pi it})$, $a : [0,1] \to [0,1]$, denoting $b_1(t) = \int_0^t u_1(e^{2\pi it})\,dt$ and $b_2(t) = \int_0^t u_2(e^{2\pi it})\,dt$, noting that by $h(1) = 1$ we have $a(0) = 0$, we get for all $t : 0 \le t \le 1$

$$b_1(t) = b_2(a(t)).$$

The functions b_i are real-analytic and invertible, since u_i are positive. Therefore we can write inverse functions and conclude that $a = b_2^{-1} \circ b_1$ is real-analytic. Hence h is real-analytic.

The function h extends to a holomorphic function on a neighbourhood of S^1, and we can replace R_2 by $R_3 = R_2 \circ h$ in a neighbourhood of S^1 in D_2. By definition, R_1 considered on $\operatorname{cl} D_1$ and R_3 outside D_1 coincide on S^1. So by Peinleve's Lemma they glue together to a holomorphic mapping R on a neighbourhood of S^1. So $R(S^1)$ is a real-analytic curve, and the proof of the first part is complete.

Suppose now that f extends to a rational function on $\overline{\mathbb{C}}$, and Ω is completely invariant. Then Ω^* is also completely invariant, and both domains are basins of attraction to sinks, $p_1 \in \Omega$ and $p_2 \in \overline{\mathbb{C}} \setminus \operatorname{cl}\Omega$ respectively (use the Brouwer Theorem and the Schwarz Lemma). Let R_i defined above satisfy $R_1(0) = p_1$ and $R_2(\infty) = p_2$. The maps $g_i = R_i^{-1} \circ f \circ R_i$ preserving D_i and S^1 must be Blaschke products. Let $a_1, \ldots, a_d$ be the zeros of g_1 in D_1 with counted multiplicities. Their number is d, since d is the degree of f, and $R_1(a_i)$ are f-pre-images of p. Denote $B_1(z) = \prod_{i=1}^d \frac{z-a_i}{1-\bar{a}_i z}$. Each factor $\frac{z-a_i}{1-\bar{a}_i z}$ is a homography preserving S^1, so their product also preserves S^1.

For B_1, as above, g_1/B_1 is holomorphic on D_1, has no zeros there, and its continuous extension to S^1 (see Section 9.4) preserves S^1. Hence by the Maximum Principle applied to g_1/B_1 and B_1/g_1 the function g_1/B_1 is a constant λ_1. So $g_1 = \lambda_1 B$ for $|\lambda_1| = 1$. In fact, as one of the zeros of B is 0, as $0 = R_1^{-1}(p_1)$ is a fixed point for g_1, we can write

$$g_1(z) = \lambda_1 z \prod_{i=2}^{d} \frac{z - a_i}{1 - \bar{a}_i z}.$$

Similarly we prove that

$$g_2(z) = \lambda_2 z \prod_{i=2}^{d} \frac{z - a_i'}{1 - \bar{a}_i' z}.$$

for $1/\bar{a}_i'$ the poles of g_2 in D_2.

Note that each Blaschke product B, for which 0 is a fixed point, preserves the length measure l on $\partial\mathbb{D}$. Indeed, let ϕ be an arbitrary real continuous function on $\partial\mathbb{D}$ and $\tilde{\phi}$ its harmonic extension to $\mathbb{D}$. Then

$$\int \phi\, dl = \tilde{\phi}(0) = \tilde{\phi}(B(0)) = \int \phi \circ B\, dl, \tag{9.5.3}$$

since $\tilde{\phi} \circ B$ is harmonic as a composition of a holomorphic mapping with a harmonic function. We conclude that both g_1 and g_2 preserve the length measure l. Hence $h = \mathrm{id}$ and $R_1 = R_2$ on S^1 glue together to a homography R on $\overline{\mathbb{C}}$, g_1 and g_2 extend each other holomorphically to $g := g_1 = g_2$ on $\overline{\mathbb{C}}$, and $f = R \circ g \circ R^{-1}$.

Finally, if f is a polynomial, then ∞ is a pole of multiplicity d: hence $g(z) = z^d$. ♣

Example 9.5.6. In Section 6.1, Example 6.1.10, we provided an example of an expanding repeller, an invariant Jordan curve for $f_c(z) = z^d + c$ for $d = 2$ and $c \approx 0$. Similarly, for any $d \geq 2$ there exists an invariant Jordan curve J_c, being a Julia set for f_c, cutting the Riemann sphere $\bar{\mathbb{C}}$ into two components, Ω and Ω^*, which are basins of attraction to a fixed point p_c near 0 and to the fixed point at ∞. The existence of the expanding repeller J_c follows from Proposition 6.1.7. The rest of the scenario is an easy exercise. We can conclude from Theorem 9.5.5 that $c \neq 0$ implies $\mathrm{HD}(J_c) > 1$.

Now we present another proof of Theorem 9.5.5, avoiding the Riesz Theorem, and so more applicable in other situations: see for example Exercise 9.14.

Proof of Theorem 9.5.5, a second method. It is convenient now to use the half-plane rather than a disc, so we consider a univalent conformal map $R : \{z \in \mathbb{C} : \Im z > 0\} \to \Omega$ extending to a homeomorphism $R : \mathrm{cl}\{\Im z \geq 0\} \cup \infty \to \mathrm{cl}\, \Omega$.

By our assumptions, R is absolutely continuous on the real axis $\mathbb{R}$. Denote the restriction of R to this axis by Ψ. Then $\Psi(x)$ is differentiable a.e., and it is equal to the integral of its derivative: see [Pommerenke, 1992, Chapter 6.3].

Therefore $\Psi' \neq 0$ on a set of positive Lebesgue measure in $\mathbb{R}$. By Egorov's Theorem $\frac{\Psi(x+h)-\Psi(x)}{h} - \Psi'(x) \to 0$ uniformly for $|h| \to 0, h \neq 0$, except for a set of an arbitrarily small measure (for a finite measure equivalent to Lebesgue). To be concrete, there exists $c > 0$ and a sequence of numbers $\varepsilon_n \searrow 0$ such that the following set has positive Lebesgue measure:

$$Q = \{x \in \mathbb{R} : c \leq |\Psi'(x)| \leq 1/c,\ |\Psi(x+h) - \Psi(x) - \Psi'(x)h| \leq |h|/n \,\text{if}\, |h| \leq \varepsilon_n\}.$$

Let μ denote, as before, the probability g-invariant measure on $\mathbb{R}$ equivalent to Lebesgue (remember that we have replaced the unit disc by the upper half-plane, but we use the same notation R, g and μ). We shall prove that R extends to a holomorphic map on a neighbourhood in $\mathbb{C}$ of any point in $\mathbb{R}$ – that is, it is real-analytic on $\mathbb{R}$ – by using the formula

$$R = f^n \circ R \circ g^{-n}.$$

The point is to choose the right backward branches g^{-n} so that R in the centre in the above identity is almost affine. We shall use the natural extension $(\tilde{\mathbb{R}}, \tilde{\mu}, \tilde{g})$ (see Section 2.8), where $\tilde{\mathbb{R}}$ can be understood as the space of g-trajectories and $\pi = \pi_0$ maps $\tilde{\mathbb{R}}$ to $\mathbb{R}$ and is defined by $\pi(x_n, n = \dots, -1, 0, 1, \dots) = x_0$. (We shall use this method more extensively in Chapter 11.)

Since g is ergodic, $\tilde{g}^{-1}$ is ergodic (see Section 2.2 and Exercise 2.14), and in conclusion there exists $x_0 \in \mathbb{R}$ and a sequence G_j of backward branches of g^{-n_j} defined on an interval I with x_0 in the middle, $r := |I|/2$, such that $x_j := G_j(x_0) \in Q$ for all $j = 1, 2, \dots$. Define affine maps

$$A_j(y) := \Psi'(x_j)(y - x_j) + \Psi(x_j) \tag{9.5.4}$$

from $\mathbb{R}$ to $\mathbb{C}$.

First we show that we have uniform convergence on I as $j \to \infty$:

$$\Psi_j := f^{n_j} \circ A_j \circ G_j \to \Psi. \tag{9.5.5}$$

With fixed G_j and $x \in I$, denote $y := g_j(x_0)$ and $y + h := G_j(x)$. If $\lambda_g^{-n_j} \operatorname{diam}_s(I) < \varepsilon_n$ for λ_g the expanding constant for g, where diam_s denotes the diameter in the spherical metric, then by the definition of the set Q, taking into account that $\Psi(y) = A_j(y)$, we obtain

$$\left| \frac{\Psi(y+h) - \Psi(y)}{A_j(y+h) - \Psi(y)} - 1 \right| \le \frac{(1/n_j)|h|}{|\Psi'(y)h|} \le \frac{1/n_j}{c}. \tag{9.5.6}$$

Then by bounded distortion for iterates of f, Lemma 6.2.2, we obtain for a constant $C \ge 1$ depending on f

$$\left| \frac{f^{n_j}\Psi(y+h) - f^{n_j}\Psi(y)}{f^{n_j}A_j(y+h) - f^{n_j}\Psi(y)} - 1 \right| \le e^{Cr}/n_j c. \tag{9.5.7}$$

To use Lemma 6.2.2 we need to check its assumptions (we consider $x = \Psi(y), y_1 = \Psi(y+h), y_2 = A_j(y+h)$ in the notation of Lemma 6.2.2): that is, to check that for all $k = 0, 1, \dots, n_j$

$$|f^k\Psi(y+h) - f^k\Psi(y)| < r \text{ and } |f^k A_j(y+h) - f^k\Psi(y)| < r. \tag{9.5.8}$$

The first estimate follows immediately from the expanding property of f, that is, the estimate $|f^k\Psi(y+h) - f^k\Psi(y)| \le \lambda_f^{n_j - k} \operatorname{diam} \Psi(I) < r$, where λ_f is the expanding constant for f.

The second estimate can be proved by induction, jointly with (9.5.7) for all $f^k, k = 0, 1, \dots, n_j$ in place of f^{n_j} in (9.5.7). For each k_0, having assumed (9.5.8) for all $k \le k_0$, we obtain (9.5.7) for f^{k_0} in place of f^{n_j}, by Lemma 6.2.2. In particular, the bound is by $1 - \lambda_f^{-1}$ if n_j is large enough.

Hence in the fraction, writing $k = k_0$, we get in the denominator $|f^k A_j(y+h) - f^k\Psi(y)| \le \operatorname{diam} \Psi(I)$, since the numerator is bounded by $\lambda_f^{-(n_j - k)} \operatorname{diam} \Psi(I) \le \lambda_f^{-1} \operatorname{diam} \Psi(I)$. Hence $|f^{k+1} A_j(y+h) - f^{k+1}\Psi(y)| < K \operatorname{diam} \Psi(I) = r$.

Note that in the course of induction we verify that the consecutive points $f^k A_j(y+h)$, as being close to $\operatorname{Fr}\Omega$, belong to the domain of f.

Now we calculate, using $f^{n_j} \circ \Psi \circ G_j = \Psi$ for all j and (9.5.7), that

$$\begin{aligned}|\Psi_j(x) - \Psi(x)| &= |\Psi_j(x) - \Psi_j(x_0) - (\Psi(x) - \Psi(x_0))| \\ &= \left| \frac{f^{n_j}\Psi(y+h) - f^{n_j}\Psi(y)}{f^{n_j}A_j(y+h) - \Psi(x_0)} - 1 \right| \cdot |f^{n_j}A_j(y+h) - \Psi(x_0)| \\ &\leq e^{Cr}/n_j c)|f^{n_j}A_j(y+h) - \Psi(x_0)|,\end{aligned}$$

which tends to 0 for $j \to \infty$.

Now consider Ψ_j as defined on a complex neighbourhood of I. To this end consider A_j as affine maps of $\mathbb{C}$, given by the same formula (9.5.4) as before. By (9.5.8), considered for complex x and consequently complex h, the maps Ψ_j are well defined and uniformly bounded. Thus we can apply the Montel Theorem and choose a convergent sub-sequence from Ψ_j. The limit must be a holomorphic extension of Ψ by uniqueness, because it is equal to Ψ on I.

Finally, Ψ extends holomorphically to a neighbourhood of every $z \in \mathbb{R} \cup \infty$, since by the topological exactness of g there exists $x \in I$ and an integer $n \geq 0$ such that $g^n(x) = z$. So, on a neighbourhood of z, we define the extension $\Psi = f^n \circ \Psi \circ g_x^{-n}$, where Ψ in the centre has been already defined in a neighbourhood of I. ♣

Now we shall prove the following corresponding fact on the radial behaviour of Riemann mapping.

Theorem 9.5.7. *Let $f : \operatorname{Fr}\Omega \to \operatorname{Fr}\Omega$ be a conformal expanding repeller with Ω a Jordan domain. Depending on whether $c(\omega) = 0$ or $c(\omega) \neq 0$, either $\partial\Omega$ is real-analytic and the Riemann map $R : D^1 \to \Omega$ and its derivative R' extend holomorphically beyond ∂D^1, or for almost every $z \in \partial D^1$,*

$$\limsup_{r \to 1} |R'(rz)| \exp c\sqrt{\log(1/1-r)\log_3(1/1-r)} = \begin{cases} \infty & \text{if } c \leq c(\omega) \\ 0 & \text{if } c > c(\omega) \end{cases} \tag{9.5.9}$$

and

$$\limsup_{r \to 1} \left(|R'(rz)| \exp c\sqrt{\log(1/1-r)\log_3(1/1-r)}\right)^{-1} = \begin{cases} \infty & \text{if } c \leq c(\omega) \\ 0 & \text{if } c > c(\omega) \end{cases} \tag{9.5.10}$$

Moreover, the radial limsup can be replaced by the non-tangential one.

Proof. Let $n > 0$ be the least integer for which $g^n(rz) \in B(0, r_0)$ for some fixed $r_0 < 1$. We have $R'(rz) = ((f^n)'(R(rz)))^{-1} \cdot R'(g^n(rz)) \cdot (g^n)'(rz)$. Hence, for some constant $K > 0$ independent of r and z,

$$K^{-1} \leq \frac{|R'(rz)|}{|((f^n)'(R(rz)))^{-1}| \cdot |R'(g^n(rz))|} \leq K.$$

By the Bounded Distortion Theorem the rz in the denominator can be replaced by z, and n depends on r as described by (9.1.2) with r replaced by $1-r$. Now we proceed as in the proof of Theorem 9.3.1, replacing deviations of $S_n(\phi) - P(\phi)n + \kappa \log|(f^n)'(x)|$ by the deviations of $\log|(g^n)'(x)| - \log|(f^n)'(x)|$. The proof is complete. ♣

9.6 Pressure versus integral means of the Riemann map

In this section we establish a close relation between the integral means of derivatives of the Riemann map to a domain Ω and the topological pressure of the function $-t\log|f'|$ for a mapping f on the boundary of Ω. This links holomorphic dynamics with analysis, because in the notion of β below f is not involved. Given $t \in \mathbb{R}$, define

$$\beta(t) = \limsup_{r\to 1} \frac{\log \int_{\partial\mathbb{D}} |R'(rz)|^t dl(z)}{-\log(1-r)}, \tag{9.6.1}$$

the integral with respect to the length measure. We shall prove the following.

Theorem 9.6.1. *Assume that* $(\mathrm{Fr}\,\Omega, f)$ *is a conformal expanding repeller (as in Theorem 9.4.7). If the lifted (desingularized) map* $g: \partial\mathbb{D} \to \partial\mathbb{D}$ *is of the form* $z \mapsto z^d$, $d \geq 2$, *then*

$$\beta(t) = t - 1 + \frac{\mathrm{P}(f, -t\log|f'|)}{\log d}. \tag{9.6.2}$$

In particular, in (9.6.1) limsup can be replaced by lim.

Proof. Fix $0 < r < 1$. Fix $n = n(r)$ to be the first integer for which $|g^n(rz)| < r_0$ for $z \in \partial\mathbb{D}$, where $r_0 < 1$ is a constant such that f is defined on a neighbourhood of $\mathrm{cl}\, R(\{r_0 \leq |w| \leq 1\})$. Note that n is independent of z, and that there exists a constant $A \geq 1$ such that $A^{-1} \leq |R'(w)| \leq A$ for all $w \in B(0, r_0)$.

Then, for all $z \in \partial\mathbb{D}$,

$$|R'(rz)|^t = |R'(g^n(rz))|^t \frac{(|g^n)'(rz)|^t}{|(f^n)'(R(rz))|^t}.$$

Divide $\partial\mathbb{D}$ into d^n arcs $I_j, j = 0, \dots, d^n - 1$ with the end points $z_j := e^{(2\pi i)j/d^n}$ and z_{j+1}. Note that $\{z_j := j = 0, \dots, d^n - 1\} = g^{-n}(\{1\})$.

By Hölder continuity of the continuous extension of R to $\mathrm{cl}\,\mathbb{D}$ (see Theorem 9.4.7), $f' \circ R$ is Hölder continuous on $\mathrm{cl}\,\mathbb{D}$. Hence there is a constant $K > 0$ such that the ratio $|(f^n)'(R(w_1))/(f^n)'(R(w_2))|$ is bounded by K for all n all j and $w_1, w_2 \in rI_j$, (see Chapter 4). Hence

$$\int_{I_j} |(f^n)'(R(rz))|^{-t} dl(z) \asymp (2\pi r d^{-n} |(f^n)'(R(rz_j))|^{-t}|,$$

where $\asymp$ means the equality up to a bounded factor.

By our definition of n, $r_0^d \le r^{d^n} \le r_0$: hence $d\log(r_0) \le d^n \log r \le \log(r_0)$. Since there exists a constant $B \ge 1$ such that $B^{-1}(1-r) \le -\log r \le B(1-r)$ for all r sufficiently close to 1, we get $B^{-1}\log 1/r_0 \le d^n(1-r) \le Bd\log 1/r_0$. Therefore $-\log B + \log\log 1/r_0 \le n\log d + \log(1-r) \le \log B + \log\log 1/r_0 + \log d$. Hence $n\log d - C \le -\log(1-r) \le n\log d + C$ for some constant C. Thus, using $|g^n)'(rz)| = d^n|rz|^{d^n-1} \asymp d^n$,

$$\lim_{r\to 1}\frac{\log\int_{\partial\mathbb{D}}|R'(rz)|^t dl(z)}{-\log(1-r)} = \lim_{n\to\infty}\frac{1}{n\log d}\log\Big(\sum_{j=0}^{d^n-1}2\pi r d^{-n}d^{nt}|(f^n)'(R(rz_j))|^{-t}\Big)$$

$$= -1+t+\sum\lim_{n\to\infty}\frac{1}{n\log d}\log\sum_j|f^n)'(R(rz_j))|^{-t}$$

$$= t-1+\frac{\mathrm{P}(g,-t\log|f'|\circ R)}{\log d} = t-1+\frac{\mathrm{P}(f,-t\log|f'|)}{\log d}.$$

Above, to get pressures, we use the equalities

$$|(f^n)'(R(rz_j))|^{-t} = \exp S_n(-\log|f'|\circ R)(rz_j),$$

where $S_n(\phi) = \sum_{k=0}^{n-1}\phi\circ g^k$ with $\phi = -t\log|f'|\circ R$, and apply the definition of pressure $\mathrm{P}_x(T,\phi)$ provided in Proposition 4.4.3. To get $\mathrm{P}(g,-t\log|f'|\circ R)$ we replace n-th pre-images rz_j of the point $g^n(rz_j)$ (not depending on j) by pre-images of $g^n(z_j)=1$, therefore computing $\mathrm{P}_1(g,\phi)$. As ϕ is Hölder continuous we can apply Lemma 4.4.2, so the latter pressure is indeed $\mathrm{P}(g,\phi)$.

Now replace 1 by an arbitrary $r_0 : 0 < r_0 < 1$ close to 1 so that ϕ is defined on its all g^n-pre-images, $n = 0,1,\ldots$. Then $\mathrm{P}_1(g,\phi) = \mathrm{P}_{r_0}(g,\phi)$, the latter defined by the same formula as in Proposition 4.4.3 (though $r_0 \notin \partial\mathbb{D}$, our repeller for g), since R hence ϕ are Hölder continuous.

To get $\mathrm{P}(f,-t\log|f'|)$ we replace pre-images of r_0 by pre-images of $R(r_0)$ using the fact that R is injective on $\mathbb{D}$. We obtain $\mathrm{P}_{r_0}(g,\phi) = \mathrm{P}_{R(r_0)}(f,-t\log|f'|)$, and the latter expression can be replaced by $\mathrm{P}_{R(1)}(f,-t\log|f'|)$. This is equal to $\mathrm{P}(f,-t\log|f'|)$, owing to the Hölder continuity of $-t\log|f'|$. The topological transitivity of f on $\mathrm{Fr}\,\Omega$ assumed in Proposition 4.4.3 used here follows from the topological transitivity of g on $\partial\mathbb{D}$.

Finally, limsup can be replaced by lim in $\beta(t)$, since lim in $\mathrm{P}_x(T,\phi)$ exists in Proposition 4.4.3. The proof is complete. ♣

Remark 9.6.2. The equality (9.2.7) holds even if we do not assume that f is expanding on $\mathrm{Fr}\,\Omega$; it is sufficient to assume boundary repelling to the side of Ω, as in Proposition 9.4.5. To this end we need to define pressure appropriately. The above proof works for $P_x(f,-t\log|f'|)$ for an arbitrary $x\in\Omega$ close to $\mathrm{Fr}\,\Omega$: see also [Binder, Makarov & Smirnov 2003, Lemma 2].

This pressure does not depend on $x\in\Omega$ by the Koebe Distortion Lemma for iteration of branches of f^{-1} in Ω: see Section 6.2. This notion makes sense, and is also independent of x for $x\in\mathrm{Fr}\,\Omega$ for 'most' x: see [Przytycki 1999] for the case where Ω is a basin of infinity for a polynomial. Compare Section 12.5.

Remark 9.6.3. If f is of degree d on Ω simply connected and f expanding on $\mathrm{Fr}\,\Omega$, then

$$F(\alpha) := \inf_{t\in\mathbb{R}}\left(t + \frac{\alpha P(t)}{\log d}\right) \tag{9.6.3}$$

for $P(t) := P(f, -t\log|f'|)$ is the spectrum of dimensions of measure with maximal entropy $F_{\mu_{\max}}(\alpha)$: see the beginning of Section 9.2 and Exercise 9.6.

If f is a polynomial and Ω basin of ∞, then measure with maximal entropy is the harmonic measure ω (from ∞: see [Brolin 1965]): hence (9.6.3) is the formula for the spectrum of dimensions of harmonic measure related to the Hausdorff dimension.

One can ask under what conditions the same formula would hold for a simply connected Ω in the absence of f, where in place of $P(t)/\log d$ one puts $\beta(t)-t+1$: compare (9.6.1).

Remark 9.6.4. The following conjecture is of interest. For $B(t) := \sup\beta(t)$, the supremum being taken over all simply connected domains with a boundary consisting of more than one point, and for $B_{\mathrm{poly}}(t) := \sup_\Omega \beta(t)$ the supremum taken over Ω being simply connected basins of attraction to ∞ for polynomials,

$$B(t) = B_{\mathrm{poly}}(t).$$

It is known that $B_t = B_{\mathrm{snowflake}}(t)$, where $B_{\mathrm{snowflake}}(t)$ is defined as the $\sup\beta(t)$ with supremum taken over Ω being complements of Carleson's snowflakes: see Section 9.7.

Remark 9.6.5. The following is called the Brennan conjecture: $B_{\mathrm{BSC}}(-2) = 1$ (BSC means the supremum over bounded simply connected domains).

This has been verified for Ω simply connected basins of ∞ for quadratic polynomials in [Barański, Volberg, & Zdunik 1998], the variant saying that

$$\iint_{\mathbb{D}} |R'|^{-2+\varepsilon}\,|dz|^2 < \infty.$$

A stronger conjecture is that

$$B(t) = |t|^2/4 \quad \text{for} \quad |t| \le 2 \qquad \text{and} \quad |t|-1 \quad \text{for} \quad |t| \ge 2.$$

9.7 Geometric examples: snowflake and Carleson's domains

This last section of this chapter is devoted to applying the results of preceding sections to geometric examples such as Koch's snowflake and Carleson's example. Following the idea of the proof of Theorems 9.3.1, 9.5.4 and 9.5.5, and coping with additional technicalities (see [Przytycki, Urbański & Zdunik, 1991, Theorem C, Section 6]), one can prove the following.

Theorem 9.7.1. *Let Ω be a simply connected domain in $\mathbb{C}$, with the boundary $\operatorname{Fr}\Omega = \partial\Omega$ being a Jordan curve. Let $\eth_j$, $j = 1, 2\ldots, k$ be a finite family of compact arcs in $\partial\Omega$ with pairwise disjoint interiors. Denote $\bigcup \eth_j$ by $\eth$ (we do not assume that this curve is connected). Assume that there exists a family of conformal maps f_j, $j = 1, \ldots, k$ (which may reverse the orientation on $\mathbb{C}$) on neighbourhoods U_j of $\eth_j$. For every j assume that $f_j(\Omega \cap U_j) \subset \Omega$, $|f_j'| > 1$ on U_j, and*

$$f_j(\partial\Omega \cap U_j) \subset \partial\Omega. \tag{9.7.1}$$

Assume also the Markov partition property: for every $j = 1, \ldots k$, $f_j(\eth_j) = \bigcup_{s \in I_j} \eth_s$ for some subset $I_j \subset \{1, 2, \ldots, k\}$. Consider the $k \times k$ matrix $A = A_{jk}$, where $A_{jk} = 1$ if $k \in I_j$ and $A_{jk} = 0$ if $k \notin I_j$. Assume that A is aperiodic: that is, there exists n such that all the entries of A^n are positive (compare Section 4.3). Then there exists a transition parameter $c(\omega, \eth) \geq 0$ such that, for the harmonic measure ω on $\partial\Omega$ viewed from Ω, restricted to $\eth$, the claims of Theorem 9.5.4 and Theorem 9.5.5 (the analyticity of $\eth$ in the case $c(\omega, \eth) = 0$) hold for $\eth$.

Here (9.7.1) is a crucial assumption, allowing us to prove Theorem 9.7.1. To have it satisfied, one sometimes needs to construct a sophisticated Markov partition of $\partial\Omega$ rather than a natural one: see the snowflake example below (Figure 9.5) and Chapter 1. See also the discussion in [Makarov 1986].

Example 9.7.2 (the snowflake). To every side of an equilateral triangle, in the middle we glue from outside a three times smaller triangle. To every side of the resulting polygon we glue again an equilateral triangle three times smaller, and so on infinitely many times. The triangles do not overlap in this construction and the boundary of the resulting domain Ω is a Jordan curve. This Ω is called Koch's snowflake (Figure 9.5). It was first described by Helge von Koch in 1904.

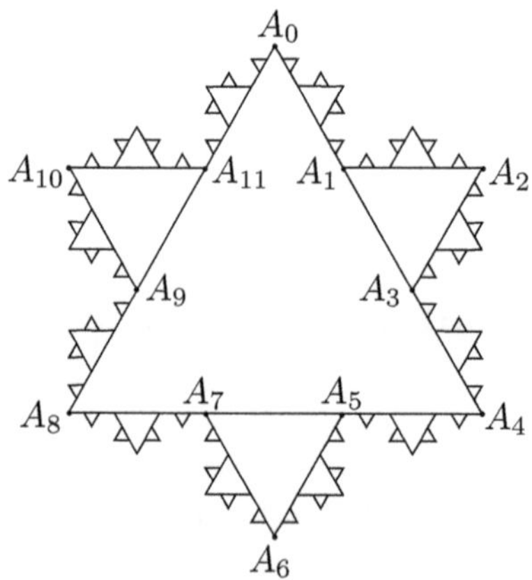

Figure 9.5 Snowflake.

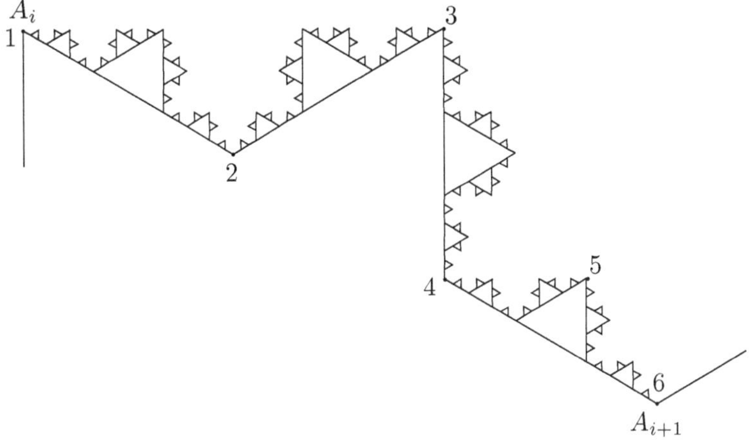

Figure 9.6 A fragment of the snowflake.

Denote the curve in $\partial\Omega$ joining a point $x \in \partial\Omega$ to $y \in \partial\Omega$ in the clockwise direction just by xy. For every $\partial_i := A_iA_{i+1(mod 12)} \subset \partial\Omega$, $i = 0, 1, \ldots, 11$, we consider its covering by the curves 12, 23, 34, 45, 56 in Ω: see Figure 9.6. This covering, together with the affine maps

$$\begin{aligned} &12,\ 34 \to 16 \text{ (preserving orientation on } \partial\Omega) \\ &23 \to 61 \text{ (reversing orientation)} \\ &56 \to 36 \text{ (preserving orientation)} \\ &45 \to 63 \text{ (reversing orientation)} \end{aligned}$$

gives a Markov partition of ∂_i satisfying the assumptions of Theorem 9.7.1.

Since $\partial\Omega$ (and all its subcurves) is definitely not real-analytic ($\mathrm{HD}(\partial\Omega) = \log 4/\log 3$), the assertion of Theorem 9.7.1 is valid with $c(\omega, \partial_i) > 0$. We may denote $c(\omega, \partial_i)$, by $c(\omega)$ since it is independent of ∂_i by symmetry.

Example 9.7.3 (Carleson's domain). We recall Carleson's construction from [Carleson 1985]. We fix a broken line γ with the first and last segment lying in the same straight line in $\mathbb{R}^2$, with no other segments intersecting the segment $\overline{1, d-1}$ (see Figure 9.7).

Then we take a regular polygon Ω^1 with vertices $T_0, T_1, \ldots, T_n$ and glue to every side of it, from outside, the rescaled, not mirror-reflected, curve γ so that the ends of the glued curve coincide with the ends of the side. The resulting curve bounds a second polygon Ω^2. Denote its vertices by $A_0, A_1, \ldots$ (Figure 9.8). Then we again glue the rescaled γ to all sides of Ω^2 and obtain a third-order polygon Ω^3 with vertices $B_0, B_1, \ldots$. Then we build Ω^4 with vertices $C_0, C_1, \ldots \Omega^5$ with $D_0, D_1, \ldots$ etc.

Assume that there is no self-intersecting of the curves $\partial\Omega^n$ in this construction. Moreover, assume that in the limit we obtain a Jordan curve $\mathcal{L} = \mathcal{L}(\Omega^1, \gamma) =$

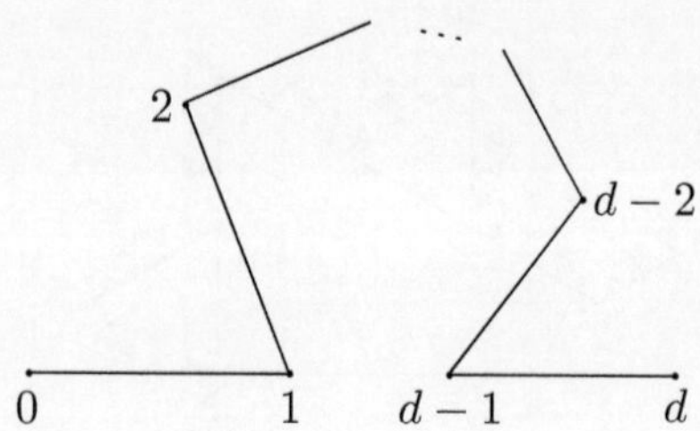

Figure 9.7 Construction of Carleson's domain.

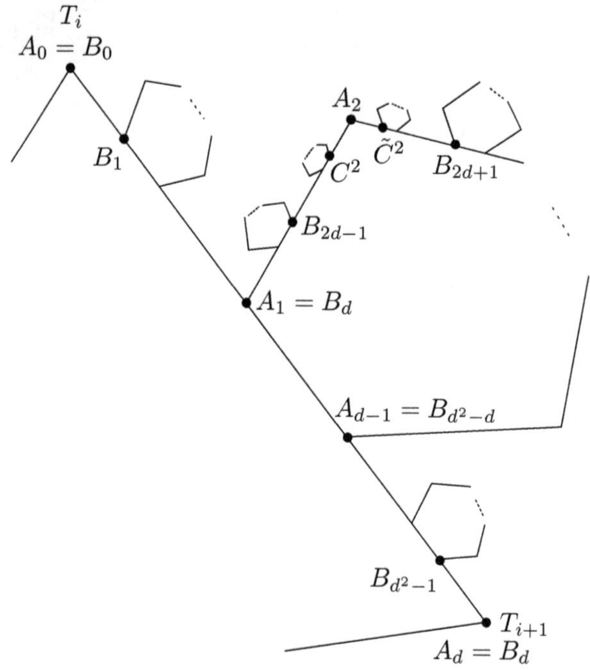

Figure 9.8 Carleson's domain.

$\partial\Omega$. The natural Markov partition of each curve T_iT_{i+1}in $\mathcal{L}$ into curves A_jA_{j+1} with $f(A_jA_{j+1}) = T_iT_{i+1}$, considered by Carleson, does not satisfy the property (9.7.1), so we cannot succeed with it. Instead we proceed as follows. Define in an affine fashion

$$f(B_{d(j-1)+1}B_{dj-1}) = A_1A_{d-1}$$

for every $j = 1, 2 \ldots, d$. Now divide every arc $B_{dj-1}A_j$ for $j = 1, 2 \ldots, d$ and A_jB_{dj+1}, $j = 1, 2 \ldots, d$ into curves with ends in the vertices of the polygon $\Omega^4 : C^j \in B_{dj-1}A_j$, $\tilde{C}^j \in A_jB_{dj+1}$ respectively, the closest to $A_j(\neq A_j)$. Let, for $j = 1, 2, \ldots, d-1$,

$$f(C^jA_j) = B_{dj-1}A_j, \quad f(B_{dj-1}C^j) = A_{d-1}B_{d^2-1},$$
$$f(A_j\tilde{C}^j) = A_jB_{dj+1}, \quad f(\tilde{C}^jB_{dj+1}) = B_1A_1.$$

This gives a Markov partition of $B_1B_{d^2-1}$ with aperiodic transition matrix: see the discussion after Definition 4.3.3 and Theorem 4.5.7. Instead of the broken line γ in the construction of Ω, we can consider the line $\gamma^{(2)}$, consisting of d^2 segments, which arises by glueing to every side of γ a rescaled γ. Consecutive gluing of the rescaled $\gamma^{(2)}$ to the polygon Ω^1 gives consecutively Ω^3, Ω^5 etc. The same construction as above gives a Markov partition of $D_1D_{d^4-1}$ in T_iT_{i+1}. By continuing this procedure we approximate T_iT_{i+1}, so from Theorem 9.7.1 and from the symmetry we deduce that there exists a transition parameter $c(\omega)$ such that the assertion of Theorem 9.5.4(b) is satisfied. Observe that Carleson's assumption that the broken line $1, 2, \ldots, d-1$ does not intersect $\overline{1, d-1}$ has not been needed in these considerations. Also, the assumption that Ω^1 is a regular polygon can be omitted; one can prove that $c(\omega)$ does not depend on T_iT_{i+1} by considering a Markov partition with aperiodic transition matrix, which involves all the sides of Ω^1 simultaneously.

Exercises

Multifractal analysis

9.1. Prove the equalities of Rényi and Hentschel–Procaccia spectra.

9.2. Prove Proposition 9.2.4 about Legendre transform pairs and the remarks preceding and following it.

9.3. Prove for $\alpha = -T'(1)$ that $F(\alpha) = \alpha$ and $F'(\alpha) = 1$ and $F'(-T'(\pm\infty)) = \pm\infty$ (see Figure 9.3).

9.4. Prove that if ϕ is not co-homologous to $-\operatorname{HD}(X)\log|f'|$ then the singular part $\hat{X}$ of X is non-empty. Moreover, $\operatorname{HD}(\hat{X}) = \operatorname{HD}(X)$.

Hint: Using the Shadowing Lemma from Chapter 4, find trajectories that have blocks close to blocks of trajectories typical for $\mu_{-\operatorname{HD}(X)\log|f'|}$ of length N interchanging with blocks close to blocks typical for μ_ϕ of length εN, for N arbitrarily large and $\varepsilon > 0$ arbitrarily small.

9.5. Define the *lower* and *upper information dimension* $\underline{I}(\nu)$ and $\overline{I}(\nu)$, replacing, in the definition of $I(\nu)$, the limit $\lim_r$ by the lower and upper limits respectively. Prove that $\operatorname{HD}_\star(\nu) \le \underline{I}(\nu) \le \overline{I}(\nu) \le \operatorname{PD}^\star(\nu)$: see (9.2.1).

Sketch of the proof. For an arbitrary $\varepsilon > 0$ there exist $C > 0$ and $A \subset X$, with $\nu(X\backslash A) \le \varepsilon$ such that for all r small enough there exists a partition $\mathcal{F}_r$ of A, satisfying $H_\nu(r) + \varepsilon \ge -\sum_{B\in\mathcal{F}_r} \nu(B)\log\nu(B) \ge \sum_{B\in\mathcal{F}_r} \nu(B)\operatorname{HD}_\star(\nu)\log\frac{1}{C\operatorname{diam}B} \ge \operatorname{HD}_\star(\nu)(1-\varepsilon)\log\frac{1}{Cr}$.

On the other hand, for the partition $\mathcal{B}_r$ of X into intersections with boxes (cubes) of sides of length r (compare Proposition 8.4.6 and the partition involved in the definition of Rényi dimension, but consider here disjoint cubes, that is, open from one side), we have

$$\begin{aligned}\overline{I}(\nu) &= \limsup_{r\to 0} \frac{H_\nu(r)}{-\log r} \le \limsup_{r\to 0} \frac{-\sum_{B\in\mathcal{B}_r} \nu(B)\log\nu(B)}{-\log r} \\ &\le \limsup_{r\to 0} \frac{\int \log\nu(B_r(x))\,d\nu(x)}{\log r} \le \int \Big(\limsup_{r\to 0} \frac{\log\nu(B_r(x))}{\log r}\Big) d\nu(x) \le \mathrm{PD}^\star(\nu),\end{aligned}$$

where $B_r(x)$ denotes the cube of side r containing x.

Prove that it has been valid here to use cubes instead of balls in the definition of $\overline{d}_\nu(x)$. To this end prove that for ν-a.e. $x \in X$, we have $\lim \frac{\log\nu(B_r(x))}{\log\nu(B(x,r))} = 1$. Use the Borel–Cantelli Lemma.

Prove that we could use Fatou's Lemma (changing the order of limsup and integral), owing to the existence of a ν-integrable function that bounds from above all the functions $\log\nu(B(x,r))/\log r$ (or $\log\nu(B_r(x))/\log r$). Use the Borel–Cantelli Lemma again for, say, $r = 2^{-k}$.

9.6. Let $\mu = \mu_\phi$ be a measure of maximal entropy on a topologically exact conformal expanding repeller X such that every point $x \in X$ has exactly d pre-images (so $\phi = -\log d$). Prove (deduce from Theorem 9.2.5) that $F(\alpha) = \inf_{t\in\mathbb{R}}\Big(t + \frac{\alpha\,\mathrm{P}(t)}{\log d}\Big)$, or more concretely $F(\alpha) = T + \frac{\alpha\,\mathrm{P}(T)}{\log d}$, where $\alpha = -\frac{\log d}{P'(T)}$.

9.7. Let $\phi_i : X \to \mathbb{R}$ be Hölder continuous functions for $i = 1,\dots,k$ and μ_{ϕ_i} their Gibbs measures. Define $X_{\alpha_1,\dots,\alpha_k} = \{x \in X : d_{\mu_i}(x) = \alpha_i \text{ for all } i = 1,\dots,k\}$. Define $\phi_{q_1,\dots,q_k,t} = -t\log|f'| + \sum_i q_i\phi_i$ and $T(q_1,\dots,q_k)$ as the only zero of the function $t \mapsto \mathrm{P}(\phi_{q_1,\dots,q_k,t})$. Prove the same properties of T as in Theorem 9.2.5, and in particular that

$$\mathrm{HD}(X_{\alpha_1,\dots,\alpha_k}) = \inf_{(q_1,\dots,q_k)\in\mathbb{R}^k} \sum_i q_i\alpha_i + T(q_1,\dots,q_k),$$

wherever the infimum is finite.

Fluctuations for Gibbs measures

9.8. Prove $\mu_\phi \perp \Lambda_{\alpha_{c_0}}$ in case (b) of Theorem 9.3.1.

Hint: Use a function more refined than $\sqrt{2\sigma^2 n\log\log n}$: see the Kolmogorov test after Theorem 2.11.1. Use LIL (upper bound) for $S_n(\log|\phi'| - \chi_{\mu_\phi})$ (the Birkhoff Ergodic Theorem as used above is not suficient). For details see [Przytycki, Urbański & Zdunik 1989].

9.9. Prove a theorem analogous to Theorem 9.3.1, comparing μ_ϕ with packing measures. In particular, prove that, for ψ not a co-boundary, for the gauge function $\alpha_c(r) = r^\kappa \exp(-c\sqrt{\log 1/r \log_3 1/r})$ and $c_0 = \sqrt{2\sigma^2_{\mu_\phi}(\psi)/\chi_{\mu_\phi}(f)}$ it holds that $\mu_\phi \ll \Pi_{\alpha_c}$ for all $0 < c < c_0$, and $\mu_\phi \perp \Pi_{\alpha_c}$ for all $c > c_0$.

Harmonic measure

9.10. Prove (9.5.2) – that is, $\mathrm{h}_{R_*(\mu)}(f) = \mathrm{h}_\mu(g)$ – in the case where f is expanding, but not assuming that Ω is Jordan. To this end prove that R is finite-to-one on $\partial\mathbb{D}$.

9.11. Prove that if Ω is a Jordan domain with boundary preserved by a conformal expanding map f defined on its neighbourhood, and harmonic measures ω_Ω and $\omega_{\overline{\mathbb{C}}\setminus \mathrm{cl}\,\Omega}$ on $\mathrm{Fr}\,\Omega$ (that is, harmonic measures on $\mathrm{Fr}\,\Omega$ viewed from inside and outside) are equivalent, then they are equivalent to the Hausdorff measure Λ_1 (and hence $\mathrm{Fr}\,\Omega$ is real-analytic).

Remark. A part of this theorem holds without assuming the existence of f: see Bishop *et al.* [1989]. It has an important intuitive meaning. Harmonic measure is supported on a set exposed to the side from which it is defined, easily accessible by Brownian motion. These sets in $\mathrm{Fr}\,\Omega$ viewed from inside and outside are very different, except for the case, where $\mathrm{Fr}\,\Omega$ is rectifiable.

9.12. Prove that if $(\mathrm{Fr}\,\Omega, f)$ is an expanding conformal repeller for a rational function f, $\mathrm{Fr}\,\Omega$ is an analytic Jordan curve, and Ω is a basin of attraction to a sink, then $\mathrm{Fr}\,\Omega$ is a geometric circle. (The assumption that Ω is a basin of attraction is weaker than the assumption that Ω is completely invariant in Theorem 9.5.5.)

Hint: Owing to the analyticity of $\mathrm{Fr}\,\Omega$, a Riemann map $R : \mathbb{D} \to \Omega$ extends holomorphically to a neighbourhood U of $\mathrm{cl}\,\mathbb{D}$. Consider g a Blaschke product extending $R^{-1}fR$ defined on $\mathbb{D}$. We can assume that g has a sink at ∞. Next, extend R to $\mathbb{C}$ holomorphically by $f^n \circ R \circ g^{-n}$, with branches g^{-n} and n large enough that $g^{-n}(z) \in U$. Check that the extension does not depend on the choice of the branches g^{-n}. If g is not of the form $g(z) = Az^d$, then the above formula defines R on $\overline{\mathbb{C}}$. If $g(z) = Az^d$, prove separately that R does not have an essential singularity at ∞. Finally prove that the extended R is invertible. For details see [Brolin 1965, Lemma 9.1].

If we do not assume anything about the f-invariance of Ω or Ω^*, then Jordan $\mathrm{Fr}\,\Omega$ need not be a geometric circle. Consider, for example, the mapping $F(x,y) = (4x, 4y)$ on the 2-torus $\mathbb{R}^2/\mathbb{Z}^2$ and its factor, the so-called Lattés map, $f := PFP^{-1}$ on the Riemann sphere, where P is the Weierstrass elliptic function. Then $P(\{y = 1/4 + \mathbb{Z}\})$ is an f-invariant expanding repelling Jordan curve, but it is not a geometric circle (we owe this example to A. Eremenko).

9.13. Prove that if for two conformal expanding repellers (J_1, f_1) and (J_2, f_2) in $\mathbb{C}$ being Jordan curves, the multipliers at all periodic orbits in J corresponding by a conjugating homeomorphism h, coincide, that is, for each periodic point $q \in J_1$ of period n we have $|(f_1^n)'(q)| = |(f_2^n)'(h(q))|$, then the conjugacy extends to a conformal map to neighbourhoods.

9.14. Let $A : \mathbb{R}^d/Z^d \to \mathbb{R}^d/Z^d$ be a hyperbolic toral automorphism given by an integer matrix of determinant 1. Let $\Phi(x_1, \dots, x_d) = (\varepsilon^{2\pi i x_1}, \dots, \varepsilon^{2\pi i x_d})$ map this torus to the torus $T^d = \{|z_1| = \cdots = |z_d| = 1\} \subset \mathbb{C}^d$. It extends to $\mathbb{C}^d/\mathbb{Z}^d$. Define $B = \Phi A \Phi^{-1}$. Let f be a holomorphic perturbation of B on a neighbourhood of T^d. Prove that close to T^d there is a topological torus S invariant for f such that

A on T^d and f on S are topologically conjugate by a homeomorphism h close to identity. Prove that if for each A-periodic orbit $p, A(p), \ldots, A^{n-1}(p)$ of period n absolute values of eigenvalues of differentials $DA^n(p)$ and of $Df^n(h(p))$ coincide (one says that Lyapunov spectra of periodic orbits coincide), then h extends to a holomorphic mapping on a neighbourhood of T^d.

Bibliographical notes

The fact that the zero of the geometric pressure is equal to the Hausdorff dimension, for any topologically mixing CER (see Corollary 9.1.7), is often called *Bowen's formula*; compare [Bowen 1979]. See [Rugh 2008] for an interesting generalization for a larger class of C^1-*conformal repellers.*

The section on multifractal analysis relies mainly on the monographs by Y. Pesin [Pesin 1997] and K. Falconer [Falconer 1997] (although the details are modified: for example, we do not use Markov partition). The reader can find there comprehensive expositions and further references. The development of this theory has been stimulated by physicists: the paper often quoted is [Hasley *et al.* 1986].

Proofs of Propositions 9.4.3 and 9.4.5 are slight modifications of proofs in [Przytycki 1986a, Chapter 7]. For Proposition 9.4.3 see also [Ghys 1984]. It is commonly used in the study of Siegel discs (Herman) and hedgehogs for Cremer points (Perez-Marco). The assumption that f is expanding, for the Riemann map R to be Hölder, is not necessary. A non-uniform hyperbolicity – that is, Collet–Eckmann or topological Collet–Eckmann – is sufficient: see [Graczyk & Smirnov 1998], [Przytycki & Rohde 1998] or [Przytycki 2000, Proposition 5.2]. A condition intermediate between expanding and non-uniform hyperbolicity (all critical points in $\mathrm{Fr}\,\Omega$ eventually periodic) in the case where Ω is a basin of attraction of a rational map is equivalent to a John property of Ω (stronger than Hölder): see [Carleson, Jones & Yoccoz 1994].

The formula for $\chi_\mu(R)$ in Theorem 9.4.7 holds for $(\partial\Omega, f)$ repelling to the side of Ω, as in Proposition 9.4.5, provided μ has positive entropy. The expanding assumption for f is not needed. See [Przytycki 1986a, Theorem 1].

The proofs of Theorem 9.5.4 and Theorem 9.5.5 in full generality can be found in [Przytycki 1986a], [Przytycki, Urbański & Zdunik 1989] and [Przytycki, Urbański & Zdunik 1991]. The proof of Theorem 9.5.5 is essentially taken from [Sullivan 1982]: see also [Przytycki 1986b]. The idea of the proof is similar to quasi-Fuchsian groups (either the Hausdorff dimension of the limit quasicircle is bigger than 1, or the group is Fuchsian) in [Bowen 1979]. It is a simple example of the strategy used in a proof of the Mostow Rigidity Theorem: see for example [Sullivan 1982]. In the general polynomial f case, with the basin Ω of attraction to ∞ simply connected, the dichotomy that either $\mathrm{HD}(\mathrm{Fr}\,\Omega) > 1$ or $f(z) = z^d$ was proved by A. Zdunik in [Zdunik 1990] and [Zdunik 1991]. A more careful look proves that for any Ω with f defined on a neighbourhood of $\mathrm{Fr}\,\Omega$ with boundary repelling to the side Ω as in Propositions 9.4.5 and Corollary 9.5.2, either $\mathrm{HD}(\mathrm{Fr}\,\Omega) > 1$, and in fact even the hyperbolic dimension $\mathrm{HyD}(X) > 1$

(see [Przytycki 2006]: for the definition see Section 12.2), or $\mathrm{Fr}(\Omega)$ is a real-analytic Jordan curve or interval: see [Zdunik 1991]. In the case where f extends to a rational mapping of $\overline{\mathbb{C}}$ it is either a finite Blaschke product in appropriate coordinates, as in Theorem 9.5.5, or a 2:1 factor of a finite Blaschke product (a Tchebyshev polynomial if f is a polynomial).

The references for Remarks 9.6.2–9.6.4 include [Carleson & Jones 1992], [Makarov 1999], [Binder, Makarov & Smirnov 2003] and [Beliaev & Smirnov 2005], where further references are provided.

10

Sullivan's classification of conformal expanding repellers

This chapter relies on ideas of the proof of the rigidity theorem drafted by D. Sullivan in the Proceedings of Berkeley's International Congress of Mathematicians in 1986: see [Sullivan 1986]. In Chapter 7, Example 7.1.10 shows that two expanding repellers can be Lipschitz conjugate, but not analytically (nor even differentially) conjugate.

So in Chapter 7 we provided an additional invariant, the scaling function for an expanding repeller in the line, taking 'gaps' into account, and proved that it determined the $C^{1+\varepsilon}$-structure.

In this chapter, following Sullivan, we distinguish a class of conformal expanding repellers (CERs) called *non-linear*, and prove that the class of equivalence of the geometric measure, and in particular the class of Lipschitz conjugacy, determines the conformal structure.

This is amazing: a holomorphic structure preserved by a map is determined by a measure.

10.1 Equivalent notions of linearity

Definition 10.1.1. Consider a CER (X, f) for compact $X \subset \mathbb{C}$. Denote by Jf the Jacobian of f with respect to the Gibbs measure μ_X equivalent to a geometric measure m_X on X. We call (X, f) *linear* if one of the following conditions holds:

(a) The Jacobian Jf, is locally constant.

(b) The function $\mathrm{HD}(X) \log |f'|$ is co-homologous to a locally constant function on X.

(c) The conformal structure on X admits a conformal affine refinement so that f is affine (that is, there exists an atlas $\{\varphi_t\}$ that is a family of conformal injections $\phi_t : U_t \to \mathbb{C}$ where $\bigcup_t U_t \supset X$ such that all the maps $\phi_t \phi_s^{-1}$ and $\phi_t f \phi_s^{-1}$ are affine).

Recall that as the conformal map f may change the orientation of $\mathbb{C}$ on some components of its domain, we can write $|f'|$ but not f' unless f is holomorphic.

Proposition 10.1.2. *The conditions (a), (b) and (c) are equivalent.*

Before we prove this proposition we single out from CER's, *real-analytic repellers* (this repeats Definition 6.3.1).

Definition 10.1.3. We call (X, f) *real-analytic* if X is contained in the union of a finite family of real-analytic open arcs and closed curves with pairwise disjoint closures.

Lemma 10.1.4. *If there exists a connected open domain U in $\mathbb{C}$ intersecting X for a CER (X, f), and if there exists a real-analytic function k on it equal identically 0 on $U \cap X$ but not on U, then (X, f) is real-analytic.*

Proof. Pick an arbitrary $x \in U \cap X$. Then in a neighbourhood V of x the set $E = \{k = 0\}$ is a finite union of pairwise disjoint real-analytic curves and of the point x. This follows from the existence of a finite decomposition of the germ of E at x into irreducible germs and from the form of each such germ: see for example Proposition 5.8 in [Malgrange 1967]. As the sets $f^n(X \cap V)$, $n \geq 0$ cover X, X is compact, and f is open on X, we conclude that X is contained in a finite union of real-analytic curves γ_j and a finite set of points A such that the closures of γ_j can intersect only in A.

Suppose that there exists a point $x \in X$ such that X is not contained in any real-analytic curve in every neighbourhood of x. Then the same is true for every point $z \in X \cap f^{-n}\{x\}$, $n \geq 0$, and hence for an infinite number of points (because pre-images of x are dense in X by the topological exactness of f: see Chapter 4). But we proved above that the number of such points is finite, so we have arrived at a contradiction. We conclude that X is contained in a one-dimensional real-analytic submanifold of $\mathbb{C}$. ♣

Proof of Proposition 10.1.2.

(a)⇒ (b). Let u be the eigenfunction $\mathcal{L}u = u$ for the transfer operator $\mathcal{L} = \mathcal{L}_\phi$ for the function $\phi = -\kappa \log |f'|$, where $\kappa = \mathrm{HD}(X)$, as in Section 5.3. Here the eigenvalue $\lambda = \exp \mathrm{P}(f, \phi)$ is equal to 1: see Section 9.1.

For an arbitrary $z \in X$ we have, in its neighbourhood in X,

$$\mathrm{Const} = \log Jf = \kappa \log |f'(x)| + \log u(f(x)) - \log u(x) \qquad (10.1.1)$$

(b)⇒ (c). The function u extends to a real-analytic function $u_\mathbb{C}$ in a neighbourhood of X (see Section 6.3), so the function $\log Jf$ extends to a real-analytic function $\log Jf_\mathbb{C}$ by the right-hand-side equality in the formula (10.1.1), for $u_\mathbb{C}$ instead of u. We have two cases: either $\log Jf_\mathbb{C}$ is not locally constant on every neighbourhood of X, and then by Lemma 10.1.4 (X, f) is real-analytic, or $\log Jf_\mathbb{C}$ is locally constant. Let us first consider the latter case.

Fix $z \in X$. Choose an arbitrary sequence of points $z_n \in X$, $n \geq 0$ such that $f(z_n) = z_{n-1}$, and choose branches f_ν^{-n} mapping z to z_n. Because of the expanding property of f they are all well defined on a common domain around z. For every x close to z denote $x_n = f_\nu^{-n}(x)$. We have $\operatorname{dist}(x_n, z_n) \to 0$, so by (10.1.1) for $\log Jf_{\mathbb{C}}$

$$\begin{aligned} &\sum_{n=1}^{\infty} \kappa(\log|f'(x_n)| - \log|f'(z_n)|) \\ &= \log u_{\mathbb{C}}(x) - \log u_{\mathbb{C}}(z) + \lim_{n\to\infty}(\log u_{\mathbb{C}}(z_n) - \log u_{\mathbb{C}}(x_n)) \\ &= \log u_{\mathbb{C}}(x) - \log u_{\mathbb{C}}(z). \end{aligned} \tag{10.1.2}$$

We conclude that $\log u_{\mathbb{C}}(x)$ is a harmonic function in a neighbourhood of z in $\mathbb{C}$ as the limit of a convergent series of harmonic functions; we use the fact that the compositions of harmonic functions with the conformal maps f_ν^{-n} are harmonic. Close to z we take a so-called *harmonic conjugate function* h so that $\log u(x) + ih(x)$ is holomorphic.

Write $F_z = \exp(\log u + ih)$, and denote by $\tilde{F}_z$ a primitive function for F_z in a neighbourhood of z. This is a chart, because $F_z(z) \neq 0$. The *atlas* given by the charts $\tilde{F}_z$ is affine (conformal) by the construction. We have, owing to (10.1.1) for the extended u,

$$|(\tilde{F}_{f(z)} \circ f \circ \tilde{F}_z^{-1})'(F_z(x))| = u_{\mathbb{C}}(f(x))|f'(x)|/u_{\mathbb{C}}(x) = \text{Const},$$

so the differential of f is locally constant in our atlas.

In the case where (X, f) is real-analytic we consider just the charts ϕ_t, being primitive functions of u on real-analytic curves containing X into $\mathbb{R}$ with unique complex extensions to neighbourhoods of these curves into a neighbourhood of $\mathbb{R}$ in $\mathbb{C}$. The equality $\log Jf_{\mathbb{C}} = \text{Const}$ holds on these curves, so the derivatives of $\phi_t f \phi_s^{-1}$ are locally constant.

(c)⇒ (a). Denote the maps $\phi_t f \phi_s^{-1}$ by $\tilde{f}_{t,s}$. In a neighbourhood (in X) of an arbitrary $z \in X$ we have

$$\begin{aligned} u(x) &= \lim_{n\to\infty} \mathcal{L}^n(1)(x) = \lim_{n\to\infty} \sum_{y\in f^{-n}(x)} |(f^n)'(y)|^{-\kappa} \\ &= \lim_{n\to\infty} |\phi'(x)|^\kappa \sum_y |\phi'(y)|^{-\kappa} |\tilde{f}'(y)|^{-n\kappa} \\ &= \text{Const} \lim_{n\to\infty} |\phi'(x)|^\kappa \sum_y \tilde{f}'(y)|^{-n\kappa} = |\phi'(x)|^\kappa \,\text{Const}\,. \end{aligned} \tag{10.1.3}$$

To simplify the notation we have omitted the indices at ϕ and $\tilde{f}$ here; of course, they depend on z and y's more precisely on the branches of f^{-n} on our neighbourhood of z mapping z to y's . Const also depends on z. We could omit the functions $\phi'(y)$ in the last line of (10.1.3), because the diameters of the domains of $\phi'(y)$ that were involved converged to 0 when $n \to \infty$ owing to the expanding property of f, so these functions were almost constant.

Hence, owing to (10.1.3), in a neighbourhood of every $x \in X$ we get

$$Jf(x) = \text{Const } u(f(x))|f'(x)|^\kappa / u(x) = \text{Const } |\tilde{f}'(x)|^\kappa = \text{Const}. \qquad \clubsuit$$

Remark 10.1.5. In the (b)⇒(c) part of the proof of Proposition 10.1.2, as $-\kappa \log |f'|$ is harmonic we do not need to refer to Section 5.4 for the real-analyticity of u. The formula (10.1.2) gives a harmonic extension of u to a neighbourhood of an arbitrary $z \in X$, depending on the choice of the sequence (z_n). If two extensions u_1, u_2 do not coincide on a neighbourhood of z, then in a neighbourhood of z, $X \subset \{u_1 - u_2 = 0\}$.

If equation (10.1.1) does not extend to a neighbourhood of z, then again $X \subset \{v = \text{Const}\}$ for a harmonic function v extending the right-hand side of (10.1.1).

In both cases (X, f) happens to be real-analytic, and to prove it we do not need to refer to Malgrange's book as in the proof of Lemma 10.1.4. Indeed, for any non-constant harmonic function k on a neighbourhood of $x \in X$ such that $X \subset \{k = 0\}$ we consider a holomorphic function F such that $k = \Re F$ and $F(x) = 0$. Then $E = \{k = 0\} = \{\Re F = 0\}$. If F has a d-multiple zero at x then it is a standard fact that E is a union of d analytic curves intersecting at x within the angle $\frac{\pi}{d}$.

We end this section by giving one more condition implying linearity.

Lemma 10.1.6. *Suppose for a CER (X, f) that there exists a Hölder continuous line field in the tangent bundle on a neighbourhood of X, invariant under the differential of f. In other words, there exists a complex-valued, nowhere-zero Hölder continuous function α such that, for every x in a neighbourhood of X,*

$$\operatorname{Arg} \alpha(x) + \operatorname{Arg} f'(x) = \operatorname{Arg} \alpha(f(x)) + \varepsilon(x)\pi \qquad (10.1.4)$$

where $\varepsilon(x)$ is a locally constant function equal to 0 or 1. This is in the case where f preserves the orientation at x; if it reverses the orientation we replace $\operatorname{Arg} f'$ *in (10.2.1) by* $-\operatorname{Arg} \bar{f}'$.

Then (X, f) is linear.

Proof. As in the proof of Proposition 10.1.2, calculation (10.1.2), if f is holomorphic we have for x in a neighbourhood of $z \in X$ in $\mathbb{C}$

$$\operatorname{Arg} \alpha(z) - \operatorname{Arg} \alpha(x) = \sum_{n=1}^{\infty} (\operatorname{Arg}(f'(z_n)) - \operatorname{Arg}(f'(x_n))).$$

If we allow f to reverse the orientation, then we replace $\operatorname{Arg} f'$ by $-\operatorname{Arg} \bar{f}'$ in the above formula for such n that f changes the orientation in a neighbourhood of x_n. So $\operatorname{Arg} \alpha(x)$ is a harmonic function. Close to z we find a conjugate harmonic function h, so we get a family of holomorphic functions $F_z = \exp(-h + i \operatorname{Arg} \alpha$, which primitive functions give the atlas we have looked for.

Remark 10.1.7. The condition for (X, f) in Lemma 10.1.6 is stronger than the linearity property. Indeed, we can define f on the union of the discs $D_1 = \{|z| < 1\}$ and $D_2 = \{|z-3| < 1\}$ by $f(z) = 5\exp 2\pi\vartheta i$ on D_1, where ϑ is irrational, and $f(z) = 5(z-3)$ on D_2. This is an example of an iterated function system from Section 5.5. We get a CER (X, f), where $X = \bigcap_{n=0}^{\infty} f^{-n}(\{|z| < 5\})$. It is linear because it satisfies condition (c). Meanwhile $0 \in X, f(0) = 0$ and $f'(0) = 5\exp 2\pi\vartheta i$, so equation (10.1.4) has no solution at $x = 0$ for any iterate of f.

Remark 10.1.8. If we assume, in place of (10.1.4), that $\operatorname{Arg} f'(x) - \operatorname{Arg} \alpha(f(x)) - \operatorname{Arg} \alpha(x)$ is locally constant, then we get the condition equivalent to linearity.

10.2 Rigidity of non-linear CERs

In this section we shall prove the main theorem of Chapter 10.

Theorem 10.2.1. *Let $(X, f), ((Y, g)$ be two non-linear conformal expanding repellers in $\mathbb{C}$. Let h be an invertible mapping from X onto Y, preserving Borel σ-algebras and conjugating f to g, $h \circ f = g \circ h$. Suppose that one of the following assumption is satisfied:*

1. h and h^{-1} are Lipschitz continuous.

2. h and h^{-1} are continuous and preserve so-called Lyapunov spectra: that is, for every periodic $x \in X$ and integer n such that $f^n(x) = x$ we have $|(f^n)'(x)| = |(g^n)'(h(x))|$.

3. h_* maps a geometric measure m_X on X to a measure equivalent to a geometric measure m_Y on Y.

Then h extends from X (or from a set of full measure m_X in case 3) to a conformal homeomorphism on a neighbourhood of X.

First we discuss the assumptions. The equivalence of conditions 1 and 2 easily follows from Proposition 4.4.5, with the use of 'bounded distortion' (exercise). Condition 1 implies 3 by the definition of geometric measures 6.6.5. One of the steps of the proof of the theorem will assert that 3 implies 1 under the non-linearity assumption. Without this assumption the assertion may be false. A positive result is that if h is continuous, then, for a constant $C > 0$ and every $x_1, x_2 \in X$,

$$C < \frac{|h(x_1) - h(x_2)|^{\mathrm{HD}(Y)}}{|x_1 - x_2|^{\mathrm{HD}(X)}} < C^{-1}.$$

(We leave the proof to the reader.)

It may happen that $\mathrm{HD}(X) \neq \mathrm{HD}(Y)$, for example if X is a $1/3$ – Cantor set, and for g we remove each time half of the interval from the middle.

A basic observation to prove Theorem 10.2.1 is that

$$Jg \circ h = Jf \text{ and moreover } Jg^j \circ h = Jf^j \tag{10.2.1}$$

for every integer $j > 0$. This follows from $g^j \circ h = h \circ f^j$ and $Jh \equiv 1$. Recall that we consider Jacobians with respect to the Gibbs measures equivalent to geometric measures.

Observe finally that (X, f) linear implies (Y, g) linear. Indeed, if (X, f) is linear then Jf and hence Jg admit only a finite number of values, in view of $Jg \circ h = Jf$. As Jg is continuous, this implies that Jg is locally constant: that is, (Y, g) is linear.

Lemma 10.2.2. *If a CER (X, f) is non-linear, then there exists $x \in X$ such that* $\operatorname{grad} Jf_{\mathbb{C}}(x) \neq 0$.

Proof. If $\operatorname{grad} Jf_{\mathbb{C}} \equiv 0$ on X, then as $Jf_{\mathbb{C}}$ is real-analytic we have either $\operatorname{grad} Jf_{\mathbb{C}} \equiv 0$ on a neighbourhood of X in $\mathbb{C}$, or by Lemma 10.1.4 (X, f) is real-analytic and $\operatorname{grad} Jf_{\mathbb{C}} \equiv 0$ on real-analytic curves containing X. In both cases we obtain by integration Jf locally constant on X, which contradicts the non-linearity assumption.

Now we shall prove the theorem in the simplest case to show the reader the main idea, working later also in the general case.

Proposition 10.2.3. *The assertion of Theorem 10.2.1 holds if we suppose additionally that (X, f) and $Y, g)$ are real-analytic and the conjugacy h is continuous.*

Proof. Let M, N be real-analytic manifolds containing X, Y respectively. By the non-linearity of X and Lemma 10.2.2 there exists $x \in X$ and its neighbourhood U in M such that $F := Jf_{\mathbb{C}}|_U : U \to \mathbb{R}$ has a real-analytic inverse $F^{-1} : F(U) \to U$. Then in view of (10.2.1) $h^{-1} = F^{-1} \circ Jg_{\mathbb{C}}$ on $h(U \cap X)$, so h^{-1} on $h(U \cap X)$ extends to a real-analytic map on a neighbourhood of $h(U \cap X)$ in N.

Now we use the assumption that h^{-1} is continuous, so $h(U \cap X)$ contains an open set v in Y. There exists a positive integer n such that $g^n(V) = Y$: hence for every $y \in Y$ there exists a neighbourhood W of y in N such that a branch g_ν^{-n} of g^{-n} mapping y and even $W \cap Y$ into V is well defined. So we have $h^{-1} = f^n \circ h^{-1} \circ g_\nu^{-n}$ extended on W to a real-analytic map. This gives a real-analytic extension of h^{-1} on a neighbourhood of Y, because two such extensions must coincide on the intersections of their domains by the real-analyticity and the fact that Y has no isolated points.

Similarly, using the non-linearity of (Y, g) and the continuity of h we prove that h extends analytically. By the analyticity and again the lack of isolated points in X and Y the extensions are inverse to each other, so h extends even to a biholomorphic map.

Now we pass to the general case.

Lemma 10.2.4. *Suppose that there exists $x \in X$ such that* $\operatorname{grad} Jf_{\mathbb{C}}(x) \neq 0$ *in the case where X is real-analytic, or there exists an integer $k \geq 1$ such that* $\det(\operatorname{grad} Jf_{\mathbb{C}}, \operatorname{grad}(Jf_{\mathbb{C}} \circ f^k)) \neq 0$ *in the other case.*

(In other words, we suppose that $Jf_{\mathbb{C}}$, respectively $(Jf_{\mathbb{C}}, Jf_{\mathbb{C}} \circ f^k)$, gives a coordinate system on a real (complex) neighbourhood of x.)

Assume the analogous property for (Y, g).

Let $h : X \to Y$ satisfy property 3 assumed in Theorem 10.2.1. Then h extends from a set of full geometric measure in X to a bi-Lipschitz homeomorphism of X onto Y conjugating f with g.

Proof. We can suppose that $\mathrm{HD}(X) \geq \mathrm{HD}(Y)$: recall that HD denotes Hausdorff dimension. Pick x with the property assumed in the lemma. Let U be its neighbourhood in M (as in the proof of Proposition 10.2.3) or in $\mathbb{C}$ if (X, f) is not real-analytic, so that $F := (Jf_{\mathbb{C}}, Jf_{\mathbb{C}} \circ f^k)$ is an embedding on U. Let $y \in Y$ be a density point of the set $h(U \cap X)$ with respect to the Gibbs measure μ_Y equivalent to the geometric measure m_Y. (Recall that we have proved that almost every point is a density point for an arbitrary Borel probability measure on a Euclidean space in Chapter 8, Theorem 8.5.4, relying on Besicovitch's Theorem.) So if we denote $(Jg_{\mathbb{C}}, Jg_{\mathbb{C}} \circ g^k)$ in a neighbourhood (real or complex) of y by G, we have for every $\delta > 0$ such $\varepsilon_0 = \varepsilon_0(\delta) > 0$ that for every $0 < \varepsilon < \varepsilon_0$:

$$\frac{\mu_Y(B(y,\varepsilon) \cap h(U \cap X))}{\mu_Y(B(y,\varepsilon))} > 1 - \delta$$

and

$$h^{-1} = F^{-1} \circ G \qquad \text{on } h(U \cap X).$$

(Observe that the last equality may be false outside $h(U \cap X)$, even very close to y, because h^{-1} may map such points to $(Jf_{\mathbb{C}}, Jf_{\mathbb{C}} \circ f^k)^{-1} \circ G$ with a branch of $(Jf_{\mathbb{C}}, Jf_{\mathbb{C}} \circ f^k)^{-1}$ different from F^{-1}.)

Now for every $\varepsilon > 0$ small enough there exists an integer n such that $\operatorname{diam} g^n B(y, \varepsilon)$ is greater than a positive constant, $g^n|_{B(y,\varepsilon)}$ is injective, and the distortion of g^n on $B(y,\varepsilon)$ is bounded by a constant C, both constants depending only on (Y, g). Then if $\varepsilon < \varepsilon_0(\delta)$ we obtain for $Y_\delta := g^n(h(U \cap X) \cap B(y, \varepsilon))$

$$\frac{\mu_Y(g^n(B(y,\varepsilon)) \setminus Y_\delta)}{\mu_Y(g^n(B(y,\varepsilon)))} < C \frac{\mu_Y(B(y,\varepsilon) \setminus h(U \cap X))}{\mu_Y(B(y,\varepsilon))} < C\delta.$$

So

$$\frac{\mu_Y(Y_\delta)}{\mu_Y(g^n(B(y,\varepsilon)))} > 1 - C\delta. \tag{10.2.2}$$

We have

$$\begin{aligned} |(f^n)'(h^{-1}(y))|^{\mathrm{HD}(X)} &\leq \mathrm{Const}\, Jf(h^{-1}(y)) \\ &= \mathrm{Const}\, Jg(y) \leq \mathrm{Const}\, |(f^n)'(y)|^{\mathrm{HD}(Y)}. \end{aligned}$$

As we assumed $\mathrm{HD}(X) \geq \mathrm{HD}(Y)$ we obtain

$$|(f^n)'(h^{-1}(y))| \leq \mathrm{Const}\, |(f^n)'(y)|^{\mathrm{HD}(Y)/\,\mathrm{HD}(X)} \leq \mathrm{Const}\, |(f^n)'(y)|. \tag{10.2.3}$$

Then, owing to the bounded distortion property for iteration of f and g, we obtain that $h^{-1} = f^n h^{-1} g^{-1}$ is Lipschitz on Y_δ with Lipschitz constant independent of δ, or more precisely bounded by $\mathrm{Const} \sup \|D(F^{-1} \circ G\|$, where $F^{-1} \circ G$ is considered on a real (complex) neighbourhood of y and Const is that from (10.2.3).

There exists an integer $K>0$ such that for every n, $g^K g^n B(y,\varepsilon(n))$ covers Y. Because Jg is bounded, separated from 0, this gives that h^{-1} on $g^K(Y_\delta)$ is Lipschitz with a Lipschitz constant independent from δ and $\mu(g^K(Y_\delta))>1-$ Const δ for δ arbitrarily small. We conclude that h^{-1} is Lipschitz on a set of full measure μ_Y, so it has a Lipschitz extension to Y.

We conclude also that $\mathrm{HD}(X)=\mathrm{HD}(Y)$. Otherwise $\operatorname{diam} h^{-1}(Y_\delta)\to 0$, so because $\operatorname{supp}\mu_X=X$ we would get $\operatorname{diam} X=0$. So we can interchange above the roles of (X,f) and (Y,g) and prove that h is Lipschitz.

The next step will assert that for non-linear repellers the assumptions of Lemma 10.2.4 about the existence of coordinate systems are satisfied.

Lemma 10.2.5. *If (X,f) is a non-linear CER then there either exists $x\in X$ such that* $\operatorname{grad} Jf_{\mathbb{C}}(x)\neq 0$ *in the case where X is real-analytic, or there exists an integer $k\geq 1$ such that* $\det(\operatorname{grad} Jf_{\mathbb{C}},\operatorname{grad}(Jf_{\mathbb{C}}\circ f^k))\neq 0$ *in the case where (X,f) is not real-analytic.*

Proof. We know already from Lemma 10.2.2 that there exists $\hat{x}\in X$ such that $\operatorname{grad} Jf_{\mathbb{C}}(z)\neq 0$, so we may restrict our considerations to the case where (X,f) is not real-analytic.

Suppose the lemma is false. Then for all $k>0$ the functions

$$\Phi_k:=\det(\operatorname{grad} Jf_{\mathbb{C}},\operatorname{grad}(Jf_{\mathbb{C}}\circ f^k))$$

are identically equal to 0 on X. Let W be a neighbourhood of $\hat{x}$ in $\mathbb{C}$, where $\operatorname{grad} Jf_{\mathbb{C}}\neq 0$.

Let us consider on W the line field $\mathcal{V}$ orthogonal to $\operatorname{grad} Jf_{\mathbb{C}}$. Because of the topological exactness of f on X for every $x\in X$, there exists $y\in W\cap X$ and $n\geq 0$ such that $f^n(y)=x$.

Thus define at x

$$\mathcal{V}_x:=Df^n(\mathcal{V}_y). \tag{10.2.4}$$

We shall now prove that if $x=f^k(y)=f^l(z)$ for some $y,z\in W\cap X,\ k,l\geq 0$, then

$$Df^k(\mathcal{V}_y)=Df^l(\mathcal{V}_z). \tag{10.2.5}$$

If (10.2.5) is false, then close to x there exist $x'\in X$ and $m\geq 0$ such that $f^m(x')\in W$ (we again refer to the topological exactness of f) and $Df^k(\mathcal{V}_{y'})\neq Df^l(\mathcal{V}_{z'})$, where $f^k(y')=f^l(z')=x'$, $y'\in X$ is close to y, and $z'\in X$ is close to z. We obtain $Df^{k+m}(\mathcal{V}_{y'})\neq Df^{l+m}(\mathcal{V}_{z'})$, so either $Df^{k+m}(\mathcal{V}_{y'})\neq\mathcal{V}_{f^m(x')}$ or $Df^{l+m}(\mathcal{V}_{z'})\neq\mathcal{V}_{f^m(x')}$. Consider the first case (the second is of course similar). We obtain that Jf and $Jf\circ f^{k+m}$ give a coordinate system in a neighbourhood of y': that is, $\Phi_{k+m}(y')\neq 0$, contrary to the supposition.

Thus formula (10.2.4) defines a line field at all points of X that is Df-invariant. Observe, however, that the same formula defines a real-analytic extension of the line field to a neighbourhood of x in $\mathbb{C}$, because $\mathcal{V}$ is real-analytic on a neighbourhood of $y\in W$, and f is analytic. Each two such germs of extensions related to two different pre-images of x must coincide, because they coincide on X; otherwise (X,f) would be real-analytic. Now we can choose a finite cover $B_j=B(x_j,\delta_j)$ of a neighbourhood of X with discs, $x_j\in X$, so that for the respective F_j-branches of f^{-n_j} leading x_j into W we have $F_j(3B_j) \subset W$, where $3B_j := B(x_j, 3\delta_j)$. Hence the formula (10.2.4) defines $\mathcal{V}$ on $3B_j$. So if $B_i \cap B_j \neq \emptyset$, then we have $3B_i \subset B_j$ or vice versa. So $3B_i \cap 3B_j \cap X \neq 0$, and hence the extensions of $\mathcal{V}$ on $3B_i$ and $3B_j$, and in particular on B_i and on B_j, coincide on the intersection. This is so because they coincide on the intersection with X, and (X, f) is not real-analytic.

(We used 3δ because it can happen that $B_i \cap B_j \neq \emptyset$ but $B_i \cap B_j \cap X = \emptyset$.)

Thus $\mathcal{V}$ extends real-analytically to a neighbourhood of X. This field is Df-invariant on a neighbourhood of X, because we can define it in a neighbourhood of $x \in X$ and $f(x)$ by (10.2.4), taking the same $y \in W \cap X$, where $f^n(y) = x$, $f^{n+1}(y) = f(x)$. So by Lemma 10.1.6 (X, f) is linear, which contradicts the assumption that (X, f) is non-linear.

Corollary 10.2.6. *If for $(X, f), (Y, g)$ the assumptions of Theorem 10.2.1 are satisfied, and if (Y, g) is real-analytic, then (X, f) is real-analytic too.*

Proof. Because of Lemma 10.2.5 the assumptions of Lemma 10.2.4 are satisfied. So $h^{-1} = F^{-1} \circ G$ on a neighbourhood of $y \in Y$ by the continuity of h^{-1} (see notation in the proof of Lemma 10.2.4). Denote a real-analytic manifold that Y is contained in by N. Then $Jg_{\mathbb{C}} \neq \text{Const}$ on any neighbourhood of y in N. Otherwise h^{-1} would be constant, but y is not isolated in Y, and so h^{-1} would not be injective.

Recall that we can consider $F^{-1} \circ G$ as a real-analytic extension of h^{-1} to a neighbourhood V of y in N. So the differential of $F^{-1}G$ is 0 at most at isolated points, and so different from 0 at a point $y' \in V \cap Y$. We conclude, owing to the continuity of h, that in a neighbourhood of $h^{-1}(y')$, X is contained in a real-analytic curve. So (X, f) is a real-analytic repeller.

Now we shall collect together what we have done and make a decisive step in proving Theorem 10.2.1: that is, we shall prove that the conjugacy extends to a real-analytic diffeomorphism.

Proof of Theorem 10.2.1. If both (X, f) and (Y, g) are real-analytic, then the conjugacy extends real-analytically to a real-analytic manifold and so complex analytically to its neighbourhood by Proposition 10.2.3. Its assumptions hold by Lemmas 10.2.4 and 10.2.2. If both (X, f) and (Y, g) are not real-analytic (a mixed situation is excluded by Corollary 10.2.6), then by Lemma 10.2.4, which assumptions hold because of Lemma 10.2.5, we can assume that the conjugacy h is a homeomorphism of X onto Y. But h^{-1} extends to a neighbourhood of $y \in Y$ in $\mathbb{C}$ to a real-analytic map. We use here again the notation of Lemma 10.2.4 and proceed precisely as in Proposition 10.2.3, Lemma 10.2.4 and Corollary 10.2.6 by writing $h^{-1} = F^{-1} \circ G$. This gives a real-analytic extension of h^{-1} to a neighbourhood of an arbitrary $y \in Y$ by the formula $f^n \circ h^{-1} \circ g_\nu^{-1}$, precisely as in the proof of Proposition 10.2.3.

For two different branches F_1, F_2 of g^{-n_1}, g^{-n_2} respectively, mapping y into the domain of $F^{-1} \circ G$, germs of the extensions must coincide because they coincide on the intersection with Y: see Lemma 10.1.4.

Now we build a real-analytic extension of h^{-1} to a neighbourhood of Y in the same way as we extended $\mathcal{V}$ in the proof of Lemma 10.2.5, again using the assumption (Y, g) is not real-analytic. Similarly we extend h.

Denote the extensions by $\tilde{h}, \widetilde{h^{-1}}$. We have $\widetilde{h^{-1}} \circ \tilde{h}$ and $\tilde{h} \circ \widetilde{h^{-1}}$ equal to the identity on X, Y respectively. Then these compositions extend to the identities to neighbourhoods, otherwise (X, f) or (Y, g) would be real-analytic. We conclude that $\tilde{h}$ is a real-analytic diffeomorphism. Finally, observe that $g\tilde{h} = \tilde{h}f$ on a neighbourhood of X, because this equality holds on X itself, and our functions are real-analytic; otherwise (X, f) would be real-analytic.

The only thing we still need to prove is the following:

Lemma 10.2.7. *If (X, f) is a non-linear CER, not real-analytic, and there is a real-analytic diffeomorphism h on a neighbourhood of X to a neighbourhood of Y for another CER (Y, g) such that $h(X) = Y$ and h conjugates f with g in a neighbourhood of X, then h is conformal.*

Proof. Suppose for simplification that f, g and h preserve the orientation of $\mathbb{C}$; we shall comment on the general case at the end.

For any orientation-preserving diffeomorphism Φ of a domain in $\mathbb{C}$ into $\mathbb{C}$ denote the complex dilatation function by ω_Φ. Recall that $\omega_\Phi := \frac{d\Phi}{d\bar{z}} / \frac{d\Phi}{dz}$. (The reader not familiar with the complex dilatation and its properties is advised to read the first 10 pages of the classical Ahlfors book [Ahlfors 1966].) The geometric meaning of the argument of $\omega_\Phi(z)$ may be explained by the equality $\frac{1}{2}\omega_\Phi = \alpha$, where α corresponds to the direction in which the differential $D\Phi$ at z attains its maximum. In other words, it is the direction of the smaller axis of the ellipse in the tangent space at z which is mapped by $D\Phi$ to the unit circle. Of course, this makes sense if $\omega(z) \neq 0$. Observe finally that $\omega(z) = 0$ if and only if $\frac{d\Phi}{d\bar{z}} = 0$. Let us now return to our concrete maps.

If $\frac{dh}{d\bar{z}} \equiv 0$ on X, then as $\frac{dh}{d\bar{z}}$ is a real-analytic function we have $\frac{dh}{d\bar{z}} \equiv 0$ on a neighbourhood of X; otherwise (X, f) would be real-analytic. But this means that h is holomorphic, which proves our lemma. It remains to prove that the case $\frac{dh}{d\bar{z}} \not\equiv 0$ on X is impossible.

Observe that if $\frac{dh}{d\bar{z}}(x) = 0$ then $\frac{dh}{d\bar{z}}(f(x)) = 0$, because $h = ghf_\nu^{-1}$ on a neighbourhood of $f(x)$ for the branch f_ν^{-1} of f^{-1} mapping $f(x)$ to x, and because g and f_ν^{-1} are conformal. So if there exists $x \in X$ such that $\frac{dh}{d\bar{z}}(x) \neq 0$, then this also holds for all x's from a neighbourhood, and as a consequence of the topological exactness of f for all x in a neighbourhood of X. Thus we have a complex-valued function ω_h that is nowhere zero on a neighbourhood of X.

Recall now that for any two orientation-preserving diffeomorphisms Φ and Ψ, if Ψ is holomorphic then

$$\omega_{\Psi \circ \Phi} = \omega_\Phi$$

and if Φ is conformal then

$$\omega_\Psi \circ \Phi = \left(\frac{\Phi'}{|\Phi'|} \right)^2 \omega_{\Psi \circ \Phi} = \omega_\Phi.$$

Applying this to the equation $h \circ f = g \circ h$ we obtain

$$\omega_h \circ f = \left(\frac{f'}{|f'|}\right)^2 \omega_{h \circ f} = \left(\frac{f'}{|f'|}\right)^2 \omega_{g \circ h} = \left(\frac{f'}{|f'|}\right)^2 \omega h.$$

Thus $\alpha(x) := \frac{1}{2}\omega_h(x)$ satisfies equation (10.1.4), and by Lemma 10.1.6 (X, f) is linear, which contradicts our assumption that it is non-linear.

In the case where a diffeomorphism reverses the orientation, we write everywhere above $\omega_{\bar{\Phi}}$ instead of ω_Φ, and if Φ is conformal-reversing orientation we write $\overline{\bar{\Phi}'}$ instead of Φ'. Additionally, some omegas should be conjugated in the formulae above. We also arrive at (10.1.4). (In this situation the complex notation is not convenient. Everything becomes trivial if we act with differentials on line fields. We leave writing this down to the reader.)

Example 10.2.8. If $f_c(z) = z^2 + c$ for $c \in \mathcal{M}_0$ (see Example 6.1.10 and Example 9.5.6 (for $z^d + c$)), then the Julia set $J(f_c) = X_{f_c}$ is a Jordan curve and (X_{f_c}, f_c) is non-linear, except for $c = 0$.

Indeed, if it is linear, then by Definition 10.1.1 (a) the function $-\mathrm{HD}(X_{f_c}) \log |f'|$ is co-homologous to constant on X_{f_c}, because this set is connected. Hence, by Theorem 9.5.5, $f_c(z) = z^2$, that is, $c = 0$.

In fact $(J(f), f|_{J(f)})$ is non-linear for every rational map f without critical points in its Julia set $J(f)$, and in particular f expanding on $J(f)$, except for $f(z) = z^d$, $|d| \geq 2$. This follows from [Zdunik 1990]: compare [Przytycki & Urbański 1999, Section 3].

Example 10.2.9. Let X be a Cantor set in the line $\mathbb{R}$, which is an image by h of Σ^d as in Section 7.1: that is, $h \in \mathcal{H}$. Consider the map $h \circ s \circ h^{-1}$, where s is the shift to the left on Σ^d. Suppose that this map extends to s_h, which is locally affine: that is, the scaling function stabilizes, $S_n/S_{n+1} \equiv 1$ for all n large enough (compare Theorem 7.2.4). Then the repeller (X, s_h) is linear, by Definition 10.1.1 (c).

Remark 10.2.10. In the presence of critical points in $J(f)$ for f non-exceptional (that is, with parabolic orbifold) $J(f)$ contains non-linear invariant expanding repellers for f. See [Przytycki & Urbański 1999, Section 3], [Zdunik 1990] and [Prado 1997].

Bibliographical notes

As we have already mentioned, this chapter relies on ideas by Dennis Sullivan: see [Sullivan 1986]. Written in 1991, this was followed by many papers applying its ideas: see for example [Przytycki & Urbański 1999], [Mauldin, Przytycki & Urbański 2001] and [Urbański 2001] in $\mathbb{R}^d, d \geq 3$. See also [Mauldin & Urbański 2003, section 7.3]. In recent years this rigidity has been intensively applied in studies of the iterations of entire and meromorphic maps.

11

Holomorphic maps with invariant probability measures of positive Lyapunov exponent

11.1 Ruelle's inequality

Let X be a compact subset of the Riemann sphere $\overline{\mathbb{C}}$, and let $\mathcal{A}(X)$ denote the set of all continuous maps $f : X \to X$ that can be analytically extended to an open neighbourhood $U(f)$ of X. In this section we work only with the standard spherical metric on $\overline{\mathbb{C}}$, normalized so that the area of $\overline{\mathbb{C}}$ is 1. In particular, all the derivatives are computed with respect to this metric.

Let us recall and extend Definition 9.1.2. Let μ be an f-invariant Borel probability measure on X. Since $|f'|$ is bounded, the integral $\int \log |f'| \, d\mu$ is well defined, and moreover $\int \log |f'| \, d\mu < +\infty$. The number

$$\chi_\mu = \chi_\mu(f) = \int \log |f'| \, d\mu$$

is called the *Lyapunov characteristic exponent* of μ and f. Note that $\int \log |f'| d\mu = -\infty$ is not excluded. In fact it is possible, for example if $X = \{0\}$ and $f(z) = z^2$.

On the other hand, for every rational function $f : \overline{\mathbb{C}} \to \overline{\mathbb{C}}$ and every f-invariant μ supported on the Julia set $J(f)$ (see Chapter 1, Example 1.6), it holds that $\chi_\mu \geq 0$. For the proof see [Przytycki 1993]. We shall often assume $\chi_\mu > 0$; compare Definition 9.1.2. We say, then, that μ is a *hyperbolic measure* (following [Katok & Hasselblatt 1995]).

By the Birkhoff Ergodic Theorem (Theorem 2.2.5) the *Lyapunov characteristic exponent*

$$\chi_\mu(x) = \lim_{n\to\infty} \frac{1}{n} \log |(f^n)'(x)|$$

exists for a.e. x and $\int \chi_\mu(x)\, d\mu(x) = \chi_\mu$. (In fact one allows $\log |f'|$ with integral $-\infty$ here, so one needs to extend Theorem 2.2.5 slightly. This is not difficult.)

This section is devoted to prove the following.

Theorem 11.1.1 (Ruelle's inequality). *If $f \in \mathcal{A}(X)$, then $\mathrm{h}_\mu(f) \le 2\int \max\{0, \chi_\mu(x)\}\, d\mu$. For ergodic μ this yields $\mathrm{h}_\mu(f) \le 2\max\{0, \chi_\mu\}$.*

Proof. Consider a sequence of positive numbers $a_k \searrow 0$, and $\mathcal{P}_k, k = 1, 2, \ldots$ an increasing sequence of partitions of the sphere $\overline{\mathbb{C}}$ consisting of elements of diameters $\le a_k$ and of (spherical) areas $\ge \frac{1}{4}a_k^2$. Check that such partitions exist.

For every $g \in \mathcal{A}(X)$, $x \in X$ and $k \ge 1$ let

$$N(g, x, k) = \#\{P \in \mathcal{P}_k : g(P_k(x) \cap U(g)) \cap P \ne \emptyset\}.$$

Our first aim is to show that, for every $k > k(g)$ large enough,

$$N(g, x, k) \le 4\pi(|g'(x)| + 2)^2. \qquad (11.1.1)$$

Fix $x \in X$ and consider k so large that $\mathcal{P}_k(x) \subset U(g)$ and a Lipschitz constant of $g|_{\mathcal{P}_k(x)}$ does not exceed $|g'(x)| + 1$. Thus the set $g(\mathcal{P}_k(x))$ is contained in the ball $B(g(x), (|g'(x)| + 1)a_k)$. Therefore, if $g(\mathcal{P}_k(x)) \cap P \ne \emptyset$, then

$$P \subset B(g(x), (|g'(x)| + 1)a_k + a_k) = B(g(x), (|g'(x)| + 2)a_k).$$

Hence $N(g, x, k) \le \pi(|g'(x)| + 2)^2 a_k^2 / \frac{1}{4}a_k^2 = 4\pi(|g'(x)| + 2)^2$, and (11.1.1) is proved.

Let $N(g, x) = \sup_{k > k(g)} N(g, x, k)$. In view of (11.1.1) we get

$$N(g, x) \le 4\pi(|g'(x)| + 2)^2. \qquad (11.1.2)$$

Now note that for every finite partition $\mathcal{A}$ one has

$$\begin{aligned} \mathrm{h}(g, \mathcal{A}) &= \lim_{n\to\infty} \frac{1}{n+1} \mathrm{H}(\mathcal{A}^n) \\ &= \lim_{n\to\infty} \frac{1}{n+1} \Big(\mathrm{H}(g^{-n}(\mathcal{A})|\mathcal{A}^{n-1}) + \cdots + \mathrm{H}(g^{-1}(\mathcal{A})|\mathcal{A}) + \mathrm{H}(\mathcal{A})\Big) \\ &\le \lim_{n\to\infty} \frac{1}{n} \Big(\mathrm{H}(g^{-n}(\mathcal{A})|g^{-(n-1)}(\mathcal{A})) + \cdots + \mathrm{H}(g^{-1}(\mathcal{A})|\mathcal{A})\Big) \\ &= \mathrm{H}(g^{-1}(\mathcal{A})|\mathcal{A}). \end{aligned} \qquad (11.1.3)$$

(Compare this computation with that done in Theorem 2.4.5 or in the proof of Theorem 2.5.4, which would result in $\mathrm{h}(g, \mathcal{A}) \le \mathrm{H}(\mathcal{A}|g^{-1}(\mathcal{A}))$.) Going back to our situation, since

$$\mathrm{H}_{\mu_{\mathcal{P}_k(x)}}(g^{-1}(\mathcal{P}_k)|\mathcal{P}_k(x)) \le \log \#\{P \in \mathcal{P}_k : g^{-1}(P) \cap \mathcal{P}_k(x) \ne \emptyset\} = \log N(g, x, k)$$

and by Theorem 2.8.7a, we obtain

$$\begin{aligned} \mathrm{h}_\mu(g) &\le \limsup_{k\to\infty} \mathrm{H}_\mu(g^{-1}(\mathcal{P}_k)|\mathcal{P}_k) = \limsup_{k\to\infty} \int \mathrm{H}_{\mu_{\mathcal{P}_k(x)}}(g^{-1}(\mathcal{P}_k)|\mathcal{P}_k(x))\, d\mu(x) \\ &\le \limsup_{k\to\infty} \int \log N(g, x, k)\, d\mu(x) \le \int \log N(g, x)\, d\mu(x). \end{aligned}$$

Applying this inequality to $g = f^n$ ($n \geq 1$ an integer) and employing (10.1.2) we get

$$\begin{aligned} \mathrm{h}_\mu(f) &= \frac{1}{n}\mathrm{h}_\mu(f^n) \leq \frac{1}{n}\int \log N(f^n,x)\,d\mu(x) = \int \frac{1}{n}\log N(f^n,x)\,d\mu(x) \\ &\leq \int \frac{1}{n}\log 4\pi(|(f^n)'(x)|+2)^2\,d\mu(x). \end{aligned}$$

Since $0 \leq \frac{1}{n}\log(|(f^n)'(x)|+2)^2 \leq 2(\log(\sup_X|f'|)+1)$ and $\lim_{n\to\infty}\frac{1}{n}\log(|(f^n)'(x)|+2) = \max\{0,\chi_\mu(x)\}$ for μ-a.e $x \in X$, it follows from the Dominated Convergence Theorem (Section 2.1) that

$$h_\mu(f) \leq \lim_{n\to\infty}\int \frac{1}{n}\log(|(f^n)'(x)|+2)^2\,d\mu(x) = \int \max\{0,2\chi_\mu(x)\}\,d\mu.$$

The proof is complete. ♣

11.2 Pesin's theory

In this section we work in the same setting and follow the same notation as in Section 11.1.

Lemma 11.2.1. *If μ is a Borel finite measure on $\mathbb{R}^n$, $n \geq 1$, a is an arbitrary point of $\mathbb{R}^n$ and the function $z \mapsto \log|z-a|$ is μ-integrable, then for every $C > 0$ and every $0 < t < 1$,*

$$\sum_{n\geq 1}\mu(B(a,Ct^n)) < \infty.$$

Proof. Since μ is finite and since, given $t < s < 1$, there exists $q \geq 1$ such that $Ct^n \leq s^n$ for all $n \geq q$, without losing generality we may assume that $C = 1$. Recall that given $b \in \mathbb{R}^n$, and two numbers $0 \leq r < R$, $R(b,r,R) = \{z \in \mathbb{C} : r \leq |z-b| < R\}$. Since $-\log(t^n) \leq -\log|z-a|$ for every $z \in B(a,t^n)$ we get the following:

$$\begin{aligned} \sum_{n\geq 1}\mu(B(a,t^n)) &= \sum_{n\geq 1} n\mu(R(a,t^{n+1},t^n)) = \frac{-1}{\log t}\sum_{n\geq 1} -\log(t^n)\mu(R(a,t^{n+1},t^n)) \\ &\leq \frac{-1}{\log t}\int_{B(a,t)} -\log|z-a|\,d\mu(z) < +\infty. \end{aligned}$$

The proof is complete. ♣

Lemma 11.2.2. *If μ is a Borel finite measure on $\overline{\mathbb{C}}$, $n \geq 1$, and $\log|f'|$ is μ integrable, then the function $z \mapsto \log|z-c| \in L^1(\mu)$ for every critical point c of f. If additionally μ is f-invariant, then also the function $z \mapsto \log|z-f(c)| \in L^1(\mu)$.*

Proof. That $\log|z-c| \in L^1(\mu)$ follows from the fact that near c we have $C^{-1}|z-q|^{q-1} \leq |f'(z)| \leq C|z-c|^{q-1}$, where $q \geq 2$ is the order of the critical point c and

$C \geq 1$ is a universal constant, and since out of any neighbourhood of the set of critical points of f, $|f'(z)|$ is uniformly bounded away from zero and infinity. In order to prove the second part of the lemma consider a ray R emanating from $f(c)$ such that $\mu(R) = 0$ and a disc $B(f(c), r)$ such that $f_c^{-1} : B(f(c), r) \setminus R \to \overline{\mathbb{C}}$, an inverse branch of f sending $f(c)$ to c, is well defined. Let $D = B(f(c), r) \setminus R$. We may additionally require $r > 0$ to be so small that $|z - f(c)| \asymp |f_c^{-1}(z) - c|^q$. It suffices to show that the integral $\int_D \log|z - f(c)|\, d\mu(z)$ is finite. And indeed, by f-invariance of μ we have

$$\begin{aligned}
\int_D \log|z - f(c)|\, d\mu(z) &= \int_X 1_D(z) \log|z - f(c)|\, d\mu(z) \\
&\asymp \int_X 1_D(z) \log|f_c^{-1}(z) - c|^q\, d\mu(z) \\
&= \int_X (1_D \circ f)(z) \log|z - c|^q\, d\mu(z) \\
&= \int_X 1_{f^{-1}(D)} \log|z - c|^q\, d\mu(z).
\end{aligned}$$

Note here that the function $1_D(z) \log|f_c^{-1}(z) - c|^q$ is indeed well defined on X, and that, unlike most of our comparability signs, the sign in the formula above means an additive comparability. The finiteness of the last integral follows from the first part of this lemma. ♣

Theorem 11.2.3. *Let $(Z, \mathcal{F}, \nu)$ be a measure space with an ergodic measure-preserving automorphism $T : Z \to Z$. Let $f : X \to X$ be a continuous map from a compact set $X \subset \overline{\mathbb{C}}$ onto itself, having a holomorphic extension onto a neighbourhood of X ($f \in \mathcal{A}(X)$). Suppose that μ is an f-invariant ergodic measure on X with positive Lyapunov exponent. Suppose also that $h : Z \to X$ is a measurable mapping such that $\nu \circ h^{-1} = \mu$ and $h \circ T = f \circ h$ ν-a.e.. Then for ν-a.e. $z \in Z$ there exists $r(z) > 0$ such that the function $z \mapsto r(z)$ is measurable, and the following is satisfied.*

For every $n \geq 1$ there exists $f_{x_n}^{-n} : B(x, r(z)) \to \overline{\mathbb{C}}$, an inverse branch of f^n sending $x = h(z)$ to $x_n = h(T^{-n}(z))$. In addition, for an arbitrary χ, $-\chi_\mu(f) < \chi < 0$ (not depending on z) and a constant $K(z)$,

$$|(f_{x_n}^{-n})'(y)| < K(z)\, \mathrm{e}^{\chi n} \quad \textit{and} \quad \frac{|(f_{x_n}^{-n})'(w)|}{|(f_{x_n}^{-n})'(y)|} \leq K$$

for all $y, w \in B(x, r(z))$. K is here the Koebe constant corresponding to the scale $1/2$.

Proof. Suppose first that $\mu\big(\bigcup_{n \geq 1} f^n(\mathrm{Crit}(f))\big) > 0$. Since μ is ergodic this implies that μ must be concentrated on a periodic orbit of an element $w \in \bigcup_{n \geq 1} f^n(\mathrm{Crit}(f))$. This means that $w = f^q(c) = f^{q+k}(c)$ for some $q, k \geq 1$ and $c \in \mathrm{Crit}(f)$, and

$$\mu(\{f^q(c), f^{q+1}(c), \dots, f^{q+k-1}(c)\}) = 1.$$

Since $\int \log|f'|\,d\mu > 0$, $|(f^k)'(f^q(c))| > 1$. Thus the theorem is obviously true for the set $h^{-1}(\{f^q(c), f^{q+1}(c), \ldots, f^{q+k-1}(c)\})$ of ν measure 1.

So, suppose that $\mu\big(\bigcup_{n\ge1} f^n(\mathrm{Crit}(f))\big) = 0$. Set $R = \min\{1, \mathrm{dist}(X, \overline{\mathbb{C}}\backslash U(f))\}$ and fix $\lambda \in (e^{\frac{1}{4}\chi}, 1)$. Consider $z \in Z$ such that $x = h(z) \notin \bigcup_{n\ge1} f^n(\mathrm{Crit}(f))$,

$$\lim_{n\to\infty} \frac{1}{n} \log|(f^n)'(h(T^{-n}(z))| = \chi_\mu(f),$$

and $x_n = h(T^{-n}(z)) \in B(f(\mathrm{Crit}(f)), R\lambda^n)$ only for finitely many n's. We shall first demonstrate that the set of points satisfying these properties is of full measure ν. Indeed, the first requirement is satisfied by our hypothesis, and the second is due to Birkhoff's Ergodic Theorem. In order to prove that the set of points satisfying the third condition has ν measure 1, not that

$$\begin{aligned}\sum_{n\ge1} \nu\big(T^n(h^{-1}(B(f(\mathrm{Crit}(f)), R\lambda^n)))\big) &= \sum_{n\ge1} \nu\big(h^{-1}(B(f(\mathrm{Crit}(f)), R\lambda^n))\big)\\ &= \sum_{n\ge1} \mu(B(f(\mathrm{Crit}(f)), R\lambda^n)) < \infty,\end{aligned}$$

where the last inequality is due to Lemmas 11.2.2 and 11.2.1. The application of the Borel–Cantelli Lemma now finishes the demonstration. Now fix an integer $n_1 = n_1(z)$ so large that $x_n = h(T^{-n}(z)) \notin B(f(\mathrm{Crit}(f)), R\lambda^n)$ for all $n \ge n_1$. Note that because of our choices there exists $n_2 \ge n_1$ such that $|(f^n)'(x_n)|^{-1/4} < \lambda^n$ for all $n \ge n_2$. Finally, set $S = \sum_{n\ge1} |(f^n)'(x_n)|^{-1/4}$, $b_n = \frac{1}{2} S^{-1} |(f^{n+1})'(x_{n+1})|^{\frac{-1}{4}}$, and

$$\Pi = \Pi_{n=1}^{\infty}(1 - b_n)^{-1},$$

which converges since the series $\sum_{n\ge1} b_n$ converges. Choose now $r = r(z)$ so small that $16r(z)\Pi K S^3 \le R$, all the inverse branches $f_{x_n}^{-n} : B(x_0, \Pi r(z)) \to \overline{\mathbb{C}}$ are well defined for all $n = 1, 2, \ldots, n_2$ and $\mathrm{diam}\big(f_{x_{n_2}}^{-n_2}(B\big(x_0, r\Pi_{k\ge n_2}(1 - b_k)^{-1}\big)\big) \le \lambda^{n_2} R$. We shall show by induction that for every $n \ge n_2$ there exists an analytic inverse branch $f_{x_n}^{-n} : B\big(x_0, r\Pi_{k\ge n}(1 - b_k)^{-1}\big) \to \mathbb{C}$, sending x_0 to x_n and such that

$$\mathrm{diam}\big(f_{x_n}^{-n}(B\big(x_0, r\Pi_{k\ge n}(1 - b_k)^{-1}\big)\big) \le \lambda^n R.$$

Indeed, for $n = n_2$ this follows immediately from our requirements imposed on $r(z)$. So, suppose that the claim is true for some $n \ge n_2$. Since $x_n = f_{x_n}^{-n}(x_0) \notin B(\mathrm{Crit}(f), R\lambda^n)$, and since $\lambda^n R \le R$, there exists an inverse branch $f_{x_{n+1}}^{-1} : B(x_n, \lambda^n R) \to \mathbb{C}$ sending x_n to x_{n+1}. Since $\mathrm{diam}\big(f_{x_n}^{-n}(B\big((x_0, r\Pi_{k\ge n}(1 - b_k)^{-1}\big)\big) \le \lambda^n R$, the composition $f_{x_{n+1}}^{-1} \circ f_{x_n}^{-n} B\big(x_0, r\Pi_{k\ge n}(1 - b_k)^{-1}\big) \to \mathbb{C}$ is well defined and forms the inverse branch of f^{n+1} that sends x_0 to x_{n+1}. By the Koebe distortion theorem we now estimate

$$\begin{aligned}
\operatorname{diam}\bigl(f_{x_{n+1}}^{-(n+1)}(B(x_0, r\Pi_{k\ge n+1}(1-b_k)^{-1}))\bigr)
&\le 2r\Pi_{k\ge n+1}(1-b_k)^{-1}|(f^{n+1})'(x_{n+1})|^{-1}Kb_n^{-3} \\
&\le 16r\Pi K S^3 |(f^{n+1})'(x_{n+1})|^{-1}|(f^{n+1})'(x_{n+1})|^{\frac{3}{4}} \\
&= 16r\Pi K S^3 |(f^{n+1})'(x_{n+1})|^{-\frac{1}{4}} \\
&\le R\lambda^{n+1},
\end{aligned}$$

where we wrote the last inequality sign because of our choice of r and the number n_2. Putting $r(z) = r/2$, the second part of this theorem now follows as a combined application of the equality $\lim_{n\to\infty}\frac{1}{n}\log|(f^n)'(x_n)| = \chi_\mu(f)$ and the Koebe distortion theorem. ♣

As an immediate consequence of Theorem 11.2.3 we get the following.

Corollary 11.2.4. *Assume the same notation and asumptions as in Theorem 11.2.3. Fix $\varepsilon > 0$. Then there exist a set $Z(\varepsilon) \subset Z$, the numbers $r(\varepsilon) \in (0,1)$ and $K(\varepsilon) \ge 1$ such that $\mu(Z(\varepsilon)) > 1-\varepsilon$, $r(z) \ge r(\varepsilon)$ for all $z \in Z(\varepsilon)$ and with $x_n = h(T^{-n}(z))$*

$$\begin{aligned}
K(\varepsilon)^{-1}\exp(-(\chi_\mu+\varepsilon)n) &\le |(f_{x_n}^{-n})'(y)| \\
&\le K(\varepsilon)\exp(-(\chi_\mu-\varepsilon)n) \ \text{ and } \ \frac{|(f_{x_n}^{-n})'(w)|}{|(f_{x_n}^{-n})'(y)|} \le K
\end{aligned}$$

for all $n \ge 1$, all $z \in Z(\varepsilon)$ and all $y, w \in B(x_0, r(\varepsilon))$. K is here the Koebe constant corresponding to the scale $1/2$.

Remark 11.2.5. In our future applications the system (Z, f, ν) will be usually given by the natural extension of the holomorphic system (f, μ).

11.3 Mañé's partition

In this section, basically following Mañé's book [Mañé 1987], we construct the so-called *Mañé's partition*, which will play an important role in the proof of a part of the Volume Lemma given in the next section. We begin with the following elementary fact.

Lemma 11.3.1. *If $x_n \in (0,1)$ for every $n \ge 1$ and $\sum_{n=1}^\infty nx_n < \infty$, then $\sum_{n=1}^\infty -x_n\log x_n < \infty$.*

Proof. Let $S = \{n : -\log x_n \ge n\}$. Then

$$\sum_{n=1}^\infty -x_n\log x_n = \sum_{n\notin S} -x_n\log x_n + \sum_{n\in S} -x_n\log x_n \le \sum_{n=1}^\infty nx_n + \sum_{n\in S} -x_n\log x_n.$$

Since $n \in S$ means that $x_n \le e^{-n}$, and since $\log t \le 2\sqrt{t}$ for all $t \ge 1$, we have

$$\sum_{n\in S} x_n\log\frac{1}{x_n} \le 2\sum_{n=1}^\infty x_n\sqrt{\frac{1}{x_n}} \le 2\sum_{n=1}^\infty e^{-\frac{1}{2}n} < \infty.$$

The proof is complete. ♣

The next lemma is the main and simultaneously the last result of this section.

Lemma 11.3.2. *If μ is a Borel probability measure concentrated on a bounded subset M of a Euclidean space, and $\rho : M \to (0,1]$ is a measurable function such that $\log \rho$ is integrable with respect to μ, then there exists a countable measurable partition, called Mañé's partition, $\mathcal{P}$ of M, such that $\mathrm{H}_\mu(\mathcal{P}) < \infty$ and*

$$\operatorname{diam}(\mathcal{P}(x)) \le \rho(x)$$

for μ-almost every $x \in M$.

Proof. Let q be the dimension of the Euclidean space containing M. Since M is bounded, there exists a constant $C > 0$ such that for every $0 < r < 1$ there exists a partition $\mathcal{P}_r$ of M of diameter $\le r$ and which consists of at most Cr^{-q} elements. For every $n \ge 0$ put $U_n = \{x \in M : e^{-(n+1)} < \rho(x) \le e^{-n}\}$. Since $\log \rho$ is a non-positive integrable function, we have

$$\sum_{n=1}^{\infty} -n\mu(U_n) \ge \sum_{n=1}^{\infty} \int_{U_n} \log \rho \, d\mu = \int_M \log \rho \, d\mu > -\infty$$

so that

$$\sum_{n=1}^{\infty} n\mu(U_n) < +\infty. \tag{11.3.1}$$

Now define $\mathcal{P}$ as the partition whose atoms are of the form $Q \cap U_n$, where $n \ge 0$ and $Q \in \mathcal{P}_{r_n}$, $r_n = e^{-(n+1)}$. Then

$$\mathrm{H}_\mu(\mathcal{P}) = \sum_{n=0}^{\infty} \Big(-\sum_{U_n \supset P \in \mathcal{P}} \mu(P) \log \mu(P)\Big).$$

But for every $n \ge 0$

$$\begin{aligned}
-\sum_{U_n \supset P \in \mathcal{P}} \mu(P) \log \mu(P) &= \mu(U_n) \sum_P -\frac{\mu(P)}{\mu(U_n)} \log\Big(\frac{\mu(P)}{\mu(U_n)}\Big) \\
&\quad - \mu(U_n) \sum_P \frac{\mu(P)}{\mu(U_n)} \log(\mu(U_n)) \\
&\le \mu(U_n)(\log C - q \log r_n) - \mu(U_n) \log \mu(U_n) \\
&\le \mu(U_n) \log C + q(n+1)\mu(U_n) - \mu(U_n) \log \mu(U_n).
\end{aligned}$$

Thus, summing over all $n \ge 0$, we obtain

$$\mathrm{H}_\mu(\mathcal{P}) \le \log C + q + q \sum_{n=0}^{\infty} n\mu(U_n) + \sum_{n=0}^{\infty} -\mu(U_n) \log \mu(U_n).$$

Therefore, looking at (11.3.1) and Lemma 10.3.1, we conclude that $\mathrm{H}_\mu(\mathcal{P})$ is finite. Also, if $x \in U_n$, then the atom $\mathcal{P}(x)$ is contained in some atom of $\P_{r_n}$ and therefore

$$\operatorname{diam}(\mathcal{P}(x)) \le r_n = e^{-(n+1)} < \rho(x).$$

Now the remark that the union of all the sets U_n is of measure 1 completes the proof. ♣

11.4 Volume Lemma and the formula $\mathrm{HD}(\mu) = \mathrm{h}_\mu(f)/\chi_\mu(f)$

In this section we keep the notation of Sections 11.1 and 11.2, and our main purpose is to prove the following two results, which generalize the respective results in Chapter 9.

Theorem 11.4.1. *If $f \in \mathcal{A}(X)$ and μ is an ergodic f-invariant measure with positive Lyapunov exponent, then* $\mathrm{HD}(\mu) = \mathrm{h}_\mu(f)/\chi_\mu(f)$.

Theorem 11.4.2 (Volume Lemma). *With the assumptions of Theorem 11.4.1*

$$\lim_{r\to 0} \frac{\log(\mu(B(x,r)))}{\log r} = \frac{\mathrm{h}_\mu(f)}{\chi_\mu(f)}$$

for μ-a.e. $x \in X$.

In view of Theorem 8.6.5, Theorem 11.4.1 follows from Theorem 11.4.2, and we need only prove the latter one. Let us first prove

$$\liminf_{r\to 0} \frac{\log(\mu(B(x,r)))}{\log r} \geq \frac{\mathrm{h}_\mu(f)}{\chi_\mu(f)} \tag{11.4.1}$$

for μ-a.e. $x \in X$. By Corollary 9.1.10 there exists a finite partition $\mathcal{P}$ such that for an arbitrary $\varepsilon > 0$ and every x in a set X_o of full measure μ there exists $n(x) \geq 0$ such that, for all $n \geq n(x)$,

$$B(f^n(x), e^{-\varepsilon n}) \subset \mathcal{P}(f^n(x)). \tag{11.4.2}$$

From now on let us work in the natural extension $(\tilde{X}, \tilde{f}, \tilde{\mu})$. Let $\tilde{X}(\varepsilon)$ and $r(\varepsilon)$ be given by Corollary 11.2.4: that is, $\tilde{X}(\varepsilon) = Z(\varepsilon)$. In view of Birkhoff's Ergodic Theorem there exists a measurable set $\tilde{F}(\varepsilon) \subset \tilde{X}(\varepsilon)$ such that $\tilde{\mu}(\tilde{F}(\varepsilon)) = \tilde{\mu}(\tilde{X}(\varepsilon))$ and

$$\lim_{n\to\infty} \frac{1}{n} \sum_{j=1}^{n-1} \chi_{\tilde{X}(\varepsilon)} \circ \tilde{f}^n(\tilde{x}) = \tilde{\mu}(\tilde{X}(\varepsilon))$$

for every $\tilde{x} \in \tilde{F}(\varepsilon)$. Let $F(\varepsilon) = \pi(\tilde{F}(\varepsilon))$. Then $\mu(F(\varepsilon)) = \tilde{\mu}(\pi^{-1}(F(\varepsilon)) \geq \tilde{\mu}(\tilde{F}(\varepsilon)) = \tilde{\mu}(\tilde{X}(\varepsilon))$ converges to 1 if $\varepsilon \searrow 0$. Consider now $x \in F(\varepsilon) \cap X_o$, and take $\tilde{x} \in \tilde{F}(\varepsilon)$ such that $x = \pi(\tilde{x})$. Then by the above there exists an increasing sequence $\{n_k = n_k(x) : k \geq 1\}$ such that $\tilde{f}^{n_k}(\tilde{x}) \in \tilde{X}(\varepsilon)$ and

$$\frac{n_{k+1} - n_k}{n_k} \leq \varepsilon \tag{11.4.3}$$

for every $k \geq 1$. Moreover, we can assume that $n_1 \geq n(x)$. Consider now an integer $n \geq n_1$ and the ball $B\big(x, Cr(\varepsilon)\exp(-(\chi_\mu + (2 + \log\|f'\|)\varepsilon)n)\big)$, where $0 < C < (Kr(\varepsilon))^{-1}$ is a constant (possibly depending on x) so small that

$$f^q\big(B(x, Cr(\varepsilon)\exp -(\chi_\mu + (2 + \log\|f'\|)\varepsilon)n)\big) \subset P(f^q(x)) \tag{11.4.4}$$

for every $q \le n_1$, and $K(\varepsilon) \ge 1$ is the constant appearing in Corollary 11.2.4. Take now any q, $n_1 \le q \le n$, and associate k such that $n_k \le q \le n_{k+1}$. Since $\tilde{f}^{n_k}(\tilde{x}) \in \tilde{X}(\varepsilon)$, and since $\pi(\tilde{f}^{n_k}(\tilde{x})) = f^{n_k}(x)$, Corollary 11.2.4 produces a holomorphic inverse branch $f_x^{-n_k} : B(f^{n_k}(x), r(\varepsilon)) \to \overline{\mathbb{C}}$ of f^{n_k} such that $f_x^{-n_k}(f^{n_k}(x)) = x$ and

$$f_x^{-n_k}(B(f^{n_k}(x), r(\varepsilon))) \supset B\big(x, K(\varepsilon)r(\varepsilon)^{-1}\exp(-(\chi_\mu + \varepsilon)n_k)\big).$$

Since $B(x, Cr(\varepsilon)\exp-(\chi_\mu + (2+\log\|f'\|)\varepsilon)n) \subset B\big(x, K(\varepsilon)^{-1}r(\varepsilon)\exp-(\chi_\mu + \varepsilon)n_k)\big)$, it follows from Corollary 11.2.4 that

$$f^{n_k}\big(B(x, Cr(\varepsilon)\exp-(\chi_\mu + (2+\log\|f'\|)\varepsilon)n)\big) \subset\subset B\big(f^{n_k}(x), \\ CKr(\varepsilon)e^{-\chi_\mu(n-n_k)}\exp(\varepsilon(n_k - (2+\log\|f'\|)n))\big).$$

Since $n \ge n_k$, and since by (11.4.3) $q - n_k \le \varepsilon n_k$, we therefore obtain

$$\begin{aligned}
&f^q\big(B(x, Cr(\varepsilon)\exp-(\chi_\mu + (2+\log\|f'\|)\varepsilon)n)\big) \subset \\
&\quad \subset B(f^q(x), CK(\varepsilon)r(\varepsilon)e^{-\chi_\mu(n-n_k)}\exp(\varepsilon(n_k - (2+\log\|f'\|)n)) \\
&\qquad \times \exp((q-n_k)\log\|f'\|) \\
&\quad \subset B(f^q(x), CK(\varepsilon)r(\varepsilon)\exp\big(\varepsilon(n_k\log\|f'\| + n_k - 2n - n\log\|f'\|)\big) \\
&\quad \subset B(f^q(x), CK(\varepsilon)r(\varepsilon)e^{-\varepsilon n}) \subset B(f^q(x), e^{-\varepsilon q}).
\end{aligned}$$

Combining this, (11.4.2) and (11.4.4), we get

$$B\big(x, Cr(\varepsilon)\exp-(\chi_\mu + (2+\log\|f'\|)\varepsilon)n)\big) \subset \bigvee_{j=0}^{n} f^{-j}(\mathcal{P})(x).$$

Therefore, applying Theorem 2.5.5 (the Shannon–McMillan–Breiman Theorem), we have

$$\liminf_{n\to\infty} -\frac{1}{n}\log\mu\big(B(x, Cr(\varepsilon)\exp-(\chi_\mu + (2+\log\|f'\|)\varepsilon)n)\big) \ge \mathrm{h}_\mu(f, \mathcal{P}) \ge \mathrm{h}_\mu(f) - \varepsilon.$$

This means that, denoting the number $Cr(\varepsilon)\exp-(\chi_\mu + (2+\log\|f'\|)\varepsilon)n)$ by r_n, we have

$$\liminf_{n\to\infty} \frac{\log\mu(B(x, r_n)}{\log r_n} \ge \frac{\mathrm{h}_\mu(f) - \varepsilon}{\chi_\mu(f) + (2+\log\|f'\|)\varepsilon}.$$

Now, since $\{r_n\}$ is a geometric sequence, and since $\varepsilon > 0$ can be taken arbitrarily small, we conclude that for μ-a.e. $x \in X$

$$\liminf_{n\to\infty} \frac{\log\mu(B(x, r)}{\log r} \ge \frac{\mathrm{h}_\mu(f)}{\chi_\mu(f)}.$$

This completes the proof of (11.4.1). ♣

Remark. Since here $X \subset \mathbb{C}$, we could have considered a partition $\mathcal{P}$ of a neighbourhood of X in $\mathbb{C}$, where $\partial_{\mathcal{P},a}$ would have a more standard sense: see remark after Corollary 9.1.9.

Now let us prove that

$$\limsup_{r\to 0} \frac{\log(\mu(B(x,r)))}{\log r} \le \mathrm{h}_\mu(f)/\chi_\mu(f) \tag{11.4.5}$$

for μ-a.e. $x \in X$.

In order to prove this formula we again work in the natural extension $(\tilde{X}, \tilde{f}, \tilde{\mu})$, and we apply Pesin theory. In particular, the sets $\tilde{X}(\varepsilon)$, $\tilde{F}(\varepsilon) \subset \tilde{X}(\varepsilon)$ and the radius $r(\varepsilon)$, produced in Corollary 11.2.4, have the same meaning as in the proof of (11.4.1). To begin with, note that there exist two numbers $R > 0$ and $0 < Q < \min\{1, r(\varepsilon)/2\}$ such that the following two conditions are satisfied:

$$\text{If } z \notin B(\mathrm{Crit}(f), R), \text{ then } f|_{B(z,Q)} \text{ is injective.} \tag{11.4.6}$$

$$\text{If } z \in B(\mathrm{Crit}(f), R), \text{ then } f|_{B(z,Q\,\mathrm{dist}(z,\mathrm{Crit}(f)))} \text{ is injective.} \tag{11.4.7}$$

Observe also that if z is sufficiently close to a critical point c, then $f'(z)$ is of order $(z-c)^{q-1}$, where $q \ge 2$ is the order of the critical point c. In particular, the quotient of $f'(z)$ and $(z-c)^{q-1}$ remains bounded away from 0 and ∞, and therefore there exists a constant number $B > 1$ such that $|f'(z)| \le B\,\mathrm{dist}(z, \mathrm{Crit}(f))$. So, in view of Lemma 11.2.2, the logarithm of the function $\rho(z) = Q\min\{1, \mathrm{dist}(z, \mathrm{Crit}(f))$ is integrable, and consequently Lemma 11.3.2 applies. Let $\mathcal{P}$ be the Mañé's partition produced by this lemma. Then $B(x, \rho(x)) \supset \mathcal{P}(x)$ for μ-a.e. $x \in X$, say for a subset X_ρ of X of measure 1. Consequently

$$B_n(x,\rho) = \bigcap_{j=0}^{n-1} f^{-j}\big(B(f^j(x), \rho(f^j(x)))\big) \supset \mathcal{P}_0^n(x) \tag{11.4.8}$$

for every $n \ge 1$ and every $x \in X_\rho$. By our choice of Q and the definition of ρ, the function f is injective on all balls $B(f^j(x), \rho(f^j(x)))$, $j \ge 0$, and therefore f^k is injective on the set $B_n(x,\rho)$ for every $0 \le k \le n-1$. Now, let $x \in F(\varepsilon) \cap X_\rho$ and let k be the greatest subscript such that $q = n_k(x) \le n-1$. Denote by f_x^{-q} the unique holomorphic inverse branch of f^q produced by Corollary 11.2.4, which sends $f^q(x)$ to x. Clearly $B_n(x,\rho) \subset f^{-q}(B(f^q(x), \rho(f^q(x))))$, and since f^q is injective on $B_n(x,\rho)$ we even have

$$B_n(x,\rho) \subset f_x^{-q}(B(f^q(x), \rho(f^q(x)))).$$

By Corollary 11.2.4, $\mathrm{diam}\big(f_x^{-q}(B(f^q(x), \rho(f^q(x))))\big) \le K\exp(-q(\chi_\mu - \varepsilon))$. Since, by (11.4.3), $n \le q(1+\varepsilon)$, we finally deduce that

$$B_n(x,\rho) \subset B\left(x, K\exp\left(-n\frac{\chi_\mu - \varepsilon}{1+\varepsilon}\right)\right).$$

Thus, in view of (11.4.8),

$$B\left(x, K\exp\left(-n\frac{\chi_\mu - \varepsilon}{1+\varepsilon}\right)\right) \supset \mathcal{P}_0^n(x).$$

Therefore, denoting the radius of the ball above by r_n, it follows from the Shanon–McMillan–Breiman theorem that, for μ-a.e $x \in X$,

$$\limsup_{n\to\infty} -\frac{1}{n}\log\mu(B(x,r_n) \le \mathrm{h}_\mu(f,\mathcal{P}) \le \mathrm{h}_\mu(f).$$

So

$$\limsup_{n\to\infty} \frac{\log\mu(B(x,r_n)}{\log r_n} \le \frac{\mathrm{h}_\mu(f)}{\chi_\mu(f)-\varepsilon}(1+\varepsilon).$$

Now, since $\{r_n\}$ is a geometric sequence, and since ε can be taken arbitrarily small, we conclude that for μ-a.e. $x \in X$

$$\limsup_{n\to\infty} \frac{\log\mu(B(x,r)}{\log r} \le \frac{\mathrm{h}_\mu(f)}{\chi_\mu(f)}.$$

This completes the proof of (11.4.5) and, because of (11.4.1), also the proof of Theorem 11.4.2. ♣ ♣

11.5 Pressure-like definition of the functional $\mathrm{h}_\mu + \int \phi\, d\mu$

In this section we prepare some general tools used in the next section to approximate topological pressure on hyperbolic sets. No smoothness is assumed here; we work in a purely metric setting only. Our exposition is similar to that contained in Chapter 3.

Let $T : X \to X$ be a continuous map of a compact metric space (X, ρ), and let μ be a Borel probability measure on X. Given $\varepsilon > 0$ and $0 \le \delta \le 1$, a set $E \subset X$ is said to be $\mu - (n, \varepsilon, \delta)$-*spanning* if

$$\mu\Big(\bigcup_{x\in E} B_n(x,\varepsilon)\Big) \ge 1-\delta.$$

Let $\phi : X \to \mathbb{R}$ be a continuous function. We define

$$Q_\mu(T,\phi,n,\varepsilon,\delta) = \inf_E \Big\{\sum_{x\in E} \exp S_n\phi(x)\Big\},$$

where the infimum is taken over all $\mu-(n, \varepsilon, \delta)$-spanning sets E. The main result of this section is the following.

Theorem 11.5.1. *For every $0 < \delta < 1$ and every ergodic measure μ,*

$$\begin{aligned}\mathrm{h}_\mu(T) + \int \phi\, d\mu &= \lim_{\varepsilon\to 0}\liminf_{n\to\infty} \frac{1}{n}\log Q_\mu(T,\phi,n,\varepsilon,\delta)\\ &= \lim_{\varepsilon\to 0}\limsup_{n\to\infty} \frac{1}{n}\log Q_\mu(T,\phi,n,\varepsilon,\delta).\end{aligned}$$

Proof. Denote the number following the first equality sign by $\underline{\mathrm{P}}_\mu(T,\phi,\delta)$, and the number following the second equality sign by $\overline{\mathrm{P}}_\mu(T,\phi,\delta)$. First, following essentially the proof of Part I of Theorem 3.4.1, we shall show that

$$\underline{\mathrm{P}}_\mu(T,\phi,\delta) \ge \mathrm{h}_\mu(T) + \int \phi\,d\mu. \tag{11.5.1}$$

Indeed, as in that proof, consider a finite partition $\mathcal{U} = \{A_1,\ldots,A_s\}$ of X into Borel sets and compact sets $B_i \subset A_i$, $i = 1,2,\ldots,A_s\}$, such that for the partition $\mathcal{V} = \{B_1,\ldots,B_s, X\setminus(B_1\cup\ldots\cup B_s)\}$ we have $H_\mu(\mathcal{U}|\mathcal{V}) \le 1$. For every $\theta > 0$ and $q \ge 1$, set

$$X_q = \Big\{x \in X : -\frac{1}{n}\log\mu(\mathcal{V}^n(x)) \ge \mathrm{h}_\mu(T,\mathcal{V}) - \theta \;\text{ for all }\; n \ge q$$
$$\frac{1}{n}S_n\phi(x) \ge \int \phi\,d\mu - \theta \;\text{ for all }\; n \ge q\Big\}.$$

Now fix $0 \le \delta < 1$. It follows from the Shannon–McMillan–Breiman Theorem and Birkhoff's Ergodic Theorem that for q large enough $\mu(X_q) > \delta$. Take $0 < \varepsilon < \frac{1}{2}\min\{\rho(B_i,B_j) : 1 \le i < j \le s\} > 0$ so small that

$$|\phi(x) - \phi(y)| < \theta$$

if $\rho(x,y) \le \varepsilon$. Since for every $x \in X$ the set $B_n(x,\varepsilon)\cap X_q$ can be covered by at most 2^n elements of $\mathcal{V}^n$,

$$\mu(B_n(x,\varepsilon)\cap X_q) \le \exp\big(n(\log 2 - \mathrm{h}_\mu(T,\mathcal{V}) + \theta)\big).$$

Now let E be a $\mu-(n,\varepsilon,\delta)$-spanning set for $n \ge q$, and consider the set $E' = \{x \in E : B_n(x,\varepsilon)\cap X_q \ne \emptyset\}$. Take any point $y(x) \in B_n(x,\varepsilon)\cap X_q$. Then by the choice of ε, $S_n\phi(x) - S_n\phi(y) > -n\theta$. Therefore we have

$$\begin{aligned}
&\sum_{x\in E} \exp S_n\phi(x)\exp\Big(-n\Big(\mathrm{h}_\mu(T,\mathcal{V}) + \int\phi\,d\mu - 3\theta - \log 2\Big)\Big)\\
&\quad\ge \sum_{x\in E'} \exp S_n\phi(x)\exp\Big(-n\Big(\mathrm{h}_\mu(T,\mathcal{V}) + \int\phi\,d\mu - 3\theta - \log 2\Big)\Big)\\
&\quad= \sum_{x\in E'} \exp(S_n\phi(x) - n\int\phi\,d\mu)\exp\Big(-n(\mathrm{h}_\mu(T,\mathcal{V}) - 3\theta - \log 2)\Big)\\
&\quad= \sum_{x\in E'} \exp\Big(S_n\phi(x) - S_n\phi(y) + S_n\phi(y) - n\int\phi\,d\mu\Big)\\
&\qquad\times\exp\big(-n\big(\mathrm{h}_\mu(T,\mathcal{V}) - 3\theta - \log 2\big)\big)\\
&\quad\ge \sum_{x\in E'} \exp(-n\theta)\exp(-n\theta)\exp(2n\theta)\exp\big(-n(\mathrm{h}_\mu(T,\mathcal{V}) - \theta - \log 2)\big)\\
&\quad= \sum_{x\in E'} \exp\big(n(\log 2 - \mathrm{h}_\mu(T,\mathcal{V}) + \theta)\big)\\
&\quad\ge \sum_{x\in E'} \mu(B_n(x,\varepsilon)\cap X_q) \ge \mu(X_q) - \delta > 0,
\end{aligned}$$

which implies that

$$Q_\mu(T,\phi,n,\varepsilon,\delta) \geq \mathrm{h}_\mu(T,\mathcal{V}) + \int \phi\,d\mu - 3\theta - \log 2.$$

Since $\theta > 0$ is an arbitrary number, and since $\mathrm{h}_\mu(T,\mathcal{U}) \leq \mathrm{h}_\mu(T,\mathcal{V}) + H_\mu(\mathcal{U}|\mathcal{V}) \leq \mathrm{h}_\mu(T,\mathcal{V}) + 1$, letting $\varepsilon \to 0$, we get

$$\underline{\mathrm{P}}_\mu(T,\phi,\delta) \geq \mathrm{h}_\mu(T,\mathcal{U}) - 1 + \int \phi\,d\mu - \log 2.$$

Therefore, by the definition of entropy of an automorphism, $\underline{\mathrm{P}}_\mu(T,\phi,\delta) \geq \mathrm{h}_\mu(T) + \int \phi\,d\mu - \log 2 - 1$. Using now the standard trick, always applied in this setting, whose point is to replace T by its arbitrary iterates T^k and ϕ by $S_k\phi$, we obtain $k\underline{\mathrm{P}}_\mu(T,\phi,\delta) \geq k\,\mathrm{h}_\mu(T) + k\int \phi\,d\mu - \log 2 - 1$. So, dividing this inequality by k, and letting $k \to \infty$, we finally obtain

$$\underline{\mathrm{P}}_\mu(T,\phi,\delta) \geq \mathrm{h}_\mu(T) + \int \phi\,d\mu.$$

Now let us prove that

$$\overline{\mathrm{P}}_\mu(T,\phi,\delta) \leq \mathrm{h}_\mu(T) + \int \phi\,d\mu, \tag{11.5.2}$$

where $\overline{\mathrm{P}}_\mu(T,\phi,\delta)$ denotes limsup appearing in the statement of Theorem 11.5.1. Fix $0 < \delta < 1$; then $\varepsilon > 0$ and $\theta > 0$. Let $\mathcal{P}$ be a finite partition of X of diameter $\leq \varepsilon$. By the Shannon–McMillan–Breiman Theorem and Birkhoff's Ergodic Theorem there exists a Borel set $Z \subset X$ such that $\mu(Z) > 1 - \delta$ and

$$\frac{1}{n}S_n\phi(x) \leq \int \phi\,d\mu + \theta \quad\text{and}\quad -\frac{1}{n}\log\mu(\mathcal{P}^n(x)) \leq \mathrm{h}_\mu(T) + \theta \tag{11.5.3}$$

for every n large enough and all $x \in Z$. From each element of $\mathcal{P}^n$ having a non-empty intersection with Z choose one point obtaining, say, a set $\{x_1, x_2, \ldots, x_q\}$. Then $B_n(x_j,\varepsilon) \supset \mathcal{P}^n(x_j)$ for every $j = 1, 2, \ldots, q$, and therefore the set $\{x_1, x_2, \ldots, x_q\}$ is $\mu - (n,\varepsilon,\delta)$-spanning. By the second part of (10.5.3) we have $q \leq \exp(n(\mathrm{h}_\mu(T) + \theta))$. Using also the first part of (10.5.3), we get

$$\sum_{j=1}^{q} \exp S_n\phi(x_j) \leq \exp(n(\mathrm{h}_\mu(T) + \theta + \int \phi\,d\mu + \theta)).$$

Therefore $Q_\mu(T,\phi,n,\varepsilon,\delta) \leq \exp(n(\mathrm{h}_\mu(T) + \theta + \int \phi\,d\mu + \theta))$, and letting consequtively $n \to \infty$ and $\varepsilon \to 0$, we obtain $\overline{\mathrm{P}}_\mu(T,\phi,\delta) \leq \mathrm{h}_\mu(T) + \int \phi\,d\mu + 2\theta$. Since θ is an arbitrary positive number, (11.5.2) is proved. This and (11.5.1) complete the proof of Theorem 11.5.1. ♣

11.6 Katok's theory: hyperbolic sets, periodic points, and pressure

In this section we return to the setting of Section 10.1. Let X be a compact subset of the closed complex plane $\overline{\mathbb{C}}$, and let $f : X \to X$ be a continuous map that can be analytically extended to an open neighbourhood $U = U(f)$ of X.

Let μ be an f-invariant ergodic measure on X with positive Lyapunov exponent, and let $\phi : \overline{U} \to \mathbb{R}$ be a real continuous function. Our first aim is to show that the number $\mathrm{h}_\mu(f)+\int \phi\, d\mu$ can be approximated by the topological pressures of ϕ on hyperbolic subsets of U, and then as a straightforward consequence we shall obtain the same approximation for the topological pressure $\mathrm{P}(f,\phi)$.

Theorem 11.6.1. *If μ is an f-invariant ergodic measure on X with positive Lyapunov exponent χ_μ, and if $\phi : \overline{U} \to \mathbb{R}$ is a real-valued continuous function, then there exists a sequence X_k, $k = 1, 2, \ldots$, of compact f-invariant subsets of U, (topologically) Cantor sets, such that for every k the restriction $f|_{X_k}$ is a conformal expanding repeller,*

$$\liminf_{k\to\infty} \mathrm{P}(f|_{X_k}, \phi) \geq \mathrm{h}_\mu(f) + \int \phi\, d\mu, \tag{11.6.1}$$

and if μ_k is any ergodic f-invariant measure on X_k, then the sequence μ_k, $k = 1, 2, \ldots$, converges to μ in the weak--topology on $\overline{U}$. Moreover $\chi_{\mu_k}(f|_{X_k}) = \int \log|f'|\, d\mu_k \to \int \log|f'|\, d\mu = \chi_\mu(f)$. If X is repelling then one finds $X_k \subset X$.*

In particular, μ_k can be supported by individual periodic orbits in X_k. For more properties of X_k see the remarks after the proof.

Proof. Since $\mathrm{P}(f|_{X_k}, \phi + c) = \mathrm{P}(f|_{X_k}, \phi) + c$, and since $\mathrm{h}_\mu(f) + \int(\phi + c)\, d\mu = \mathrm{h}_\mu(f)+\int \phi\, d\mu+c$, adding a constant if necessary, we can assume that ϕ is positive: that is, that $\inf \phi > 0$. As in Section 10.2, we work in the natural extension $(\tilde{X}, \tilde{f}, \tilde{\mu})$. Given $\delta > 0$, let $\tilde{X}(\delta)$ and $r(\delta)$ be produced by Corollary 11.2.4. The set $\pi(\tilde{X}(\delta))$ is assumed to be compact. This corollary implies the existence of a constant $\chi' > 0$ (possibly with a smaller radius $r(\delta)$) such that

$$\mathrm{diam}\big(f_{x_n}^{-n}(B(\pi(\tilde{x}), r(\delta)))\big) \leq e^{-n\chi'} \tag{11.6.2}$$

for all $\tilde{x} \in \tilde{X}(\delta)$ and $n \geq 0$. Fix a countable basis $\{\psi_j\}_{j=1}^{\infty}$ of the Banach space $C(\overline{U})$ of all continuous real-valued functions on $\overline{U}$. Fix $\theta > 0$ and an integer $s \geq 1$. In view of Theorem 11.5.1 and continuity of functions ϕ and ψ_i there exists $\varepsilon > 0$ so small that

$$\liminf_{n\to\infty} \frac{1}{n} \log Q_\mu(T, \phi, n, \varepsilon, \delta) - (\mathrm{h}_\mu(f) + \int \phi\, d\mu) > -\theta, \tag{11.6.3}$$

if $|x - y| < \varepsilon$, then

$$|\phi(x) - \phi(y)| < \theta \tag{11.6.4}$$

and

$$|\psi_i(x) - \psi_i(y)| < \frac{1}{2}\theta \tag{11.6.5}$$

for all $i = 1, 2, \ldots, s$.

Set $\beta = r(\delta)/2$, and fix a finite $\beta/2$-spanning set of $\pi(\tilde{X}(\delta))$, say $\{x_1, \ldots, x_t\}$. That is, $B(x_1, \beta/2) \cup \ldots \cup B(x_t, \beta/2) \supset \pi(\tilde{X}(\delta/2))$. Let $\mathcal{U}$ be a finite partition of X with diameter $< \beta/2$, and let n_1 be sufficiently large that

$$\exp(-n_1\chi') < \min\{\beta/3, K^{-1}\}. \tag{11.6.6}$$

Given $n \geq 1$, define

$$\begin{aligned}\tilde{X}_{n,s} = \{\tilde{x} \in \tilde{X}(\delta) : \tilde{f}^q(\tilde{x}) \in \tilde{X}(\delta) \ \text{ and } \ \pi(\tilde{f}^q(\tilde{x})) \in \mathcal{U}(\pi(\tilde{x}))\\ \text{for some } \ q \in [n+1, (1+\theta)n]\\ \left|\frac{1}{k}S_k(\psi_i)(\pi(\tilde{x})) - \int \psi_i \, d\mu\right| < \frac{1}{2}\theta\\ \text{for every } k \geq n \ \text{ and all } \ i = 1, 2, \ldots, s\}.\end{aligned}$$

By Birkhoff's Ergodic Theorem, $\lim_{n\to\infty} \mu(\tilde{X}_{n,s}) = \mu(\tilde{X}(\delta)) > 1 - \delta$. Therefore there exists $n \geq n_1$ so large that $\mu(\tilde{X}_{n,s}) > 1 - \delta$. Let $X_{n,s} = \pi((\tilde{X}_{n,s}))$. Then $\mu(X_{n,s}) > 1 - \delta$, and let $E_n \subset X_{n,s}$ be a maximal (n, ε)-separated subset of $X_{n,s}$. Then E_n is a spanning set of $X_{n,s}$, and therefore it follows from (11.6.3) that, for all n large enough,

$$\frac{1}{n}\log \sum_{x\in E_n} \exp S_n\phi(x) - (\mathrm{h}_\mu(f) + \int \phi \, d\mu) > -\theta.$$

Equivalently,

$$\sum_{x\in E_n} \exp(S_n\phi(x)) > \exp(n(\mathrm{h}_\mu(f) + \int \phi \, d\mu - \theta)).$$

For every $q \in [n+1, (1+\theta)n]$, let

$$V_q = \{x \in E_n : f^q(x) \in \mathcal{U}(x)\},$$

and let $m = m(n)$ be a value of q that maximizes $\sum_{x\in V_q} \exp(S_n\phi(x))$. Since $\bigcup_{q=n+1}^{(1+\theta)n} V_q = E_n$, we thus obtain

$$\begin{aligned}\sum_{x\in V_m} \exp S_n\phi(x) &\geq (n\theta)^{-1} \sum_{q=n+1}^{(1+\theta)n} \sum_{x\in V_q} \exp S_n\phi(x)\\ &\geq (n\theta)^{-1} \sum_{x\in E_n} \exp(S_n\phi(x)) \geq \exp(n(\mathrm{h}_\mu(f) + \int \phi \, d\mu - 2\theta)).\end{aligned}$$

Consider now the sets $V_m \cap B(x_j, \beta/2)$, $1 \leq j \leq t$, and choose the value $i = i(m)$ of j that maximizes $\sum_{x\in V_m\cap B(x_j,\beta/2)} \exp(S_n\phi(x))$. Thus, writing D_m for $V_m \cap B(x_{i(m)}, \beta/2)$, we have $V_m = \bigcup_{j=1}^t V_m \cap B(x_i, \beta/2)$ and

$$\sum_{x\in D_m} \exp S_n\phi(x) \geq \frac{1}{t}\exp(n(\mathrm{h}_\mu(f) + \int \phi\, d\mu - 2\theta)).$$

Since ϕ is positive, this implies that

$$\sum_{x\in D_m} \exp S_m\phi(x) \geq \frac{1}{t}\exp(n(\mathrm{h}_\mu(f) + \int \phi\, d\mu - 2\theta)). \tag{11.6.7}$$

Now, if $x \in D_m$, then $|f^m(x) - x_i| \leq |f^m(x) - x| + |x - x_i| < \beta/2 + \beta/2 = \beta$, and therefore

$$f^m(x) \in B(x_i, \beta) \subset B(f^m(x), 2\beta).$$

Thus, by (11.6.2) and as $m \geq n \geq n_1$, we have $\mathrm{diam}\big(f_{\tilde{x}}{-m}(B(f^m(x), 2\beta)\big) \leq \exp(-m\chi') < \beta/3$, where $\tilde{x} \in \pi^{-1}(x) \cap \tilde{X}_{n,s}$. Therefore

$$f_{\tilde{x}}^{-m}(B(x_i, \beta)) \subset B\left(x_i, \frac{\beta}{2} + \frac{\beta}{3}\right) = B\left(x_i, \frac{5}{6}\beta\right).$$

In particular,

$$\overline{f_{\tilde{x}}^{-m}(B(x_i, \beta))} \subset B(x_i, \beta). \tag{11.6.8}$$

Consider now two distinct points $y_1, y_2 \in D_m$. Then $f_{\tilde{y}_2}^{-m}(B(x_i, \beta)) \cap f_{\tilde{y}_1}^{-m}(B(x_i, \beta)) = \emptyset$, and decreasing β slightly, if necessary, we may assume that

$$f_{\tilde{y}_2}^{-m}(\overline{B(x_i, \beta)}) \cap f_{\tilde{y}_1}^{-m}(\overline{B(x_i, \beta)}) = \emptyset.$$

Let

$$\xi = \min\left\{\beta, \min\left\{\mathrm{dist}\big(f_{\tilde{y}_2}^{-m}(\overline{B(x_i, \beta)}), f_{\tilde{y}_1}^{-m}(\overline{B(x_i, \beta)}\big) : y_1, y_2 \in D_m, y_1 \neq y_2\right\}\right\}.$$

Now define inductively the sequence of sets $\{X^{(j)}\}_{j=0}^\infty$ contained in $U(f)$ by setting

$$X^{(0)} = \overline{(B(x_i, \beta)} \text{ and } X^{(j+1)} = \bigcup_{x\in D_m} f_{\tilde{x}}^{-m}(X^{(j)}).$$

By (11.6.8),$\{X^{(j)}\}_{j=0}^\infty$ is a descending sequence of non-empty compact sets, and therefore the intersection

$$X^* = X^*(\theta, s) = \bigcap_{j=0}^\infty X^{(j)}$$

is also a non-empty compact set. Moreover, by the construction $f^m(X^*) = X^*$, $f^m|_{X^*}$ is topologically conjugate to the full one-sided shift generated by an alphabet consisting of $\#D_m$ elements, and it immediately follows from Corollary 11.2.4 that $f^m|_{X^*}$ is an expanding map. Since $f^m|_{X^*}$ is an open map, by Lemma 6.1.2 the triple (f^m, X^*, U_m) is a conformal expanding repeller with a sufficiently small neighborhood U_m of X^*. Thus $(f, X(\theta, s), W_s)$, is a conformal expanding set, where

$$X(\theta, s) = \bigcup_{l=0}^{m-1} f^l(X^*) \text{ and } W_s = \bigcup_{l=0}^{m-1} f^l(U_m).$$

This can be extended to a conformal expanding repeller $\hat{X}(\theta, s)$ in W_s by Proposition 4.5.6.

Now fix an integer $j \geq 1$. For any j-tuple $(z_0, z_1, \ldots, z_{j-1})$, $z_l \in D_m$ choose exactly one point y from the set $f_{z_{j-1}}^{-m} \circ f_{z_{j-2}}^{-m} \circ \ldots \circ f_{z_0}^{-m}(X^*)$, and denote the made-up set by A_j. Since, by (11.6.4) and (11.6.6), $S_{jm}\phi(y) \geq \sum_{l=0}^{j-1} S_m\phi(z_l) - jm\theta$, we see that

$$\sum_{y \in A_j} \exp S_{jm}\phi(y) \geq \Big(\sum_{x \in D_m} \exp S_m\phi(x) \Big)^j \exp(-jm\theta)$$

and

$$\frac{1}{j} \log \sum_{y \in A_j} \exp S_{jm}\phi(y) \geq \log \sum_{x \in D_m} \exp S_m\phi(x) - m\theta.$$

In view of the definition of ξ, the set A_j is (j, ξ)-separated for f^m, and ξ is an expansive constant for f^m. Hence, letting $j \to \infty$, we obtain

$$\begin{aligned} \mathrm{P}(f^m|_{X^*}, S_m\phi) &\geq \log \sum_{x \in D_m} \exp S_m\phi(x) - m\theta \\ &\geq n\big(\mathrm{h}_\mu(f) + \int \phi \, d\mu - 2\theta\big) - \log t - m\theta, \end{aligned}$$

where the last inequality was written in view of (11.6.7). Since $n + 1 \leq m \leq n(1 + \theta)$, and since $\inf \phi > 0$ (and consequently $\mathrm{h}_\mu(f) + \int \phi \, d\mu > 0$), we get

$$\begin{aligned} \mathrm{P}(f|_{\hat{X}(\theta,s)}, \phi) \geq \mathrm{P}(f|_{X(\theta,s)}, \phi) &= \frac{1}{m} \mathrm{P}(f^m|_{X(\theta,s)}, S_m\phi) \geq \frac{1}{m} \mathrm{P}(f^m|_{X^*}, S_m\phi) \\ &\geq \frac{1}{1+\theta} \left(\mathrm{h}_\mu(f) + \int \phi \, d\mu - 2\theta \right) - \frac{\log t}{m} - \theta. \end{aligned}$$

Supposing now that n (and consequently also m) was chosen sufficiently large, we get

$$\mathrm{P}(f|_{X(\theta,s)}, \phi) \geq \frac{1}{1+\theta} (\mathrm{h}_\mu(f) + \int \phi \, d\mu) - 4\theta.$$

If now ν is any ergodic f-invariant measure on $\hat{X}(\theta, s)$, then it follows from the definition of the set $\tilde{X}_{n,s}$, the construction of the set $X(\theta, s)$ and since $\hat{X}(\theta, s)$ is arbitrarily close to it, and else by the Birkhoff Ergodic Theorem, that $| \int \psi_i \, d\nu - \int \psi_i \, d\mu| < \theta$ for every $i = 1, 2, \ldots, s$. A similar estimate for $\log |f'|$ follows from the definition of $\tilde{X}(\delta)$ and Corollary 11.2.4. Therefore, for example, the sets $X_k = \hat{X}(1/k, k)$ satisfy the assertions of Theorem 11.6.1.

Finally, if the set X is repelling, that is if $\bigcap_{n \geq 0} f^{-n}(U) = X$, then the sets X_k are all contained in X. ♣

Remark 11.6.2. In fact the sets X_k in Theorem 11.6.1 can be found independent of ϕ.

To see this, set $\phi \equiv 0$. Find X_k for this function. We get

$$\limsup_{k \to \infty} \mathrm{h}_{\mathrm{top}}(f|X_k) \geq \mathrm{h}_\mu(f).$$

Let μ_k be a measure of maximal entropy on X_k, for $k = 1, 2, ...$: that is, $\mathrm{h}_{\mu_k}(f) = \mathrm{h}_{\mathrm{top}}(f|_{X_k})$. Consider an arbitrary continuous function $\phi : \overline{U} \to \mathbb{R}$. Then $\mu_k \to \mu$ weakly*: hence $\int \phi \, d\mu_k \to \int \phi \, d\mu$. Hence, with the use of the Variational Principle,

$$\liminf_{k\to\infty} \mathrm{P}(f|_{X_k}, \phi) \geq \liminf_{k\to\infty} (\mathrm{h}_{\mu_k}(f) + \int \phi \, d\mu_k) \geq \mathrm{h}_\mu(f) + \int \phi \, d\mu.$$

Note also that although for the maximal measures μ_k we have

$$\liminf_{k\to\infty} \mathrm{h}_{\mu_k}(f) \geq \mathrm{h}_\mu(f),$$

this need not be true for all sequences μ_k.

It is possible to find the sets X_k with f topologically mixing on them, common for (finite) families of measures μ, and thus common for families of ϕ in Corollary 11.6.4, by building 'bridges'. For details see [Gelfert, Przytycki & Rams 2009] (based on [Prado 1997]).

Remark 11.6.3. One can find (correct) X_k above so that each f_{X_k} is topologically transitive, and even topologically mixing.

This follows from the general Theorem 4.3.8 on the existence of spectral decomposition. It implies that for each k there exists $\Omega_k \subset X_k$ such that $f|_{\Omega_k}$ is open (see Lemma 4.3.10), topologically transitive, and satisfying $\mathrm{h}_{\mathrm{top}}(f|_{\Omega_k}) = \mathrm{h}_{\mathrm{top}}(f|_{X_k})$ (see Exercise 4.4). Hence, using μ_k measures of maximal entropy on Ω_k, we obtain (11.6.1) as in Remark 11.6.2.

In fact one can prove that (f, X_k), found in the proof of Theorem 11.6.1, are already topologically transitive. Indeed, $(f^n, X^*(\theta, s))$ are topologically mixing, since by construction they are topologically conjugate with one-sided shifts. Hence each $(f, X(\theta, s))$ is topologically transitive. So the transition matrix $A = (a_{i,j})$, considered in the proof of Proposition 4.5.6, defined by $a_{i,j} = 1$ if there exists $g_{i,j} : U_j \to U_i$, a branch of f^{-1} with non-empty $g(U_j) \cap X$, and 0 otherwise, is irreducible. This follows from the existence of a trajectory dense in X. The same matrix A is the transition matric of a topological Markov chain Σ_A topologically conjugate to the resulting (f, X_k).

This (f, X_k) extends to a topologically mixing Cantor expanding repeller by adding a 'bridge' of length mutually prime with respect to m. We leave this to the reader as an exercise.

Corollary 11.6.4. *If* $\mathrm{P}(f, \phi) > \sup \phi$, *then there exists a sequence* X_k, $k = 1, 2, \ldots$, *of compact f-invariant subsets of an arbitrarily small neighbourhood of* X *such that for every* k, $(X_k, f|_{X_k})$ *is a Cantor conformal expanding repeller satisfying*

$$\liminf_{k\to\infty} \mathrm{P}(f|_{X_k}, \phi) \geq \mathrm{P}(f, \phi). \tag{11.6.9}$$

If X *is repelling, then one finds* $X_k \subset X$ *and*

$$\lim_{k\to\infty} \mathrm{P}(f|_{X_k}, \phi) = \mathrm{P}(f, \phi).$$

Proof. By the Variational Principle $\mathrm{P}(f,\phi) = \lim_{k\to\infty}(\mathrm{h}_{\nu_k}(f) + \int \phi\, d\nu_k)$ for a sequence of Borel probability measures ν_k on X. Because $\mathrm{P}(f,\phi) > \sup\phi$ we have $\mathrm{h}_{\nu_k}(f) > 0$ for k large enough. Hence, owing to Ruelle's inequality (Theorem 11.1.1), $\chi_{\nu_k} > 0$. Now we apply Theorem 11.6.1 and for each k large enough find X_k satisfying the assertion of the theorem for $\mu = \nu_k$ such that

$$\mathrm{P}(f|_{X_k},\phi) \geq \mathrm{h}_{\nu_k}(f) + \int \phi\, d\nu_k - 1/k.$$

Any limit for $k \to \infty$ satisfies (11.6.9). In the case where X is repelling, the estimate from the other side follows immediately from $X_k \subset X$.

Our last immediate conclusion concerns periodic points.

Corollary 11.6.5. *If $f : X \to X$ is repelling and $\mathrm{h_{top}}(f) > 0$, then f has infinitely many periodic points. Moreover, the number of periodic points of period n grows exponentially fast with n.*

Exercises

11.1. Prove the following general version of Theorem 11.1.1. Let X be a compact f-invariant subset of a smooth Riemannian manifold for a C^1 mapping $f : U \to M$, defined on a neighbourhood U of X. Let μ be an f-invariant Borel probability measure X. Then

$$h_\mu(f) \leq \int_X \max\{0, \chi_\mu^+(x)\}\, d\mu(x),$$

where $\chi_\mu^+(x) = \lim_{n\to\infty} \frac{1}{n} \log \|(Df^n)^\wedge\|$. Here Df^n is the differential and $(Df^n)^\wedge$ is the exterior power, the linear operator between the exterior algebras generated by the tangent spaces at x and $f^n(x)$. The norm is induced by the Riemann metric. $\|(Df^n)^\wedge\|$ is the supremum of the volumes of Df^n-images of unit cubes in k-dimensional subspaces of T_xM with $k = 0, 1, \ldots, \dim M$.

Bibliographical notes

Theorem 11.1.1 relies on [Ruelle 1978b].

The content of Sections 11.2, 11.5 and 11.6 corresponds to facts from Pesin's and Katok's theories for diffeomorphisms [Katok & Hasselblatt 1995, Supplement 5]. For Theorem 11.2.3 see for example [Przytycki, Urbański & Zdunik 1989]. Mañé's partition for diffeomorphisms was discussed in [Mañé 1987]. References to the Volume Lemma are for example [Mañé 1988], [Przytycki 1985] and [Ledrappier 1984].

The problem of constructing $X_k \subset X$ in the case where (X, f) is not a repeller, in Theorem 11.6.1, was recently dealt with in [Przytycki 2005a]

The theorem in Exercise 11.1 is due to Ruelle: see Ruelle [1978b]. Compare Theorem 11.1.1.

12

Conformal measures

12.1 General notion of conformal measures

Let $T : X \to X$ be a continuous map of a compact metric space (X, ρ), and let $g : X \to \mathbb{R}$ be a non–negative measurable function. A Borel probability measure m on X is said to be *g–conformal* for $T : X \to X$ if

$$m(T(A)) = \int_A g\, dm \tag{12.1.1}$$

for any Borel set $A \subset X$, such that $T|_A$ is injective and $T(A)$ is Borel measurable. Sets with this property will be called *special sets.*

If $g > 0$, then T is backward quasi-invariant (non-singular) with respect to the g-conformal measure m: see Chapter 5, Section 5.2.

Now consider an arbitrary Borel probability measure m on X, backward quasi-invariant for T. Assume that T is uniformly bounded to one or countable to one, that is, $X = \bigcup X_j$, where X_j are measurable, pairwise disjoint, and for each j the map $T|_{X_j} \to T(X_j)$ is a measurable isomorphism, as in Section 5.2. Denote $\hat{g} := d(m \circ (T|_{X_j})^{-1})/dm$.

Consider, as in Section 5.2, the operator $\mathcal{L}_m : L^1(m) \to L^1_m$, defined in the present notation and the notation of (5.2.8) by

$$\mathcal{L}_m(u)(x) = \mathcal{L}_{\log \hat{g}}(x) = \sum_{T(y)=x} u(y)\hat{g}(y).$$

So, for all $u \in L^1(m)$,

$$\int \mathcal{L}_m^*(1\!\!1) u\, dm = \int 1\!\!1 \mathcal{L}_m(u)\, dm = \int u\, dm$$

(see (5.2.4)). We conclude that, by Proposition 5.2.1, if m is a g-conformal measure and $g > 0$, then $\hat{g} = 1/g$ and

$$\mathcal{L}^*_{-\log g}(1\!\!1) = \mathcal{L}^*_m(1\!\!1) = 1\!\!1. \tag{12.1.2}$$

Conversely, if m is backward quasi-invariant, $\hat{g} > 0$ and (12.1.2) holds, then for $g = 1/\hat{g}$ the measure m is g-conformal.

Note that even if T is continuous, $\mathcal{L}_m$ need not map $C(X)$ into $C(X)$, unlike for T open, continuous. However, if we assume $\mathcal{L}_m : C(X) \to C(X)$ and T being uniformly bounded to one, then $\mathcal{L}_m^* : C^*(X) \to C^*(X)$. Then, under the above constraints concerning positivity, we conclude with the following proposition.

Proposition 12.1.1. *A probability measure m is g-conformal if and only if*

$$\mathcal{L}^*_{-\log g}(m) = m.$$

Now, since we can have troubles with the operator $\mathcal{L}^*$ for T not open, we shall provide another general method of constructing conformal measures, called the *Patterson–Sullivan method.* The construction will make use of the following simple fact. For a sequence $\{a_n : n \geq 1\}$ of reals the number

$$c = \limsup_{n\to\infty} \frac{a_n}{n} \tag{12.1.3}$$

will be called the *transition parameter* of $\{a_n : n \geq 1\}$. It is uniquely determined by the property that

$$\sum_{n\geq 1} \exp(a_n - ns)$$

converges for $s > c$ and diverges for $s < c$. For $s = c$ the sum may converge or diverge. By a simple argument one obtains the following.

Lemma 12.1.2. *There exists a sequence $\{b_n : n \geq 1\}$ of positive reals such that*

$$\sum_{n=1}^{\infty} b_n \exp(a_n - ns) \begin{cases} < \infty & s > c \\ = \infty & s \leq c \end{cases}$$

and $\lim_{n\to\infty} \frac{b_n}{b_{n+1}} = 1$.

Proof. If $\sum \exp(a_n - nc) = \infty$, put $b_n = 1$ for every $n \geq 1$. If $\sum \exp(a_n - nc) < \infty$, choose a sequence $\{n_k : k \geq 1\}$ of positive integers such that $\lim_{k\to\infty} n_k n_{k+1}^{-1} = 0$ and $\varepsilon_k := a_{n_k} n_k^{-1} - c \to 0$. Setting

$$b_n = \exp\left(n\left(\frac{n_k - n}{n_k - n_{k-1}}\varepsilon_{k-1} + \frac{n - n_{k-1}}{n_k - n_{k-1}}\varepsilon_k\right)\right) \quad \text{for } n_{k-1} \leq n < n_k,$$

it is easy to check that the lemma follows. ♣

Getting back to dynamics, let $\{E_n\}_{n=1}^{\infty}$ be a sequence of finite subsets of X such that

$$T^{-1}(E_n) \subset E_{n+1} \qquad \text{for every} \quad n \geq 1. \tag{12.1.4}$$

Let $\phi : X \to \mathbb{R}$ be an arbitrary measurable function of bounded absolute value.

Functions of the form $-\phi + \text{Const}$ will play the role of 'potential' functions; $\exp(-\phi + \text{Const})$ corresponds to the Jacobian g discussed above.

Let

$$a_n = \log\Big(\sum_{x \in E_n} \exp(S_n\phi(x))\Big),$$

where $S_n\phi = \sum_{0 \le k < n} \phi \circ T^k$. Denote by c the transition parameter of this sequence. Choose a sequence $\{b_n : n \ge 1\}$ of positive reals as in Lemma 11.1.2 for the sequence $\{a_n : n \ge 1\}$. For $s > c$ define

$$M_s = \sum_{n=1}^{\infty} b_n \exp(a_n - ns) \tag{12.1.5}$$

and the normalized measure

$$m_s = \frac{1}{M_s} \sum_{n=1}^{\infty} \sum_{x \in E_n} b_n \exp(S_n\phi(x) - ns)\delta_x, \tag{12.1.6}$$

where δ_x denotes the unit mass at the point $x \in X$. Let A be a special set. Using (12.1.4) and (12.1.6) it follows that

$$\begin{aligned} m_s(T(A)) &= \frac{1}{M_s} \sum_{n=1}^{\infty} \sum_{x \in E_n \cap T(A)} b_n \exp(S_n\phi(x) - ns) \\ &= \frac{1}{M_s} \sum_{n=1}^{\infty} \sum_{x \in A \cap T^{-1}E_n} b_n \exp(S_n\phi(T(x)) - ns) \\ &= \frac{1}{M_s} \sum_{n=1}^{\infty} \sum_{x \in A \cap E_{n+1}} b_n \exp[S_{n+1}\phi(x) - (n+1)s] \exp(s - \phi(x)) \\ &\quad - \frac{1}{M_s} \sum_{n=1}^{\infty} \sum_{x \in A \cap (E_{n+1} \setminus T^{-1}E_n)} b_n \exp(S_n\phi(T(x)) - ns). \end{aligned} \tag{12.1.7}$$

Set

$$\begin{aligned} \Delta_A(s) = \Bigg| &\frac{1}{M_s} \sum_{n=1}^{\infty} \sum_{x \in A \cap E_{n+1}} b_n \exp[S_{n+1}\phi(x) - (n+1)s] \exp(s - \phi(x)) \\ &- \int_A \exp(c - \phi)\, dm_s \Bigg| \end{aligned}$$

and observe that

$$\begin{aligned} \Delta_A(s) = \frac{1}{M_s} \Bigg| &\sum_{n=1}^{\infty} \sum_{x \in A \cap E_{n+1}} \exp[S_{n+1}\phi(x) - (n+1)s] \\ &\times \exp(-\phi(x)) \big[b_n e^s - b_{n+1} e^c\big] - b_1 \sum_{x \in A \cap E_1} e^{c-s} \Bigg| \end{aligned}$$

$$\leq \frac{1}{M_s}\sum_{n=1}^{\infty}\sum_{x\in A\cap E_{n+1}}\left|\frac{b_n}{b_{n+1}}-e^{c-s}\right|b_{n+1}\exp(s-\phi(x))$$
$$\times \exp[S_{n+1}\phi(x)-(n+1)s]+\frac{1}{M_s}b_1\exp(c-s)\,\sharp(A\cap E_1)$$
$$\leq \frac{1}{M_s}\sum_{n=1}^{\infty}\sum_{x\in E_{n+1}}\left|\frac{b_n}{b_{n+1}}-e^{c-s}\right|b_{n+1}\exp(s-\phi(x))$$
$$\times \exp[S_{n+1}\phi(x)-(n+1)s]+\frac{1}{M_s}b_1\exp(c-s)\,\sharp E_1.$$

By Lemma 11.1.2 we have $\lim_{n\to\infty} b_{n+1}/b_n = 1$ and $\lim_{s\downarrow c} M_s = \infty$. Therefore

$$\lim_{s\downarrow c}\Delta_A(s)=0 \tag{12.1.8}$$

uniformly for all special sets A.

Any weak accumulation point, when $s \searrow c$, of the measures $\{m_s : s > c\}$ defined by (12.1.6) will be called a *limit measure* (associated to the function ϕ and the sequence $\{E_n : n \geq 1\}$).

In order to find conformal measures among the limit measures, it is necessary to examine (12.1.7) in greater detail. To begin with, for a Borel set $D \subset X$, consider the following condition:

$$\lim_{s\downarrow c}\frac{1}{M_s}\sum_{n=1}^{\infty}\sum_{x\in D\cap(E_{n+1}\setminus T^{-1}E_n)} b_n\exp[S_n\phi(T(x))-ns]=0. \tag{12.1.9}$$

We shall need the following definitions.

A point $x \in X$ is said to be singular for T if at least one of the following two conditions is satisfied:

$$\text{There is no open neighbourhood } U \text{ of } x \text{ such that } T|_U \text{ is injective.} \tag{12.1.10}$$
$$\forall_{\varepsilon>0}\exists_{0<r<\varepsilon} \text{ such that } T(B(x,r)) \text{ is not an open subset of } X. \tag{12.1.11}$$

The set of all singular points is denoted by $\mathrm{Sing}(T)$, the set of all points satisfying condition (12.1.10) is denoted by $\mathrm{Crit}(T)$, and the set of all points satisfying condition (12.1.11) is denoted by $X_0(T)$.

It is easy to give examples where $X_0 \cap \mathrm{Crit}(T) \neq \emptyset$. If $T : X \to X$ is an open map, no point satisfies condition (12.1.11): that is, $X_0(T) = \emptyset$.

Lemma 12.1.3. *Let m be a Borel probability measure on X, and let Γ be a compact set containing* $\mathrm{Sing}(T)$. *If (12.1.1) for g integrable holds for every special set A whose closure is disjoint from Γ and such that $m(\partial A) = m(\partial T(A)) = 0$, then (12.1.1) continues to hold for every special set A disjoint from Γ.*

Proof. Let A be a special set disjoint from Γ. Fix $\varepsilon > 0$. Since on the complement of Γ the map T is open, for each point $x \in A$ there exists an open neighbourhood

$U(x)$ of x such that $T|_{U(x)}$ is a homeomorphism, $m(\partial U(x)) = m(\partial T(U(x))) = 0$, $\overline{U(x)} \cap \Gamma = \emptyset$, and such that

$$\int_{\cup U(x) \setminus A} g\, dm < \varepsilon.$$

Choose a countable family $\{U_k\}$ from $\{U(x)\}$ that covers A, and define recursively $A_1 = U_1$ and $A_n = U_n \setminus \bigcup_{k<n} U_k$. By the assumption of the lemma, each set A_k satisfies (11.1.1) and hence

$$\begin{aligned} m(T(A)) &= m\Big(\bigcup_{k=1}^{\infty} T(A \cap A_k)\Big) \le \sum_{k=1}^{\infty} m(T(A_k)) \\ &= \sum_{k=1}^{\infty} \int_{A_k} g\, dm = \int_A g\, dm + \sum_{k=1}^{\infty} \int_{A_k \setminus A} g\, dm \\ &\le \int_A g\, dm + \varepsilon. \end{aligned}$$

If $\varepsilon \to 0$, it follows that

$$m(T(B)) \le \int_B g\, dm$$

for any special set B disjoint from Γ. Using this fact, the lower bound for $m(T(A))$ is obtained from the following estimate, if $\varepsilon \to 0$:

$$\begin{aligned} m(T(A)) &= m\Big(\bigcup_{k=1}^{\infty} T(A \cap A_k)\Big) = \sum_{k=1}^{\infty} m(T(A \cap A_k)) \\ &= \sum_{k=1}^{\infty} (m(T(A_k)) - m(T(A_k \setminus A))) \ge \sum_{k=1}^{\infty} \int_{A_k} g\, dm - \int_{A_k \setminus A} g\, dm \\ &= \int_{\bigcup_{k \ge 1} A_k} g\, dm - \int_{\bigcup_{k\ge 1} A_k \setminus A} g\, dm \ge \int_A g\, dm - \varepsilon. \end{aligned}$$

This proves the lemma. ♣

Lemma 12.1.4. *Let $\phi : X \to \mathbb{R}$ be a function of bounded absolute value and m be a limit measure as above, and let Γ be a compact set containing* Sing(T). *Assume that every special set $D \subset X$ with $m(\partial D) = m(\partial T(D)) = 0$ and $\bar{D} \cap \Gamma = \emptyset$ satisfies condition (12.1.9). Then $m(T(A)) = \int_A \exp(c - \phi)\, dm$ for every special set A disjoint from Γ.*

Proof. Let $D \subset X$ be a special set such that $\bar{D} \cap \Gamma = \emptyset$ and $m(\partial D) = m(\partial T(D)) = 0$. It follows immediately from (12.1.7)–(12.1.9) that $m(T(D)) = \int_D \exp(c - \phi)\, dm$. Applying Lemma 12.1.3 completes the proof. ♣

Lemma 12.1.5. *Let m be a limit measure. If condition (12.1.9) is satisfied for $D = X$, then $m(T(A)) \ge \int_A \exp(c - f)\, dm$ for every special set A disjoint from* Crit(T).

Proof. Suppose first that A is compact and $m(\partial A) = 0$. From (11.1.7), (12.1.8) and the assumption one obtains

$$\lim_{s \in J} |m_s(T(A)) - \int_A \exp(c - \phi)\, dm_s| = 0,$$

where J denotes the sub-sequence along which m_s converges to m. Since $T(A)$ is compact, this implies

$$m(T(A)) \geq \liminf_{s \in J} m_s(T(A)) = \lim_{s \in J} \int_A \exp(c - \phi)\, dm_s = \int_A \exp(c - \phi)\, dm.$$

Now, drop the assumption that $m(\partial A) = 0$ but keep A compact, and assume additionally that for some $\varepsilon > 0$ the ball $B(A, \varepsilon)$ is also special. Choose a descending sequence A_n of compact subsets of $B(A, \varepsilon)$ whose intersection equals A and $m(\partial A_n) = 0$ for every $n \geq 0$. By what has already been proved,

$$m(T(A)) = \lim_{n \to \infty} m(T(A_n)) \geq \int_{A_n} \exp(c - \phi)\, dm = \int_A \exp(c - \phi)\, dml.$$

The next step is to prove the lemma for A, an arbitrary open special set disjoint from $\mathrm{Crit}(T)$, by partitioning it by countably many compact sets. Then one approximates from above special sets of sufficiently small diameters by special open sets, and the last step is to partition an arbitrary special set disjoint from $\mathrm{Crit}(T)$ by sets of such small diameters that the lemma holds. ♣

Lemma 12.1.6. *Let Γ be a compact subset of X containing* $\mathrm{Sing}(T)$. *Suppose that for every integer $n \geq 1$ there are a continuous function $g_n : X \to X$ and a measure m_n on X satisfying (12.1.1) for $g = g_n$ and for every special set $A \subset X$ with*

$$\overline{A} \cap \Gamma = \emptyset \tag{a}$$

and satisfying

$$m_n(B) \geq \int_B g_n\, dm_n$$

for any special set $B \subset X$ such that $B \cap \mathrm{Crit}(T) = \emptyset$. Suppose, moreover, that the sequence $\{g_n\}_{n=1}^{\infty}$ converges uniformly to a continuous function $g : X \to \mathbb{R}$. Then for any weak accumulation point m of the sequence $\{m_n\}_{n=1}^{\infty}$ we have

$$m(T(A)) = \int_A g\, dm \tag{b}$$

for all special sets $A \subset X$ such that $A \cap \Gamma = \emptyset$ and

$$m(T(B)) \geq \int_B g\, dm \tag{c}$$

for all special sets $B \subset X$ such that $B \cap \mathrm{Crit}(T) = \emptyset$.

Moreover, if (a) is replaced by

$$\overline{A} \cap (\Gamma \setminus (\mathrm{Crit}(T) \setminus X_0(T))) = \emptyset, \tag{a'}$$

then for any $x \in \mathrm{Crit}(T) \setminus X_0(T)$

$$m(\{T(x)\}) \le g(x) m(\{x\}) \le q(x) m(\{T(x)\}), \tag{d}$$

where $q(x)$ *denotes the maximal number of pre-images of single points under the transformation* T *restricted to a sufficiently small neighbourhood of* x.

The proof of property (b) is a simplification of the proof of Lemma 12.1.4, and the proof of property (c) is a simplification of the proof of Lemma 12.1.5. The proof of (d) uses the same techniques and is left for the reader.

12.2 Sullivan's conformal measures and dynamical dimension: I

As in Chapter 11, let X denote a compact subset of the Riemann sphere $\overline{\mathbb{C}}$, and let $f \in \mathcal{A}(X)$, which means that $f : X \to X$ is a continuous map that can be analytically extended to an open neighbourhood $U(f)$ of X.

Let $t \ge 0$. Any $|f'|^t$-conformal measure for $f : X \to X$ is called a t-conformal Sullivan's measure or, even shorter, a t-conformal measure. Rewriting (12.1.1) this means that

$$m(f(A)) = \int_A |f'|^t \, dm \tag{12.2.1}$$

for every special set $A \subset X$. An obvious but important property of conformal measures is formulated in the following.

Lemma 12.2.1. *If* $f : X \to X$ *is topologically exact, then every Sullivan's conformal measure is positive on non-empty open sets of* X.

In particular, it follows from this lemma that if f is topologically exact, then for every $r > 0$

$$M(r) = \inf\{m(B(x,r)) : x \in X\} > 0 \tag{12.2.2}$$

Denote by $\delta(f)$ the infimum over all exponents $t \ge 0$ for which a t-conformal measure for $f : X \to X$ exists. Call $\delta(f)$ the *conformal dimension* of X.

Our aim in the next two sections is to show the existence of conformal measures and, moreover, to establish more explicit dynamical characterization of the number $\delta(f)$. We are going to prove that, under some additional assumptions, $\delta(f)$ coincides with the dynamical dimension $\mathrm{DD}(X)$ of X and the hyperbolic dimension $\mathrm{HyD}(X)$ of X, which are defined as follows:

$$\mathrm{DD}(X) = \sup\{\mathrm{HD}(\mu) : \mu \in M_e^+(f)\}$$
$$\mathrm{HyD}(X) = \sup\{\mathrm{HD}(Y) : f|_Y \text{ is a conformal expanding repeller}\}.$$

In HyD one can even restrict Y to being topological Cantor sets.

In this section we shall prove the following two results.

Lemma 12.2.2. *If $f: X \to X$ is topologically exact, then* $\mathrm{DD}(X) \le \delta(f)$.

Proof. Our main idea 'to get to a large scale' is the same as in [Denker & Urbański 1991b]. However, to carry it out we use Pesin's theory, described in Section 10.2, instead of Mañé's partition, applied in [Denker & Urbański 1991b]. So, let $\mu \in M_e^+(f)$ and let m be a t-conformal measure. We again work in the natural extension $(\tilde{X}, \tilde{f}, \tilde{\mu})$. Fix $\varepsilon > 0$, and let $\tilde{X}(\varepsilon)$ and $r(\varepsilon)$ be given by Corollary 11.2.4. In view of the Birkhoff Ergodic Theorem there exist a measurable set $\tilde{F}(\varepsilon) \subset \tilde{X}(\varepsilon)$ such that $\tilde{\mu}(\tilde{F}(\varepsilon)) = \tilde{\mu}(\tilde{X}(\varepsilon))$ and an increasing sequence $\{n_k = n_k(\tilde{x}) : k \ge 1\}$ such that $\tilde{f}^{n_k}(\tilde{x}) \in \tilde{X}(\varepsilon)$ for every $k \ge 1$. Let $F(\varepsilon) = \pi(\tilde{F}(\varepsilon))$. Then $\mu(F(\varepsilon)) = \tilde{\mu}(\pi^{-1}(F(\varepsilon)) \ge \tilde{\mu}(\tilde{F}(\varepsilon)) \ge 1 - \varepsilon$. Consider now $x \in F(\varepsilon)$, and take $\tilde{x} \in \tilde{F}(\varepsilon)$ such that $x = \pi(\tilde{x})$. Since $\tilde{f}^{n_k}(\tilde{x}) \in \tilde{X}(\varepsilon)$, and since $\pi(\tilde{f}^{n_k}(\tilde{x}) = f^{n_k}(x)$, Corollary 11.2.4 produces a holomorphic inverse branch $f_x^{-n_k} : B(f^{n_k}(x), r(\varepsilon)) \to \overline{\mathbb{C}}$ of f^{n_k} such that $f_x^{-n_k} f^{n_k}(x) = x$ and

$$f_x^{-n_k}\big(B(f^{n_k}(x), r(\varepsilon))\big) \subset B\big(x, K|(f^{n_k})'(x)|^{-1} r(\varepsilon)\big).$$

Set $r_k(x) = K|(f^{n_k})'(x)|^{-1} r(\varepsilon)$. Then by Corollary 11.2.4 and the t-conformality of m,

$$\begin{aligned} m(B(x, r_k(x))) &\ge K^{-t}|(f^{n_k})'(x)|^{-t} m\big(B(f^{n_k}(x), r(\varepsilon))\big) \\ &\ge M(r(\varepsilon))^{-1} K^{-2t} r(\varepsilon)^{-t} r_k(x)^t. \end{aligned}$$

Therefore it follows from Theorem 8.5.1 (the Besicovitch Covering Theorem) that $\Lambda_t(F(\varepsilon)) \le M(r(\varepsilon)) K^{2t} r(\varepsilon)^t b(2) < \infty$. Hence $\mathrm{HD}(F(\varepsilon)) \le t$. Since $\mu\big(\bigcup_{n=1}^\infty F(1/n)\big) = 1$, it implies that $\mathrm{HD}(\mu) \le t$. This completes the proof. ♣

Theorem 12.2.3. *If $f: X \to X$ is topologically exact and X is a repelling set for f, then* $\mathrm{HyD}(X) = \mathrm{DD}(X)$.

Proof. In order to see that $\mathrm{HyD}(X) \le \mathrm{DD}(X)$ note only that in view of Theorems 5.3.2 and 9.1.6 and Corollary 9.1.7 there exists $\mu \in M_e^+(f|_Y) \subset M_e^+(f)$ such that $\mathrm{HD}(\mu) = \mathrm{HD}(Y)$. In order to prove that $\mathrm{DD}(X) \le \mathrm{HyD}(X)$ we shall use Katok's theory from Section 11.6 applied to μ, an arbitrary ergodic invariant measure of positive entropy. First, for every integer $n \ge 0$, define on X a new continuous function

$$\phi_n = \max\{-n, \log|f'|\}.$$

Then $\phi_n \ge \log|f'|$ and $\phi_n \searrow \log|f'|$ pointwise on X. Since, in addition, $\phi_n \le \log||f'||_\infty$, it follows from the Lebesgue Monotone Convergence Theorem that $\lim_{n\to\infty} \int \phi_n \, d\mu = \chi_\mu(f) = \int \log|f'| \, d\mu > 0$. Fix $\varepsilon > 0$. Then for all n sufficiently large, say $n \ge n_0$, $\int \phi_n d\mu \le \chi_\mu/(1-\varepsilon)$, which implies that

$$\mathrm{h}_\mu(f) = \mathrm{HD}(\mu)\chi_\mu \ge (1-\varepsilon)\,\mathrm{HD}(\mu) \int \phi_n \, d\mu. \tag{12.2.3}$$

Fix such $n \ge n_0$. Let $X_k \subset X$, $k \ge 0$, be the sequence of conformal expanding repellers produced in Theorem 11.6.1 for the measure μ and the function

$-\mathrm{HD}(\mu)\phi_n$, and let μ_k be an equilibrium state of the map $f|_{X_k}$ and the potential $-\mathrm{HD}(\mu)\phi_n$ restricted to X_k. It follows from the second part of Theorem 11.6.1 that $\lim_{k\to\infty}\int \phi_n\,d\mu_k = \int \phi_n\,d\mu > 0$. Thus, by Theorem 11.6.1 and (12.2.3),

$$\begin{aligned}\liminf_{k\to\infty}\Big(\mathrm{h}_{\mu_k} - \mathrm{HD}(\mu)\int \phi_n\,d\mu_k\Big) &= \liminf_{k\to\infty}\mathrm{P}\big(f|_{X_k}, -\mathrm{HD}(\mu)\phi_n\big)\\ &\geq \mathrm{h}_\mu(f) - \mathrm{HD}(\mu)\int \phi_n\,d\mu \geq -\varepsilon\,\mathrm{HD}(\mu)\int \phi_n\,d\mu.\end{aligned}$$

Hence, for all k large enough,

$$\begin{aligned}\mathrm{h}_{\mu_k} &\geq \mathrm{HD}(\mu)\int \phi_n\,d\mu_k - 2\varepsilon\,\mathrm{HD}(\mu)\int \phi_n\,d\mu\\ &\geq \mathrm{HD}(\mu)\int \phi_n\,d\mu_k - 3\varepsilon\,\mathrm{HD}(\mu)\int \phi_n\,d\mu_k\\ &= (1-3\varepsilon)\,\mathrm{HD}(\mu)\int \phi_n\,d\mu_k \geq (1-3\varepsilon)\,\mathrm{HD}(\mu)\int \log|f'|\,d\mu_k.\end{aligned}$$

Thus

$$\mathrm{HD}(X_k) \geq \mathrm{HD}(\mu_k) = \frac{\mathrm{h}_{\mu_k}(f)}{\chi_{\mu_k}} \geq (1-3\varepsilon)\,\mathrm{HD}(\mu).$$

So, letting $\varepsilon \to 0$ completes the proof. ♣

12.3 Sullivan's conformal measures and dynamical dimension: II

In this section $f : \overline{\mathbb{C}} \to \overline{\mathbb{C}}$ is assumed to be a rational map of degree ≥ 2, and X is its Julia set $J(f)$. Nevertheless, it is worth mentioning that some results proved here continue to hold under the weaker assumption that $f|_X$ is open or X is a perfect locally maximal set for f. By $\mathrm{Crit}(f)$ we denote the set of all critical points contained in the Julia set $J(f)$.

Lemma 12.3.1. *If $z \in J(f)$ and $\overline{\{f^n(z) : n \geq 0\}} \cap \mathrm{Crit}(f) = \emptyset$, then the series $\sum_{n=1}^{\infty} |(f^n)'(z)|^{\frac{1}{3}}$ diverges.*

Proof. Putting ∞ in Crit (f), we can assume that the forward orbit of z is bounded away from ∞, and use the Euclidean metric on $\mathbb{C}$. By the assumption there exists $\varepsilon > 0$ such that for every $n \geq 0$ the map f restricted to the ball $B(f^n(z), \varepsilon)$ is injective. Since f is uniformly continuous, there exists $0 < \alpha < 1$ such that, for every $x \in \overline{\mathbb{C}}$,

$$f(B(x, \alpha\varepsilon)) \subset B(f(x), \varepsilon). \tag{12.3.1}$$

Suppose that the series $\sum_{n=1}^{\infty}|(f^n)'(z)|^{\frac{1}{3}}$ converges. Then there exists $n_0 \geq 1$ such that $\sup_{n\geq n_0}(2|(f^n)'(z)|)^{\frac{1}{3}} < 1$. Choose $0 < \varepsilon_1 = \varepsilon_2 = \ldots = \varepsilon_{n_o} < \alpha\varepsilon$ so small that for, every $n = 1, 2, \ldots, n_0$,

$$f^n \text{ restricted to the ball } B(z, \varepsilon_n) \text{ is injective} \tag{12.3.2}$$

and

$$f^n(B(z, \varepsilon_n)) \subset B(f^n(z), \varepsilon). \tag{12.3.3}$$

For every $n \geq n_0$ define ε_{n+1} inductively by

$$\varepsilon_{n+1} = (1 - (2|(f^n)'(z)|)^{\frac{1}{3}})\varepsilon_n. \tag{12.3.4}$$

Then $0 < \varepsilon_n < \alpha\varepsilon$ for every $n \geq 1$. Assume that (12.3.2) and (12.3.3) are satisfied for some $n \geq n_0$. Then by the Koebe Distortion Lemma 6.2.4 and (12.3.4) the set $f^n(B(z, \varepsilon_{n+1}))$ is contained in the ball centred at $f^n(z)$ and of radius

$$\varepsilon_{n+1}|(f^n)'(z)|\frac{2}{(1 - \varepsilon_{n+1}/\varepsilon_n)^3} = \frac{2\varepsilon_{n+1}|(f^n)'(z)|}{2|(f^n)'(z)|} = \varepsilon_{n+1} < \alpha\varepsilon.$$

Therefore, since f is injective on $B(f^n(z), \varepsilon)$, formula (12.3.2) is satisfied for $n + 1$, and using also (12.3.1) we get

$$f^{n+1}(B(z, \varepsilon_{n+1})) = f\big(f^n(B(z, \varepsilon_{n+1}))\big) \subset f(B(f^n(z), \alpha\varepsilon)) \subset B(f^{n+1}(z), \varepsilon).$$

Thus (12.3.3) is satisfied for $n + 1$.

Let $\varepsilon_n \searrow \varepsilon_0$. Since the series $\sum_{n=1}^{\infty}|(f^n)'(z)|^{\frac{1}{3}}$ converges, it follows from (11.3.4) that $\varepsilon_0 > 0$. Clearly, (12.3.2) and (12.3.3) remain true with ε_n replaced by ε_0. It follows that the family $\{f^n|_{B(z,\frac{1}{2}\varepsilon_0)}\}_{n=1}^{\infty}$ is normal, and consequently $z \notin J(f)$. This contradiction completes the proof. ♣

As an immediate consequence of this lemma and of Birkhoff's Ergodic Theorem we get the following.

Corollary 12.3.2. *If μ is an ergodic f-invariant measure for which there exists a compact set $Y \subset J(f)$ such that $\mu(Y) = 1$ and $Y \cap \mathrm{Crit}(f) = \emptyset$, then $\chi_\mu \geq 0$.*

In fact the assumption $Y \cap \mathrm{Crit}(f) = \emptyset$ is not needed: see [Przytycki 1993]. Compare Theorem 12.3.10.

Now let Ω be a finite subset of $\overline{\bigcup_{n=1}^{\infty} f^n(\mathrm{Crit}(f))}$ such that

$$\Omega \cap \overline{\{f^n(c) : n = 1, 2 \ldots\}} \neq \emptyset \qquad \text{for every } c \in \mathrm{Crit}(f) \tag{12.3.5}$$

and

$$\Omega \cap \mathrm{Crit}(f) = \emptyset. \tag{12.3.6}$$

Sets satisfying these conditions exist, since no critical point of f lying in $J(f)$ can be periodic. Now let $V \subset J(f)$ be an open neighbourhood of Ω, and define

$K(V)$ to be the set of those points of $J(f)$ whose forward trajectory avoids V. Equivalently this means that

$$K(V) = \{z \in J(f) : f^n(z) \notin V \text{ for every } n \geq 0\} = \bigcap_{n=0}^{\infty} f^{-n}(J(f) \setminus V).$$

Hence $K(V)$ is a compact subset of $J(f)$ and $f(K(V)) \subset K(V)$. Consequently we can consider the dynamical system $f|_{K(V)} : K(V) \to K(V)$. Note that $f(K(V)) = K(V)$ does not hold for all sets V, and that usually $f^{-1}(K(V)) \not\subset K(V)$. Simple considerations based on (11.3.5) and the definition of sets $K(V)$ give the following.

Lemma 12.3.3. $\mathrm{Crit}(f|_{K(V)}) \subset \mathrm{Crit}(f) \cap K(V) = \emptyset$, $K(V)_0(f) = \mathrm{Sing}(f) \subset \partial V$, *and* $-t\log|f'|$ *is a well-defined continuous function on* $K(V)$.

Now fix $z \in K(V)$ and set $E_n = f|_{K(V)}^{-n}(z)$, $n \geq 0$. Then $E_{n+1} = f|_{K(V)}^{-1}(E_n)$, and therefore the sequence $\{E_n\}$ satisfies (11.1.9) with $D = K(V)$. Take $t \geq 0$ and let $c(t, V)$ be the transition parameter associated to this sequence and the function $-t\log|f'|$. Put $\mathrm{P}(t, V) = \mathrm{P}(f|_{K(V)}, -t\log|f'|)$. We shall prove the following.

Lemma 12.3.4. $c(t, V) \leq \mathrm{P}(t, V)$.

Proof. Since $K(V)$ is a compact set disjoint from $\mathrm{Crit}(f)$, the map $f|_{K(V)}$ is locally 1-to-1, which means that there exists $\delta > 0$ such that $f|_{K(V)}$ restricted to any set with diameter $\leq \delta$ is 1-to-1. Consequently, all the sets E_n are (n, ε)-separated for $\varepsilon < \delta$. Hence the required inequality $c(t, V) \leq \mathrm{P}(t, V)$ follows immediately from Theorem 3.3.2. ♣

The standard straightforward arguments showing continuity of topological pressure also prove the following.

Lemma 12.3.5. *The function* $t \mapsto c(t, V)$ *is continuous.*

Set

$$s(V) = \inf\{t \geq 0 : c(t, V) \leq 0\} < +\infty.$$

We shall prove the following.

Lemma 12.3.6. $s(V) \leq \mathrm{DD}(J(f))$.

Proof. Suppose that $\mathrm{DD}(J(f)) < s(V)$, and take $0 \leq \mathrm{DD}(J(f)) < t < s(V)$. From this choice, and by Lemma 12.3.4, we have $0 < c(t, V) \leq \mathrm{P}(t, V)$, and by the Variational Principle (Theorem 3.4.1) there exists $\mu \in M_e(f_{K(V)}) \subset M_e(f)$ such that $\mathrm{P}(t, V) \leq \mathrm{h}_\mu(f) - t\chi_\mu(f) + c(t, V)/2$. Therefore, by Corollary 12.3.2 and Lemma 12.3.3, we get $\mathrm{h}_\mu(f) \geq c(t, V)/2 > 0$, and applying additionally Theorem 11.1.1 (Ruelle's inequality), $\chi_\mu(f) > 0$. Hence it follows from Theorem 11.4.1 that

$$t \leq \mathrm{HD}(\mu) - \frac{1}{2}\frac{c(t, V)}{\chi_\mu} < \mathrm{HD}(\mu) \leq \mathrm{DD}(J(f)).$$

This contradiction completes the proof. ♣

Let m be a limit measure on $K(V)$ associated to the sequence E_n and the function $-s(V)\log|f'|$. Since $c(0,V)\geq 0$ and $s(V)<\infty$, it follows from Lemma 12.3.5 that $c(s(V),V)=0$. Therefore, applying Lemma 12.1.4 and Lemma 12.1.5 with $\Gamma=\partial V$, we see that $m(f(A))\geq\int_A|f'|^{s(V)}\,dm$ for any special set $A\subset K(V)$ and $m(f(A))=\int_A|f'|^{s(V)}\,dm$ for any special set $A\subset K(V)$ such that $A\cap\partial V=\emptyset$. Treating m as a measure on $J(T)$ and using straightforward measure-theoretic arguments we now deduce from this that

$$m(f(A))\geq\int_A|f'|^{s(V)}\,dm \tag{12.3.7}$$

for any special set $A\subset J(f)$ and

$$m(f(A))=\int_A|f'|^{s(V)}\,dm \tag{12.3.8}$$

for any special set $A\subset J(f)$ such that $A\cap\bar{V}=\emptyset$. Now we are in position to prove the following.

Lemma 12.3.7. *For every Ω there exist $0\leq s(\Omega)\leq \mathrm{DD}(J(f))$ and a Borel probability measure m on $J(f)$ such that*

$$m(f(A))\geq\int_A|f'|^{s(\Omega)}\,dm$$

for any special set $A\subset J(f)$ and

$$m(f(A))=\int_A|f'|^{s(\Omega)}\,dm$$

for any special set $A\subset J(f)$ disjoint from Ω.

Proof. For every $n\geq 1$ let $V_n=B(\Omega,\frac{1}{n})$, and let m_n be the measure on $J(f)$ satisfying (12.3.7) and (12.3.8) for the neighbourhood V_n. Using Lemma 12.1.6 we shall show that any weak* limit m of the sequence of measures $\{m_n\}_{n=1}^\infty$ satisfies the requirements of Lemma 12.3.7. First observe that the sequence $\{s(V_n)\}_{n=1}^\infty$ is non-decreasing, and denote its limit by $s(\Omega)$. Therefore the sequence of continuous functions $g_n=|f'|^{s(V_n)}$, $n=1,2,\ldots$, defined on $J(f)$, converges uniformly to the continuous function $g=|f'|^{s(\Omega)}$. Let A be a special subset of $J(f)$ such that

$$\overline{A}\cap(\mathrm{Sing}(f)\cup\Omega)=\emptyset. \tag{12.3.9}$$

Then one can find a compact set $\Gamma\subset J(f)$ disjoint from $\overline{A}$ and such that $\mathrm{Int}(\Gamma)\supset\mathrm{Sing}(f)\cup\Omega$. So, using also Lemma 12.3.3, we see that for any n sufficiently large, say $n\geq q$,

$$\overline{V_n}\subset\Gamma \qquad\text{and}\qquad \overline{V_n}\cap\mathrm{Crit}(f)=\emptyset. \tag{12.3.10}$$

Therefore, by (12.3.7) and (12.3.8), we conclude that Lemma 12.1.6 applies to the sequence of measures $\{m_n\}_{n=q}^\infty$ and the sequence of functions $\{g_n\}_{n=q}^\infty$. Hence the first property required in our lemma is satisfied for any special subset of

$J(f)$ disjoint from $\mathrm{Crit}(f)$, and since $A \cap \Gamma = \emptyset$, the second property is satisfied for the set A. So, since any special subset of $J(f)$ disjoint from $\mathrm{Sing}(f) \cup \Omega$ can be expressed as a disjoint union of special sets satisfying (12.3.9), an easy computation shows that the second property is satisfied for all special sets disjoint from $\mathrm{Sing}(f) \cup \Omega$. Therefore, in order to complete the proof, it is enough to show that the second requirement of the lemma is satisfied for every point of the set $\mathrm{Sing}(f)$. First note that by (12.3.10) and (12.3.8) formula (a′) in Lemma 12.1.6 is satisfied for every $n \geq q$ and every $x \in \mathrm{Crit}(f) \setminus J(f)_0(f)$. As $f : J(f) \to J(f)$ is an open map, the set $J(f)_0(f)$ is empty, and $\mathrm{Sing}(f) = \mathrm{Crit}(f)$. Consequently, formula (d) of Lemma 11.1.6 is satisfied for any critical point $c \in J(f)$ of f. Since $g(c) = |f'(c)|^{s(\Omega)} = 0$, this formula implies that $m(f()) \leq 0$. Thus $m(\{f(c)\}) = 0 = |f'(c)|^{s(\Omega)} m(\{c\})$. The proof is complete. ♣

Lemma 12.3.8. *Let m be the measure constructed in Lemma 12.3.7. If for some $z \in J(f)$ the series $S(t,z) = \sum_{n=1}^{\infty} |(f^n)'(z)|^t$ diverges, then $m(\{z\}) = 0$, or a positive iteration of z is a parabolic point of f. Moreover, if z itself is periodic, then $m(\{f(z)\}) = |f'(z)|^t m(\{z\})$.*

Proof. Suppose that $m(\{z\}) > 0$. Assume first that the point z is not eventually periodic. Then by the definition of a conformal measure on the complement of some finite set we get $1 \geq m(\{f^n(z) : n \geq 1\}) \geq m(\{z\}) \sum_{n=1}^{\infty} |(f^n)'(z)|^t = \infty$, which is a contradiction. Hence z is eventually periodic, and therefore there exist positive integers k and q such that $f^k(f^q(z)) = f^q(z)$. Since $f^q(z) \in J(f)$, and since the family of all iterates of f on a sufficiently small neighbourhood of an attractive periodic point is normal, this implies that $|(f^k)'(f^q(z))| \geq 1$. If $|(f^k)'(f^q(z))| = \lambda > 1$ then, again by the definition of a conformal measure on the complement of some finite set, $m(\{f^q(z)\}) > 0$ and $m(\{f^{kn}(f^q(z))\}) \geq \lambda^{nt} m(\{f^q(z)\})$. Thus $m(\{f^{kn}(f^q(z))\})$ converges to ∞, which is a contradiction. Therefore $|(f^k)'(f^q(z))| = 1$, which finishes the proof of the first assertion of the lemma. In order to prove the second assertion assume that $q = 1$. Then, again using the definition of conformal measures on the complement of some finite set, we get $m(\{f(z)\}) \geq m(\{z\})|f'(z)|^t$, and on the other hand

$$m(\{z\}) = m(\{f^{k-1}(f(z))\}) \geq m(\{f(z)\})|(f^{k-1})'(f(z))|^t = m(\{f(z)\})|f'(z)|^{-t}.$$

Therefore $m(\{f(z)\}) = m(\{z\})|f'(z)|^t$. The proof is complete. ♣

Corollary 12.3.9. *If for every $x \in \mathrm{Crit}(f)$ one can find $y(x) \in \overline{\{f^n(x) : n \geq 0\}}$ such that the series $S(t, y(x))$ diverges for every $0 \leq t \leq \mathrm{DD}(J(f))$, then there exists an s-conformal measure for $f : J(f) \to J(f)$ with $0 \leq s \leq \mathrm{DD}(J(f))$.*

Proof. Let m be the measure constructed in Lemma 12.3.7. Since $S(t, y(x))$ diverges for every $0 \leq t \leq \mathrm{DD}(J(f))$, we see that $y(x) \notin \mathrm{Crit}(f)$. If for some $x \in \mathrm{Crit}(f)$, $y(x)$ is a non-periodic point, eventually falling into a parabolic point, then let $z(x)$ be this parabolic point; otherwise put $z(x) = y(x)$. The set $\Omega = \{z(x) : x \in \mathrm{Crit}(f)\}$ meets the conditions (12.3.5) and (12.3.6), and is contained in $\overline{\bigcup_{n=1}^{\infty} f^n(\mathrm{Crit}(f))}$. Since for every $t \geq 0$ and $z \in J(f)$ the divergence of the series $S(t,z)$ implies the divergence of the series $S(t,f(z))$, it follows immediately from Lemma 12.3.7 and Lemma 12.3.8 that the measure m is s-conformal. ♣

Fortunately, the assumptions on the existence of $y(x)$ with the divergence of $(t,y(x))$ hold. They follow from the following fact, for which we refer the reader to [Przytycki 1993] and omit the proof here.

Theorem 12.3.10. *For every f-invariant probability measure μ on $J(f)$, $\int \log|f'|\,d\mu \geq 0$, in particular $\log|f'|$ is μ-integrable. For μ ergodic this reads that the Lyapunov characteristic exponent is non-negative, $\chi_\mu(f) \geq 0$. For μ a.e. y,*

$$\limsup_{n\to\infty} |(f^n)'(y)| \geq 1.$$

Now we are in position to finish the proof of the following main result of this section.

Theorem 12.3.11. $\mathrm{HyD}(J(f)) = \mathrm{DD}(J(f)) = \delta(f)$, *and there exists a $\delta(f)$-conformal measure for $f: J(f) \to J(f)$.*

Proof. For every $x \in \mathrm{Crit}(f)$ the set $\overline{\{f^n(x) : n \geq 0\}}$ is closed and forward invariant under f. Therefore, in view of Theorem 3.1.8 (the Bogolubov–Krylov Theorem), there exists $\mu \in M_e(f)$ supported on $\overline{\{f^n(x) : n \geq 0\}}$. By Theorem 12.3.10 there exists at least one point $y(x) \in \overline{\{f^n(x) : n \geq 0\}}$ such that $\limsup_{n\to\infty} |(f^n)'(y(x))| \geq 1$, and consequently the series $S(t,y(x))$ diverges for every $t \geq 0$. So, in view of Corollary 12.3.9, there exists an s-conformal measure for $f: J(f) \to J(f)$ with $0 \leq s \leq \mathrm{DD}(J(f))$. Combining this with Lemma 12.2.2 and Theorem 12.2.3 completes the proof. ♣

12.4 Pesin's formula

Theorem 12.4.1 (Pesin's formula). *Assume that X is a compact subset of the closed complex plane $\overline{\mathbb{C}}$, and that $f \in \mathcal{A}(X)$. If m is a t-conformal measure for f, and $\mu \in M_e^+(f)$ is absolutely continuous with respect to m, then $\mathrm{HD}(\mu) = t = \delta(f)$.*

Proof. In view of Lemma 12.2.2 we need only prove that $t \leq \mathrm{HD}(\mu)$, and in order to do this we essentially combine the arguments from the proof of Lemma 12.2.2 and the proof of formula (11.4.1). So, we work in the natural extension $(\tilde{X}, \tilde{f}, \tilde{\mu})$. Fix $0 < \varepsilon < \chi_\mu/3$, and let $\tilde{X}(\varepsilon)$ and $r(\varepsilon)$ be given by Corollary 11.2.4. In view of the Birkhoff Ergodic Theorem there exists a measurable set $\tilde{F}(\varepsilon) \subset \tilde{X}(\varepsilon)$ such that $\tilde{\mu}(\tilde{F}(\varepsilon)) \geq 1 - 2\varepsilon$ and

$$\lim_{n\to\infty} \frac{1}{n} \sum_{j=1}^{n-1} \chi_{\tilde{X}(\varepsilon)} \circ \tilde{f}^n(\tilde{x}) = \tilde{\mu}(\tilde{X}(\varepsilon))$$

for every $\tilde{x} \in \tilde{F}(\varepsilon)$. Let $F(\varepsilon) = \pi(\tilde{F}(\varepsilon))$. Then $\mu(F(\varepsilon)) = \tilde{\mu}(\pi^{-1}(F(\varepsilon)) \geq \tilde{\mu}(\tilde{F}(\varepsilon)) \geq 1 - 2\varepsilon$. Consider now $x \in F(\varepsilon) \cap X_o$, and take $\tilde{x} \in \tilde{F}(\varepsilon)$ such that $x = \pi(\tilde{x})$. Then by the above there exists an increasing sequence $\{n_k = n_k(x) : k \geq 1\}$ such that $\tilde{f}^{n_k}(\tilde{x}) \in \tilde{X}(\varepsilon)$ and

$$\frac{n_{k+1} - n_k}{n_k} \leq \varepsilon \tag{12.4.1}$$

for every $k \geq 1$. Moreover, Corollary 11.2.4 produces holomorphic inverse branches $f_x^{-n_k} : B(f^{n_k}(x), r(\varepsilon)) \to \overline{\mathbb{C}}$ of f^{n_k} such that $f_x^{-n_k} f^{n_k}(x) = x$ and

$$f_x^{-n_k}(B(f^{n_k}(x), r(\varepsilon))) \subset B(x, K|(f^{n_k})'(x)|^{-1} r(\varepsilon)).$$

Set $r_k = r_k(x) = K^{-1}|(f^{n_k})'(x)|^{-1} r(\varepsilon)$. By Corollary 11.2.4 $r_k \leq K^{-2} \exp(-(\chi_\mu - \varepsilon)n_k) r(\varepsilon)$. So, using Corollary 11.2.4 again and (12.4.1) we can estimate

$$\begin{aligned} r_k = r_{k+1}|(f^{n_{k+1}-n_k})'(f^{n_k}(x))| &\leq r_{k+1} K \exp(\chi_\mu + \varepsilon)(n_{k+1} - n_k)) \\ &\leq r_{k+1} K \exp(\chi_\mu + \varepsilon) n_{k+1}\varepsilon) \leq K r_{k+1} \exp(\chi_\mu - \varepsilon) 2 n_{k+1} \varepsilon) \\ &\leq r_{k+1} K (K^{-2} r(\varepsilon) r_{k+1}^{-1})^{2\varepsilon} = K^{1-4\varepsilon} r(\varepsilon)^{2\varepsilon} r_{k+1}^{1-2\varepsilon}. \end{aligned}$$

Take now any $0 < r \leq r_1$, and find $k \geq 1$ such that $r_{k+1} < r \leq r_k$. Then using this estimate, t-conformality of m, and invoking Corollary 11.2.4 once more, we get

$$\begin{aligned} m(B(x,r)) \leq m(B(x, r_k)) &\leq K^t |(f^{n_k})'(x)|^{-t} m(B(x, r(\varepsilon))) \\ &\leq K^{2t} r(\varepsilon)^{-t} r_k^t \leq K^{(3-4\varepsilon)t} r(\varepsilon)^{2\varepsilon t} r^{(1-2\varepsilon)t}. \end{aligned}$$

So, by Theorem 8.5.1 (the Besicovitch Covering Theorem), $\Lambda_{(1-2\varepsilon)t}(X) \geq \Lambda_{(1-2\varepsilon)t}(F(\varepsilon)) > 0$, whence $\mathrm{HD}(X) \geq (1 - 2\varepsilon)t$. Letting $\varepsilon \to 0$ completes the proof. ♣

Remark 12.4.2. For m being the Riemann measure on $\overline{\mathbb{C}}$, which is 2-conformal by definition, $\mathrm{HD}(m) = 2$ is obvious, even without assuming the existence of μ.

Of course, there exist 2-conformal measures for which no $\mu \in M_e^+(f)$ with $\mu \ll m$ exists. Take, for example, $f(z) = z^2 + 1/4$. It has a parabolic fixed point $z = 1/2$, as $f'(1/2) = 1$. Put $m(1/2) = 1$, and for each $n \geq 0$ and $w \in f^{-n}(1/2)$ put $m(w) = |(f^n)'(w)|^{-t}$. For $t \geq 2$ the series $\Sigma := \sum_{n,w} |(f^n)'(w)|^{-t}$ converges (Exercise; use the Koebe Distortion Theorem). Normalize m by dividing by Σ. Check that there is no $\mu \in M_e^+$ with $\mu \ll m$. In this example, for $t = \delta(f)$, the measure μ exists. However, this is not always the case. Consider $f(z) = z^2 - 3/4$ and m built as above, starting from the fixed point $-1/2$. See [Aaronson, Denker & Urbański 1993].

Other nice examples and estimates, for ∞-renormalizable polynomials, can be found in [Avila & Lyubich 2008] and [Levin & Świątek 2009], with other references therin.

For an arbitrary 2-conformal m the equality $\mathrm{h}_\mu(f)/\chi_\mu(f) = \mathrm{HD}(\mu) = 2$, that is, $h_\mu(f) = 2\chi_\mu(f)$, is non-trivial. For m Riemann measure, the first equality is non-trivial. In higher dimensions its analogue is usually called the Pesin formula: see [Mañé 1987]. It corresponds to Rokhlin's equality in Theorem 2.9.7.

The following theorem, converse to Theorem 12.4.1, holds. We formulate it for f a rational function on $\overline{\mathbb{C}}$ and X its Julia set. We shall not prove it here. We refer to [Ledrappier 1984] and the recent [Dobbs 2008].

Theorem 12.4.3. *If m is a t-conformal measure supported on $J(f)$ for $f : \overline{\mathbb{C}} \to \overline{\mathbb{C}}$ a rational function of degree at least 2 on the Riemann sphere, and μ is an f-invariant ergodic probability measure on $J(f)$ of positive Lyapunov exponent such that $\mathrm{HD}(\mu) \geq t$, then $\mu \ll m$. Moreover, the density $d\mu/dm$ is bounded away from 0. In particular, μ is unique satisfying these properties.*

12.5 More about geometric pressure and dimensions

Here we provide a simple proof of $\mathrm{HyD}(J(f)) = \delta(f)$ (see Theorem 12.3.11), omitting the construction via the sets $K(V)$ and omitting Pesin's theory.

Let $f : \overline{\mathbb{C}} \to \overline{\mathbb{C}}$ be a rational mapping of degree $d \geq 2$ on the Riemann sphere $\overline{\mathbb{C}}$. Here we denote by $\mathrm{Crit}(f)$ the set of all critical points in $\overline{\mathbb{C}}$: that is, $f'(x) = 0$ for $x \in \mathrm{Crit}(f)$. As before, the symbol $J = J(f)$ stands for the Julia set of f. Absolute values of derivatives and distances are considered with respect to the standard Riemann sphere metric. We consider pressures below for all $t > 0$. All the pressures will occur to coincide, giving rise to a generalization of the geometric pressure $\mathrm{P}(t)$ introduced in Section 9.1 in the uniformly expanding case.

Definition 12.5.1 (Tree pressure). For every $z \in \overline{\mathbb{C}}$ define

$$\mathrm{P}_{\mathrm{tree}}(z,t) := \limsup_{n\to\infty} \frac{1}{n} \log \sum_{f^n(x)=z} |(f^n)'(x)|^{-t}.$$

Definition 12.5.2 (Hyperbolic pressure).

$$\mathrm{P}_{\mathrm{hyp}}(t) := \sup_X \mathrm{P}(f|_X, -t\log|f'|),$$

where the supremum is taken over all compact f-invariant (that is, $f(X) \subset X$) Cantor-repelling hyperbolic (expanding) subsets of J. The property of X being a Cantor set can be skipped, giving the same definition: compare Theorem 11.6.1.

$\mathrm{P}(f|_X, -t\log|f'|)$ denotes the standard topological pressure for the continuous mapping $f|_X : X \to X$ and continuous real-valued potential function $-t\log|f'|$ on X, as in the previous sections.

Note that these definitions imply that $\mathrm{P}_{\mathrm{hyp}}(t)$ is a continuous monotone decreasing function of t.

In the definition of the hyperbolic pressure one can restrict the supremum to be over Cantor-repelling hyperbolic sets X such that $f|_X$ is topologically transitive: see Remark 11.6.3.

Definition 12.5.3 (Conformal pressure)**.** Set $\mathrm{P}_{\mathrm{Conf}}(t) := \log \lambda(t)$, where

$$\lambda(t) = \inf\{\lambda > 0 : \text{there is } \mu, \text{ a } \lambda|f'|^t\text{– conformal probability measure on } J(f)\}.$$

We know that the set of λ's above is non-empty, from Section 12.3. However, we want this section to be independent. So the existence of $\lambda(t)$ will be proved again later, more directly.

In the sequel we shall call any $\lambda|f'|^t$-conformal probability measure on $J(f)$ a (λ, t)-*conformal measure* for f, and call a $(1,t)$-conformal measure a t-*conformal measure* for f.

Proposition 12.5.4. *For each $t > 0$ the number $\mathrm{P}_{\mathrm{Conf}}(t)$ is attained: that is, there exists a (λ, t)-conformal measure with $\log \lambda = \mathrm{P}_{\mathrm{Conf}}(t)$.*

This proposition follows from the following (compare the proof of Lemma 12.3.7).

Lemma 12.5.5. *If μ_n is a sequence of (λ_n, t)-conformal measures for f on $J(f)$ for an arbitrary $t > 0$, weakly* convergent to a measure μ, and $\lambda_n \to \lambda$, then μ is a (λ, t)-conformal measure.*

Proof. Let $E \subset J$ be a Borel set on which f is injective. Then E can be decomposed into a countable union of critical points and sets E_i pairwise disjoint and such that f is injective on a neighbourhood V of $\operatorname{cl} E_i$. For every ε there exist compact set K and open U such that $K \subset E_i \subset U \subset V$ and $\mu(U) - \mu(K) < \varepsilon$ and $\mu(f(U)) - \mu(f(K)) < \varepsilon$. Consider an arbitrary continuous function $\chi : J \to [0,1]$ so that χ is 1 on K and 0 on $J \setminus U$. Then there exists $s : 0 < s < 1$ such that for $A = \chi^{-1}([s,1])$, $\mu(\partial f(A)) = 0$. Then the weak* convergence of μ_n implies $\mu_n(f(A)) \to \mu(f(A))$, as $n \to \infty$: see Theorem 3.1.4. Moreover, this weak* convergence and $\lambda_n \to \lambda$ imply $\int \chi \lambda_n |f'|^t d\mu_n \to \int \chi \lambda |f'|^t d\mu$. Therefore from $\mu_n(f(A)) = \int_A \lambda_n |f'|^t d\mu_n$, letting $\varepsilon \to 0$, we obtain $\mu(f(E_i)) = \int_{E_i} \lambda |f'|^t d\mu$.

If $E = \{c\}$, where $c \in \operatorname{Crit}(f) \cap J(f)$, then for every $r > 0$ small enough and for all n, we have $\mu_n(f(B(c,r))) \leq 2(\sup_k \lambda_k)(2r)^t$, and since the bound is independent of n we get $\mu(f(c)) = 0$: hence $\mu(f(c)) = \int_c |f'|^t d\mu$, as $f'(c) = 0$. ♣

Remark 12.5.6. For a continuous map $T : X \to X$ of a compact metric space X, for an integrable function $g : X \to \mathbb{R}$, and for an arbitrary $\varepsilon \geq 0$, a probability measure m on X is said to be ε-g-*conformal* if for every special set $A \subset X$ we have

$$|m(T(A)) - \int_A g \, dm| \leq \varepsilon.$$

Compare (12.1.1). Then, in Lemma 12.5.5, it is sufficient to assume that μ_n is a sequence of ε_n-$\lambda_n|f'|^t$-conformal measures, with $\varepsilon_n \to 0$.

Definition 12.5.7. We call $z \in \overline{\mathbb{C}}$ *safe* if

(1) $z \notin \bigcup_{j=1}^{\infty} f^j(\operatorname{Crit}(f))$ and

(2) $\liminf_{n\to\infty} \frac{1}{n} \log \operatorname{dist}(z, f^n(\operatorname{Crit}(f))) = 0$.

Remark 12.5.8. For every safe $z \in \overline{\mathbb{C}}$ and every $t > 0$ the pressure $\mathrm{P}_{\mathrm{tree}}(z,t)$ is finite. Indeed, if $z \notin B(f^n(\mathrm{Crit}(f)), \varepsilon\lambda^{-n})$ for all $n = 1, 2, ...$ and some $\varepsilon > 0$ and $\lambda > 1$, then for each $x \in f^{-n}(z)$ the mapping f^n is univalent on $\mathrm{Comp}_x f^{-n}B(f^n(\mathrm{Crit}(f)), \frac{\varepsilon}{2}\lambda^{-n})$, with distortion bounded by a constant $C > 0$: see the Koebe Distortion Lemma 6.2.3. Recall that Comp_x denotes the component containing x. Hence

$$|(f^n)'(x)| \geq C^{-1} \frac{\frac{\varepsilon}{2}\lambda^{-n}}{\mathrm{diam}\, \mathrm{Comp}_x f^{-n}B(z, \frac{\varepsilon}{2}\lambda^{-n}} \geq C^{-1}\frac{\varepsilon}{2}\lambda^{-n}.$$

Summing up over x, and letting $\lambda \to 1$ and $n \to \infty$, we obtain

$$\mathrm{P}_{\mathrm{tree}}(z,t) \leq \log \deg f. \tag{12.5.1}$$

Definition 12.5.9. We call a point $z \in \overline{\mathbb{C}}$ *expanding* (or *hyperbolic*) if there exist $\Delta > 0$ and $\lambda = \lambda_z > 1$ such that for all n large enough f^n is univalent on $\mathrm{Comp}_z f^{-n}(B(f^n(z), \Delta))$ and $|(f^n)'(z)| \geq \lambda^n$.

Proposition 12.5.10. *The set S of expanding safe points in J is non-empty. Moreover,* $\mathrm{HD}(S) \geq \mathrm{HyD}(J)$.

Proof. The set NS of non-safe points is of zero Hausdorff dimension. This follows from $NS \subset \bigcup_{j=1}^{\infty} f^j(\mathrm{Crit}(f)) \cup \bigcup_{\xi<1} \bigcap_{n=1}^{\infty} \bigcup_{j=n}^{\infty} B(f^j(\mathrm{Crit}(f)), \xi^j)$, finiteness of $\mathrm{Crit}(f)$, and from $\sum_n (\xi^n)^t < \infty$ for every $0 < \xi < 1$ and $t > 0$. Therefore the existence of expanding safe points in J follows from the existence of hyperbolic sets $X \subset J$ with $\mathrm{HD}(X) > 0$. Note that every point in a hyperbolic set X is expanding. ♣

Theorem 12.5.11. *For all $t > 0$, all expanding safe $z \in J$ and all $w \in \overline{\mathbb{C}}$*

$$\mathrm{P}_{\mathrm{tree}}(z,t) \leq \mathrm{P}_{\mathrm{hyp}}(t) \leq \mathrm{P}_{\mathrm{Conf}}(t) \leq \mathrm{P}_{\mathrm{tree}}(w,t).$$

We shall provide a proof later on. Now let us state corollaries.

Corollary 12.5.12. *For all $t > 0$ $\mathrm{P}_{\mathrm{hyp}}(t) = \mathrm{P}_{\mathrm{Conf}}(t)$ and* $\mathrm{HyD}(J) = \delta(f)$.

Proof. The first equality follows from Theorem 12.5.11 and the existence of expanding safe points in J, and the second from the fact that both quantities are first zeros of $\mathrm{P}_{\mathrm{hyp}}(t)$ and $\mathrm{P}_{\mathrm{Conf}}(t)$. We shall prove the latter, including the existence of a finite zero.

First notice that $\mathrm{P}_{\mathrm{hyp}}(t)$ is monotone decreasing, which follows immediately from the monotone decreasing of $\mathrm{P}(X,t) := \mathrm{P}(f|_X, -t\log|f'|)$ for every expanding repeller $X \subset J$: see for example the discussion after Theorem 9.1.4 and Definition 12.5.2. Continuity follows from the equicontinuity of the family $\mathrm{P}(X,t)$ following, using the definition of pressure, from its uniform Lipschitz continuity with the Lipschitz constant $\sup \log|f'|$. (In fact, by the Variational Principle the Lipschitz constant of all $\mathrm{P}(X,t)$ is bounded by $\sup_\mu \chi_\mu(f)$, the supremum over all probability f-invariant measures on J.) If t_0 is the first zero of $\mathrm{P_{hyp}}(t)$ (we have not yet excluded the case $\mathrm{P_{hyp}}(t) > 0$ for all t; in such a case write $t_0 = \infty$) and $t_0(X)$ is zero of $\mathrm{P}(X,t)$, then $\mathrm{P}(X,t) \to \mathrm{P_{hyp}}(t)$ for all t implies $t_0(X) \to t_0$. Since $t_0(X) = \mathrm{HD}(X) \le 2$ (see Corollary 9.1.7), t_0 is finite.

Observe finally that $\delta(f)$ is also the first zero t_0 of $\mathrm{P_{Conf}}(t)$ (which we know already to be equal to $\mathrm{P_{hyp}}(t)$). It cannot be larger, because there exists a t_0-conformal measure, owing to Proposition 12.5.4. It cannot be smaller, since $\mathrm{P_{Conf}}(t) > 0$ for $t < t_0$. ♣

We obtain also a simple proof of the following.

Corollary 12.5.13. $\mathrm{P_{tree}}(z,t)$ *does not depend on* z *for* $z \in J$ *expanding safe.*

Proof of Theorem 12.5.11. 1. We prove first that $\mathrm{P_{tree}}(z,t) \le \mathrm{P_{hyp}}(t)$. Fix expanding safe $z = z_0 \in J$ and $\lambda = \lambda_{z_0} > 1$ according to Definition 12.5.9. Since z_0 is expanding, we have for $\delta = \Delta/2$, $l = 2\alpha n$ and all n large enough

$$W := \mathrm{Comp}_{z_0} f^{-l} B(f^l(z_0), 2\delta) \subset B(z, \varepsilon\lambda^{-\alpha n}),$$

and f^l is univalent on W. Since z_0 is safe we have

$$B(z_0, \lambda^{-\alpha n}) \cap \bigcup_{j=1}^{2n} f^j(\mathrm{Crit}(f)) = \emptyset$$

for arbitrary constants $\varepsilon, \alpha > 0$.

By the Koebe Distortion Lemma for ε small enough, for every $1 \le j \le 2n$ and $z_j \in f^{-j}(z_0)$ we have

$$\mathrm{Comp}_{z_j} f^{-j} B(z_0, \varepsilon\lambda^{-\alpha n}) \subset B(z_j, \delta).$$

Let $m = m(\delta)$ be such that $f^m(B(y,\delta/2)) \supset J$ for every $y \in J$. Then, putting $y = f^l(z_0)$, for every $z_n \in f^{-n}(z_0)$ we find $z'_n \in f^{-m}(z_n) \cap f^m(B(y,\delta/2))$. Hence the component W_{z_n} of $f^{-m}(\mathrm{Comp}_{z_n} f^{-n}(B(z_0, \varepsilon\lambda^{-\alpha n}))$ containing z'_n is contained in $\subset B(y, \frac{3}{2}\delta))$, and f^{m+n} is univalent on W_{z_n} (provided $m \le n$).

Therefore f^{m+n+l} is univalent from $W'_{z_n} := \mathrm{Comp}(f^{-(m+n+l)}(B(y,2\delta)) \subset W_{z_n}$ onto $B(y,2\delta)$. The mapping

$$F = f^{m+n+l} : \bigcup_{z_n \in f^{-n}(z_0)} W'_{z_n} \to B(y, 2\delta)$$

has no critical points: hence $Z := \bigcap_{k=0}^{\infty} F^{-k}(B(y,2\delta))$ is a repelling expanding F-invariant Cantor subset of J.

We obtain for a constant $C > 0$ resulting from distortion and $L = \sup|f'|$,

$$\begin{aligned} \mathrm{P}(F|_Z, -t\log|F'|) &\ge \log\Big(C \sum_{z_n \in f^{-n}(z_0)} |(f^{m+n+l})'(z'_n)|^{-t}\Big) \\ &\ge \log\Big(C \sum_{z_n \in f^{-n}(z_0)} |(f^n)'(z_n)|^{-t} L^{-t(m+l)}\Big). \end{aligned} \tag{12.5.2}$$

Hence on the expanding f-invariant set $Z' := \bigcup_{j=0}^{m+n+l-1} f^j(Z)$ we obtain

$$\mathrm{P}(f|_{Z'}, -t\log|f'|) \geq \frac{1}{m+n+l}\mathrm{P}(F, -t\log|F'|)$$

$$\geq \frac{1}{m+n+l}\Big(\log C - t(m+l)\log L + \log \sum_{z_n\in f^{-n}(z_0)} |(f^n)'(z_n)|^{-t}\Big).$$

Passing with n to ∞ and next letting $\alpha \searrow 0$ we obtain

$$\mathrm{P}(f|_{Z'}, -t\log|f'|) \geq \mathrm{P}_{\mathrm{tree}}(z_0, t).$$

Finally one can find an f-invariant repelling expanding Cantor set Z'' containing Z', contained in J as in the proof of Theorem 11.6.1, relying on Proposition 4.5.6. The latter inequality for Z'' in place of Z' is of course satisfied.

Note that we proved by the way that $P(z_0, t) < \infty$ for z_0 safe and expanding. This is, however, weaker than (12.5.1), proved for all z safe.

2. $\mathrm{P}_{\mathrm{hyp}}(t) \leq \mathrm{P}_{\mathrm{Conf}}(t)$. Let μ be an arbitrary (λ, t)-conformal measure on J. From the *topological exactness* of f on J (see [Carleson & Gamelin 1993]) we get $\int_U \lambda^N |(f^N)'|^t d\mu \geq 1$. Hence $\mu(U) > 0$ (compare Lemma 12.2.1).

Let X be an arbitrary f-invariant non-empty isolated hyperbolic subset of J. Then, for U small enough, $(\exists C)(\forall x_0 \in X)(\forall n \geq 0)(\forall x \in X \cap f^{-n}(x_0))$ f^n maps $U_x = \mathrm{Comp}_x f^{-n}(U)$ onto U univalently with distortion bounded by C. So, for every n,

$$\mu(U) \cdot \sum_{x\in f^{-n}(x_0)\cap X} \lambda^{-n}|(f^n)'(x)|^{-t} \leq C \sum_{x\in f^{-n}(x_0)\cap X} \mu(U_x) \leq C.$$

Hence

$$\mathrm{P}(f|_X, -\log\lambda - t\log|f'|) \leq 0, \ \text{ hence } \ \mathrm{P}(f|_X, -t\log|f'|) \leq \log\lambda.$$

3. Now we prove $\mathrm{P}_{\mathrm{Conf}}(t) \leq \mathrm{P}_{\mathrm{tree}}(w, t)$, and in particular that the definition of $\mathrm{P}_{\mathrm{Conf}}(t)$ makes sense. The proof is via the Patterson–Sullivan construction, as started in Section 11.5.1, but it is much simpler and direct, omitting approximation via $K(V)$'s in the following sections. We can assume that $\mathrm{P}_{\mathrm{tree}}(w, t) < \infty$, otherwise there is nothing to prove.

Let us assume first that w is such that for any sequence $w_n \in f^{-n}(w)$ we have $w_n \to J$. This means that w is not in an attracting periodic orbit, nor in a Siegel disc, nor in a Herman ring: see [Carleson & Gamelin 1993]. Assume also that w is not periodic. Let $\mathrm{P}_{\mathrm{tree}}(w, t) = \lambda$. Then for all $\lambda' > \lambda$

$$\sum_{x\in f^{-n}(w)} (\lambda')^{-n}|(f^n)'(x)|^{-t} \to 0$$

exponentially fast, as $n \to \infty$. We find a sequence of numbers $\phi_n > 0$ such that $\lim_{n\to\infty} \phi_n/\phi_{n+1} \to 1$, and for $A_n := \sum_{x\in f^{-n}(w)} \lambda^{-n}|(f^n)'(x)|^{-t}$ the series

$\sum_n \phi_n A_n$ is divergent: compare Lemma 12.1.2. For every $\lambda' > \lambda$ consider the measure

$$\mu_{\lambda'} = \sum_{n=0}^{\infty} \sum_{x \in f^{-n}(w)} D_x \cdot \phi_n \cdot (\lambda')^{-n} |(f^n)'(x)|^{-t} / \Sigma_{\lambda'},$$

where D_x is the Dirac delta measure at x, and $\Sigma_{\lambda'}$ is the sum over all x of the weights at D_x, so that $\mu_{\lambda'}(J) = 1$. Notice that m'_λ is $(1/\Sigma_{\lambda'})$-$(\lambda'|f'|^t)$-conformal.

Indeed, the only point where this purely atomic measure is not conformal is w. But $f(w)$ does not belong to $\bigcup_{n \geq 0} f^{-n}(\{w\})$, since w is not periodic: hence $\mu_{\lambda'}(\{f(w)\}) = 0$.

Finally we find a (λ, t)-conformal measure μ as a weak* limit of a convergent sub-sequence of $\mu_{\lambda'}$ as $\lambda' \searrow \lambda$: see Lemma 12.5.5 and Remark 12.5.6.

If w is in an attracting periodic orbit that is one of at most two exceptional fixed points (∞ for polynomials, 0 or ∞ for $z \mapsto z^k$, in adequate coordinates), then it is a critical value, so $\mathrm{P}_{\mathrm{tree}}(w, t) = \infty$. If w is in a non-exceptional periodic orbit or in a Siegel disc or Herman ring S, take $w' \in f^{-1}(w)$ neither in the periodic orbit of w, nor in the periodic orbit of S in the latter cases. Then for w' we have the first case: hence $\mathrm{P}_{\mathrm{Conf}}(f) \leq \mathrm{P}_{\mathrm{tree}}(w', t) \leq \mathrm{P}_{\mathrm{tree}}(w, t)$. The latter inequality follows from

$$\mathrm{P}_{\mathrm{tree}}(w', t) = \limsup_{n \to \infty} \frac{1}{n-1} \sum_{x \in f^{-(n-1)}(w')} |(f^{-(n-1)})'(x)|^{-t} \leq$$

$$\leq \limsup_{n \to \infty} \frac{1}{n} \sum_{x \in f^{-(n-1)}(w')} |(f^n)'(x)|^{-t} \sup_{z \in \overline{\mathbb{C}}} |f'|^t \leq \mathrm{P}_{\mathrm{tree}}(w, t).$$

♣

Remark 12.5.14. There is a direct simple proof of $\mathrm{P}_{\mathrm{tree}}(z, t) \leq \mathrm{P}_{\mathrm{Conf}}(t)$ for μ-a.e. z, using the Borel–Cantelli Lemma: see [Przytycki 1999, Theorem 2.4].

Remark 12.5.15. In [Przytycki 1999, Th.3.4] a stronger completing Corollary 11.5.13 has been proved, also by elementary means, namely that $\mathrm{P}_{\mathrm{tree}}(z, t)$ does not depend on $z \in \overline{\mathbb{C}}$ except for a zero Hausdorff dimension set of z's.

To complete this section it is worth mentioning one more definition of pressure: see [Przytycki 1999] and [Przytycki, Rivera-Letelier & Smirnov 2004].

Definition 12.5.16.

$$\mathrm{P}_{\mathrm{varhyp}}(t) = \sup\{\mathrm{h}_\mu(f) - t\chi_\mu(f)\},$$

the supremum taken over all f-invariant probability ergodic measures on J with positive Lyapunov exponent: that is, over all hyperbolic f-invariant measures.

The inequalities $\mathrm{P}_{\mathrm{hyp}}(t) \geq \mathrm{P}_{\mathrm{varhyp}}(t) \geq \mathrm{P}_{\mathrm{hyp}}(t)$ hold by Theorem 11.6.1 and the Variational Principle, Theorem 3.4.1, respectively.

Remark 12.5.17. In conclusion we can denote all the pressures above by $\mathrm{P}(t)$, as anticipated at the beginning of this section, and call it *geometric pressure.*

A remarkable dichotomy holds for rational maps: either $\mathrm{P}(t)$ is strictly decreasing to $-\infty$ as $t \nearrow \infty$, or $\mathrm{P}(t) \equiv 0$ for all $t \geq t_0 = \mathrm{HyD}(J)$. The first happens precisely for so-called Topological Collet–Eckmann maps (abbr. TCE maps). Here is one characterization of this class, which explains another name: *non-uniformly hyperbolic.* A rational map $f : \overline{\mathbb{C}} \to \overline{\mathbb{C}}$ is TCE if and only if

$$\inf_{\mu} \chi_\mu(f) > 0,$$

the infimum taken over all probability f-invariant measures on J. For details of this theory see [Przytycki, Rivera-Letelier & Smirnov 2003].

Remark 12.5.18. In the definition of tree pressure lim sup can be replaced by lim, which happens to exist (compare Proposition 4.4.3). Indeed, writing for any $z \in X$, for any topologically transitive expanding repeller $X \subset J(f)$

$$\sum_{f^n(x)=z, x\in X} |(f^n)'(x)|^{-t} \leq \sum_{f^n(x)=z, x\in J(f)} |(f^n)'(x)|^{-t},$$

and applying lim inf $\frac{1}{n}$ on both sides, one obtains $\mathrm{P}_{\mathrm{hyp}}(t) \leq \mathrm{P}^{-}_{\mathrm{tree}}(t)$, where the superscript 'minus' means we take lim inf instead lim sup in the definition of the tree pressure. Note that for the left-hand sum the limit exists, by Proposition 4.4.3.

Bibliographical notes

Section 12.1 roughly follows [Denker & Urbański 1991a]. However, here the set Sing need not be finite; this is the version introduced and used in [Denker & Urbański 1991b]. Sections 12.2 and 12.3 follow [Denker & Urbański 1991b], with some simplifications. For example, the proof of Lemma 12.2.2 is much simpler.

The construction of conformal measures was first sketched in [Sullivan 1983], and followed an analogous notion and construction by S.J. Patterson on the limit sets of a Kleinian group.

The content of Section 12.5 has been extracted from [Przytycki, Rivera-Letelier & Smirnov 2004]: see [Przytycki 2005b].

We owe Remark 12.5.18 to K. Barański and A. Zdunik.

References

[Aaronson, Denker & Urbański 1993] J. Aaronson, M. Denker, M. Urbański: Ergodic theory for Markov fibered systems and parabolic rational maps. *Trans. Am. Math. Soc.* 337 (1993), 495–548.

[Adler, Konheim & McAndrew 1965] R. L. Adler, A. G. Konheim, M. H. McAndrew: Topological entropy. *Trans. Am. Math. Soc.* 39 (1964), 37–56.

[Ahlfors 1966] L. V. Ahlfors: *Lectures on Quasiconformal Mappings.* D. van Nostrand Company Inc., Princeton, NJ, 1966.

[Anosov 1970] D. V. Anosov: On a class of invariant sets of smooth dynamical systems. *Proceedings 5th Int. Conf. on Nonlin. Oscill. 2.* Kiev 1970, 39–45 (in Russian).

[Astala, Iwaniec & Martin 2009] K. Astala, T. Iwaniec, G. Martin: *Elliptic Partial Differential Equations and Quasiconformal Mappings in the Plane.* Princeton University Press, 2009.

[Avila & Lyubich 2008] A. Avila, M. Lyubich: Hausdorff dimension and conformal measures of Feigenbaum Julia sets. *J. Am. Math. Soc.* 21(2) (2008), 305–363.

[Avila, Lyubich & de Melo 2003] A. Avila, M. Lyubich, W. de Melo: Regular or stochastic dynamics in real analytic families of unimodal maps. *Invent. Math.* 154 (2003), 451–550.

[Avila, Martens & de Melo 2001] A. Avila, M. Martens, W. de Melo: On the dynamics of the renormalization operator. In *Global Analysis of Dynamical Systems: Festschrift dedicated to Floris Takens for his 60th birthday*, Ed. H. Broer, B. Krauskopf, G. Vegter, Institute of Physics Publishing, Bristol and Philadelphia, pp. 449–460, 2001.

[Baladi 2000] V. Baladi: *Positive Transfer Operators and Decay of Correlations.* World Scientific, Singapore, 2000.

[Barański, Volberg, & Zdunik 1998] K. Barański, A. Volberg, A. Zdunik: Brennan's conjecture and the Mandelbrot set. *Int. Math. Res. Notices* 12 (1998), 589–600.

[Barnsley 1988] M. Barnsley: *Fractals Everywhere.* Academic Press, Boston, 1988.

[Beardon 1991] A. F. Beardon: *Iteration of Rational Functions.* Springer-Verlag, New York, 1991.

[Beliaev & Smirnov 2005] D. Beliaev, S. Smirnov: Harmonic measure on fractal sets. In A. Laptev (ed.), *European Congress of Mathematics, Stockholm, Sweden*, pp. 41–59. European Mathematical Society, Zurich, 2005.

[Bielefeld 1990] B. Bielefeld: Conformal dynamics problem list. Preprint SUNY Stony Brook IMS 1990/1.

[Billingsley 1965] P. Billingsley: *Ergodic Theory and Information.* Wiley, New York, 1965.

[Billingsley 1968] P. Billingsley: *Convergence of Probability Measures.* Wiley, New York, 1968.

[Billingsley 1979] P. Billingsley: *Probability and Measure.* Wiley, New York, 1979.

[Billingsley 1999] P. Billingsley: *Convergence of Probability Measures*, 2nd edn. Wiley, New York, 1999.

[Binder, Makarov & Smirnov 2003] I. Binder, N. Makarov, S. Smirnov: Harmonic measure and polynomial Julia sets. *Duke Math. J.* 117(2) (2003), 343–365.

[Bishop & Phelps 1963] E. Bishop, R. R. Phelps: The support functionals of a convex set. *AMS Proc. Symp. Pure Math.* 7 (1963), 27–35.

[Bishop *et al.* 1989] C. J. Bishop, L. Carleson, J. B. Garnett, P. W. Jones: Harmonic measures supported on curves. *Pac. J. Math.* 138(2) (1989), 233–236.

[Bogolyubov & Hacet 1949] N. N. Bogolyubov, B. M. Hacet: On some mathematical questions of the theory of statistical equilibrium. *Dokl. Akad. Nauk SSSR* 66 (1949), 321–324.

[Bourbaki 1981] N. Bourbaki: *Espaces Vectoriels Topologiques.* Masson, Paris, 1981.

[Bowen 1970] R. Bowen: Markov partitions for Axiom A diffeomorphisms. *Am. J. Math.* 92 (1970), 725–749.

[Bowen 1975] R. Bowen: *Equilibrium States and the Ergodic Theory of Anosov Diffeomorphisms.* Lecture Notes in Math. 470, Springer Verlag, Berlin, 1975.

[Bowen 1979] R. Bowen: Hausdorff dimension of quasi-circles. *Publ. Math. IHES* 50 (1979), 11–26.

[Bowen & Series 1979] R. Bowen, C. Series: Markov maps associated with Fuchsian groups. *Publ. Math. IHES* 50 (1979), 153–170.

[Boyarsky & Góra 1997] A. Boyarsky, P. Góra: *Laws of Chaos, Invariant Measures and Dynamical Systems in One Dimension.* Birkhäuser, Boston, 1997.

[Brolin 1965] H. Brolin: Invariant sets under iteration of rational functions. *Ark. Mat.* 6 (1965), 103–144.

[Buff & Cheritat 2008] X. Buff, A. Cheritat: Quadratic Julia sets with positive area, arXiv:math/0605514v2.

[Carathéodory 1927] C. Carathéodory: *Vorlesungen über reelle Funktionen.* Leipzig-Berlin, 1927.

[Carleson 1985] L. Carleson: On the support of harmonic measure for sets of Cantor type. *Ann. Acad. Sci. Fenn.* 10 (1985), 113–123.

[Carleson & Gamelin 1993] L. Carleson, T. W. Gamelin: *Complex Dynamics.* Springer-Verlag, New York, 1993.

[Carleson & Jones 1992] L. Carleson, P. Jones: On coefficient problems for univalent functions and conformal dimension. *Duke Math. J.* 66(2) (1992), 169–206.

[Carleson, Jones & Yoccoz 1994] L. Carleson, P. Jones & J.-Ch. Yoccoz: Julia and John. *Bol. Soc. Bras. Mat.* 25 (1994), 1–30.

[Collet & Eckmann 1980] P. Collet, J.-P. Eckmann: *Iterated Maps of the Interval as Dynamical Systems.* Birkhäuser, Boston, 1980.

[Collingwood & Lohwater 1966] E. F. Collingwood, A. J. Lohwater: *The Theory of Cluster Sets.* Cambridge University Press, 1966.

[Conway 1997] J. B. Conway: *A Course in Functional Analysis.* Springer, New York, 1997.

[Cornfeld, Fomin & Sinai 1982] I. P. Cornfeld, S. V. Fomin, Ya. G. Sinai: *Ergodic Theory.* Springer-Verlag, New York, 1982.

[Coven & Keane 2006] E. M. Coven, M. Keane: Every compact metric space that supports a positively expansive homeomorphism is finite. In *Dynamics & Stochastics*, 304–305. IMS Lecture Notes Monogr. Ser., 48, Inst. Math. Statist., Beachwood, OH, 2006.

[Crovisier 2002] S. Crovisier: Une remarque sur les ensembles hyperboliques localement maximaux. *C. R. Acad. Sci. I* 334(5) (2002), 401–404.

[Cui 1996] G. Cui: On the smoothness of conjugacy for circle covering maps. *Acta Math. Sin. (N.S.)* 12(2) (1996) 122–125.

[Denker, Grillenberger & Sigmund 1976] M. Denker, Ch. Grillenberger, K. Sigmund: *Ergodic Theory on Compact Spaces.* Lecture Notes in Mathematics, 527, Springer-Verlag, Berlin, 1976.

[Denker & Heinemann 1998] M. Denker, S.-M. Heinemann: Jordan tori and polynomial endomorphisms in $\mathbb{C}^2$. *Fund. Math.* 157(2-3) (1998), 139–159.

[Denker & Urbański 1991a] M. Denker, M. Urbański: On the existence of conformal measures. *Trans. Am. Math. Soc.* 328(2) (1991), 563–587.

[Denker & Urbański 1991b] M. Denker, M. Urbański: On Sullivan's conformal measures for rational maps of the Riemann sphere. *Nonlinearity* 4 (1991), 365–384.

[Dinaburg 1971] E. I. Dinaburg: On the relations among various entropy characteristics of dynamical systems. *Math. USSR Izv.* 5 (1971), 337–378.

[Dobbs 2008] N. Dobbs: Measures with positive Lyapunov exponent and conformal measures in rational dynamics. arXiv:0804.3753v1 .

[Doob 1953] J. L. Doob: *Stochastic Processes.* Wiley, New York, 1953.

[Douady, Sentenac & Zinsmeister 1997] A. Douady, P. Sentenac, M. Zinsmeister: Implosion parabolique et dimension de Hausdorff. *CR Acad. Sci. I* 325 (1997), 765–772.

[Dunford & Schwartz 1958] N. Dunford, J. T. Schwartz: *Linear Operators I.* Wiley, New York, 1958.

[Edwards 1995] R. E. Edwards: *Functional Analysis: Theory and Applications,* Dover Books on Mathematics. Dover, New York, 1995.

[Ellis 1985] R. S. Ellis: *Entropy, Large Deviations and Statistical Mechanics.* Springer, New York, 1985.

[Falconer 1985] K. Falconer: *The Geometry of Fractal Sets.* Cambridge University Press, 1985.

[Falconer 1997] K. Falconer: *Techniques in Fractal Geometry.* Wiley, Chichester, 1997.

[de Faria *et al.* 2006] E. de Faria, W. de Melo, A. Pinto: Global hyperbolicity of renormalization for C^r unimodal mappings. *Ann. Math.* 164(3) (2006) 731–824.

[Farrell & Jones 1979] F. T. Farrell, L. E. Jones: Markov cell structures for expanding maps in dimension 2. *Trans. Am. Math. Soc.* 255 (1979), 315–327.

[Farrell & Jones 1993] F. T. Farrell, L. E. Jones: *Markov Cell Structures near a Hyperbolic Set.* Memoirs of the AMS 491 (1993).

[Fatou 1919–1920] P. Fatou: Sur les equationes fonctionelles. *Bull. Soc. Math. Fr.* 47 (1919), 161–271; 48 (1920), 33–94, 208–314.

[Fisher 2006] T. Fisher: Hyperbolic sets that are not locally maximal. *Ergod. Theor. Dyn. Syst.* 26(5) (2006), 1491–1509.

[Frink 1937] A. H. Frink: Distance functions and metrization problem. *Bull. Am. Math. Soc.* 43 (1937), 133–142.

[Gelfert, Przytycki & Rams 2009] K. Gelfert, F. Przytycki, M. Rams: Lyapunov spectrum for rational maps. arXiv:0809.3363. To appear in *Math. Annalen.*

[Ghys 1984] E. Ghys: Transformations holomorphes au voisinage d'une curbe de Jordan. *C. R. Acad. Sci. I* 298 (1984), 385–388.

[Goodman 1971] T. Goodman: Relating topological entropy and measure entropy. *Bull. Lond. Math. Soc.* 3 (1971), 176–180.

[Goodwyn 1969] L. W. Goodwyn: Topological entropy bounds measure-theoretic entropy. *Proc. Am. Math. Soc.* 23 (1969), 679–688.

[Gordin 1969] M. I. Gordin: The central limit theorem for stationary processes. *Dokl. Akad. Nauk SSSR* 188 (1969), 739–741. English transl.: *Sov. Math. Dokl.* 10 (1969), 1174–1176.

[Graczyk & Świątek 1998] J. Graczyk, G. Świątek: *The Real Fatou Conjecture.* Annals of Mathematical Studies, Princeton University Press, Princeton, 1998.

[Graczyk & Smirnov 1998] J. Graczyk, S. Smirnov: Collet, Eckmann and Hölder. *Invent. Math.* 133 (1998), 69–96.

[Halmos 1950] P. Halmos: *Measure Theory.* Van Nostrand, New York, 1950.

[Hasley *et al.* 1986] T. C. Hasley, M. Jensen, L. Kadanoff, I. Procaccia, B. Shraiman: Fractal measures and their singularities: the characterization of strange sets. *Phys. Rev. A* 33(2) (1986), 1141–1151.

[Hayman & Kennedy 1976] W. K. Hayman, P. B. Kennedy: *Subharmonic Functions*, Vol. I. Academic Press, London, 1976.

[Hoover 1991] D. N. Hoover: Convergence in distribution and Skorokhod convergence for the general theory of processes. *Probab. Theory Rel.* 89 (1991), 239–259.

[Ibragimov 1962] I. A. Ibragimov: Some limit theorems for stationary processes. *Theor. Probab. Appl.* 7 (1962), 349–382.

[Iglesias, Portela & Rovella 2009] J. Iglesias, A. Portela, A. Rovella: Stability modulo singular sets. *Fund. Math.* 204(2) (2009), 155–175.

[Ionescu Tulcea & Marinescu 1950] C. T. Ionescu Tulcea, G. Marinescu: Theorie ergodique pour des classes d'operations completement continues. *Ann. Math.* 52(1) (1950), 140–147.

[Israel 1979] R. Israel: *Convexity in the Theory of Lattice Gases.* Princeton University Press, 1979.

[Jiang 1996] Y. Jiang: *Renormalization and Geometry in One-Dimensional and Complex Dynamics.* Advanced Series in Nonlinear Dynamics 10, World Scientific, Singapore, 1996.

[Jordan et al. 2010] T. Jordan, M. Kesseböhmer, M. Pollicott, B. O. Stratmann: Sets of non-differentiability for conjugacies between expanding interval maps. To appear in *Fund. Math.*

[Julia 1918] G. Julia: Mémoire sur l'iteration des fonctions rationelles. *J. Math. Pures Appl.* 8(1) (1918), 47–245.

[Kato 1966] T. Kato: *Perturbation Theory for Linear Operators.* Springer, Berlin, 1966.

[Katok & Hasselblatt 1995] A. Katok, B. Hasselblatt: *Introduction to the Modern Theory of Dynamical Systems.* Cambridge University Press, 1995.

[Keller 1998] G. Keller: *Equilibrium States in Ergodic Theory.* Cambrige University Press, 1998.

[Kelley 1955] J. L. Kelley: *General Topology.* Van Nostrand, Princeton, NJ, 1955.

[Kolmogoroff 1933] A. N. Kolmogoroff: *Grundbegriffe der Wahrscheinlichkeitsrechnung.* Springer-Verlag, Berlin, 1933.

[Koopman 1931] B. O. Koopman: Hamiltonian systems and transformations in Hilbert space. *Proc. Natl Acad. Sci. USA* 17 (1931), 315–318.

[Kornfeld, Fomin & Sinai 1982] I. P. Kornfeld, S. V. Fomin, Ya. G. Sinai: *Ergodic Theory.* Springer-Verlag, New York, 1982.

[Krzyżewski 1982] K. Krzyżewski: On analytic invariant measures for expanding mappings. *Coll. Math.* 46(1) (1982), 59–65.

[Kushnirenko 1972] A. G. Kushnirenko: Problems in dynamical systems theory on manifolds. *IX Mathematical Summer School*, Kiev 1972 (in Russian).

[Lanford 1982] O. E. Lanford III: A computer assisted proof of of the Feigenbaum conjecture. *Bull. Am. Math. Soc.* 6 (1982), 427–434.

[Lang 1965] S. Lang: *Algebra.* Addison-Wesley, Reading, MA, 1970.

[Lasota & Mackey 1985] A. Lasota, M. Mackey: *Probabilistic Properties of Deterministic Systems.* Cambridge University Press, 1985.

[Lasota, Li & Yorke 1984] A. Lasota, T. Y. Li, J. A. Yorke: Asymptotic periodicity of the iterates of Markov operators. *Trans. Am. Math. Soc.* 286 (1984), 751–764.

[Lasota & Yorke 1973] A. Lasota, J. A. Yorke: On the existence of invariant measures for piecewise monotonic transformations. *Trans. Am. Math. Soc.* 186 (1973), 481–488.

[Ledrappier 1984] F. Ledrappier: Quelques propriétés ergodiques des applications rationnelles. *C.R. Acad. Sci. I, Math.* 299(1) (1984), 37–40.

[Ledrappier & Young 1985] F. Ledrappier, L.-S. Young: The metric entropy of diffeomorphisms, Part I and Part II. *Ann. Math.* (2) 122 (1985), 509–539 and 540–574.

[Leonov 1961] V. P. Leonov: On the dispersion of time-dependent means of stationary stochastic process. *Theor. Probab. Appl.* 6 (1961), 93–101.

[Levin & Świątek 2009] G. Levin, G. Świątek: Measure of the Julia set of the Feigenbaum map with infinite criticality. *Ergod. Th. Dynam. Sys.*, online 2009.

[Lojasiewicz 1988] S. Lojasiewicz: *An Introduction to the Theory of Real Functions.* Wiley, Chichester, 1988.

[Lyubich 1983] M. Yu. Lyubich: Entropy properties of rational endomorphisms of the Riemann sphere. *Ergod. Theor. Dyn. Syst.* 3 (1983), 351–385.

[Lyubich 1999] M. Lyubich: Feigenbaum–Coullet–Tresser universality and Milnor's hairness conjecture. *Ann. Math.* 149 (1999), 319–420.

[Lyubich & Lyubich 1986] M. Yu. Lyubich, Yu. I. Lyubich: Perron-Frobenius theory for almost periodic operators and semigroups representations. *Teoriya Funktsii* 46 (1986), 54–72.

[Lyubich & Minsky 1997] M. Lyubich, Y. Minsky: Laminations in holomorphic dynamics. *J. Differ. Geom.* 47(1) (1997), 17–94.

[Makarov 1985] N. Makarov: On the distortion of boundary sets under conformal mappings. *Proc. Lond. Math. Soc.* 51 (1985), 369–384.

[Makarov 1986] N. Makarov: Metric properties of harmonic measure. In: *Proc. Inter. Congress of Mathematicians, Berkeley 1986*, Am. Math. Soc., 1986, 766–776.

[Makarov 1999] N. Makarov: Fine structure of harmonic measure. *St. Petersbourg Math. J.* 10 (1999), 217–268.

[Malgrange 1967] B. Malgrange: *Ideals of Differentiable Functions.* Oxford University Press, 1967.

[Mandelbrot 1982] B. Mandelbrot: *The Fractal Geometry of Nature.* W. H. Freeman, San Francisco, 1982.

[Mañé 1987] R. Mañé: *Ergodic Theory and Differentiable Dynamics.* Springer 1987.

[Mañé 1988] R. Mañé: The Hausdorff dimension of invariant probabilities of rational maps. In *Dynamical Systems, Valparaiso 1986*, 86–117. Lecture Notes in Math., 1331, Springer, Berlin, 1988.

[Mañé, Sad & Sullivan 1983] R. Mañé, P. Sad, D. Sullivan: On the dynamics of rational maps. *Ann. scientifiques de l'E.N.S.* 4e *série*, 16(2) (1983), 193–217.

[Marczewski & Ryll-Nardzewski 1953] E. Marczewski, C. Ryll-Nardzewski: Remarks on the compactness and non direct products of measures. *Fund. Math.* 40 (1953), 165–170.

[Mattila 1995] P. Mattila: *Geometry of Sets and Measures in Euclidean Spaces, Fractals and Rectifiability.* Cambridge University Press, 1995.

[Mauldin & Urbanski 1996] D. Mauldin, M. Urbański: Dimensions and measures in infinite iterated function systems. *Proc. Lond. Math. Soc.* (3) 73 (1996), 105–154.

[Mauldin & Urbanski 2003] D. Mauldin, M. Urbański: *Graph Directed Markov Systems: Geometry and Dynamics of Limit Sets.* Cambridge University Press, 2003.

[Mauldin, Przytycki & Urbanski 2001] D. Mauldin, F. Przytycki & M. Urbanski: Rigidity of conformal iterated function systems. *Compos. Math.* 129 (2001), 273–299.

[McCluskey & Manning 1983] H. McCluskey, A. Manning: Hausdorff dimension for horseshoes. *Ergod. Th. and Dynam. Sys.* 3 (1983), 251–260.

[McMullen 1996] C. McMullen: *Renormalization and 3-Manifolds Which Fiber Over the Circle.* Annals of Mathematics Studies 142, Princeton University Press, 1996.

[de Melo & Pinto 1999] W. de Melo, A. Pinto: Rigidity of C^2 renormalizable unimodal maps *Commun. Math. Phys.* 208 (1999), 91–105.

[de Melo & van Strien 1993] W. de Melo, S. van Strien: *One-Dimensional Dynamics.* Springer-Verlag, Berlin, 1993.

[Milnor 1999] J. Milnor: *Dynamics in One Complex Variable.* Vieweg, Wiesbaden, 1999, 2000; Princeton University Press, 2006.

[Misiurewicz 1973] M. Misiurewicz: Diffeomorphism without any measure with maximal entropy. *Bull. Acad. Pol. Sci* 21 (1973), 903–910.

[Misiurewicz 1976] M. Misiurewicz: A short proof of the variational principle for a $\mathbb{Z}_+^N$ action on a compact space. *Astérisque* 40 (1976), 147–187.

[Misiurewicz & Przytycki 1977] M. Misiurewicz, F. Przytycki: Topological entropy and degree of smooth mappings. *Bull. Acad. Pol. Sci. Math. Astronom. Phys.* 25(6) (1977), 573–574.

[Neveu 1964] J. Neveu: *Bases Mathématiques du Calcul des Probabilités.* Masson et Cie, Paris, 1964.

[Ornstein 1970] D. Ornstein: Bernoulli shifts with the same entropy are isomorphic. *Adv. Math.* 4 (1970), 337–352.

[Paluba 1989] W. Paluba: The Lipschitz condition for the conjugacies of Feigenbaum mappings. *Fund. Math.* 132 (1989), 227–258.

[Parry 1969] W. Parry: *Entropy and Generators in Ergodic Theory.* W. A. Benjamin, Inc., New York, 1969.

[Parry 1981] W. Parry: *Topics in Ergodic Theory.* Cambridge Tracts in Mathematics, 75. Cambridge University Press, 1981.

[Pesin 1997] Ya. Pesin: *Dimension Theory in Dynamical Systems: Contemporary Views and Applications.* Chicago Lectures in Mathematics Series, University of Chicago Press, 1997.

[Petersen 1983] K. Petersen: *Ergodic Theory.* Cambridge University Press, 1983.

[Phelps 1966] R. R. Phelps: *Lectures on Choquet's Theorem.* Van Nostrand, Princeton, 1966.

[Philipp & Stout 1975] W. Philipp, W. Stout: *Almost Sure Invariance Principles for Partial Sums of Weakly Dependent Random Variables.* Memoirs of the American Mathematical Society 161 (1975).

[Pilyugin 1999] S. Yu. Pilyugin: *Shadowing in Dynamical Systems.* Lecture Notes in Mathematics 1706, Springer, Berlin, 1999.

[Pinto & Rand 1988] A. Pinto, D. Rand: Global phase space universality, smooth conjugacies and renormalization: 1. The $C^{1+\alpha}$ case. *Nonlinearity* 1 (1988), 181–202.

[Pinto & Rand 1992] A. Pinto, D. Rand: Global phase space universality, smooth conjugacies and renormalization: 2. The $C^{k+\alpha}$ case using rapid convergence of Markov families. *Nonlinearity* 4 (1992), 49–79.

[Pommerenke 1992] Ch. Pommerenke: *Boundary Behaviour of Conformal Maps.* Springer, Berlin, 1992.

[Prado 1997] E. Prado: Teichmüller distance for some polynomial-like maps. Preprint SUNY Stony Brook IMS 1996/2 (revised 1997), arXiv:math/9601214v1

[Przytycki 1976] F. Przytycki: Anosov endomorphisms. *Studia Math.* 58(3) (1976), 249–285.

[Przytycki 1977] F. Przytycki: On Ω-stability and structural stability of endomorphisms satisfying Axiom A. *Stud. Math.* 60(1) (1977), 61–77.

[Przytycki 1985] F. Przytycki: Hausdorff dimension of harmonic measure on the boundary of an attractive basin for a holomorphic map. *Invent. Math.* 80(1) (1985), 161–179.

[Przytycki 1986a] F. Przytycki: Riemann map and holomorphic dynamics. *Invent. Math.* 85 (1986), 439–455.

[Przytycki 1986b] F. Przytycki: On holomorphic perturbations of $z \mapsto z^n$. *Bull. Pol. Acad. Sci. Math.* 34(3–4) (1986), 127–132.

[Przytycki 1990] F. Przytycki: On the Perron–Frobenius–Ruelle operator for rational maps on the Riemann sphere and for Hölder continuous functions. *Bol. Soc. Bras. Mat.* 20(2) (1990), 95–125.

[Przytycki 1993] F. Przytycki: Lyapunov characteristic exponents are nonnegative. *Proc. Am. Math. Soc.* 119 (1993), 309–317.

[Przytycki 1999] F. Przytycki: Conical limit sets and Poincaré exponent for iterations of rational functions. *Trans. Am. Math. Soc.* 351(5) (1999), 2081–2099.

[Przytycki 2000] F. Przytycki: Hölder implies CE. *Astérisque* 261 (2000), 385–403.

[Przytycki 2005a] F. Przytycki: Expanding repellers in limit sets for iterations of holomorphic functions. *Fund. Math.* 186 (2005), 85–96.

[Przytycki 2005b] F. Przytycki: Hyperbolic Hausdorff dimension is equal to the minimal exponent of conformal measure on Julia set: a simple proof. RIMS Kyoto, *Kokyuroku*, Vol. 1447, 'Complex Dynamics' (2005) pp. 182–186.

[Przytycki 2006] F. Przytycki: On the hyperbolic Hausdorff dimension of the boundary of the basin of attraction for a holomorphic map and of quasirepellers. *Bull. Pol. Acad. Sci. Math.* 54(1) (2006), 41–52.

[Przytycki & Rivera-Letelier 2007] F. Przytycki, J. Rivera-Letelier: Statistical properties of topological Collet–Eckmann maps. *Ann. Sci. Éc. Norm. Sup.* 4e ser., 40 (2007), 135–178.

[Przytycki & Rivera-Letelier 2008] F. Przytycki, J. Rivera-Letelier: Nice inducing schemes and the thermodynamics of rational maps. arXiv:0806.4385

[Przytycki, Rivera-Letelier & Smirnov 2003] F. Przytycki, J. Rivera-Letelier, S. Smirnov: Equivalence and topological invariance of conditions for non-uniform hyperbolicity in the iteration of rational maps. *Invent. Math.* 151(1) (2003), 29–63.

[Przytycki, Rivera-Letelier & Smirnov 2004] F. Przytycki, J. Rivera-Letelier, S. Smirnov: Equality of pressures for rational functions. *Ergod. Th. and Dynam. Sys.* 23 (2004), 891–914.

[Przytycki & Rohde 1998] F. Przytycki, S. Rohde: Porosity of Collet–Eckmann Julia sets. *Fund. Math.* 155 (1998), 189–199.

[Przytycki & Tangerman 1996] F. Przytycki, V. Tangerman: Cantor sets in the line: scaling functions and the smoothness of the shift-map. *Nonlinearity* 9 (1996), 403–412.

[Przytycki & Urbański 1989] F. Przytycki, M. Urbański: On the Hausdorff dimension of some fractal sets. *Stud. Math.* 93 (1989), 155–186.

[Przytycki & Urbański 1999] F. Przytycki, M. Urbański: Rigidity of tame rational functions. *Bull. Pol. Acad. Sci.* 47(2) (1999), 163–182.

[Przytycki, Urbański & Zdunik 1989] F. Przytycki, M. Urbański, A. Zdunik: Harmonic, Gibbs and Hausdorff measures for holomorphic maps. Part 1. *Ann. Math.* 130 (1989), 1–40.

[Przytycki, Urbański & Zdunik 1991] F. Przytycki, M. Urbański, A. Zdunik: Harmonic, Gibbs and Hausdorff measures for holomorphic maps. Part 2. *Stud. Math.* 97 (1991), 189–225.

[Rogers 1970] C. A. Rogers: *Hausdorff Measures.* Cambridge University Press, 1970.

[Rokhlin 1949] V. A. Rokhlin: On the fundamental ideas of measure theory. *Mat. Sb.* 67(1) (1949), 107–150; *Am. Math. Soc. Trans.* (1)10 (1962), 1–54.

[Rokhlin 1967] V. A. Rokhlin: Lectures on the entropy theory of transformations with invariant measure. *Uspehi Mat. Nauk* 22(5) (1967), 3–56; *Russian Math. Surveys* 22(5) (1967), 1–52.

[Ruelle 1973] D. Ruelle: Statistical mechanics on a compact set with $\mathbb{Z}^\nu$ action satisfying expansivness and specification. *Trans. Am. Math. Soc.* 185 (1973), 237–251.

[Ruelle 1978a] D. Ruelle: *Thermodynamic Formalism.* Addison-Wesley, Reading, MA, 1978.

[Ruelle 1978b] D. Ruelle: An inequality of the entropy of differentiable maps. *Bol. Soc. Bras. Mat.* 9 (1978), 83–87.

[Rugh 2008] H. H. Rugh: On the dimensions of conformal repellers. Randomness and parameter dependency. *Ann. Math.* 168(3) (2008), 695–748.

[Shub 1969] M. Shub: Endomorphisms of compact differentiable manifolds. *Am. J. Math.* 91 (1969), 175–199.

[Shub, Sullivan 1985] M. Shub, D. Sullivan: Expanding endomorphisms of the circle revisited. *Ergod. Th. and Dynam. Sys.* 5 (1985), 285–289.

[Sinai 1982] Ya. G. Sinai: *Theory of Phase Transitions: Rigorous Results.* Akadémiai Kiadó, Budapest 1982.

[Slodkowski 1991] Z. Slodkowski: Holomorphic motions and polynomial hulls. *Proc. Am. Math. Soc.* 111 (1991), 347–355.

[Smale 1967] S. Smale: Differentiable dynamical systems. *Bull. Am. Math. Soc.* 73 (1967), 747–817.

[Smania 2005] D. Smania: On the hyperbolicity of the period-doubling fixed point. *Trans. Am. Math. Soc.* 358(4) (2005), 1827–1846.

[Stein 1970] E. M. Stein: *Singular Integrals and Differentiability Properties of Functions.* Princeton University Press, 1970.

[Steinmetz 1993] N. Steinmetz: *Rational Iteration: Complex Analytic Dynamical Systems.* Walter de Gruyter, Berlin 1993.

[Stratmann & Urbański 2003] B. Stratmann, M. Urbanski: Real analyticity of topological pressure for parabolically semihyperbolic generalized polynomial-like maps. *Indagat. Math.* 1(14) (2003), 119–134.

[Stroock 1993] D. W. Stroock: *Probability Theory: An Analytic View.* Cambridge University Press, 1993.

[Sullivan 1982] D. Sullivan: Seminar on conformal and hyperbolic geometry. Notes by M. Baker and J. Seade, Preprint IHES, 1982.

[Sullivan 1983] D. Sullivan: Conformal dynamical systems. In *Geometric Dynamics*, Lecture Notes in Math., Vol. 1007, Springer-Verlag, 1983, pp. 725–752.

[Sullivan 1986] D. Sullivan: Quasiconformal homeomorphisms in dynamics, topology, and geometry. In *Proc. Inter. Congress of Mathematicians, Berkeley 1986*, Am. Math. Soc., 1986, 1216–1228.

[Sullivan 1988] D. Sullivan: Differentiable structures on fractal-like sets determined by intrinsic scaling functions on dual Cantor sets. *The Mathematical Heritage of Hermann Weyl*, AMS Proc. Symp. Pure Math. Vol. 48 (1988), pp. 15–23.

[Sullivan 1991] D. Sullivan: Bounds, quadratic differentials and renormalization conjectures. In *Mathematics into the Twenty-First Century*, Am. Math. Soc. Centennial Publication, vol.2, Am. Math. Soc., Providence, RI, 1991, pp. 417–466.

[Tsuji 1959] M. Tsuji: *Potential Theory in Modern Function Theory.* Maruzen, Tokyo, 1959.

[Urbański 2001] M. Urbański: Rigidity of multi-dimensional conformal iterated function systems. *Nonlinearity*, 14 (2001), 1593–1610.

[Urbański & Zinsmeister 2001] M. Urbański, M. Zinsmeister: Geometry of hyperbolic Julia-Lavaurs sets. *Indagat. Math.* 12 (2001) 273–292.

[Walters 1976] P. Walters: A variational principle for the pressure of continuous transformations. *Am. J. Math.* 17 (1976), 937–971.

[Walters 1978] P. Walters: Invariant measures and equilibrium states for some mappings which expand distances. *Trans. Am. Math. Soc.* 236 (1978), 121–153.

[Walters 1982] P. Walters: *An Introduction to Ergodic Theory.* Springer-Verlag, New York, 1982.

[Walters 1992] P. Walters: Differentiability properties of the pressure of a continuous transformation on a compact metric space. *J. Lond. Math. Soc.* (2) 46 (1992), 471–481.

[Young 1999] L.-S. Young: Recurrence times and rates of mixing. *Israel J. Math.* 110 (1999), 153–188.

[Zdunik 1990] A. Zdunik: Parabolic orbifolds and the dimension of the maximal measure for rational maps. *Invent. Math.* 99 (1990), 627–649.

[Zdunik 1991] A. Zdunik: Harmonic measure versus Hausdorff measures on repellers for holomorphic maps. *Trans. Am. Math. Soc.* 326 (1991), 633–652.

[Zinsmeister 1996] M. Zinsmeister: *Formalisme thermodynamique et systèmes dynamiques holomorphes.* Panoramas et Synthèses, 4. Société Mathématique de France, Paris, 1996. English translation: *Thermodynamic Formalism and Holomorphic Dynamical Systems*, SMF/AMS Texts and Monographs, 2, 2000.

Index

For EU product safety concerns, contact us at Calle de José Abascal, 56–1°,
28003 Madrid, Spain or eugpsr@cambridge.org.

www.ingramcontent.com/pod-product-compliance
Ingram Content Group UK Ltd.
Pitfield, Milton Keynes, MK11 3LW, UK
UKHW042207080726
473066UK00007B/300

* 9 7 8 0 5 2 1 4 3 8 0 0 1 *